QUATERNARY SEA-LEVEL CHANGES
A Global Perspective

Flooding caused by a rise in global mean sea level has the potential to affect the lives of more than 1 billion people in coastal areas worldwide. There have been significant changes in sea level over the past 2 million years, both at the local and global scales, and a complete understanding of natural cycles of change as well as anthropogenic effects is imperative for future global development.

This book reviews the history of research into these sea-level changes and summarises the methods and analytical approaches used to interpret evidence for sea-level changes. It provides an overview of the changing climates of the Quaternary, examines the processes responsible for global variability of sea-level records, and presents detailed reviews of sea-level changes for the Pleistocene and Holocene. The book concludes by discussing current trends in sea level and likely future sea-level changes.

This is an important and authoritative summary of evidence for sea-level changes in our most recent geological period, and provides a key resource for academic researchers, and graduate and advanced undergraduate students, working in tectonics, stratigraphy, geomorphology and physical geography, environmental science, and other aspects of Quaternary studies.

COLIN V. MURRAY-WALLACE is a Quaternary geologist and currently a Professor and Head of the School of Earth & Environmental Sciences in the University of Wollongong. His long-standing research interests have centred on Quaternary sea-level changes, neotectonics, carbonate depositional systems and amino acid racemisation dating, and he has undertaken coastal field investigations in southern Australia, Vietnam, Hawaii, and in South Africa. Professor Murray-Wallace was project leader of IGCP (International Geological Correlation Programme) project 437 (1999–2003) 'Coastal environmental change during sea-level highstands', and leader of the INQUA (International Union for Quaternary Research) Coastal and Marine Commission (2004–2007). He has served as Editor-in-Chief of the journal *Quaternary Science Reviews* since 2008.

COLIN D. WOODROFFE is a coastal geomorphologist in the School of Earth & Environmental Sciences at the University of Wollongong. He has studied the stratigraphy and development of coasts in Australia and New Zealand, as well as on islands in the West Indies, and Indian and Pacific Oceans. He has written and co-authored three coastal studies books. Professor Woodroffe was national representative on the INQUA Quaternary Shorelines subcommission, and served on the committees of both the IGCP project 274 'Coastal Evolution in the Quaternary', and its follow-up project, and on the Scientific Steering Committee of the LOICZ (Land–Ocean Interactions in the Coastal Zone) project within IGBP. He was a lead author on the coastal chapter in the 2007 Intergovernmental Panel on Climate Change (IPCC) Fourth Assessment report. In 2012 he was awarded the R.J. Russell Award from the Coast and Marine Specialty Group of the Association of American Geographers.

QUATERNARY SEA-LEVEL CHANGES

A Global Perspective

COLIN V. MURRAY-WALLACE
AND
COLIN D. WOODROFFE

University of Wollongong, Australia

CAMBRIDGE
UNIVERSITY PRESS

University Printing House, Cambridge CB2 8BS, United Kingdom

Published in the United States of America by Cambridge University Press, New York

Cambridge University Press is part of the University of Cambridge.

It furthers the University's mission by disseminating knowledge in the pursuit of education, learning and research at the highest international levels of excellence.

www.cambridge.org
Information on this title: www.cambridge.org/9780521820837

First published 2014

Printed and bound in the United Kingdom by TJ International Ltd. Padstow Cornwall

A catalogue record for this publication is available from the British Library

Library of Congress Cataloging-in-Publication Data
Murray-Wallace, Colin V.
Quaternary sea-level changes : a global perspective / Colin V. Murray-Wallace, Colin D. Woodroffe.
pages cm.
ISBN 978-0-521-82083-7 (hardback)
1. Sea level–History. 2. Paleoceanography–Holocene. 3. Pleistocene-Holocene boundary.
I. Woodroffe, C. D. II. Title.
GC89.M87 2014
551.45'8–dc23   2013036070

ISBN 978-0-521-82083-7 Hardback

To John Chappell
in appreciation of his truly outstanding contributions
to the study of Quaternary sea-level changes

Contents

Preface

When the work of the geologist is finished and the final comprehensive report is written, the longest and most important chapter will be upon the latest and shortest of the geological periods.

(Grove Karl Gilbert, *1890, p. 1*)

The investigation of relative sea-level changes has a rich intellectual lineage and has played a central role in development of the Earth Sciences. With the emergence of empirically based explanations for field observations of natural phenomena, an increasing appreciation that the ocean surface has not remained constant, relative to land, prompted several lines of geological inquiry. Myths and legends, as well as religious explanations for Earth surface events, such as flood legends, provided an early framework, examined by the emerging discipline of Earth Science. Extensive fieldwork in the nineteenth century, culminating in the Glacial Theory, and technological developments in geochemistry and geochronology in the latter twentieth century led to a geologically coherent explanation for fluctuations in relative sea level during the Quaternary, the past 2.59 Ma.

Responses of the Earth's surface environments to Quaternary sea-level changes are complex and far-reaching. The effects of Quaternary sea-level changes extend from upper reaches of near-coastal drainage basins to the edge of continental shelves. Sea-level changes have amplified the effects of local climate changes, such as enhanced aridity due to increased continentality at times of lower sea level, and promoted marine abrasion along many bedrock-dominated coastlines. During the Quaternary, relative sea-level changes have exposed continental shelves and in places created land bridges which have acted as dispersal routes for a diverse range of biota, including early humans. Relative sea-level changes since the Last Glacial Maximum (LGM) 21,000 years ago, have resulted in the modern configuration of coastlines, and the coastal landforms we see today have developed in the past few millennia. The significance of sea-level changes in coastal landscape evolution is likely to have increased following the middle Pleistocene transition, approximately 1.2 Ma ago, when the amplitude of sea-level changes between glacials and interglacials, as well as the duration of glacial cycles, increased.

A paper by Rhodes Fairbridge, entitled 'Eustatic changes in sea level', published in *Physics and Chemistry of the Earth* in 1961, provided a stimulus for both of us. The earliest glimpses of this book began when one of us (CVM-W) was fossicking in the basement stacks of the Barr Smith Library in the University of Adelaide. There, he unearthed this paper which helped establish a nascent appreciation of the significance of relative sea-level changes in the geological record and represented a wonderful source for reflection, generating many questions. What were the causes of Quaternary sea-level changes? Why do different geographical regions appear to have contrasting relative sea-level histories? Were these seemingly contrasting records real or an illusory artefact of sampling? Could the contrasting records be explained within a coherent geohistorical framework? Would future technological improvements in aspects of geochemistry and geochronology begin to address many of the issues raised in the paper? Is the record of Quaternary sea-level changes relevant to an understanding of future sea-level changes?

The paper also focused the research and thinking behind the sea-level research that CDW undertook across the tropics, following his discovery of an original reprint of the paper hidden away in a cupboard in the Department of Geography at the University of Cambridge. In many respects the questions raised by Fairbridge's paper continue to be the focus of ongoing research, and are addressed in this book.

The paper by Fairbridge led to the realisation that Australian coastal landforms contributed significantly to a global perspective on Quaternary sea-level changes. Showing a high degree of tectonic stability due to several cratons, and located in the far-field of continental icesheets with minimal direct glaciation, Australia represents an ideal location to undertake palaeo-sea-level investigations. One person who seized on that opportunity, and has made an unparalleled contribution to the science of sea-level changes, is John Chappell, to whom we dedicate this book. Through meticulous research, John unravelled the complex series of changes in relative sea level from palaeo-sea-level evidence on the Huon Peninsula in Papua New Guinea, on the northern geological margin of the Australian continent, from the later middle Pleistocene through to the late Holocene. He complemented these studies with investigations more widely across the Australian continent. John has profoundly influenced our research, and the research of many other investigators around the world; it has been a pleasure and privilege to work alongside him, to learn from him, and to share our ideas with him.

Many people have influenced us over the years and we are particularly grateful for the lengthy and valuable discussions on aspects of Quaternary sea-level changes, as well as field excursions, with Tony Belperio, David Bowen, Robert Bourman, John Cann, Bill Carter, Peter Cowell, Patrick De Deckker, Robert Devoy, Charles Fletcher III, Don Forbes, Roland Gehrels, Victor Gostin, Nick Harvey, David Hopley, Kurt Lambeck, Antony Long, Roger McLean, Dan Muhs, Robert Nicholls, Richard Peltier, Paolo Pirazzoli, Peter Roy, Yoshi Saito, Ian Shennan, Andy Short, David Smith, Tom Spencer, David Stoddart, Bruce Thom, Sandy Tudhope,

Masatomo Umitsu, and the late Orson van de Plassche, amongst others. We have each pursued these interests through the activities, in particular, of successive projects of IGCP (the International Geological Correlation Programme, now called the International Geoscience Programme), through which we have met a wider sphere of sea-level researchers. In addition, numerous 'Time Lords' (geochronologists) have provided inspiration and support over the years, including Mike Barbetti, Steve Eggins, Stewart Fallon, David Fink, Rainer Grün, Quan Hua, David Price, Ulrich Radtke, John Wehmiller, Jian-xin Zhao, and the late John Prescott. Several colleagues read drafts of chapters, or responded to questions to assist our understanding, or clarify and supply data or illustrative material, including Robert Bourman, Allan Chivas, Tim Cohen, Zenobia Jacobs, Richard Roberts, and David Smith. We are grateful for their comments, but accept responsibility for any errors or omissions. Colleagues in the School of Earth & Environmental Sciences and the GeoQuest Research Centre at the University of Wollongong who have provided inspiration over the years include Ted Bryant, Allan Chivas, Lesley Head, Brian Jones, Gerald Nanson, David Price, and Bob Young. We have also had the privilege of supervising several gifted postgraduate students who are continuing the Australian academic tradition in sea-level research, including Brendan Brooke, Mack Dixon, David Kennedy, Craig Sloss, Scott Smithers, and Adam Switzer.

We would particularly like to thank Peter Johnson, cartographer *par excellence*, for preparing all the figures. We have a long working relationship with Peter and sincerely thank him for his dedication to this project. We also express our gratitude to the editors at Cambridge University Press; in particular, we thank Laura Clark, Susan Francis, Abigail Jones, Caroline Mowatt, and Sara Brunton for their help and encouragement. Finally, we thank our partners Gemma and Salwa for their forbearance during this project.

Abbreviations

AAR	amino acid racemisation
ABOX	acid–base–wet oxidation
AHD	Australian Height Datum
AMS	accelerator mass spectrometry
AOGCM	Atmosphere–Ocean General Circulation Model
APSL	above present sea level
AR4	Fourth Assessment Report
BAU	business as usual
BP	Before Present (radiocarbon terminology – before 1950 AD)
BPSL	below present sea level
cal. yr BP	calendar years before present (where present is 1950 AD)
CCD	carbonate compensation depth
CLIMAP	Climate: Long-Range Investigation, Mapping and Prediction
CMAT	current mean annual air temperature
De	equivalent dose
DORIS	Doppler Orbitography and Radiopositioning Integrated Satellite
DSDP	Deep Sea Drilling Project
EDT	effective diagenetic temperature
EMIC	Earth Models of Intermediate Complexity
ENSO	El Niño–Southern Oscillation
EPA	Environmental Protection Agency
EPICA	European Project for Ice Coring in Antarctica
EPILOG	Environmental Processes of the Ice-Age: Land, Oceans, Glaciers
ESA	European Space Agency
ESR	electron spin resonance
FAD	First Appearance Datum
FAR	First Assessment Report
FBI	fixed biological indicator
Ga	giga anna (billions of years; American billion)
GI	Greenland Interstadial

GIA	glacial isostatic adjustment
GNSS	Global Navigation Satellite System
GOCE	Gravity field and Ocean Circulation Explorer
GPR	ground-penetrating radar
GPS	Global Positioning System
GRACE	Gravity Recovery And Climate Experiment
GS	Greenland Stadial
Gt	gigatonnes
Gy	Gray (absorbed dose of ionising radiation equal to 1 J/kg)
HAT	highest astronomical tide
ICP-MS	inductively coupled plasma mass spectrometry
IGCP	International Geosciences Programme (was International Geological Correlation Programme)
INQUA	International Union for Quaternary Research
IODP	Integrated Ocean Drilling Program
IPCC	Intergovernmental Panel on Climate Change
IRD	Ice Rafted Debris
IRSL	infrared-stimulated luminescence
ITRF	International Terrestrial Reference Frame
ka	kilo anna (thousands of years)
kg	kilogram (1000 grams)
LA-ICP-MS	laser-ablation inductively coupled plasma mass spectrometry
LAT	lowest astronomical tide
LGM	Last Glacial Maximum
LOICZ	Land–Ocean Interactions in the Coastal Zone
m	metre
Ma	mega anna (millions of years)
MC-ICP-MS	multi-collector, inductively coupled plasma mass spectrometry
MHW	mean high water
MHWN	mean high water neaps
MIS	Marine Isotope Stage
MLWN	mean low water neap
MLWS	mean low water springs
mm	millimetre (one-thousandth of a metre)
MPT	middle Pleistocene Transition
MSL	mean sea level
MTL	mean tide level
NADW	North Atlantic Deep Water
NGA	National Geospatial-Intelligence Agency
NGMS	noble gas mass spectrometry
NGRIP	North Greenland Ice Core Project

NSW	New South Wales
ODP	ocean drilling project
OSL	optically stimulated luminescence
Pa	Pascal (unit of pressure equal to $1\,\text{N/m}^2$; 1 newton per square metre)
PDB	Peedee Belemnite
PMS	palaeo-marsh surface
ppmv	parts per million by volume
PSMSL	Permanent Service for Mean Sea Level
RCP	representative concentration pathways
RMS	root-mean square
SAR	Second Assessment Report
SLE	sea-level equation
SLIP	sea-level index point
SMOW	Standard Mean Ocean Water
SRES	Special Report on Emissions Scenarios
SSH	sea-surface height
SST	sea-surface temperature
T/P	TOPEX/Poseidon
TAR	Third Assessment Report
TIMS	thermal ionisation mass-spectrometry
TL	thermoluminescence
TT-OSL	thermally transferred optically stimulated luminescence
VLM	vertical land movement
WAIS	West Antarctic Icesheet
WOCE	World Ocean Circulation Experiment
XBT	expendable bathythermograph

1

Sea-level changes: the emergence of a Quaternary perspective

And so the relation of land to sea changes too and a place does not always remain land or sea throughout all time, but where there was dry land there comes to be sea, and where there is now sea, changes to follow some order and cycle.

(Aristotle *in Barnes, 1984, p. 572*)

1.1 Introduction

The sea is rarely level. Its surface is perturbed in many ways over contrasting spatial and temporal scales. The wind ruffles the surface of the water generating waves, which travel across the oceans as swell. Groups of waves propagating from one storm interact with wind waves and swell from other directions generated by other storms, and account for the irregular surf where they meet the shoreline. Superimposed on these most obvious of oscillations are longer-term variations in water level. Tides are the most regular of these variations, represented by a rise and fall of the surface of the sea as a consequence of the gravitational forces imposed by the Moon and the Sun. Tidal range is generally small in the open ocean; it is negligible at nodal points (called amphidromic points), but increases with distance from them. It is amplified in shallow coastal regions to reach more than 15 m during spring tides in a few embayments around the world. The tidal stage can be predicted with considerable accuracy because the orbital relationships between Sun, Earth, and Moon, and the harmonics which give rise to tidal variations, are known. However, calculation of mean sea level, the average level of the sea without wave or tide, requires 18.6 years of observations to account for the lunar nodal cycle (Pugh, 1987). The sea surface is subject to other perturbations; it responds to atmospheric conditions, such as barometric pressure, regional wind field, and dynamic oceanographic factors that vary with ocean currents. Of particular concern are unusually high water levels, such as storm surges which inundate coastal settlements, and tsunami, which are long wavelength waves caused by earthquakes, submarine landslides involving large sediment masses, or extraterrestrial impacts from space (Bryant, 2001; Owen *et al.*, 2007). The Aceh tsunami of 26 December 2004, which resulted in the death of more than 250,000 people, and the Tohoku tsunami of 11 March 2011, which killed more than 25,000 people, are tragic reminders of the destructive power of these extreme inundation events.

Sea level determines the present geographical outline of continents. The shoreline occurs where the sea meets the land. The shoreline is characterised by distinctive coastal landforms shaped by a range of processes that extend beyond the upper and lower limits of the water surface. Cliffs, beaches, and muddy shorelines undergo gradual changes in response to the oscillatory action of waves as they approach the shoreline and subaerial processes above the water line, such as swash, the run-up of the waves, salt spray, wind action that builds dunes, and other processes. Coastlines are some of the most dynamic places on Earth, as captured in the quotation from Aristotle above, not only because of the variability in water level from day to day, and the suite of associated coastal processes, but also because the level of the sea has undergone changes over longer timescales.

The level of the sea represents the ultimate base level determining the lower limit of continental denudation, to which fluvial and other weathering processes can erode landscapes. Before it was realised that sea level itself had varied, landscapes were viewed in terms of a cycle of erosion (the geographical cycle) culminating in a broad plain, termed a peneplain, at or close to sea level (Davis, 1899). Landscapes were thought to be infrequently rejuvenated by tectonic uplift. These erosional processes determine the volume and nature of terrestrially derived sediment that may be deposited within coastal environments. Geologists have understood for a long time that many sedimentary rocks accumulated in marine or coastal basins when the sea was at a different level to that which it occupies at present. Palaeoenvironmental inferences about global sea-level changes derived from the pre-Quaternary record have provided a coherent and integrated basis for developing models for hydrocarbon exploration, termed sequence stratigraphy, based on mapping of unconformity-bounded sedimentary sequences (Van Wagoner *et al.*, 1988; Catuneanu, 2006). Many principles derived from pre-Quaternary sedimentary successions also apply to the Quaternary Period.

Sea-level changes have far-reaching effects on both coastal and terrestrial environments. During successive Pleistocene glaciations, at times of maximum icesheet development, major falls in sea level (lowstands) exposed continental shelves. This was associated with lateral shifts in vegetation communities, and created land bridges enabling migration of taxa between formerly isolated regions; for example, Britain was connected to continental Europe and the island of New Guinea was connected to Australia. During these periods when base level was lowered, rivers eroded more vigorously in headwater and other regions. The significantly larger land areas of many continents at times of lower sea levels also led to greater aridity in the interior of Australia, Africa, the Americas, and Eurasia. As rivers extended across continental shelves during the Last Glacial Maximum (LGM) many followed significantly different courses in their lower reaches. For example, the island of Borneo was connected with Peninsular Malaysia forming a broad Sunda Shelf which was drained into the South China Sea by a river system called the Molengraaff River (Tjia, 1980), and the Rhine and Thames joined as tributaries to the Channel River in northwestern Europe (Coles, 2000). Terrigenous sediments, the product of prolonged continental denudation, were carried onto, or well beyond the continental shelves and deposited in deeper water, such as the Bengal Fan in the Bay of Bengal (Kuehl *et al.*, 1989) and the River Murray at the shelf edge in southern Australia (Hill *et al.*, 2009; Schmidt *et al.*, 2010).

Sea level has been close to, or above, present levels for $< 15\%$ of the past 128,000 years (the last glacial cycle). Sea levels fluctuated considerably for the majority of this period and were significantly below present reaching a maximum of about -120 m during the LGM (Lambeck and Chappell, 2001). Sea level reoccupied its present position (the present highstand being close to the position of past highstands) only recently, during the past 6,000 years when the northern hemisphere polar icesheets that were kilometres thick during the glaciations had almost completely melted, except for Greenland. The past few thousand years have been relatively stable in terms of climate, in contrast to much of the last glacial cycle, during which sea level has oscillated significantly in response to repeated climate changes. There have also been subtle geographical variations in relative sea-level change which have been a function of the complex crustal response to changing ice and water loads on the surface of the Earth (glacio-hydro-isostatic adjustment processes). Sea-level studies provide independent evidence for these geophysical adjustments, and are an invaluable means of quantifying mantle rheology (Lambeck and Johnson, 1998).

Viewed at a global scale, redistribution of water masses accompanying glacio-eustatic sea-level changes affects the Earth's rotation as well as perturbing near-Earth satellites in an otherwise time-constant gravity field. Earth rotational changes involve variations in angular velocity about the rotation axis (length of day) and the orientation of the rotation axis (polar motion). Rates of weathering, planetary albedo at a global scale, as well as the rotational behaviour of the planet are also influenced by changes in sea level. In addition, the rapid flooding of continental shelves has been linked to some episodes of volcanism (Nakada and Yokose, 1992; McGuire *et al.*, 1997; Church *et al.*, 2005).

Quaternary highstands of sea level have coincided with interglacials, the warmer periods during the Ice Ages. The vestiges of past periods of high sea level are recorded in the varied landforms and sedimentary deposits preserved along the world's diverse coastlines. Fluctuations in relative sea level can also shift laterally the area affected by hazards such as storm surge and tsunami, especially along low gradient coastlines.

In recent years, the realisation that human activities might be warming the planet and the prospect of future enhanced greenhouse-induced sea-level rise have raised considerable concern. Climate modelling and associated projections of a possible range of sea-level rise scenarios indicate that globally mean sea level might rise by as much as 80 cm or more by 2100 (IPCC, Third Assessment Report, 2001, and Fourth Assessment Report, 2007 which produced slightly different projections, see Chapter 8). These projections have caused concern for people living in coastal lowland regions and oceanic islands, exacerbating the impact of coastal hazards to which they are already exposed (Nicholls *et al.*, 2007). A disproportionate percentage of the world's population resides in coastal lowland regions close to present sea level.

1.2 The Quaternary Period

The Quaternary Period was characterised by repeated growth and decay of continental icesheets and substantial fluctuations in sea level at a global scale. The stratigraphical record from the Quaternary Period has been subdivided on the basis of palaeoclimatic inferences derived from sedimentary successions (Bowen, 1985; Lowe and Walker, 1997; Bradley, 1999; Cronin, 2010). It is the evidence for sea-level changes during this period, their nature, timing and geomorphological significance, and the methods used to study them that are the principal focus of this book.

Consistent with the terminology of the earlier Phanerozoic record, the Quaternary refers to the most recent interval of Earth history (Figure. 1.1), following the Tertiary Period (now re-named Neogene and Paleogene). The Quaternary has been assigned Period status; *periods* are chronostratigraphically defined intervals of time intermediate in status between Era and Epoch (Salvador, 1994). The Quaternary was defined by Desnoyers (1829) based on a flat-lying succession of sediments in the Paris Basin that overlie Tertiary strata. Charles Lyell subsequently defined the Pleistocene epoch in reference to marine strata containing up to 70% fossils represented by modern equivalents (Lyell, 1839). The Quaternary Period has become largely synonymous with the term Ice Age, as corroborated by oxygen-isotope evidence from deep-sea (Shackleton *et al.*, 1990; Raymo, 1992) and ice cores (Barbante *et al.*, 2010).

The Pliocene–early Pleistocene boundary has been antedated from 1.806 Ma to 2.59 Ma (Ogg *et al.*, 2008). This is an outcome of much discussion (Gradstein *et al.*, 2004; Pillans and Naish, 2004), and the recognition, based on a range of palaeoclimatic evidence, that the onset of northern hemisphere glaciation commenced earlier than originally appreciated. In assigning an older age to the base of the Pleistocene, the Gelasian Stage (formerly a terminal Pliocene Stage) has been reassigned to the Quaternary Period. The former base of the Quaternary was defined by a stratigraphical section of marine strata at Vrica in Calabria, southern Italy, with an assigned age of 1.806 Ma (Aguirre and Pasini, 1985). It was marked by the first appearance of elements of an Arctic fauna within the Mediterranean Basin including the marine mollusc *Arctica islandica* and the ostracod *Cytheropteron testudo*. The redefined base, coinciding with Marine Isotope Stage (MIS) 103, a slightly warmer interval with a basal age of 2.59 Ma, occurs at a globally recognisable geomagnetic reversal, the onset of the Matuyama Chron (interval of reversed geomagnetic polarity). This is stratigraphically more easily identified than the colder MIS 110 at 2.73 Ma.

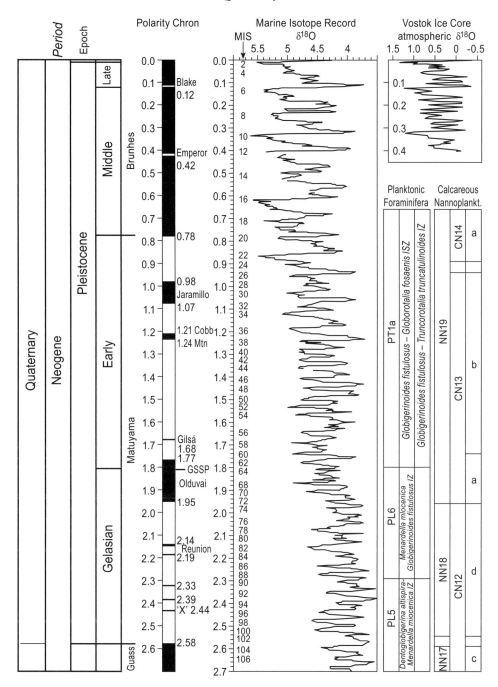

Figure 1.1 Global correlation time chart of the Quaternary Period showing the principal subdivisions of Quaternary time. The chart includes the geomagnetic polarity timescale, marine oxygen-isotope record, Vostock Ice Core, and marine biostratigraphy based on planktonic foraminifera and calcareous nanoplankton (*source*: modified after Gradstein *et al.*, 2004 to accommodate the revision to the Quaternary boundary).

The early–middle Pleistocene boundary is commonly defined by the Brunhes–Matuyama geomagnetic polarity reversal at 780 ka, although this has not been formally described. The middle–late Pleistocene boundary has been placed at about 126–128 ka (near the base of the Eemian Interglacial in Europe), although also the subject of renewed discussion (Gibbard, 2003; Pillans and Naish, 2004). The Holocene series (epoch) refers to the past 11,500 years (^{14}C years) (Figure 1.1). The Holocene has also been defined as a time of climatic amelioration following the Younger Dryas, which was a period of cooling during an overall trend of deglaciation following the LGM; it lasted approximately 1,500 years and ended some 11,500 years ago. The redefined Quaternary Period represents approximately 0.06% of Earth history. The subdivisions of the Pleistocene series (epochs in geochronological parlance) represent different percentages of Quaternary time, namely: early Pleistocene ~70%; middle Pleistocene ~25%; late Pleistocene ~4.5%, and the Holocene 0.5%.

1.3 Sea-level changes: historical development of ideas

Many cultures have recorded legends of floods, some of which were attributed to changes in sea level. More than 500 flood myths are known (Oppenheimer, 1998). Many of them are marine flood legends or descriptions of deluges (Ryan and Pitman, 1998; Burroughs, 2005). Few enable an estimation of the apparent timing of such flood events. Three examples are considered below.

In Australia, numerous Aboriginal dreamtime legends account for an episode of marine inundation which is most likely associated with the most recent deglaciation and early Holocene sea-level rise. Approximately one-seventh of the continent ($> 2,500,000\,km^2$) was flooded resulting in a loss of low gradient, foraging land and the flooding of former river valleys (Cane, 2001). A dreamtime legend describes the filling of what is now Spencer Gulf, an elongate shallow marine embayment that extends some 300 km northwards into semi-arid southern Australia. According to the legend, a kangaroo-man used a thigh bone of a mythical ancestor to open an isthmus which separated the ocean from the valley and its protected shallow-water lagoons in order to establish harmony between birds and land-dwellers over disputed access to lagoons for freshwater (Roberts and Mountford, 1969). Palaeoenvironments inferred from benthic foraminifera provide scientific evidence for the marine flooding of northern Spencer Gulf, consistent with the dreamtime legend. A calibrated radiocarbon age of $12,630 \pm 230$ years BP (Before Present, which in the case of radiocarbon is before 1950 AD) on a specimen of the intertidal to shallow subtidal cockle *Katelysia* sp. from a vibrocore of estuarine carbonate lagoon sediments collected in a modern water depth of 20 m indicates marine flooding of the northern gulf by that time (Cann et al., 2000), and attests to the antiquity of the dreamtime legend.

Evidence for abrupt drowning of the Black Sea has been examined by Ryan et al. (1997) and Ryan and Pitman (1998), who argue that around 7,600 years ago, close to the culmination of the postglacial rise in sea level, seawater burst through the narrow straits of the Bosporus Valley. They argued that this event was responsible for many of the long-standing myths concerning floods, including the Epic of Gilgamesh. Before the flooding event, the Black Sea was a large freshwater lake of significantly smaller extent. Ryan and Pitman (1998) suggested that the lake level was up to 100 m below the sill which divides the Bosporus from the present Black Sea. They inferred that inflow of water from the Bosporus to the present Black Sea was of the order of 50–100 km^3/day (greater than 200 times the flux over Niagara Falls). A counter-argument was presented by Görür *et al.* (2001) suggesting that the event was not as dramatic as suggested by Ryan and Pitman (1998) and that the change in water level was only about 18 m. A large literature has emerged on the topic and some of the controversies surrounding the Black Sea flooding event have been exacerbated by uncertainties in the value of the marine reservoir effect adopted in radiocarbon dating (Nicholas *et al.*, 2011).

The theme of flood legends is briefly explored by Lambeck (1996a) in the context of postglacial sea-level rise within the Persian Gulf and the emergence of civilisation in the Fertile Crescent. The present shoreline of the Persian Gulf was established sometime shortly before 6,000 years BP (^{14}C years). The rising sea inundated extensive low-lying areas of Lower Mesopotamia, and is likely to have constrained migration of people and establishment of the earliest settlements in Lower Mesopotamia. It is speculated that the gulf floor before sea-level rise would have represented a logical route for people travelling westward from the east of Iran during the late Palaeolithic ($> 10,000$ years BP) to early Neolithic (6,000 years BP). Excavations at the city of Ur have provided evidence for a flooding event some 5,000–6,000 years BP; the water god Enki (or Ea) appears to have been particularly honoured in the holy city of Eridu, which was situated at the mouth of the Apsu (Guirand, 1996), and it is possible that the Sumerian flood legend actually documents the culmination of postglacial sea-level rise.

1.4 Observations from classical antiquity until the nineteenth century

Since earliest antiquity, there have been suggestions of changes in the relative positions of land and sea. In his second book entitled *Euterpé*, Herodotus [c. 485–425 BC] refers to marine shells in the sediments beneath the plains of the Nile, as well as the hillslopes bounding the wide Nile Valley, and considered that the region had once been covered by the sea (Komroff, 1947). The eighteenth-century development of geology as a formal scientific discipline saw a series of refinements as observations in the field began to play a more significant role and replace a strict interpretation of biblical teaching. The Neptunist and Vulcanist schools of thought attempted to describe many seemingly disparate geological observations. The Neptunists emphasised the significance of marine agencies for the origin of rocks (exogenetic processes) whilst the Vulcanists appreciated the importance of igneous processes operative within the Earth's crust (endogenetic processes). Thus, relative changes in sea level were attributed to sea-surface variations by the Neptunists, and to crustal processes by the Vulcanists.

1.4.1 Early Mediterranean studies

Aristotle [384–322 BC] discussed the emergence of the coastal plains of Libya and noted that low-lying lands formed in response to formation of a 'barrier of silt' leading to the desiccation of lakes, and in his work on *Meteorology* was clearly aware of relative sea-level changes, as apparent from the quotation that heads this chapter. The geographer and historian Strabo [c. 64 BC–c. 23 AD] sought to explain by what processes marine shells became buried at high elevations in the landscape. He suggested that changes in land level were simultaneously accompanied by sea-level changes, such that the rise in the seafloor heralded a rise in sea level.

Leonardo da Vinci [1452–1519] described shell deposits at high elevations near Monferrato in Italy in his notebooks written sometime between 1470 and 1480. He reasoned that:

From the two lines of shells we are forced to say that the earth indignantly submerged under the sea and so the first layer was made; and then the deluge made the second.

(da Vinci, *VI Geological Problems*, in Richter, 1970)

Elsewhere, Leonardo had argued against a deluge for the origin of accumulations of shell; however, he may have made the reference above to maintain a religious correctness, as at the time it was unclear what exactly the impressions of fossils within rocks actually represented. There was a view that they had been placed within the rocks by a divine creator to test the religious faith of mortals.

1.4.2 Eighteenth-century writings on universal changes to the Earth

Although numerous workers articulated theories of the Earth in what is now regarded as the Neptunist School of thought, history has accorded Abraham Gottlob Werner [1750–1817] as the strongest proponent of the movement. As with other Neptunists, Werner was of the view that a universal ocean had formerly covered the entire surface of the Earth, and that land had periodically been exposed as a result of a progressive lowering of the world ocean (eustatic sea level in modern parlance). Werner suggested that granite and other crystalline rocks had been precipitated from the universal ocean giving rise to *primitive* rocks. A *transitional series* of strata were then deposited by subaqueous depositional processes, resulting in formation of steeply dipping strata on crystalline rock. With the progressive lowering of sea level, emergence of islands enabled subaerial processes to weather the primitive rocks and promote deposition of *secondary* formations. A continued fall in the level of the universal ocean, as receding waters were sequestered in subterranean caverns, resulted in formation of alluvial successions (Tertiary strata) flanking the crystalline massifs. A consensus emerged that there had been an overall fall in the level of the universal ocean throughout Earth history (Rudwick, 2005), although some scientists believed there had been periodic short-term rises.

Evidence for emergence of land, such as shell beds at high elevations in the landscape, and the stranding of harbours inland from the coastline, was linked to a gradual fall in the level of the 'universal ocean', termed the 'diminution of the sea' by the French diplomat and naturalist Benoît de Maillet [1656–1738]. In a book entitled *Telliamed* (published posthumously in 1748, the title being the author's surname spelt in reverse), de Maillet (1748) described a dialogue between a fictional Indian scholar named Telliamed and a French missionary (Carozzi, 1969, 1992; Oldroyd, 1996). A scholarly translation of the complete work is given by Carozzi (1968) with detailed re-interpretations of the observations made by de Maillet.

Fennoscandia features prominently in the early scientific literature on relative sea-level changes, because there are former shorelines now found elevated well above sea level (Celsius, 1743; Linné, 1745). During his travels, Celsius [1701–1744] made numerous observations of these former shorelines and the shallowing of formerly navigable channels. He concluded that sea level had fallen (the 'diminution' of water following the universal deluge) by up to 4.5 feet (1.37 m) during the previous 100 years. This value was subsequently revised to 4 feet and became widely known as the 'Celsius-value' (Wegmann, 1969). Celsius was instrumental in initiating the practice of recording water levels on rocks marking them with a line, inscribed with the date when the recording was made. This practice continued after Celsius and represents an important source of historical information on relative sea-level changes.

In his *Oratio de Telluris Habitabilis Incremento*, published in 1743, Linné envisaged dry land appearing from a once almost universal ocean. Following on from the work of Celsius and Linné, the Swedish historiographer, Olof Dahlin, had caused political commotion with his reconstruction of the former shape of the Swedish kingdom (Wegmann, 1969). He concluded that the country was formerly an archipelago, and had existed in the form it did at the time at which he wrote for only a third of history since creation. This apparently caused scandal because people '... believed, the dignity of the kingdom had been challenged' (Wegmann, 1969, p. 391).

1.4.3 Diluvial Theory – the universal flood

A central tenet of the Diluvial Theory as propounded by Reverend William Buckland [1784–1856], Reader in Geology at Oxford University, was the notion that major changes on the Earth's surface could be readily understood through literal interpretation of the Bible, particularly the deluge recorded in Genesis (Buckland, 1823). This represented a significant intellectual constraint for the development of ideas about sea level, and environmental changes in general. Accordingly, in the eighteenth and early nineteenth centuries, many geomorphological features, such as isolated glacial erratics perched on hilltops, boulder clay (till), and related superficial moraine deposits, as well as erosional landforms, were attributed to the actions of a universal deluge (Conybeare and Phillips, 1822). Some attempted to explain the origin of poorly sorted superficial deposits mantling the landscape and erosional features such as striated bedrock surfaces as the product of a mega-tsunami (Diluvian waves) (Hall, 1815). The term 'drift', introduced by Roderick Murchison (1839) in his book, *The Silurian System*, was subsequently invoked to describe seemingly chaotic assemblages of poorly sorted, unconsolidated sediments formerly termed *diluvium* by Buckland, who believed they formed during Noah's Flood (the term drift is still used by the British Geological Survey for the mapping of unconsolidated deposits of Quaternary age irrespective of their mode of formation).

Timing of the deluge was constrained by literal interpretation of the Bible. Based on a combination of astronomical evidence, historical information and genealogical studies of the Bible, James Ussher, Archbishop of Armagh in Ireland, concluded that the '. . . beginning of time according to our Chronologie, fell upon the entrance of the night of the Julian Calendar, 710 (4004 BC)' (Ussher, 1658, p. 1) and accordingly, the 'flood deposits' must have been significantly younger. In 1647 John Lightfoot, Vice-Chancellor of the University of Cambridge, suggested that the moment of creation (i.e. age of the Earth) was on 26 October 4004 BC at 9 a.m. Ussher, however, is remembered for this calculation as it was included as a side note in the authorised King James I version of the Bible (Brice, 1982; Dalrymple, 1991).

The literalist interpretation of scriptures imposed a rigid explanation for human origins and an unrealistically short time interval for the development of landforms, the Earth's major relief and the sedimentary fill of large depositional basins. In addition, emergent shell beds were commonly regarded as contemporaneous with Noah's flood rather than as evidence for relative changes in sea level.

1.4.4 The Temple of Serapis: a compelling case for relative sea-level change

In the ninth edition of his celebrated work, *Principles of Geology*, published in 1853, Charles Lyell [1797–1875] described a market place, popularly known as the Temple of Serapis at Pozzuoli, near Naples in southern Italy, which provides evidence of a recent relative change in sea level during historical times. The monument has become immortalised as the foremost example of recent relative sea-level changes through its inclusion in the frontispiece, and as a gold embossed inlay on the cover of various editions of Lyell's *Principles of Geology*. A particularly detailed description of the site was also presented by the mathematician Charles Babbage [1792–1871] (Babbage, 1847), inventor of the adding machine and author of *Reflections on the decline of science in England* published in 1830. At a regional landscape scale, the Roman marketplace at Pozzuoli is situated within the centre of the Phlegrean Fields caldera in which Monte Nuovo and Solfatara occur, a feature about 13 km in diameter whose earliest phase of volcanism extends back around 60 ka (Morhange *et al.*, 2006). The Roman marketplace was excavated in 1750 by removing volcanic ash which had partially buried the structure (see Figure 3.13a). Three vertical marble columns are covered in the markings made by marine bivalves, *Lithodomus* sp., the upper limit of which ranges between 5.68 and 5.98 m above present sea level

(Babbage, 1847). Babbage and Lyell reasoned that the degree of destruction of the lower portions of the pillars implied a long immersion in seawater, and subsequent re-emergence. Lyell attributed the initial subsidence of the area due to the eruption of the Solfatara volcanic vent in 1198 AD and subsequent uplift in 1538 AD due to the formation of Monte Nuovo. Babbage suggested that the rocks under the columns had expanded due to heating and subsequently contracted.

There has been much speculation as to the cause of the relative changes in sea level at this site. Radiocarbon dating of subfossil marine molluscs from the ancient columns, coastal cliffs and an excavated Roman cave within the Pozzuoli region has revealed evidence for three phases of relative sea-level changes in historical times between the fifth and fifteenth centuries AD (Morhange *et al.*, 2006) rather than a single phase of submergence and subsequent fall in relative sea level as originally considered by Babbage and Lyell. The first inundation of the Roman market ended at c. 400–500 AD after its last restoration in 394 AD. A second oscillation in relative sea level occurred in the early Middle Ages (c. 700–900 AD). A further submersion occurred in the late Middle Ages. This latter event occurred after a protracted phase of uplift which culminated in the eruption of Monte Nuovo in 1538, the last recorded eruption within the caldera. Coastal uplift during this last phase resulted in newly emerged land being released by two royal acts in 1503 and 1511 (Morhange *et al.*, 2006). It is intriguing to note the discrepancy in the elevation of the upper marine limit on the pillars (5.68–5.98 m above present sea level (APSL)) reported by Babbage (1847) compared with 7 m APSL reported by Morhange *et al.* (2006). Given the meticulous observations by Babbage (1847) it is likely that the discrepancy relates to recent co-seismic uplift of the Pozzuoli region rather than to uncertainties of measurement.

1.4.5 Lavoisier and the concepts of transgression and regression

The French chemist Antoine Lavoisier [1743–1794] examined the Tertiary successions of the Paris Basin and made several profound observations about relative sea-level changes (Carozzi, 1965; Friedman *et al.*, 1992; Gould, 2000). Lavoisier's memoir represents the first formal description of transgressive and regressive deposits (Lavoisier, 1789). He explained the relative stratigraphical relationships of *pelagic* shelly limestones (*bancs calcaires*), *littoral* sands and gravel and clay (*galets, sable grossier, sable fin, argile*), and pelagic chalk with flint (*craie avec cailloux*) in the context of relative sea-level changes, a concept that was revolutionary for its time (Figure 1.2). This elegant explanation provided a mechanism for superposition of deeper-water sediments (*bancs pelagiens*) over sediments formerly deposited in shallow-water environments (*bancs littoraux*). The sequence of changes involved (a) erosion of chalk coastal cliffs and deposition of gravel, sand and clay in the littoral zone, not necessarily under conditions of rising sea level; (b) a period of slow relative sea-level rise, coastal erosion and shoreline retreat accompanied by

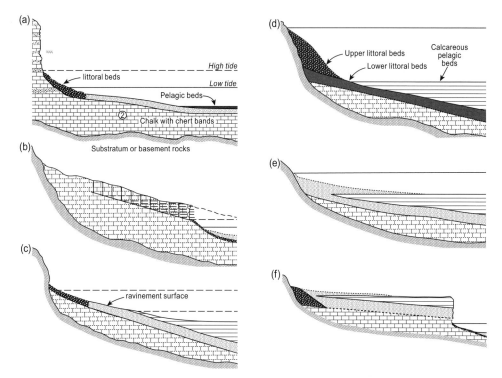

Figure 1.2 Lavoisier's conceptual schema of relative sea-level changes to explain the stratigraphical relationships of Tertiary marine strata within the Paris Basin (see text for discussion, adapted from Friedman *et al.*, 1992).

the deposition of a laterally persistent seaward-dipping, littoral deposit; and (c) continued relative sea-level rise resulting in the shoreline forming near mountains. A prominent seaward-dipping, transgressive marine erosion surface (ravinement surface in sequence stratigraphy terminology) is developed on the upper surface of the chalk by erosional processes associated with the rising sea surface, and sediments deposited in deeper water are progressively deposited on the littoral sediments. During a sea-level highstand (d), terrigenous clastic sediment is shed from the mountain range and deposited within the coastal environment resulting in a younger succession of littoral sediments (Upper littoral beds), which overlie an older succession of littoral sediments (Lower littoral beds) with development of a thicker succession of calcareous pelagic beds in deeper water. Accompanying a relative fall in sea level (e), sediments within the littoral environment are progressively deposited on the deeper-water calcareous pelagic beds, resulting in a well-defined sediment wedge; and (f) represents the final product of this sequence of depositional episodes as described by Lavoisier (1789) following a period of subaerial erosion (Carozzi, 1965). Although Lavoisier was unable to identify the causes of relative sea-level changes responsible for the successions and appealed to the physicists for an explanation, he did invoke modern sedimentary processes based on his under-standing of the coastline of northern France to explain the sequence of events (Rudwick, 2005).

Lavoisier's field observations preceded the contributions of Johannes Walther [1860–1937], who is generally credited for formulation of the concept of *lateral facies succession*: the notion that adjacent sedimentary environments and their preserved sediments, when viewed in a vertical stratigraphical profile, are seen to be super-posed. Such a study of rock relationships of pre-Quaternary sedimentary successions has contributed to interpretation of depositional sequences of Quaternary age, which represents an example of 'inverted uniformitarianism' whereby knowledge of older geological events is used to interpret more recent changes. Lavoisier's contribution is significant for recognising the possibility of cyclic sedimentation in response to sea-level changes (Carozzi, 1965; Gould, 2000), and also introduced the concept of an equilibrium cross-shore profile in response to sea-level changes (i.e. erosion, shoreline retreat and the landward translation of coastal sediments accompanying a relative rise in sea level).

1.5 Glacial action and recognition of the Ice Ages

Recognising the role of glaciations and sequence of ice ages that dominated the Quaternary was a major development which provided the theoretical framework within which to explain successive sea-level changes. The Glacial Theory, once it had been proposed and following energetic advocacy by Louis Agassiz, was rapidly adopted in the mid-nineteenth century. It provided an intellectual perspective that did not require religious dictates to explain scientific observations (i.e. in terms of sedimentary deposits that had been attributed previously to the deluge). A mechanism to explain repeated glaciations was at first elusive, but the orbital theory was proposed by James Croll, and once Milutin Milankovitch meticulously calculated the consequences of known periodicities of orbital parameters, enabled development of what has become known as the astronomical theory of ice ages.

1.5.1 Louis Agassiz and the Glacial Theory

Louis Agassiz [1807–1873] was Professor of Natural History at the University of Neuchâtel. He gained a reputation as a fish palaeontologist on the basis of a multi-volume monograph on fish fossils produced in the 1830s in which he considered that there had been several times at which taxa had experienced mass extinctions followed by new episodes of creation. Agassiz combined these ideas of catastrophism and gradualism, which enabled him to reconcile his view both with religious tradition envisaging several episodes in which there had been divine creation of new species (creationism) and the emerging view of uniformitarianism that had been initiated by James Hutton and promulgated by Charles Lyell. Initially it was not clear what could have caused the periods of extinction.

In 1836 Agassiz spent time with Johann de Charpentier who recognised evidence for more extensive glacial action in and around the Alps. He became persuaded that glaciations could explain features such as striated bedrock surfaces and boulders, as well as isolated rocks (erratics) that were scattered beyond their source region. Agassiz considered that successive glaciations, to which he gave the term *Eiszeit* after discussions with his colleague Karl Schimper, could account for the extinction events (he called the glaciers 'God's great plough'). His influential work *Etudes sur les glaciers* (1840) presented a synthesis of evidence in support of the former, wider extent of glaciers and icesheets in northern Europe and their extension south towards the Mediterranean. This explained the origin of many geomorphological and sedimentary features such as isolated dropstones and erratics of exotic lithologies, moraines, and a variety of erosional landforms now accepted as the product of glaciers and icesheets. Agassiz persuaded many other geologists of the significance of these former glacial advances. He was particularly successful in demonstrating that glaciations explained many of the landforms in Scotland; in particular he proposed that the 'parallel roads of Glen Roy', terraces along the hillside which Darwin had thought to be uplifted marine shorelines, were actually the shorelines of glacially impounded lakes (see Chapter 5, and Figure 2.7a). Several authors have presented a history of the adoption of the Glacial Theory (Chorley *et al.*, 1964; Imbrie and Imbrie, 1979; Hallam, 1983; Oldroyd, 1996; Rudwick, 2005).

The water which fell as precipitation (largely as snow) to produce extensive icesheets up to 2–3 km thick had evaporated from the oceans, and this would necessitate a substantial fall in sea level. As early as 1842, only two years after publication of Agassiz's treatise, Maclaren calculated that sea level at time of maximum icesheet development was likely to have been between 107 and 213 m below present sea level (BPSL) (Maclaren, 1842). However, it was some time before other researchers attempted to estimate the extent of sea-level lowering during maximum glaciation, and there was considerable variation in early estimates, ranging between 40 and 610 m (Daly, 1925, 1934; Guilcher, 1969; Milne *et al.*, 2002).

Early observations for changes in the relative position of land and sea in the British Isles, and principally in Scotland, were reported by Smith (1838). Smith described widespread elevated shelly deposits at a range of heights up to > 130 m APSL. He noted that some of the mollusc species within the elevated marine strata no longer live within the region. Smith also concluded that these apparent changes in relative sea level post-dated deposition of the 'diluvial' sediments and pre-dated the Roman period. Working at much the same time, the geologists John Phillips [1800–1874], Adam Sedgewick [1785–1873], Henry Thomas De la Beche [1796–1855], and Roderick Impey Murchison [1792–1871] made similar observations for relative changes in sea level in England (see references cited by Smith, 1838).

By the end of the nineteenth century, the Glacial Theory was widely accepted in the Earth Sciences. For example, a major monograph had appeared, entitled *The Great Ice Age*, by James Geike (1894) and the theory was covered in the popular works of Charles Lyell (1853, 1855). The former extent of ice masses had been determined based on careful mapping of glacial end moraines, trimlines, and related geomorphological features, but the cause of glaciations and their timing remained unclear (Ball, 1891; Lugg, 1978; Oldroyd, 1996). James Croll [1821–1890] had suggested a mechanism related to periodic variations of the Earth's orbit, which was to be championed by the Serbian mathematician, Milutin Milankovitch [1879–1958] in the early twentieth century. Their ideas, and their gradual adoption are outlined in the next section, and excellent overviews are provided by Imbrie and Imbrie (1979), Hallam (1983), and Williams *et al.* (1998).

Evidence for multiple Quaternary glaciations began to emerge with recognition of stratigraphically superposed tills. For example, Joshua Trimmer (1850) identified two distinct tills (boulder clay) in Norfolk, England, separated by peaty muds with freshwater shells, from which he inferred that the region had experienced more than one glaciation. As discussed later in this chapter, the subsequent work by Penck and Brückner (1909) provided more compelling evidence for multiple Quaternary glaciations in Europe, which it was subsequently realised meant repeated changes in sea level.

Grabau [1870–1946] attempted to explain the pre-Quaternary stratigraphical record from several continents in terms of globally synchronous sea-level changes (Grabau, 1940). He described cyclic changes in sea level evident in the geological record in terms of a 'pulsation theory' and identified evidence for flooding of continents, particularly cratonic regions. Grabau envisaged slow pulsatory rises and falls of the sea that were eustatic (global) in nature, in response to crustal movements, and in particular, vertical and horizontal movements of the Earth's oceanic crust, then termed sima (for silica and magnesium). A positive pulse-beat resulted in widespread flooding of continents (transgression), and a negative pulse-beat induced a sea-level fall (regression). Although largely forgotten today, Grabau's work is important for anticipating the now widely accepted precept that global sea-level changes provide an important framework for regional stratigraphical correlation through sequence stratigraphy (Van Wagoner *et al.*, 1988; Catuneanu, 2006).

1.5.2 The Croll–Milankovitch Hypothesis

The French physicist Joseph Alphonse Adhémar [1797–1862] first suggested the possibility that changes in configuration of the Earth's orbit accounted for episodes of glaciation. Adhémar (1842) concluded that glaciation alternated from the northern to the southern hemispheres rather than occurring in both hemispheres at the same time. His ideas have received less attention, however, due to his notion of the gravitational attraction of waters from the northern

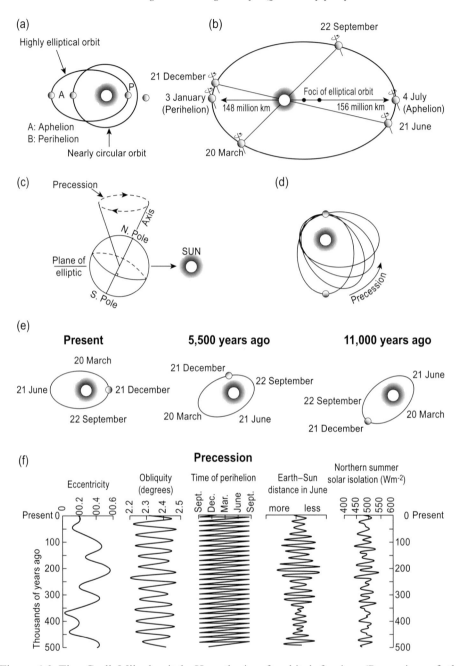

Figure 1.3 The Croll–Milankovitch Hypothesis of orbital forcing (Precession of the Equinoxes) as a mechanism of long-term global climate change; (a) changes in the *eccentricity* of the Earth's orbit around the Sun from a more circular to elliptical shape, illustrating the point at which the planet experiences aphelion (Earth farthest from the Sun) and perihelion (Earth closest to the Sun) conditions. The cycle takes 100 ka and a

hemisphere to Antarctica and episodes of catastrophic icesheet collapse with iceberg-laden tidal waves accompanying deglaciation.

James Croll systematically examined the association between long-term Quaternary climate changes and the concept of orbital forcing. Although widely credited to Milankovitch, who worked through calculations in support of periodic development of icesheets, the fundamental contributions of Croll (1875) have not been accorded the recognition they deserve. The concept of orbital forcing is therefore referred to here as the Croll–Milankovitch Hypothesis. The central premise of the Croll–Milankovitch Hypothesis is that changes in geometry of the Earth's orbit around the Sun influence the intensity of incoming solar radiation (insolation) received on the surface of the Earth, which in turn leads to long-term climate change (Milankovitch, 1941; Figure 1.3). Milankovitch suggested that three aspects of configuration of the Earth's orbit around the Sun were primarily responsible for long-term climate change. These were changes in eccentricity (100-ka cycle), obliquity (41-ka cycle), and precession (27-ka cycle) of the Earth's orbit.

The eccentricity cycle involves a change in the long axis of the ellipse of the Earth's orbit from a more circular orbit to a more pronounced ellipse, and occurs over a period of about 100 ka. A related longer cycle takes about 400 ka; the spectral peaks are actually 96, 125 and 413 ka (Maslin and Ridgwell, 2005). At present, the annual difference in distance of the Earth from the Sun between its closest and farthest positions is about 10 million km. The Earth is closest to the Sun on 3 January (*perihelion*) and farthest from the Sun on 4 July (*aphelion*). The average amount of insolation received at perihelion is 351 W/m^2 and about 329 W/m^2 at aphelion. Overall, eccentricity accounts for only about 0.03% of the change in the Earth's annual insolation budget. However, the eccentricity cycle is important for modulating effects of the precession cycle, and is also significant in influencing seasonal variations in insolation received, because the intensity of insolation reaching the Earth's surface decreases according to the square of the planet's distance from the Sun. This favours preservation of icesheets, particularly during summer months when they would otherwise experience significant ablation.

Obliquity of the Earth's plane of ecliptic involves a change in tilt of the Earth's axis of rotation which ranges from 21.8° to 24.4° and varies with a periodicity of

---

Caption for Figure 1.3 (*cont.*) longer 400-ka cycle has also been recognised; (b) the *obliquity of the ecliptic* representing the change in the tilt of the Earth's rotation axis from 21.8° to 24.4° over a period of 41 ka. The change in the angle of the rotation axis to the plane of the ecliptic amplifies seasonal changes. The current orbital conditions are illustrated, showing the position of the planet during different seasons; (c) the precession of the Earth's spin axis which, over a 27-ka period, defines a circle; (d) the precession of the Earth's orbit around the Sun over a 105-ka period; (e) the precession of the equinoxes in which perihelion occurs in each hemisphere during summer approximately every 21.7 ka; (f) resultant variations in the Earth's orbital parameters as registered at 65°N, the critical latitude for orbital forcing as identified by Milankovitch (modified after Maslin and Ridgwell, 2005).

41 ka. At present the tilt of the Earth's axis is 23.44°. Changes in obliquity enhance seasonality; the greater the obliquity, the more pronounced will be the difference between summer and winter.

The third set of orbital processes involves the Earth's precession, including changes in the elliptical orbit and the relative position of the axis of spin. The Earth's rotational axis, as defined at a fixed point such as the north geographic pole, undergoes precession over a period of 27 ka. Viewed in three-dimensional space, the axis of rotation, if considered in the context of a segment such as the radius of the planet, defines a cone of rotation, in principle analogous to a child's gyrating spinning top. In addition, a precession in the Earth's orbit around the Sun occurs over a period of 105 ka and determines the time of the year when the planet is closest to the Sun. The precessional changes result in progressive shift of the celestial position of *equinoxes* (either of the two instants at which the centre of the Sun appears coincident with the celestial equator resulting in equal lengths of day and night). Accordingly, the northern hemisphere vernal equinox (21 March) was situated a distance of approximately a quarter of the perimeter on the Earth's annual ellipse from its current position 5.5 ka ago and was a greater overall distance away from the Sun.

Collectively, changes in the Earth's orbital parameters control the long-term insolation budget of the planet. Milankovitch championed this elegant hypothesis for long-term climate change and suggested that the critical latitude for effecting changes from interglaciations to glaciations was 65°N. The land area is particularly extensive at this latitude which promotes development of icesheets, and, consequently, the curve of summer insolation at this latitude has been regarded as an indicator of the global climate curve (Hays *et al.*, 1976).

Although now seen as an elegant explanation for the origin of long-term climate changes, the Croll–Milankovitch Hypothesis fell into disfavour in the middle of the twentieth century in view of the lack of long-term proxy palaeoclimate data by which to quantitatively validate timing of glaciations. Advances in geochemistry, particularly in application of stable isotopes, and development of dating methods, represented crucial advances towards validating the Croll–Milankovitch Hypothesis. The seminal study by Hays *et al.* (1976) provided compelling evidence that variations in the Earth's orbit were the pacemaker of ice ages. Based on spectral analyses of oxygen-isotope data for the foraminifer *Globigerina bulloides* derived from two deep-sea cores (RC11–120 and E49–19) from the southern Indian Ocean, Hays *et al.* (1976) concluded that the general timing of climate changes was consistent with that predicted by Milankovitch, with spectral peaks represented at 23, 42 and 100 ka.

There is now much evidence that supports this hypothesis; although as so often is the case in the Earth Sciences, consistency between observation and predictions cannot be regarded as verification (Oreskes *et al.*, 1994). Complex feedbacks and lags modulate the influence of orbital parameters, making it more difficult to establish the direct causes of long-term climate changes (Ruddiman, 2006). Offsets between numerical ages and the mathematically modelled timing of insolation maxima, despite advances in the age-resolving power of methods such as U-series dating (see Chapter 4), may be due to factors such as sediment reworking or diagenesis in samples selected for dating, as well as systematic uncertainties in measurement techniques. Further difficulties arise in explaining aspects of the longer-term palaeo-climatic and sea-level records of the Quaternary. For example, spectral analyses of isotopic data from deep-sea cores indicates that glacial cycles lasted about 100 ka for the past 1 Ma, yet the eccentricity signal (100-ka cycle) is actually the weakest of the orbital forcing variables. Another complexity is evident in the interglacial that occurred some 400 ka ago (MIS 11); although a 400-ka cycle was identified by Milankovitch (1941) it is not expressed in the spectral records preceding this interglacial. A further difficulty relates to timing of the onset of the last interglacial maximum and the fact that it may have commenced some 10 ka before the solar forcing that is suggested to have triggered this event. Finally, the notable shift from the 41-ka cycle to the apparent 100-ka cycle (early–middle Pleistocene transition) evident in long isotopic records from deep-sea cores (Shackleton *et al.*, 1990, see Chapter 6) presents a further enigma and as yet its basis remains unexplained.

1.6 Vertical changes in land and sea level related to Quaternary climate

Voyages of exploration in the eighteenth and nineteenth centuries provided unpreced-ented opportunities for natural scientists to observe foreign lands. Many observations by scientists in previously uncharted territories were of geomorphological features developed at a scale not previously recognised in their homelands. This included instances where the land had been uplifted, such as the marine terraces of Patagonia and Chile (Darwin, 1845, 1896), and remote islands surrounded by reefs which it was inferred had experienced subsidence (Darwin, 1842; Dana, 1872).

1.6.1 Charles Darwin and James Dana

Publication of Lyell's *Principles of Geology* coincided with Charles Darwin's voyage around the world aboard HMS *Beagle*. Sailing along the South American coast, Charles Darwin [1809–1882] was impressed by the lateral persistence and constant elevation of successive marine terraces, developed on the Tertiary and Quaternary formations. Speculating on the origin of the features, Darwin (1845, p. 163) wrote:

The uprising movement has been interrupted by at least eight long periods of rest, during which the sea ate deeply back into the land, forming at successive levels the long lines of cliffs or escarpments, which separate the different plains as they rise like steps one behind the other. The elevatory movement, and the eating-back power of the sea during the periods of rest, have been equable over long lines of coast; for I was astonished to find that the step-like plains stand at nearly correspond-ing heights at far distant points.

During his voyage around the world, Darwin also described raised marine terraces of Coquimbo Bay in northern Chile (Darwin, 1845, 1896). He noted five prominent, gently seaward-sloping terraces comprising shingle, and observed marine shells scattered on the surfaces of the terraces, the highest reaching up to 80 m APSL. Darwin attributed their origin to marine agencies during progressive uplift of the land. Then on 20 February 1835 he was resting on the ground in Valdivia when a particularly severe earthquake struck, its epicentre not far away. When the *Beagle* reached the port of Concepción several days later, the devastation of the town and the suffering of its inhabitants made a deep impression on Darwin and the crew of the ship. Darwin observed a rim of molluscs and barnacles raised almost a metre out of the water at Concepción, and similar evidence for uplift of more than 2 m on the island of Santa Maria, with even greater uplift due to this event at other nearby sites. Evidence from shallowing of the harbour indicated that this pattern of uplift had been repeated episodically.

Darwin surmised that repeated uplift by small increments, such as had occurred during the Concepción earthquake, could gradually raise this mountainous coast. He extended the suggestion made by Alexander von Humboldt that earthquakes were caused by the same subterranean pressures that drove volcanoes, and inferred that this uplift could explain elevation of former shorelines and shell beds. Extrapolating from his observational evidence, this provided a fitting example of the uniformitarian approach to geology advocated by Lyell.

In a remarkable deduction, Charles Darwin extrapolated from the observation that land on the eastern margin of the Pacific was undergoing uplift, to infer that mid-ocean islands might undergo subsidence, and that this could explain the origin of coral atolls. He envisaged that fringing reefs, barrier reefs and atolls represented different stages as a result of vertical growth of reefs on subsiding volcanic foundations. The most widely accepted interpretation at the time for the numerous annular reefs in mid ocean was the theory proposed by Lyell that these were the coral-rimmed tops of volcanoes. Darwin proposed an alternative view; he suggested that there had been 'prolonged subsidence of the foundations on which the atolls were primarily based, together with the upward growth of the reef-constructing corals', and concluded that 'on this view every difficulty vanishes: ... fringing reefs are thus converted into barrier-reefs; and barrier-reefs, when encircling islands are thus converted into atolls, the instant the last pinnacle of land sinks beneath the surface of the ocean' (Darwin, 1842, p. 109). As his sketch showed (Figure 1.4), fringing reefs, barrier reefs and atolls were stages in an evolutionary sequence, driven by gradual subsidence of the volcanic island around which the reef had initially formed. Although Darwin's reasoning did not require a change in the absolute amount of water within ocean basins (i.e. such as that resulting from ice-equivalent, glacio-eustatic sea-level changes), his coral-reef theory required subsidence and a relative rise in sea level to explain the sequence of reefs. Subsequent drilling has provided support for this idea, which can be integrated into plate-tectonic theory, and can also accommodate many factors associated with changes of sea level (Guilcher, 1988).

(a)

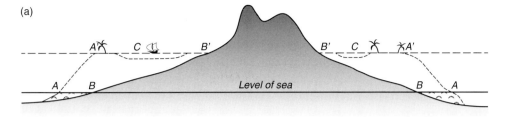

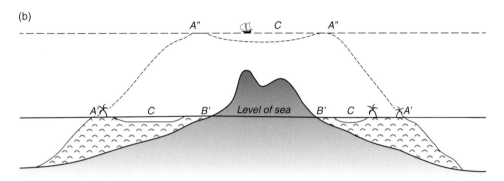

(b)

Figure 1.4 Charles Darwin's conceptual model of coral atoll formation, under conditions of relative sea-level rise, caused by subsidence of the underlying volcanic substrate. The figure illustrates the successive changes from (a) a fringing reef to a barrier reef with lagoon to (b) an atoll (Darwin, 1845: this copy redrawn from pp. 454 and 455 from the 1905 edition published by John Murray, Albermarle Street, London).

James Dwight Dana [1813–1895] read about Darwin's ideas in a Sydney newspaper, and became an ardent supporter of the subsidence theory. Dana was a member of the US Exploring Expedition to the Pacific Ocean, from 1838 to 1842, and he had many more opportunities to observe reefs than had Darwin. Dana (1872) observed that shorelines of volcanic islands behind barrier reefs comprised a series of embayments, implying that valleys had been submerged during subsidence. Later, William Morris Davis [1850–1934] extended Dana's argument that embayments in volcanic shorelines represented drowned valleys (Davis, 1928). The widespread occurrence of coral atolls in the Pacific Ocean was regarded as evidence for geographically extensive subsidence [cf. more localised subsidence within foredeeps as conceptualised by Haug (1900)], and Dana (1872) noted that the extent of subsidence was greater in the northeastern Pacific Basin than in the south.

Reginald Aldworth Daly [1871–1957] played an influential role in studies of relative sea-level changes and invoked successive glaciations to explain the sequential development of coral reefs, in what he called the Glacial Control Theory (Daly, 1915). In his benchmark book, *The Changing World of the Ice Age* (Daly, 1934),

Daly argued that atoll rims were entirely planated during the glacial maximum when sea level was around 100 m lower than present, and that the regularity to the depths of lagoons provided support for this 'glacial-control' view of reef development. However, it is clear that modern reefs have not accreted vertically through this depth of water during the postglacial, although reef growth at various depths around mid-ocean islands provides important evidence on past sea-level changes, as will be discussed in later chapters.

1.6.2 Insights from around the world

As evidence for worldwide changes in sea level became more widely accepted, it was considered that diastrophism (ocean crustal movements) was responsible. Eduard Suess [1831–1914] noted that transgressions and regressions from many widely separated regions were broadly coeval globally (Suess, 1888). On this basis, Suess formalised the notion of *eustatisch* (eustatic or worldwide) changes in sea level and considered that altered volume of the ocean basins caused sea-level changes in the longer-term geological record. He suggested that a potential mechanism was the increase in ocean volume resulting from the formation of new abyssal areas in the ocean basins, as he could not accept that all the continents had experienced vertical movements in unison. In a similar manner, Robert Chambers [1802–1871] had hypothesised that oscillations in the ocean floor were responsible for changes in sea level (Chambers, 1848). Identification of deep ocean trenches (foredeeps) adjacent to foldbelts provided scope for changes in ocean volume (Haug, 1900).

In 1801 during their voyage around the Australian coastline, the French explorers and navigators Nicholas Baudin, François Péron, and Louis Freycinet made numerous observations relating to relative changes in sea level. They found emergent shell beds within calcarenites (now termed Tamala Limestone, of middle to late Pleistocene age) along the south Western Australian coastline, and Péron was of the view that the coastal landscape had formed by the slow retreat of the sea (Mayer, 2008). In 1802, during his circumnavigation of the Australian continent, Matthew Flinders concluded that extensive Neogene limestones forming the Nullarbor Cliffs of southern Australia represented an emergent reef and had resulted from '... the gradual subsiding of the sea, or perhaps from some convulsion of nature' (Flinders, 1814, Vol. 1, p. 97).

Resulting from his extensive travels associated with his pastoral duties, the Reverend Julian Edmund Woods [1832–1889], who later published as Tenison-Woods, was the first to systematically describe the unique sequence of interglacial, limestone coastal barrier deposits of the Coorong Coastal Plain in southern South Australia (Woods, 1862; Figure 1.5). He noted the great length of the dune ranges which occur subparallel to the modern coastline of the region (e.g. he described a dune range that could be traced in excess of 80 km from Penola, northwards to Naracoorte; a feature now mapped as the West Naracoorte Range of early–middle

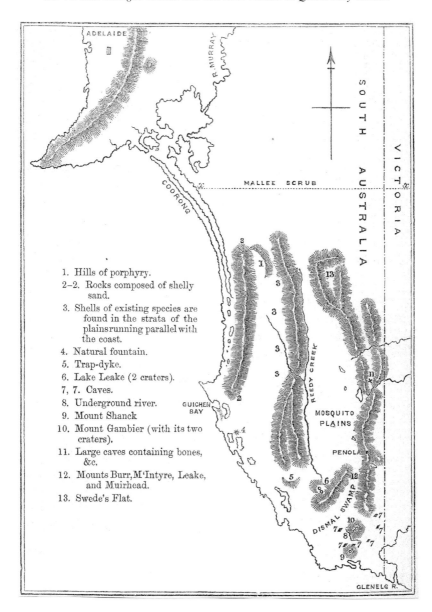

1. Hills of porphyry.
2-2. Rocks composed of shelly sand.
3. Shells of existing species are found in the strata of the plains running parallel with the coast.
4. Natural fountain.
5. Trap-dyke.
6. Lake Leake (2 craters).
7, 7. Caves.
8. Underground river.
9. Mount Shanck
10. Mount Gambier (with its two craters).
11. Large caves containing bones, &c.
12. Mounts Burr, M'Intyre, Leake, and Muirhead.
13. Swede's Flat.

Figure 1.5 A map of the Coorong Coastal Plain, southern South Australia, compiled by the Reverend Julian Edmund Tenison-Woods (1862). The pronounced lateral persistence of the coastal barrier landforms and that they signified a change in the relative positions of the land and sea were clearly appreciated by Woods.

Pleistocene age; Murray-Wallace *et al.*, 1998). Woods also suggested that (1) ongoing uplift of the coastal plain explained the origin of the raised beaches, coastal lagoons and salinas; (2) each range corresponded with a former coastline; and (3) the flats on the landward side of each barrier represented former estuaries.

1.7 Evolution of ideas in the twentieth century

The twentieth century witnessed an unparalleled explosion in knowledge about sea-level changes, particularly for the Quaternary Period. First, the more systematic examination of geological evidence in Europe provided a clearer idea of climate and sea-level changes that had occurred there. These advances were in part bolstered by important technological developments in geochemistry after the Second World War, as well as more observational fieldwork, and a proliferation in scientific publishing and dissemination of published works. Development of a range of geochronological methods, particularly radiocarbon and U-series dating, and stable isotope geo-chemistry revolutionised understanding of the rates and nature of geologically recent sea-level changes.

1.7.1 Developments in Europe

Penck and Brückner (1909) examined the stratigraphical relationships of glacial outwash terraces with terminal moraines in the Bavarian plateau south of Ulm, Augsburg, and Munich. Their work, focused on the Alps, culminated in a model of four glaciations in Europe for the Quaternary (Figure 1.6): Günz, Mindel, Riss, and Würm. During interglacials, rivers progressively cut down into previously deposited glacigene sediments and glacial deposited clasts underwent enhanced weathering. Accordingly, the duration of different interglaciations was inferred from the intensity of valley incision and the thickness of weathering rinds on glacial-deposited clasts. On this basis, Penck and Brückner (1909) concluded that in the Alps and in France, the Mindel–Riss Interglacial had the longest duration. For many years their scheme influenced palaeo-sea-level interpretations by workers in other countries including regions remote from Quaternary glaciation (Figure 1.6). Identification of multiple glaciations implied repeated relative sea-level changes during the Quaternary Period, which was subsequently established from studies of emergent coastal successions of the Mediterranean Basin.

Extensive coastal geomorphological and stratigraphical studies of the Mediterra-nean Basin resulted in the suggestion of a progressively falling sea surface during the Quaternary, embodied in the so-called 'traditional glacio-eustatic theory' (Guilcher, 1969). These ideas were reminiscent of the concepts introduced by Celsius, de Maillet and Werner, of the 'diminution of the sea'. Based on the results of numerous investi-gations in Algeria, France and Italy, it was concluded that the sea surface had oscillated in response to successive glaciations and interglaciations and concomitant ice-volume changes (De Lamothe, 1899, 1911, 1918; Gignoux, 1913; Chaput, 1917, 1927; Depéret, 1918; Daly, 1934; Baulig, 1935). It was concluded that stillstands were sufficiently long to permit the marine erosion of bedrock to form shore platforms. Specific sea-level highstand events APSL were assigned geographical names based on where the raised beaches were first formally described. The coastal highstand

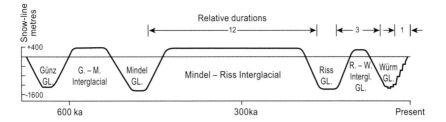

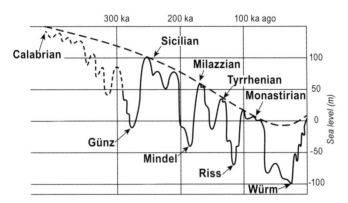

Figure 1.6 The fourfold model of Alpine Glaciation as proposed by Penck and Brückner (1909) and the glacio-eustatic sea-level oscillations as deduced from the Mediterranean Basin (after Fairbridge, 1961, with permission from Elsevier). The apparent overall decline in the elevation of the sea surface in relation to the land actually reflects a record of tectonic uplift in parts of the Mediterranean Basin rather than the progressive accumulation of ice in Antarctica. Note the compressed timescale of the Fairbridge curve compared with the timescale assumed by Penck and Brückner.

deposits and marine terraces were correlated regionally, based on palaeoshoreline elevation and with successive interglacials that included the Sicilian (80–100 m APSL), Milazzian (55–60 m APSL), Tyrrhenian (30–35 m APSL), Monastirian (15–20 m and 0–7 m APSL), and the most recent, the Flandrian (Holocene postglacial) (Zeuner, 1952; Guilcher, 1969; Figure 1.6).

The progressive decrease in altitude of individual shoreline deposits in the Mediterranean region was attributed to ongoing lowering of the ocean basins, and in particular, the Pacific Basin (Daly, 1934). To some extent this represents a surprising conclusion as the Glacial Theory was well accepted by the end of the nineteenth century, and accordingly, provided a valid mechanism *ipso facto* to explain major fluctuations in sea level. The glacio-eustatic explanation as proposed at the time by many workers generally did not consider the role of vertical tectonic movements for the Mediterranean raised shoreline deposits. In his synthesis on former sea levels in northwestern Europe, however, Dubois (1925) acknowledged that some tectonic uplift was required to explain emergent shorelines, for their elevation exceeded the potential rise in sea level from total melting of existing icesheets. Correlation of marine terraces over large distances is difficult, as many have been partially destroyed by marine and terrestrial processes. Nevertheless, in a study of the emergent shorelines of the Mediterranean Basin, Hey (1978) concluded that regional variability in elevation of individual shorelines meant that there was not a consistent basin-wide level for interglacial sea levels and that accordingly all terraces provided evidence of tectonic uplift. It is noteworthy that in the absence of tectonism, the altitude of the Milazzian shoreline would require complete deglaciation of the Antarctic (22,100 km^3) and Greenland (2,480 km^3) icesheets in order to raise sea level by as much as 66 m (Guilcher, 1969). However, the lingering problem of not being able to assign ages to these deposits represented a major impediment for understanding rates of change.

The former beaches described from southern South Australia by Woods (1862) provided a similar sequence to those in Europe, with the innermost elevated above the modern shore. In a paper presented to the 18th International Geological Congress in London, Sprigg (1948) described a record of 17 highstands at this site, which he considered the 'best preserved of the Pleistocene continental suite anywhere in the world'. He drew attention to the match with the 'sea-level oscillations, or more correctly, the variations in climate which it is assumed produced them' that resulted from the theory of Milankovitch. It was 'almost suspiciously good', and Sprigg thought it 'improbable that such good correlation would be purely fortuitous'. Again, further confirmation of the relationship was to await appropriate means of deriving ages for the landforms or the fossils they contained.

1.7.2 Advances in geochemistry and geochronology

The discovery of the isotope radiocarbon (^{14}C) by Willard F. Libby [1908–1980] and colleagues in 1946–1947 led to what has been appropriately termed a revolution in the study of recent Earth history (Renfrew, 1973). Developments in radiocarbon dating permitted carbon-bearing materials from a diverse range of geomorphological and stratigraphical contexts to be directly dated. The radiocarbon chronologies constrained the timing of geologically recent sea-level changes (for the past 50,000 years), and revealed that the most recent deglaciation occurred relatively rapidly with a concomitant rapid rise in sea level (Fairbanks, 1989; Chappell and Polach, 1991). There were some limitations to the radiocarbon studies, however; many early studies suggested the presence of high interstadial sea levels close to present sea level during the last glacial cycle (around 30,000 years ago) that are now known to be a result of sample contamination by younger carbon sources on fossil deposits actually of last interglacial age (125 ka). The contaminants could not be removed during early radiocarbon laboratory pre-treatment strategies and resulted in the derivation of many apparently 'finite' radiocarbon ages ≥ 30 ka (Thom, 1973; Giresse and Davies, 1980; Chappell, 1982, see Section 6.11).

Uranium-series dating, and in particular the activity ratios of the isotopes of uranium and thorium, represented a further important method for determining the age of coastal deposits, particularly for corals that formed during the last two interglaciations. U-series dating provided the first compelling evidence for the age of last interglacial and penultimate interglacial deposits (Barnes *et al.*, 1956; Veeh, 1966; Stearns, 1984). It also yielded preliminary numerical ages to assess the veracity of the Croll–Milankovitch Hypothesis as an explanation of long-term climate change. Since these early investigations, numerous publications have appeared on U-series dating of fossil corals from flights of emergent reef terraces, and some of the most important settings include the Huon Peninsula, Papua New Guinea (Chappell, 1974a; Chappell *et al.*, 1996a,b), Barbados (Mesolella *et al.*, 1969), and the Ryukyu Islands, Japan (Konishi *et al.*, 1974; Ota and Machida, 1987). These regions have received considerable attention because rapid rates of tectonic uplift result in clear physical separation of different sea-level highstand deposits, thus reducing potential ambiguities in geomorphological interpretations. Corals were also soon realised to be particularly suitable for dating by the U-series method, and with detailed stratigraphical analysis, confident estimates of palaeo-sea level could be inferred from the palaeoecological context of the fossil reef complexes.

1.7.3 Oxygen-isotope records from marine sediments and ice cores

Coupled with these important developments in geochronology was the advancing science of stable isotope studies. Harold Urey [1893–1981] noted that when water is heated, the stable isotopes of oxygen undergo fractionation whereby the lightest isotope (^{16}O) is preferentially selected during evaporation, such that the remaining liquid is enriched in the heavier isotopes of oxygen (^{17}O and ^{18}O), a process termed *isotopic fractionation*. This fundamentally important discovery (Urey, 1948) prompted Cesare Emiliani [1922–1995] to investigate the geochemical potential of this fractionation reaction in understanding the dynamics and nature of long-term climate change during the Pleistocene (Emiliani, 1955).

Emiliani (1955) examined the isotopic signature of fossils of the foraminifera *Globigerinoides* spp., *Globigerina* sp., and *Globoratalia* sp. from *Globigerina* ooze in 12 deep-sea cores from the Atlantic, Caribbean, and Pacific Oceans. He concluded that the down-core isotopic changes within the foraminiferal tests provided a signature of long-term ocean temperature changes, and, as a corollary, a proxy of glaciation and deglaciation (Figure 1.7). Emiliani's work revealed evidence for at least 15 glaciations during the Pleistocene, based on evidence from a Pacific deep-sea core, and therefore indicated repeated major changes in sea level at a global scale. These repeated changes in ice volume were described according to a numbering

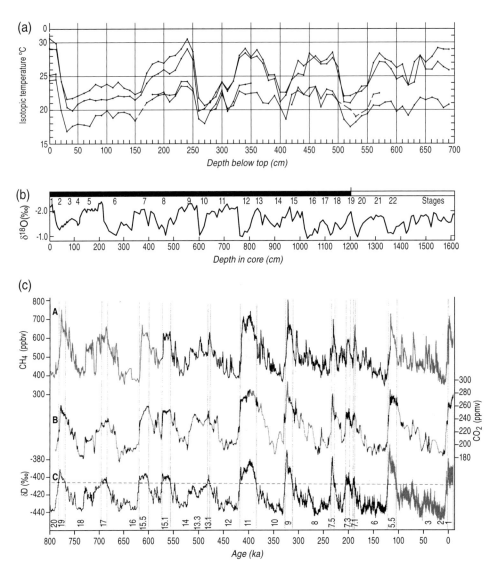

Figure 1.7 (a) Oxygen-isotope inferred palaeotemperatures from the fossil foraminifera *Globigerinoides ruber, G. sacculifera, Globigerina dubia* and *Globorotalia menardii* from marine core A179–4 (Emiliani, 1955), compared with (b) the oxygen isotopic composition of the foraminifer *Globigerinoides sacculifera* in Core V28–238 from the Solomon (Ontong Java) Plateau, Equatorial Pacific Ocean (Shackleton and Opdyke, 1973). These classical marine isotopic records of global ice-volume change, and as a corollary, Quaternary sea-level changes are compared with (c) geochemical analyses from the European Project for Ice Coring in Antarctica (EPICA) Dome C ice core that provides a palaeoclimatic record for the past 800 ka (Schilt *et al.*, 2010). The records for (A) CH_4 (ppbv), (B) CO_2 (ppmv) and (C) δD are shown. In the δD curve, the dashed line highlights peaks that represent interglacials, with the timing of interglacials represented by the portion of the peak above the dashed line.

scheme that remains in use today, whereby even numbers designate cold, and odd numbers warm stages, respectively. Full glacial lowering of sea level was inferred to be in the order of 100 m. Emiliani's 1955 paper remains very significant and provided an important framework for the subsequent development of oxygen-isotope stratigraphy (Sirocko, 1996). He attempted to derive a chronology for long-term climate change in the context of insolation changes propounded by Milankovitch (1920, 1941).

Shackleton and Opdyke (1973) continued the theme explored by Emiliani (1955) and, in a widely cited paper, they showed that deep-sea piston core Vema 28–238, from a water depth of 3120 m on the Solomon (Ontong Java) Plateau, preserves an essentially uninterrupted record spanning the past 870,000 years (Figure 1.7). Oxygen-isotope measurements were performed on specimens of the foraminifer *Globigerinoides sacculifera*. The study was important in many respects. First, it provided compelling evidence for a globally synchronous correlation in palaeoclima-tological (ice-volume) records previously inferred for the Caribbean and Atlantic by Emiliani, recording the same number of glacial cycles. The isotopic record, uncorrected for possible isostatic responses, when expressed in terms of sea-level change (i.e. 0.1‰ isotopic shift equivalent to 10 m sea-level change), also demon-strated a strong correlation with estimated sea levels for the last interglacial maximum and interstadials of the last glacial cycle from Barbados and New Guinea (Figure 1.8). These observations were particularly important because they clearly excluded the possibility of high interstadial sea levels close to present sea level within the time-range of radiocarbon dating (past 50,000 years). The longer record for the Brunhes chron from Core V28–238 suggested that, in the most general sense, sea levels attained broadly similar upper and lower levels for at least the past six interglacial and glacial maxima, respectively. This was contrary to the 'diminution of the sea' and the progressively lower interglacial levels observed over the course of the Quaternary for the Mediterranean Basin (Zeuner, 1952), indicating that these must be, in part, a record of tectonic uplift. The works of Emiliani (1955) and Shackleton and Opdyke (1973) were also significant in demonstrating many glacial–interglacial cycles, indicating that the four alpine glaciations observed by Penck and Brückner (1909) were not the only glaciations to have occurred in the Quaternary.

Noting that the isotopic changes evident within foraminifera were based on, essentially, globally synchronous events, Shackleton and Opdyke (1973, p. 48) concluded that:

... it is highly unlikely that any superior stratigraphic subdivision of the Pleistocene will ever emerge. Even more important, this subdivision is a convenient one to use because the underlying variable which we are using to correlate is the volume of terrestrially stored ice.

The ability to measure oxygen isotopes from air bubbles preserved within continental icesheets has enabled the direct measurement of stable isotope ratios of the

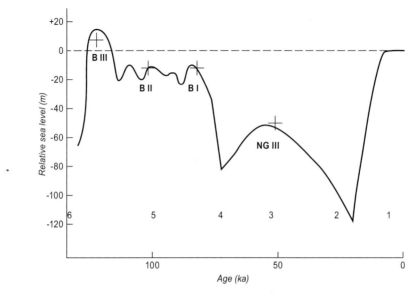

Figure 1.8 An early, generalised glacio-eustatic sea-level curve for the past 130 ka inferred from oxygen-isotope measurements of foraminifera from the deep-sea core V28–238. The data are compared with empirical observations for palaeo-sea levels from marine terraces on Barbados (BI, BII and BIII) and the Huon Peninsula, Papua New Guinea (NGIII), adapted from Shackleton and Opdyke (1973).

atmosphere at the time of ice formation, rather than indirect measurements on marine microfossils. This analytical protocol has revolutionised understanding of the rates of climate change and, as a corollary, Quaternary sea-level changes, particularly for the last glacial cycle (Dansgaard and Oeschger, 1989; Cronin, 2010). The higher resolution afforded by ice-core records has identified numerous inter-stadials involving tens of metres of relative sea-level change within a few thousand years that were not originally identified from oxygen-isotope records based on deep-sea sediments.

1.7.4 Geophysical models of sea-level changes

The formulation of geophysical models has significantly enhanced understanding of the basis for different relative sea-level histories for individual coastlines around the world, particularly in resolving contrasting records for Holocene time. From the 1950s to the early 1980s, literature on relative sea-level histories over the course of the Holocene presented what appeared to be conflicting evidence for a higher sea level in the early Holocene. The development of geophysical models provided a frame-work to explain the contrasting relative sea-level histories, spatially, in the context of glacio-hydro-isostatic models (Walcott, 1972; Farrell and Clark, 1976; Peltier, 2002a; Lambeck, 1995, 1996a,b, 2002). These models explained contrasting relative sea-level histories in terms of physical adjustments of the crust–mantle system in response to loading effects of continental icesheets and water returned to the oceans as a result of deglaciation. The development of glacio-hydro-isostatic models is examined further in Chapter 2. While geophysical models of relative sea-level changes can explain first-order changes in the sea surface in relation to the solid Earth, they have been increasingly questioned for their inability to explain local (spatial and temporal) anomalies in relative sea level associated with short-term climatic changes, as well as the effects of changing sea-surface temperatures and salinities (steric effects) (Baker and Haworth, 2000a,b). These issues related to Holocene sea-level changes are explored further in Chapter 7.

1.7.5 Sequence stratigraphy

The emergence of sequence stratigraphy in the latter twentieth century has provided an additional perspective to examine the longer-term impact of sea-level changes in the geological record (Quaternary and pre-Quaternary). In particular, sequence stratigraphical analysis provides a conceptual framework to evaluate on what basis and how sedimentary successions are ultimately preserved within the stratigraphical record. Sequence stratigraphy, a derivative of seismic stratigraphy, integrates several lines of independent stratigraphical information derived from the results of surface mapping, outcrop analysis, cores, and geophysical data, calibrated by geo-chronological and biostratigraphical data to elucidate records of relative sea-level changes (Haq *et al.*, 1987; Plint *et al.*, 1992). The ultimate goal of sequence strati-graphy is to construct long-term records of relative sea-level changes of presumed global significance, based on stratal relationships (documenting the geometry of unconformity-bounded depositional sequences). Implicit within sequence stratigraphy is the assumption that regional unconformities may be attributed to global processes such as eustatic sea-level changes.

1.7.6 International concern and a focus on current and future sea-level trends

During the late twentieth century and continuing to the present, numerous inter-national cooperative research projects have examined aspects of Quaternary sea-level changes, particularly the International Geological Correlation Programme (IGCP), funded by the International Union of Geological Sciences and UNESCO. This research effort has also been bolstered by the International Union for Quaternary Research (INQUA) subcommissions on aspects of sea-level changes and neotec-tonics. Many of these projects have derived local sea-level histories and identified the complexity in relative sea-level records between continents, which are now more clearly understood in terms of contrasting histories relating to glacio-hydro-isostatic adjustment processes, as well as contrasting neotectonic processes. A significant outcome of these projects was the production of manuals and protocols for the study of sea-level changes (van de Plassche, 1986), as well as atlases that documented the spatial and temporal variability of Holocene sea-level changes (Bloom, 1977; Pirazzoli, 1991, 1996). While these efforts continue (Shennan *et al.*, 2014), they have gained momentum and a new relevance in the context of concern about the trend of contemporary sea level, and the prospect of accelerated sea-level rise in future as a consequence of global warming.

Major developments in geodesy during the twentieth century have made a profound impact on studies of relative sea-level change by establishing a fundamentally important linkage between instrumental measurements of recent changes over timescales of years to the longer, geological record. Geographic Positioning Systems (GPS) have enabled subtle and seemingly imperceptible movements in the Earth's crust to be accurately quantified and compared with rates of change as deduced from longer-term geological evidence. The use of satellites as platforms to study the large-scale morphology of the Earth has also resulted in a more accurate interpretation of the true shape of the planet (Gaposhkin and Lambeck, 1971; King-Hele, 1972, see Section 2.3.5). Satellite altimetry offers the opportunity for near-real-time monitoring of the level of the sea across much of the ocean surface, not just around its margin. The application of data derived from satellites is discussed in Chapter 8 in the context of recording current sea-level trajectories and modelling future changes in relative sea level.

These issues are now a major focus in the debate about climate change, its impacts, and the need for adaptation on the most vulnerable shorelines around the world. Observed sea-level rise around the world over recent decades provides important independent evidence of climate change. Successive assessments by the Intergovernmental Panel on Climate Change (IPCC) have indicated that sea-level rise will accelerate in the twenty-first century, and will have serious consequences for a large proportion of the world's population living on low-lying coasts (Nicholls *et al.*, 2007). Although both tide-gauge records and satellite altimetry show a pattern of increase in global mean sea level, there remains a focus on trying to establish the relative proportions that can be attributed to each source, whether thermal expansion or sources of additional meltwater (Meehl *et al.*, 2007). It remains contentious to what extent the observed rise is an outcome of anthropogenic emissions of greenhouse gases. These issues are examined in Chapter 8.

There has been renewed interest in sea-level changes. The effort to anticipate the pattern of sea-level change in the future needs to be based on reliable estimates of the behaviour of sea level in the past. Sea-level projections, and a full consideration of the factors that will influence sea level in the future and its variation around the world, will need to be multidisciplinary in nature. The Earth Sciences offer particularly relevant insights into sea-level behaviour. In this brief overview of the history of past sea-level studies there have been several important insights, new technologies have developed, and in several instances old ideas have been revisited. In the next section some of the key concepts are summarised and the theoretical basis for their adoption is described, as background for examination of sea-level changes in subsequent chapters.

1.8 Theoretical concepts relevant to the study of Quaternary sea-level changes

Several intellectual constructs that are central to the Earth Sciences have played a major role in the study of Quaternary sea-level changes. Their implications for understanding past sea-level changes are discussed briefly in this section as a prelude to exploring the causes and trajectories of changing sea level.

The principle of *uniformitarianism*, a term introduced by Whewell (1832) in his book review of Lyell's *Principles of Geology*, is central to the Earth Sciences and has been implicitly used in the study of sea-level changes. Uniformitarianism states that geological processes and natural laws operating today to modify the Earth's crust have acted in a similar manner throughout geological history. The concept of uniformitarianism has been summarised by the popular adage that '*the present is the key to the past*'. Uniformitarianism has been discussed extensively since the term was first introduced, and has been the subject of numerous misconceptions. Uniformitarianism does not imply landscape change by only gradual processes occurring at a uniform rate. Catastrophes and rapid changes are accommodated within a uniformitarian perspective, and are embodied in the notion of rare events (Dott, 1988). The uniformitarian approach is implicit in the facies concept of stratigraphy involving comparison of ancient sedimentary structures and bedforms with their modern equivalents. Many inferences about former coastal depositional environments based on palaeoecological evidence involve uniformitarian reasoning, including sedimentary facies analysis.

A sedimentary *facies* refers to a 'body of rock characterised by a particular combination of lithology, physical and biological structures that bestow an aspect (facies) different from the bodies of rock above, below and laterally adjacent' (Walker, 1992, p. 2). Different sedimentary facies form in distinct sedimentary environments. The concept of sedimentary facies is particularly important in studies attempting to define palaeo-sea level (Figure 1.9). Commonly such studies involve analysis of lateral facies relationships as conceptualised by Johannes Walther (Blatt *et al.*, 1980). Walther's Law of *Lateral Facies Succession* states that sedimentary facies deposited in laterally adjacent sedimentary environments will, in a vertical succession, be superposed. Many palaeo-sea-level interpretations are based on inferences derived from analysis of ancient facies and uniformitarian comparisons with modern sedimentary environments. Implicit within these studies is the assumption that ancient sedimentary facies formed under essentially the same conditions as their modern counterparts (e.g. water depths, salinities, geomorphological setting). Thus, the natural field context of modern facies provides a framework to infer sea level from early Holocene or last interglacial equivalents based on a comparative study of their sedimentary

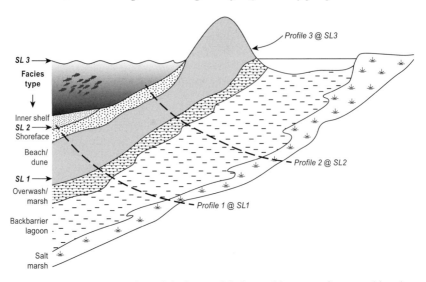

Figure 1.9 Schematic cross-section of the internal facies architecture of a coastal barrier under conditions of relative sea-level rise (transgression), illustrating the principle of *Walther's Law of Lateral Facies Succession*. In response to a relative sea-level rise, sedimentary environments shift laterally (in this case landwards) resulting in the superposition in vertical stratigraphical profile of sediments that were formerly in adjacent depositional environments. Profiles 1 and 2 represent isochronous (time-equivalent) surfaces that are equivalent in form to the modern surface (profile 3).

attributes. Many palaeoecological interpretations of ancient sedimentary successions are based on uniformitarian comparisons with *biofacies* (distinctive faunal assemblages related to specific sedimentary environments) in modern environments (e.g. use of foraminifera to infer Quaternary sea-level changes; Scott and Medioli, 1986; Cann *et al.*, 2000).

Multiple working hypotheses have long been advocated to enable a more objective evaluation of the geological processes that shape landforms (Chamberlin, 1890). Several competing hypotheses are invoked to explain the origin of a landform or suite of landforms or geomorphological processes (Gilbert, 1896; Chamberlin, 1897). Through rigorous hypothesis testing, competing hypotheses are invalidated until the remaining hypothesis hopefully represents the most appropriate explanation for the set of empirical observations. Multiple working hypotheses should ideally be tested, in parallel or in sequence, and either rejected or retained and incorporated into more detailed hypotheses (Schumm, 1991). In practice, major controversies in the Earth Sciences have seldom been resolved in this manner; one hypothesis may appear valid in one instance, but an alternative explanation may be more appropriate in other locations or at another time, and sometimes several hypotheses are amalgamated into a complex explanation. Chamberlin (1897) advocated this approach as a means of researchers maintaining a degree of objectivity in formulating explanations for the origin of landforms, rather than favouring untested, yet cherished hypotheses.

Equifinality refers to convergence in landform development in which similar morphological features may result from contrasting geomorphological processes, or may be derived from different sets of initial conditions (Haines-Young and Petch, 1983). It is not always possible to determine the specific origin of some landforms based only on an understanding of present-day evidence. If a coastal feature can be formed by more than one set of processes, then there are several explanations of its origin (multiple working hypotheses) available to a researcher. For example, a near-horizontal bench elevated a few metres above sea level might be explained as a structural feature developed on resistant lithology, or a former shore platform. Recognition of the possibility of equifinality is important in palaeo-sea-level studies as some landscape or sedimentary features may be erroneously used as evidence for former sea levels when in fact their origin is independent of sea level. Still further complexities occur when several processes act together (multicausality) or a landform is polygenetic, having been shaped by different processes at different stages in the past (Schumm, 1991). Bevin (1996, p. 298) believed that '... equifinality stems not from an inherent property of the system but from an inherent property of *the process of study of the system*'.

Ergodicity describes the concept of space–time substitution. This has been very effectively applied in coastal geomorphology, because the vast expanses of time over which many landforms develop cannot be directly observed during a human's lifetime. The ergodic hypothesis states that in certain circumstances, sampling in space may be equivalent to sampling through time (Chorley and Kennedy, 1971). Thus, by observing a series of different geomorphological features presumed to be in a genetically related continuum, it is possible to overcome the difficulty of not being able to directly observe a single landform develop over geological time. The hypothesis proposed by Charles Darwin for the origin of coral atolls illustrates the concept of ergodicity. Darwin suggested that atolls result from the upward growth of corals in response to subsidence of the ocean floor and that fringing reefs develop into barrier reefs and then progressively transform into an atoll, as the volcanic basement which was formerly encircled by reefs submerges (Darwin, 1842; see Figure 1.4). Darwin formulated his hypothesis of atoll formation based on observations of different reef forms in the Pacific and Indian Oceans, but envisaged how these individual morphologies might fit into a sequence through time. A similar approach involving ergodic assumptions to explain long-term landform development was demonstrated by Douglas Johnson in his description of the New England–Arcadian coastline (Johnston, 1925), and can be seen in relation to the formation of isolation basins, bedrock depressions in environments which initially fill with shallow-marine and coastal sediments but become isolated as a result of uplift (Shennan and Horton, 2002a,b). Geochronological methods have become important in validating ergodic hypotheses, but unfortunately some landforms presumed to be part of a temporal continuum are not directly amenable to geochronological analysis.

Systems theory is useful for understanding the interconnected and complex nature of environmental processes. The definition by Chorley and Kennedy (1971, p. 1) remains appropriate:

A system is a structured set of objects and/or attributes. These objects and attributes consist of components or variables (i.e. phenomena which are free to assume variable magnitudes) that exhibit discernible relationships with one another and operate together as a complex whole, according to some observed pattern.

Systems modelling provides a particularly useful means of conceptualising relative sea-level changes during the Quaternary, incorporating the diverse range of inter-related processes within the Earth's lithosphere, hydrosphere, pedosphere, biosphere, and atmosphere which interact to generate relative changes in sea level (van de Plassche, 1982; Figure 1.10).

Quaternary sea-level changes can be viewed as an example of a very large and complex *process–response system* operating at a global scale, but with a series of smaller subsystems at the local scale. Process–response systems refer to the relationships between a set of processes and the morphological features they form. The system involves interaction of *morphological systems* (a network of physical attributes representing the constituent parts of systems) and *cascading systems* (a chain of subsystems dynamically linked by flows of energy or mass). Flows of energy and mass within (endogenetic) and on the surface of the Earth (exogenetic) have a major role in determining the position of the sea surface in relation to the solid Earth. At a global scale, the ultimate system input for relative sea-level changes in a Quaternary context (i.e. over a timescale of thousands of years) are changes in incoming solar radiation (insolation) which is largely determined by changes in the orbital geometry of the Earth (the Croll–Milankovitch variables of eccentricity, obliquity, and precession of the Earth's orbit around the Sun). At a more local scale, plate-tectonic processes represent another system input contributing to relative sea-level changes as observed at the scale of individual shorelines (e.g. uplift at subduction zones and apparent falls in relative sea level). The Earth's geoid (overall figure or shape of the planet as represented by the projection of sea level around the globe) responds through a complex negative feedback loop to fluctuations in ice and water volumes. The rate of the litho-sphere's geoidal response is determined by rates of ice accumulation or ablation of continental icesheets, and as a corollary, rates of sea-level fall or rise, the latter returning water to ocean basins and continental shelves. Thus, the collapse of glacial forebulges that accompany and post-date deglaciation represent a form of negative feedback leading to the redistribution of water mass on the surface of the planet. Deglacial sea-level rise results in isostatic compensation on continental shelves (hydro-isostasy) and represents the process–response system

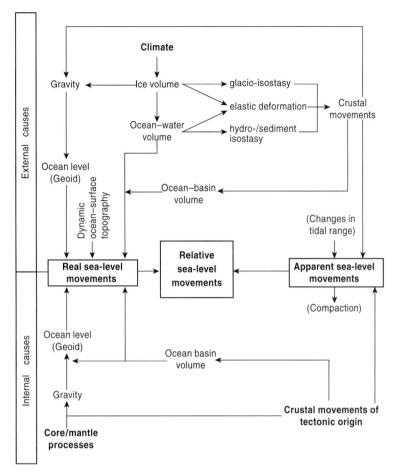

Figure 1.10 Relative sea-level changes over Quaternary timescales (ka) represented as a process–response system and illustrating the interrelationship between internal (endogenetic) and external (exogenetic) processes (after van de Plassche, 1982). The diagram reveals that numerous processes are responsible for sea-level changes.

attaining a new equilibrium; the length of time required to attain the new equilibrium is termed the *relaxation time*. These processes are examined in greater detail in Chapter 2.

1.9 Synthesis and way forward

The history of observation and theory concerning sea-level changes reveals how the emerging perspectives were constrained by the intellectual thinking at the time. Early ideas held by the Greeks were inevitably focused on the Mediterranean. The Diluvial Theory reflected biblical teaching, whereas the Glacial Theory was based on field evidence. Old problems have been revisited, but with different explanations and scientific perspectives. The evolution of ideas provides the background on which this book is founded.

1.9.1 Revisiting old ideas

In a general sense, Strabo's suggestions involving uplift of the ocean floor as a mechanism for causing concomitant sea-level changes can be seen as similar to late twentieth-century plate-tectonic concepts of changes in the volume of ocean basins as a mechanism for sea-level changes in the pre-Quaternary geological record (i.e. differential rates of seafloor spreading and associated tectono-eustatic sea-level changes; Flemming and Roberts, 1973). The early interpretations of the Mediterranean and Baltic relative sea-level records as a succession of highstands episodically falling to lower altitudes during the Pleistocene, a process considered to have been initiated during the Tertiary, is reminiscent of the diminution of the sea as conceptualised by de Maillet, Celsius, and Werner. Baulig (1935) considered that these changes were a manifestation of the general deepening of ocean basins (i.e. increasing accommodation space). Later, the concept of plate tectonics and associated major contrasts in rates of seafloor spreading would be invoked as an explanation for longer-term pre-Quaternary sea-level changes (Pitman, 1978). Accordingly, slow rates of seafloor spreading resulted in more rapid subsidence of the oceanic crust, as it cooled and moved away from the mid-ocean ridges where the crust is formed, providing greater accommodation space for seawater and lower global sea levels. In contrast, periods of Earth history characterised by faster rates of seafloor spreading gave less time for the cooling oceanic crust to subside resulting in a lower accommodation space in the oceans and higher sea levels. These relative sea-level changes occur at rates in the order of 10 mm/ka (Hsü and Winterer, 1980).

The continued acceptance of a progressively falling sea surface appears surprising when viewed from the privilege of hindsight (Rudwick, 2005). This continued to be attributed to lowering of the world ocean (eustatic fall) rather than uplift of the land, because of a failure to identify unambiguous evidence for tectonic uplift processes within regional coastal landscapes, particularly in the Mediterranean Basin. The presently accepted notion of cyclic patterns of relative sea-level changes was anticipated by Aristotle, Lavoisier, Grabau, and many others. Technological advances (particularly radiocarbon and U-series dating before the wider development of other geochronological methods) during the latter twentieth century have provided timeframes to evaluate some of the earlier, as well as current, ideas.

1.9.2 Quaternary sea-level changes: the status quo

Relative sea-level changes have fascinated Earth scientists over generations for many and varied reasons. This book examines the current understanding of Quaternary sea-level changes, from a range of geographical settings and based on several types of evidence. Knowledge of relative sea-level changes is based on direct proxies of former sea level, as well as numerous lines of indirect evidence, such as the marine oxygen-isotope record of global ice-volume changes. Further developments in oxygen-isotopic analyses of ice cores have provided unrivalled insights into the complexity of glacio-eustatic sea-level changes during the last glacial cycle, and revealed evidence for sea-level oscillations not recorded in the stratigraphical record of most coastal sedimentary successions, or in the oxygen-isotope records derived from marine sediments. The detailed nature of these relative sea-level changes is explored in Chapter 6.

In view of the limited preservation potential of older sedimentary deposits along much of the world's coastlines, the most detailed records of sea-level changes relate to the last glacial cycle (i.e. the past 128 ka). The highest resolution records of sea-level changes are after the end of the LGM, with onset of sea-level rise at approximately 19 ka ago, to the current Holocene interglacial and are based on radiocarbon dating and U-series dating. U-series dating has been particularly important in assigning ages to the two previous interglacials (last interglacial and penultimate interglacial).

Each year hundreds of journal articles are published that are relevant, and the literature on Quaternary sea-level changes is vast. Accordingly, only a partial, but hopefully representative, coverage is possible in this compilation, illustrating the historical development of the science. The causes of sea-level changes over a range of timescales, from the longer, Pleistocene record (essentially from about 2.6 million years ago to the Holocene) through Holocene time (the past 11,500 years) to the past 100 years are examined in Chapter 2.

The role of biological, geological, and geomorphological palaeo-sea-level indicators in reconstructing sea-level changes is examined in Chapter 3. The discussion focuses on a range of coastal sedimentary and erosional environments, although these are only part of the process of reliably defining sea-level changes through geological time. Geochronological methods play an essential role in assigning numerical ages to depositional events and in providing temporal frameworks for long-term environmental change. Chapter 4 examines the geochronological methods commonly applied to a variety of palaeo-sea-level indicators, describing the methods, their suitability for different materials, and their advantages and limitations.

Chapter 5 examines geologically recent vertical crustal movements within a plate-tectonic framework and their relevance to understanding Quaternary sea-level changes. Examples are presented from so-called stable trailing-edge coastlines, subduction settings, coastlines associated with transform faults and those associated with intra-plate, hotspot-related volcanism. The chapter shows how an understanding of the neotectonic characteristics of coastlines can help detailed reconstructions of Quaternary sea-level changes, and conversely, that a surprising wealth of information may be deduced about crustal movements from sea-level studies (e.g. using the last interglacial shoreline as a global geodetic datum). A detailed knowledge of neotectonic behaviour is an essential first step for evaluating the potential impacts of future sea-level changes in different geographical settings.

Pleistocene and Holocene sea-level changes are discussed in Chapters 6 and 7, respectively. These chapters are not intended as exhaustive reviews, but explore the general trends that have characterised changing sea levels during the Quaternary. Considerable concern has arisen over the past two decades about the risks associated with sea-level rise anticipated in the future, an issue examined in Chapter 8. This is a particularly significant problem for human settlements in low-lying coastal regions.

2

The causes of Quaternary sea-level changes

It belongs to the physicists who have so cleverly investigated all aspects of physics and astronomy to give us an explanation of the causes of these oscillations, to tell us if they still take place or if possibly an equilibrium might have been reached after a long sequence of centuries.

(Lavoisier, *1789*)

2.1 Introduction

Changes in relative sea level evident along coastlines result from a complex interplay of numerous processes, operating at different rates and over contrasting spatial and temporal scales. The dominant mechanism responsible for Quaternary sea-level changes has been progressive build-up of continental-scale icesheets, and their rapid decay at the end of glaciations. The history of studies of Quaternary sea-level changes, outlined in the previous chapter, reveals an increasing appreciation of processes that contribute to relative sea-level changes, and difficulties of attributing evidence observed at any single locality (e.g. distinguishing magnitude of ice-equivalent sea-level changes from variations due to glacio-hydro-isostatic adjustment processes and localised vertical tectonic movements). Geophysical modelling has provided a framework for understanding spatial and temporal variability in sea-level behaviour globally, as well as insights into relative contributions of meltwater (ice–ocean mass redistribution), solid-Earth deformation, and gravitational influences on regional and local relative sea-level histories, as foreseen in the introductory quotation from Lavoisier (1789).

This chapter explores various mechanisms responsible for relative sea-level changes during the Quaternary and their contributions to the overall magnitude of observed sea-level variation. Tectonic (endogenetic processes) and geomorphological processes affecting the Earth's surface relief (exogenetic processes) together produce particularly complex responses. Other processes, such as differential rates of sea-floor spreading, uplift of mid-ocean ridges, continental emergence, and desiccation of ocean basins, are responsible for longer-term sea-level changes, over geological timescales prior to the Quaternary; these are comprehensively reviewed elsewhere (Flemming and Roberts, 1973; Pitman, 1978; Donovan and Jones, 1979; Hallam, 1984, 1992).

The solid Earth and oceans continue to respond through isostatic adjustments to ice and complementary water loads associated with Pleistocene and early Holocene glacial cycles. Crustal uplift continues where there were former icesheets, and a worldwide sea-level signature is evident in gravitational, deformational, and rotational responses: as the viscous mantle deforms during isostatic compensation. Peripheral bulges adjacent to former icesheets experience subsidence, which, together with additions to ocean volume as a result of ongoing ice melt, produces changes to the geoid (the equipotential surface of the Earth's gravity field). Gradual subsidence of ocean basins and upward warping of continents, combined with redistribution of water into subsiding peripheral bulges, contribute to apparent relative sea-level fall in many far-field locations, known as ocean siphoning. These gravitational and deformational effects also affect rotation of the planet, which in turn feeds back into slight variations in the geoid.

2.2 Sea level and sea-level changes: some definitions

Sea level at a given point on the Earth's surface is represented by a shape called the geoid, a potentiometric surface horizontal to the local gravity field that averages variations in the ocean's surface due to tides, changes in ocean water density and atmospheric pressure, and ocean currents (Williams *et al.*, 1998; see Section 2.3.5). The sea surface at any point on the Earth's surface, called the *gauge sea level*, can be considered in relation to a fixed reference point, such as its distance from the centre of the Earth (Chappell, 1983c). This hypothetical, gauge sea level is distinct from *relative changes* in sea level as derived from geomorphological and stratigraphical evidence. The Earth's geoid provides a first-order approximation for conceptualising sea level in a global three-dimensional context, as discussed later in this chapter. When undertaking detailed surveys of sea-level evidence, it is generally necessary to refer these back to established survey datums, which in turn are expressed in relation to the origin of particular map coordinate systems. Commonly, local datums are related to ideal shapes called ellipsoids which best approximate the shape of the Earth locally. However, since the advent of GPS, it has become preferable to use a *geocentric* geoid approximation which adopts the centre of the Earth as its origin (Figure 2.1).

It is extremely difficult to quantify so-called absolute changes in ocean water volume on the surface of the Earth in view of complex relative movements of land and sea. Mean sea level refers to the time-averaged elevation of the sea surface with respect to the land, and, in particular, a fixed datum or benchmark over a period of time ranging from one month to an 18.6-year nodal cycle (Pugh, 2004). The concern which has arisen about future sea-level rise has revolved around

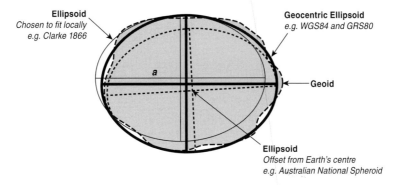

Ellipsoid	Semi-major axis *a* (m)	Inverse flattening f^{-1}
ANS	6,378,160.000	298.25
Clarke 1866	6,378,206.400	294.9786982
GRS80	6,378,137.000	298.257222101
WGS84	6,378,137.000	298.257223563

Figure 2.1 Schematic representation of the shape of the geoid, and ellipsoids that are used to approximate the Earth's surface for mapping and survey purposes. Before widespread adoption of GPS, an appropriate ellipsoid was used that represented an optimum fit on a regional or local scale, but it has now become more useful to adopt a geocentric shape, using the Earth's centre of mass as an origin. The equatorial axis, *a*, differs for ellipsoids; WGS84 uses a geocentric datum and is used in conjunction with GPS (adapted from Iliffe and Lott, 2008, and Janssen, 2009).

efforts to detect recent and ongoing changes in global mean sea level. These reflect the volume of water in the oceans, which appears to be increasing as a consequence of melting of ice and density changes as a result of warming; discussed in detail in Chapter 8. Such changes in the volume of water in the oceans are termed *eustatic*, following the adoption of the term *eustasy* by Suess (1888), as introduced in Chapter 1. In this book, much of the evidence from particular sites is interpreted in terms of relative changes in sea level, which may reflect adjustments between the land and the sea, and which can rarely be considered specifically in the context of mean sea level. The following sections examine terminology that describes aspects of sea-level changes in the context of coastal landform development and geometry of sediment bodies (lithofacies) as well as position of the sea surface relative to land.

2.2.1 Sea level and base level

In the context of long-term landscape evolution over timescales of tens of thousands to millions of years, continental denudation leads to general lowering of landscapes to a lower limit of erosion, assuming no counteracting uplift. Powell (1875) recognised that the lower limit of continental (fluvial) denudation was defined by sea level and termed this limit *ultimate base level*. The view that landscapes eroded to an ultimately near-horizontal surface was a central feature of the geographical cycle proposed by Davis, a process which he called peneplanation (Davis, 1899). In addition to ultimate base level, it is possible to recognise local, or temporary, base levels in areas distant from the coastline, such as lakes in upland settings, and at the confluences of river channels (Johnson, 1929). Changes in relative sea level will directly influence erosional processes; an increase in vigour of continental erosion (rejuvenation) accompanies a relative fall in sea level.

Defining ultimate base level has been contentious. It does not directly correspond with an averaged position of sea level, as river and shallow-water wave processes result in abrasion that extends below mean sea level (Catuneanu, 2006). Accordingly, a more realistic definition is that ultimate base level approximates mean sea level at a given point in time. Consequently, base level is a dynamic feature and long-term evolution of coastal landscapes will be affected by complex changes in sea level, rather than landscapes being shaped by a static base level.

The *graded profile* of rivers refers to a form of longitudinal equilibrium profile. Under conditions of equilibrium, fluvial sediment is deposited such that it does not change the long-term profile of the drainage system, implying an absence of major phases of sediment aggradation or erosion within valleys. In a detailed consideration of the concept of grade, Mackin (1948) proposed that rivers could aggrade or degrade. Significant changes in relative sea level (base level) will, however, lead to a change in the longitudinal profile (graded profile) of river systems (Leopold *et al.*, 1964). As an example, melting of icesheets and valley glaciers, although contributing to a relative rise in sea level, will promote localised sediment aggradation, and a fall in relative sea level will lead to river incision. More detailed examination of the longitudinal profile of river systems indicates that base level and the concept of the graded channel are not closely related (Leopold and Bull, 1979).

It has long been recognised that, during sea-level lowstands at glacial maxima, many of the world's rivers extended considerable distances across subaerially exposed continental shelves (Molengraaff, 1921; Daly, 1934). The Rhine extended by approximately 800 km and drained into the Atlantic Ocean, having adopted a course along the axis of the present English Channel (Blum and Tornqvist, 2000; Coles, 2000). Similarly, the Mekong River extended a further 400 km across the present South China Sea (the Molengraaff River of the Sunda Shelf), and the Mississippi River extended another 100 km across the Gulf of Mexico. Extension of rivers to the outer portions of continental shelves assisted in the formation of submarine canyons in many locations.

Lowstand fluvial channel networks are rejuvenated and initially incise into the continental shelf, but then aggrade in response to increased *accommodation space* (capacity of a receiving basin to act as a long-term sediment repository), producing a Type 1 sequence in the terminology of van Wagoner *et al.* (1990). If river mouths discharge over the continental shelf break they can produce a stacking of deltaic sediments at the shelf-margin rather than incision (a Type 2 sequence; van Wagoner *et al.*, 1990); shelf-margin deltas of up to 150 m thickness occur along the Gulf of Mexico (Suter, 1994). Studies of the Mississippi River and its delta by Fisk (1944) suggested that rivers incise into continental shelves during glacial phases, although the deep valley excavated during the last glaciation (termed the Wisconsin) appears to have extended only 370 km upstream of the present shoreline. Incision is not the only possible response to altered base level; rivers can adjust their pattern, roughness, or channel geometry (Wescott, 1993). Whether or not a river incises appears to depend on rate and direction of sea-level change, geological controls such as erodibility, and geomorphological characteristics such as the presence of levees which affect channel pattern (Schumm, 1993). Rivers can adjust their sinuosity, rather than responding solely by incising. Across the shelf upon which the Great Barrier Reef now occurs, rivers infilled accommodation space with extensive fluvial-deltaic deposition with only limited incision along the shelf break where there are some canyons that extended by knick-point retreat (Woolf *et al.*, 1998).

2.2.2 Relative sea-level changes

The term *relative sea-level changes* is used where either a rise or a fall in sea level is indicated, but it is not clear whether it is the land or the sea, or both, that has changed. In the strictest sense, all sea-level changes observed on the surface of the Earth should be considered *relative*, as few if any regions can be considered tectonically stable (see Chapter 5). Accordingly, it is not possible to discount even minimal land-level changes, particularly changes due to isostatic processes discussed later in this chapter. The geological record provides many examples of stratigraphical sequences that indicate relative sea-level changes, and the terms transgression and regression have been used to describe these.

A relative rise of sea level with respect to land is termed a *marine transgression*. In the geological record this is manifested by progressive encroachment of marine facies over former terrestrial environments, and the boundary between terrestrial and marine successions is represented by an unconformity of regional significance that dips in a seaward direction (Figure 2.2). Global sea-level rise will result in landward encroachment of marine environments over former terrestrial environments. However, transgressions could result from other processes, such as localised basin subsidence producing an apparent rise in sea level within a relatively restricted geographical area. A further possibility involves a rise in sea level coincident with uplift of land, but where the rate of sea-level rise is greater, resulting in a relative rise in sea level and flooding of former terrestrial environments. The depositional product is called a *transgressive facies association*, or a transgressive systems tract in the terminology of sequence stratigraphy. In the Quaternary, transgressions occurred at a relatively rapid rate during postglacial ice-melt phases. The marine transgression following the

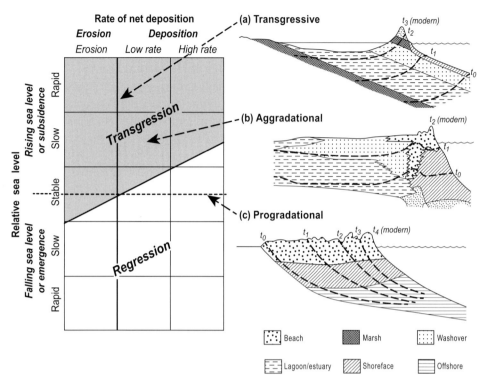

Figure 2.2 The concepts of transgression and regression as proposed by Curray (1964), who indicated that the advance or retreat of the shoreline is a function of both relative sea-level change and rate of supply or loss of sediment. Contrasting geometries and sedimentary facies of typical sand barrier coastlines are shown in profile (modified from Galloway and Hobday, 1983). The dotted lines represent isochrons (lines of time equivalence); individual isochrons mimic antecedent topography: (a) a transgressive barrier in which the shoreline moves landward in response to a relative rise in sea level; (b) an aggradational barrier which develops where vertical accretion is the dominant mode of sedimentation at a similar rate to the rate of relative sea-level rise; and (c) a regressive, or progradational, barrier that forms under conditions of a sea-level stillstand, when there is a positive supply of sediment. The arrows indicate typical location of the barrier morphologies on the 'Curray' diagram (from Woodroffe *et al.*, 2012a).

LGM at 21 ka, involved a relative sea-level rise of ~125 m over a period of about 9 ka (Fleming *et al.*, 1998), averaging around 14 mm/year.

　　Under conditions of gradual relative sea-level rise, resulting from either localised basin subsidence, glacio-eustatic sea-level rise, or a combination of these processes, sediment aggradation (sediment accumulation by vertical accretion rather than by lateral sedimentation processes) results in stacked coastal sedimentary units (Figure 2.2). In these situations, the influx of terrigenous clastic sediments is in approximate balance with the rate of relative sea-level rise so that the coastline experiences only minimal progradation and coastal barriers build vertically (Belperio *et al.*, 1988).

A relative fall in sea level is termed a *regression* and may result from a global lowering of sea level (eustatic change), or localised uplift of land, including glacio-isostasy, hydro-isostasy and a variety of tectonic processes. In the Quaternary, the dominant mechanism at a global scale responsible for marine regressions of ~125 m fall in relative sea level is the formation of continental-scale icesheets, which during glaciations approximated 3.5×10^{19} kg (Peltier, 1987). Coastal landforms and sedimentary successions formed in response to a seaward translation of the shoreline, expressed as a fall in sea level (base level), have been termed products of *forced regressions* (Posamentier *et al.*, 1990). Substantial falls of sea level may initiate erosion and fluvial incision in both terrestrial and shallow-marine environments adjacent to coastlines (continental shelves). Regressions, as manifested by a relative fall in sea level, may also occur at a time of globally rising sea level, when the rate of land uplift exceeds the rate of sea-level rise. This is evident for many regions that were occupied by former icesheets (i.e. near-field locations). In northwest Scotland, for example, an apparent fall in relative sea level is evident over the past 16 ka although for a significant portion of this time global sea level was rising (Lambeck, 1995; Shennan *et al.*, 2000a).

The term regression also refers to large-scale coastal sedimentation (progradation) resulting in apparent retreat of the sea from terrestrial environments whether or not there are changes in sea level. Lengthy periods of relatively constant sea level lasting several thousand years are termed *stillstands*. These intervals can occur during periods of high sea level (highstands) that accompany warmer intervals such as interglacials and interstadials; they may also occur during lowstands, colder intervals such as glacials and stadials when the sea level was low, but are harder to detect. In the middle and late Pleistocene, highstand durations may be as much as 10 ka, equivalent to half a precession cycle in terms of Croll–Milankovitch orbital forcing (Shackleton *et al.*, 2003), and they leave a significant imprint in the geological and geomorphological records. Given sufficient sediment supply, regressive or aggradational modes of sedimentation may occur during sea-level stillstands.

Progradation during stillstands results from large accumulations of sediment deposited in coastal environments such as deltas and barriers, but can also occur where there is a more gradual supply of sand, forming sequences of beach ridges (Figure 2.2). In these situations, within-basin sedimentary processes, such as tides, waves, and aeolian activity, are unable to rework and redistribute sediment, and accordingly the coast progrades in a seaward direction (Curray, 1964). Classic examples of this phenomenon have been described from many coastlines around the world, based on radiocarbon dating of shelly sands such as at Galveston Island on the Texas Gulf Coast (Bernard *et al.*, 1970), the Nayarit coastal plain in Mexico (Curray *et al.*, 1969), and sequences of prograded barriers along the wave-dominated coast of southeastern Australia (Roy *et al.*, 1994). Radiocarbon dating provides a first-order indication of the progressively younger beach ridges (or relict foredune ridges) that occur as these strandplains have accumulated over the mid- and late Holocene. More recently, optically stimulated luminescence (OSL) has been used to determine rates of coastal progradation (Murray-Wallace *et al.*, 2002; Ballarini *et al.*, 2003), as this method directly estimates time of deposition of quartz sand grains during beach-ridge formation, without inferring association between the age of dated materials and deposition. This is an improvement on radiocarbon dating, which determines time of cessation of ^{14}C uptake within a shell, which may significantly pre-date beach-ridge deposition (see Chapter 4).

In reviewing the interplay of relative rates of sea-level change and deposition, Curray (1964) identified several modes of transgression and regression. An *erosional regression* involved either rapid uplift or sea-level fall (or conceivably both) with low sediment supply. Rapid incision of subaerially exposed seafloor is characteristic of these conditions which are common in tectonically active areas. Uplifted terraces without a cover of coastal sediment are also formed under these conditions. A *mixed erosional and depositional regression* involved the combined processes of localised deposition as well as erosion resulting in deposition of fragmentary deposits that are susceptible to erosion at subsequent times of subaerial exposure. A *discontinuous depositional regression* is characterised by high rates of sedimentation and slow regression. Commonly under these conditions, relatively thin but laterally persistent depositional sequences form, and if subsequently subaerially exposed, are also very susceptible to erosion. A *depositional regression* refers to formation of a well-defined regressive sequence showing coastal offlap. Sediment packages showing this geometry have typically formed under stillstand conditions with minimal change in sea level, and sustained, abundant sediment supply. A *depositional transgression*, in contrast, refers to the creation of coastal onlap resulting from a relative fall in sea level which greatly outweighs the rate of sediment supply. Under these conditions the shoreline builds in a landward direction. *Coastal onlap* is expressed by low-angle strata, deposited in a shallow-water depositional environment, that rest on a higher angle surface, and indicates marine transgression. *Offlap* refers to the progressive seaward shift in the updip terminations of strata within a conformable sequence. It results in the uppermost portions of strata being exposed as successively younger strata develop. Wave action and near-shore currents rework and entrain sediment and marine biota in a landward direction producing a thin, laterally persistent *ravinement lag* deposit, commonly preserved as shelly gravels in high-energy settings or an organic-rich mud in low-energy salt-marsh or mangrove settings within a thin basal transgressive facies.

Curray (1964) also identified a *discontinuous depositional transgression* in which thin and discontinuous sedimentary units are preserved on the continental shelf and formed under conditions in which the rate of sediment supply could not keep pace with the rate of relative sea-level rise. The final two categories identified by Curray (1964) included a *rapid erosional transgression* and an *erosional transgression*. In the former, a very rapid rate of sea-level rise results in deposition of marine sediments under highstand conditions on formerly subaerially exposed sediments without preservation of shallow-water coastal facies. Finally, an *erosional transgression* refers to erosion under relatively stable sea level with minimal sediment being supplied to the coastline.

The scheme proposed by Curray (1964) for understanding interaction of net rates of coastal deposition and relative sea-level changes under either rising or falling conditions is presented in a modified form in Figure 2.2 to explain contrasting modes of coastal barrier deposition. In geological cross-sections, dashed lines represent *isochrons*, connecting equal points in time. Isochrons define former depositional surfaces and therefore former positions of barrier–surface morphologies. Under conditions of rapid sea-level rise, the coastal barrier adopts transgressive morphology characterised by a narrow barrier with finite sediment supply that progressively translates in a landward direction. Under conditions of rapid sediment supply, an aggradational barrier forms, in which the barrier accretes vertically. Progradational barriers are the classic example of a regressive facies association, formed under relatively stable sea level with abundant sediment supply. Galveston Island on the Texas Gulf Coast is an example of this stratigraphy (Bernard *et al.*, 1970), and there are many progradational barriers along the east coasts of Brazil and Australia (Thom, 1984).

2.3 Processes responsible for relative sea-level changes in the Quaternary

In the following discussion the principal causes of relative sea-level changes in the Quaternary are examined. Numerous processes, operating at various rates, are responsible for relative sea-level changes, and the magnitude of their expression observed on coastlines can vary significantly spatially.

2.3.1 Glacio-eustasy

The dominant long-term mechanism responsible for the largest component of observed relative sea-level changes for the Quaternary Period is glacio-eustasy (Bloom, 1971). Substantial changes in ice volume have occurred repeatedly throughout the Quaternary accompanying long-term climate changes, and their effects are expressed at a global scale, called *eustatic*, following introduction of the term by Suess (1888; see Chapter 1). The prefix 'glacio' denotes that the dominant cause is growth and decay of continental icesheets. During the LGM an additional 13% of the Earth's surface, excluding Antarctica, was covered in ice, corresponding to a volume of total ice cover of approximately $50 \times 10^6 \, \text{km}^3$ (Flint, 1971; Williams *et al.*, 1998). Maximum icesheet thickness exceeded 2,500 m over extensive areas of the Fennoscandian Icesheet, and in particular, the Gulf of Bothnia in northern Europe, and over 3,000 m for the Laurentide Icesheet centred over Hudson Bay on the North American continent (Boulton *et al.*, 1985; Lowe and Walker, 1997).

An extensive flight of raised shoreline deposits, represented by emergent coastal barrier successions, well-defined beach ridges, and stranded estuarine sequences, indicate uplift since melt of the Fennoscandian Icesheet (discussed in the next section). Uplift of ~250 m is evident in the Gulf of Bothnia near the centre of the former icesheet (Donner, 1995). Mörner (2003) inferred total uplift of up to 830 m since the LGM, with the locus of maximum uplift centred on Ångermanland in Sweden. Maximum rates of present-day uplift within Scandinavia are approximately 9 mm/year, located near the apex of the Gulf of Bothnia, and mountainous regions in Sweden currently experience about 7 mm/year uplift (Figure 2.3). It is not clear whether longer-term patterns of uplift were similar to present-day rates; some

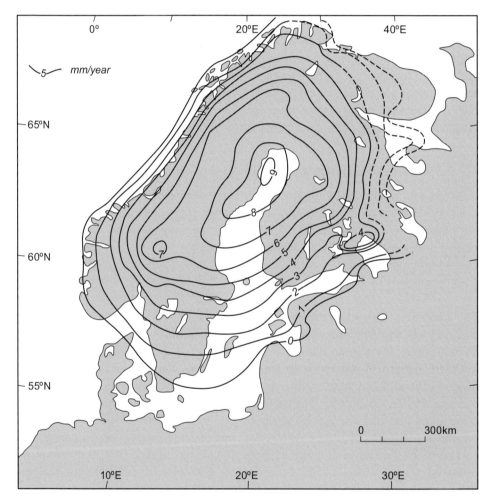

Figure 2.3 Present-day rates of isostatic uplift in Scandinavia from Donner (1995) and updated after Kakkuri (1997).

researchers have suggested that the centre of maximum uplift may have shifted with time, in part because of crustal response to readvances of ice during an overall period of deglaciation (see Donner, 1995 for a more detailed discussion).

Studies of sedimentary successions on shallow-water platforms and continental shelves in the far-field of former icesheets, such as Barbados and Australia (both Joseph Bonaparte Gulf, NW Australia, and the continental shelf off New South Wales), have indicated a maximum sea-level lowering of between 121 and 130 m during the LGM about 21 ka ago (Bard *et al.*, 1990a,b; Ferland *et al.*, 1995; Ferland and Roy, 1997; Yokoyama *et al.*, 2001). These empirically derived estimates for full glacial sea-level lowering appear to be less than would otherwise be predicted based on model calculations of ocean-volume accommodation space and estimates of water locked up in continental ice. According to Williams *et al.* (1998), eustatic sea-level lowering during the LGM should have reached about 154 m. However, the response of the ocean floor and continental shelves to differential water loads, described in more detail below, accounts for the difference between observed and predicted values. The larger water volume exerted a greater loading effect on the ocean floor, and thus the magnitude of relative sea-level rise as observed along coastlines during interglacial highstands appears reduced. Crustal isostatic loading results in different responses for ice compared to water in the liquid state. For every 100 m of ice loading, continental crust is isostatically depressed by 27 m, compared with up to 30 m in response to loading by water because of its higher density (Wilson *et al.*, 2000).

The geological record reveals that differential rates of ice-volume changes have occurred with time accompanied by glacio-eustatic sea-level changes. The last glacial cycle, for example, in a generalised sense, shows an overall gradual increase in ice volume and concomitant lowering of sea level following the last interglacial maximum (Figure 2.4). In detail, however, ice-volume changes are likely to have been more complicated, with at times quite dramatic changes in sea level evident over short time intervals, as for example, at the end of the last interglacial maximum (MIS 5e at about 116 ka in which sea level fell by as much as 20 m in only a few thousand years; Tzedakis *et al.*, 2012). Over the course of the last glacial cycle, build-up of continental icesheets and associated sea-level fall, reconstructed from isotope records from deep-sea cores and from sequences of reef terraces on uplifted coastlines, conformed to a sawtooth pattern of sequential decline (Figure 2.4). During the deglaciation that followed the LGM, however, sea-level rise was rapid and on occasions occurred at rates approximately 20 times faster than the rate of sea-level lowering during icesheet development. Based on dating of drowned coral reefs around the margin of Barbados, maximum sea level was 121 ± 5 m lower than present at 18 ka (expressed in radiocarbon years; Fairbanks, 1989; Figure 2.5). The lowstand may have persisted for several thousand years (Peltier and Fairbanks, 2006). Sea level rose slowly with onset of deglaciation increasing by 20 m in 5 ka (average rate of 4 mm/year). This was followed by a very rapid rise in sea level of up to approximately 24 m in less than 1 ka. This event, around 14 ka, when expressed as a first derivative of the sea-level record has

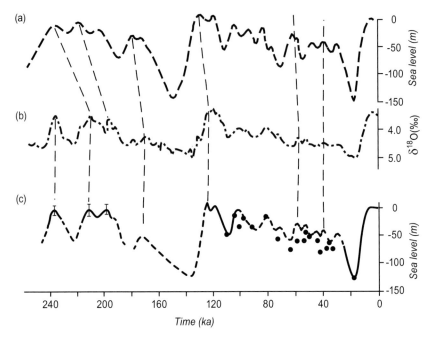

Figure 2.4 Late Quaternary sea-level variation: (a) reconstruction of sea level based on ages and uplift-corrected heights of reef terraces on Huon Peninsula; (b) oxygen-isotope record from deep-sea core V19–30 from the equatorial Pacific Ocean; and (c) reconciliation of those curves (based on Chappell and Shackleton, 1986, but incorporating revision after further dating (black dots) of several Huon terraces by Chappell *et al.*, 1996a; after Woodroffe, 2003).

been termed Meltwater pulse 1A. A slower rate of sea-level rise followed Meltwater pulse 1A (about 5.6 mm/year) for about 2.5 ka followed by Meltwater pulse 1B at ~11 ka, during which sea level rose by a further 28 m in approximately 1.5 ka (~19 mm/year; Bard *et al.*, 1990a,b; Figure 2.5, see Chapter 7, and particularly Figures 7.1 and 7.2, for a more recent interpretation of this chronology).

Sea-level curves that depict major changes in elevation of the sea surface since the start of the late Pleistocene (~128 ka) are derived from restricted regions and there-fore, strictly speaking, are unique to those regions. However, they do show broad similarities globally (Steinen *et al.*, 1973; Aharon, 1984; Lambeck and Chappell, 2001; Siddall *et al.*, 2009). Sea-level curves derived from emergent coral terraces on Barbados and the Huon Peninsula in Papua New Guinea show a striking similarity in the long-term general trend of relative sea-level changes, reflecting the dominantly glacio-eustatic cause (see Figure 7.2).

2.3.2 Isostasy

The principle of isostasy is fundamental to understanding relative changes of sea level observed along coastlines over Quaternary timescales. Isostasy refers to the tendency

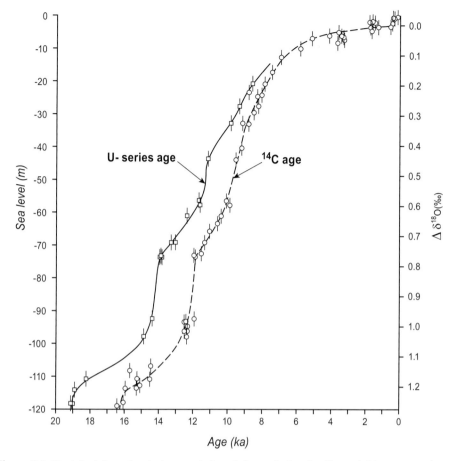

Figure 2.5 Postglacial sea-level changes inferred from dating fossil coral (*Acropora palmata*) from submerged reefs around Barbados (after Fairbanks, 1990). The right-hand curve is based on radiocarbon ages and the left-hand curve is based on $^{230}Th/^{234}U$ dated coral specimens after Bard *et al.* (1990a,b). A further interpretation of these results is shown in Figure 7.1 and discussed in Chapter 7.

of the Earth's crust–mantle system to attain a state of equilibrium with respect to its mass, thickness, and surface relief. The term *isostasy*, derived from the Greek *isostasios* (in equipose; *isos*, equal, and *stasis*, stable), was introduced by Dutton in a book review in 1882 and in an expanded form in 1889. However, although he did not use the term, the concept was first articulated by Thomas Jamieson of Ellon, in Aberdeenshire, who in his seminal paper (1865) argued that ice loading during glaciation would depress the crust, but that as icesheets decayed the land would be uplifted, drawing his evidence from eastern Scotland, largely from the Forth valley. Isostasy helps explain why oceans and continents exist as separate entities and occur at quite different elevations. The distinct bimodal distribution of the Earth's surface

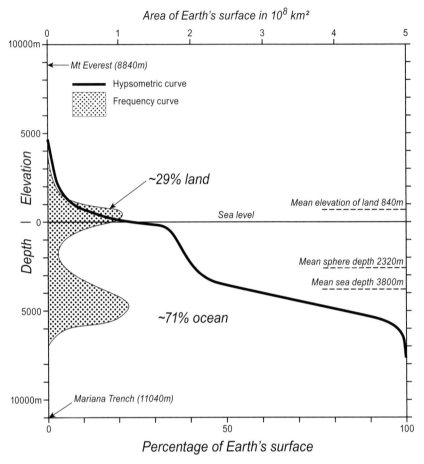

Figure 2.6 Frequency and hypsometric (cumulative frequency) curves of the Earth's major relief.

relief (land and ocean; Figure 2.6) is a manifestation of the isostatic response of the contrasting rock types and their modal densities. Continental crust, being dominated by granitic rock types has an average density (ρ) of 2,800 kg m^{-3} and an average elevation above sea level (isostatic balance) of 840 m. In contrast, oceanic crust, which is predominantly of basaltic (tholeiitic) composition, is slightly denser ($\rho = 2,900$ kg m^{-3}), and ocean floor has an average depth of 3,800 m below present sea level. Collectively, these modal crustal rock types rest on denser rocks of the upper mantle, such as peridotites ($\rho \cong 3,300$ kg m^{-3}), and have commonly been described as 'floating' on denser mantle. Continental crust averages thicknesses of 30–40 km and oceanic crust only 4–10 km, although both show considerable variation (Pinet, 1992).

At a depth of approximately 50 km below the Earth's surface, called the compensation depth, the surficial contrasts in mass and pressures are balanced and, accordingly, lithostatic pressures are equalised. Isostatic compensation of the Earth's crust–mantle system occurs in response to surface loading or unloading. Surface loads include deposition of thick piles of sediment, particularly in deltas; build-up of icesheets; the effect of water flooding continental shelves associated with sea-level rises accompanying deglaciation; and progressive volcanic outpourings of lava flows over lengthy periods of geological time. Progressive removal of surface loads by erosion of mountains, decay of continental-scale icesheets, or lowering of sea level accompanying glaciations, promotes rebound within the crust–mantle system, towards isostatic compensation. In view of the relatively high viscosity of the mantle, geologically rapid decay of continental icesheets, which occurs over periods of ~10 ka, means that rebound of crust to elevations that characterised the pre-glacial landscape is slow, and continues today long after the complete disappearance of icesheets.

Two models were advanced to explain the physical basis and geometric character-istics of isostatic compensation in the early formulation of the Principle of Isostasy, in many respects representing extensions of Archimedes' Principle of Displacement (Airy, 1855; Pratt, 1855). The Airy Hypothesis for isostasy is based on the assump-tion that the outermost shell of the Earth is of uniform density and overlies material of a higher density. Accordingly, the Earth's major surface mountain ranges are compensated by regional contrasts in thickness of the outer shell of the Earth, such that its buoyancy balances the surface load. On this basis, mountains have deep roots whereas ocean floors are significantly thinned and are said to have antiroots. In the most general sense, the base of the outer shell of the Earth can thus be regarded as an exaggerated mirror image of the surface topography. The Pratt Hypothesis of isostasy, in contrast, assumes that the base of the outermost shell of the Earth is constant and that contrasts in density of crustal rocks are associated with variations in surface topography. On this reasoning, mountains would be underlain by material of relatively low density and ocean basins by materials of higher density. It is now regarded that isostatic compensation is not confined to the Earth's crust and that compensation also occurs within the upper mantle (i.e. part of the lithosphere) as well as the underlying asthenosphere. Accompanying removal of a load, such as decay of an icesheet, isostatic rebound is initiated and is largely controlled by the viscosity of the asthenosphere. As the viscosity of asthenospheric material is approximately $1–3 \times 10^{21}$ Pa s (Pascal seconds; Lambeck and Johnston, 1998), fluid behaviour of sublithospheric material requires lengthy periods (thousands of years) to be expressed at the Earth's surface.

2.3.3 Glacial isostasy and relative sea-level changes

Many formerly glaciated regions continue to experience uplift today in response to melt of extensive icesheets, leaving shoreline features emergent as seen in Scandinavia (see Figure 2.3). The effects of glacio-isostasy and hydro-isostasy are particularly evident in the British Isles, where perhaps the greatest variations in postglacial relative sea-level histories are apparent within a relatively small geographical area. These regional contrasts have been accentuated by the former presence of an icesheet in Scotland and absence of ice in southern England. There are several former shorelines preserved around the coast of Scotland. The parallel roads of Glen Roy were initially interpreted as marine shorelines (Figure 2.7a), but are now known to be formed on the margins of ice-dammed glacial lakes (see discussion in Chapter 5). Uplifted Holocene shoreline features, such as shingle ridges (Figure 2.7b), are widespread, and inlets become increasingly isolated from marine influence as a result of uplift (Figure 2.7c), eventually forming *isolation basins* (see Chapter 3). Several

(a)   (b)

(c)   (d)

Figure 2.7 Evidence of uplifted shorelines: (a) the 'parallel roads of Glen Roy', a sequence of terraces interpreted as marine by Darwin, but shown by Agassiz to be shorelines of a former ice-dammed glacial lake, see Chapter 5 (photograph: Colin Murray-Wallace); (b) a raised shingle beach up to 8 m above sea level of Holocene age on Fladday to the north of Raasay in the Scottish Inner Hebrides (photograph: David Smith); (c) an inlet at Arisaig, Scotland, that is increasingly infrequently inundated because of uplift, and in the process of becoming an isolation basin, removed from marine influence (photograph: Colin Murray-Wallace); (d) oblique aerial photograph of the east coast of Lowther Island, central Canadian Arctic Archipelago, showing raised beach sequences formed under falling relative sea level during the late Holocene. The marine limit was established at > 100 m above sea level around 9,500 years BP, with progressively younger shorelines dated by radiocarbon ages on driftwood, whale bone, and shells (St-Hilaire-Gravel *et al.*, 2010; photograph: Dominique St-Hilaire-Gravel).

processes acting in concert are responsible for contrasting relative sea-level histories evident around the British Isles and include glacial unloading in northern Britain, variable response to meltwater loading (hydro-isostasy) of the surrounding seas and Atlantic Ocean, and to a lesser extent, effects of unloading of the Fennoscandian and North American icesheets (Lambeck, 1995). Glacial isostatic adjustment (GIA) is an important consideration in terms of current and future sea levels (see Chapter 8).

There is broad agreement between empirical evidence and theoretical model predictions for relative sea-level change for the British Isles (Lambeck, 1995; see Figure 2.8). Sites in Scotland show a consistent, overall, relative fall in sea level during postglacial times, in response to regional isostatic uplift, resulting from progressive decay of the Scottish Icesheet (Shennan *et al.*, 2000a). Geomorphological and stratigraphical evidence suggests that maximum ice thickness exceeded 950 m over northwestern Scotland based on mapping of glacial trimlines and related features (Ballantyne *et al.*, 1998). Modelling reveals that postglacial rebound between 18 and 12 ka BP (^{14}C years) largely kept pace with sea-level rise (Lambeck, 1995), resulting in a near-stationary shoreline for the North Sea coast in northern Britain. In contrast, progressive submergence was slowly commencing at this time in southern Britain with the gradual submergence of the English Channel. Between 14 and 10 ka BP (^{14}C years), an interval encompassing the Older and Younger Dryas, shorelines of the North Sea coast remained relatively unchanged in their geographical position due to the apparent dynamic equilibrium between postglacial sea-level rise and glacio-isostatic rebound (Figure 2.8). This interval witnessed progressive development of the North Sea and by 10 ka a large, shallow gulf had formed between Britain and France that was fully established by 7 ka (Lambeck, 1995).

Many case studies have documented coastal response to glacial isostasy. In Richmond Gulf, Hudson Bay, Canada, a spectacular succession of mixed sand–gravel beach ridges has developed in response to postglacial uplift (Fairbridge and Hillaire-Marcel, 1977). A similar sequence from Lowther Island in Arctic Canada is shown in Figure 2.7d, where successive shorelines have been radiocarbon-dated indicating that the 6,000-year shoreline is around 30 m above sea level and the 3,000-year shoreline is around 10 m (St-Hilaire-Gravel *et al.*, 2010).

2.3.4 Hydro-isostasy and relative sea-level changes

During the past few millennia, since the complete melt of most northern hemisphere icesheets, the eustatic component of sea-level change which dominated deglaciation has almost completely ceased. Instead, geographical variations in relative sea-level changes have become apparent at regional scales, which can be explained by models of glacio-hydro-isostasy (Nakada and Lambeck, 1989; Lambeck, 1995; Peltier, 2004; Lambeck *et al.*, 2012). These models examine the non-elastic response of the Earth's crust and upper mantle to surface loading by both ice and water.

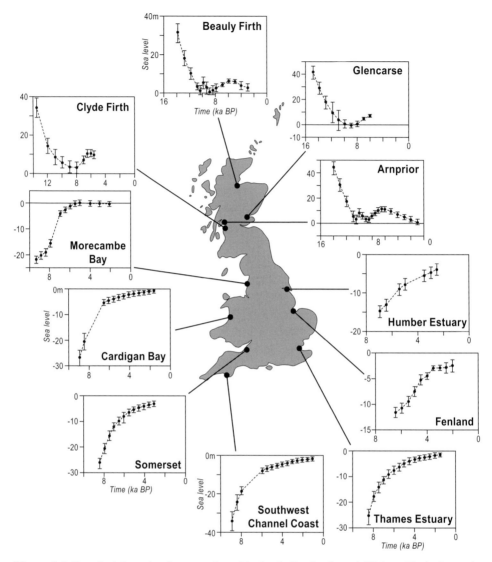

Figure 2.8 Postglacial sea-level curves from England, Scotland, and Wales. Glacio-isostatic uplift in the vicinity of the former Scottish Icesheet is reflected in progressive fall in relative sea level since the LGM. In England, at sites more distant from the Scottish Icesheet, the pattern of relative sea-level change is dominated by glacio-eustatic sea-level rise. The notable inflection in relative sea level after 8 ka for the Scottish sites is a function of changing relative rates of glacio-isostatic crustal rebound and glacio-eustatic sea-level rise (modified from Lambeck, 1995).

   Along coastlines bounded by relatively wide continental shelves (≥ 100 km), hydro-isostatic signatures are likely to be expressed in the form of raised shorelines, particularly for regions located in the far-field of former continental icesheets. The additional load of deglacial meltwater on the crust–upper mantle system results in lateral displacement of viscous mantle material away from the centre of the newly imposed load created from relative sea-level rise across formerly subaerially exposed continental shelves over several thousand years.

Located in an intra-plate setting and in the far-field of former icesheets, Australia has been largely tectonically inactive, remote from glacial isostatic adjustments, and relative sea-level histories demonstrate the spatially variable effects of hydro-isostasy (Nakada and Lambeck, 1989). Australia experienced minimal glaciation during the Quaternary, with ice cover restricted to central Tasmania and the Snowy Mountains (Bowler, 1986; Colhoun *et al.*, 2010). Neotectonic processes are limited to few parts of the Australian region and operate at slow rates (Murray-Wallace and Belperio, 1991; Bourman *et al.*, 1999; Quigley *et al.*, 2010).

The Australian postglacial record reveals a rapid sea-level rise with culmination of the transgression approximately 7,000 years ago (Belperio *et al.*, 2002; Sloss *et al.*, 2007; Lewis *et al.*, 2013). Evidence from several parts of the continent has been used to infer a steadily falling sea surface since that time (Chappell, 1983d, 1987), a function of the hydro-isostatic levering of continental margins (Nakada and Lambeck, 1989; Mitrovica and Milne, 2002). The effect of hydro-isostasy in the Australian region is particularly evident for coastlines adjacent broad continental shelves such as northern Queensland and South Australia, manifested by a seaward reduction in the amplitude of the Holocene highstand and a systematic offset in its timing. Evidence from around the Australian mainland has recently been reviewed (Lewis *et al.*, 2013); a reconsideration of this and other evidence for conflicting interpretations is summarised in Chapter 7.

Modelling of relative sea-level changes since the LGM in the Australian region has shown that following attainment of the Holocene highstand some 7,000 years ago, an additional 1.5–2 m of sea-level equivalent ice melting occurred (Nakada and Lambeck, 1989). The most likely source is the Antarctic Icesheet with a minor contribution from valley glaciers. This estimate was revised by Flemming *et al.* (1998), who suggested a eustatic sea level rise of 3–5 m between 7 ka and 1 ka. Lambeck (2002) subsequently revised the magnitude of post-highstand eustatic rise to 3 m. Relative sea-level observations from Scandinavia are best explained with 3 m of ice-equivalent sea-level rise since 6 ka (Lambeck *et al.*, 1998). These studies imply that the Antarctic Icesheet continued to provide meltwater to the global ocean following melt of the major northern hemisphere ice, but demonstrate that GIA modelling does not always produce a unique solution as it tends to rely on a relatively small subset of the global geological sea-level observations and a fit between model output and observations in one region may produce a misfit in another, as either the ice-equivalent sea-level input or the viscosity parameters can be changed to obtain similar results.

Spatial variability in relative sea-level histories is evident along the South American passive continental margin relating to contrasting crustal behaviour. In the Caribbean region, for example, geophysical modelling and field evidence indicate sea-level rise at a decelerating rate up until present with no Holocene highstand, reflecting crustal subsidence of the forebulge associated with melting of North American icesheets. In contrast, in Patagonia, a Holocene highstand is registered some 2 ka later than other sites in more northerly locations, in response to geoidal changes accompanying melting of the West Antarctic Icesheet (WAIS). The passive continental margin of much of the Atlantic coastline of South America is, like Australia, sufficiently distant from former icesheets to illustrate effects of glacio-hydro-isostatic adjustment processes (Milne *et al.*, 2005). Notwithstanding variable quality of some localised Holocene sea-level records for parts of the South American Atlantic coastline (e.g. gravel beach ridges with disarticulated shells; some shells not *in situ* as well as many reworked individuals), a review of available data suggests that in the early Holocene, a sea-level rise of 7–8 mm/year occurred until about 7 ka and was followed by an early Holocene highstand (Milne *et al.*, 2005). Regressive beach-ridge sequences have been inferred to contain evidence of oscillating sea level over the past few millennia (Angulo *et al.*, 2006).

2.3.5 The geoid and changes to its configuration

The Earth is not a perfect sphere. As recognised by Sir Isaac Newton, it is flattened at the poles and extended in its equatorial radius where centrifugal forces are stronger. The Earth's equatorial radius is 6,378 km, compared with a polar radius of 6,357 km, a flattening of approximately 1/298.25. This shape is described as an oblate ellipsoid or spheroid. In detail, however, the actual shape of the Earth is much more irregular, a manifestation of irregular distribution of mass at the surface of the planet. Irregular topography is primarily an outcome of plate-tectonic processes, with the highest point, Mount Everest in the Himalayan massif, 8,848 m above sea level, which is rapidly uplifting as a consequence of the convergence of the Indian subcontinent with Eurasia, and the lowest, 11,034 m below sea level, in the Mariana Trench, an area of subduction. However, it has become apparent that the sea surface itself is not a regular shape, but adopts a complex configuration, called the geoid.

The geoid is the shape of the Earth if it were entirely covered by ocean; it represents a configuration of equal gravitational potential, the three-dimensional form that the sea surface would adopt in the absence of tidal forcing, waves, and other disturbances, or if these variations were averaged out. It can be regarded as the theoretical shape the sea would adopt if it extended under the continents, to which gravity is always at right angles. Sea-surface height (SSH) across the oceans corresponds to the geoid (*geodetic sea level*) which is the equipotential surface, plus the mean dynamic topography (that is the average of variations caused by tides and ocean currents, and other factors), and any sea-level anomaly (such as the significant variations related to El Niño and La Niña in the equatorial Pacific; Landerer *et al.*, 2008). Irregularities of the geoid are incompletely known and cannot be observed directly, and geodesists adopt a reference ellipsoid that mathematically models the idealised shape, constrained by gravity measurements. In the past, detailed survey at regional or national level adopted an ellipsoid that provided an optimal fit for particular sections of the globe (see Figure 2.1). For example, the North American Datum 1927 (NAD27) was based on the Clarke 1866 ellipsoid with an origin in Kansas, and the European Datum 1950 (ED50) was based on the International (Hayford) Ellipsoid 1924 with an origin in Potsdam (Janssen, 2009). Since the advent of GPS which is based on satellites that orbit around the centre of the Earth, it has become necessary to adopt an ellipsoid which uses the centre of the planet as its origin. The Geodetic Reference System (GRS80) is one such *geocentric* ellipsoid, and the more recent World Geodetic System 1984 (WGS84) is another which is almost identical to GRS80 (differing in its degree of flattening only at the sixth decimal place). WGS84 is the geocentric datum (meaning its origin coincides with the Earth's centre of mass) that is used with GPS observations in the Global Navigation Satellite System (GNSS). The International Terrestrial Reference Frame (ITRF) is the most precise Earth-centred, Earth-fixed datum; it is a dynamic datum that takes into account Earth processes such as crustal motion, polar wander, and orientation of the Earth, determined through a combination of techniques such as satellite laser ranging and very long baseline interferometry (Blewitt *et al.*, 2010).

GPS receivers generally produce estimates of position that are related to the WGS84 reference ellipsoid (corresponding to a recent update of the ITRF), but this is only an approximation of the geoid, and there are significant discrepancies between the reference ellipsoid (WGS84) and the geoid that vary with location (Figure 2.9). The reference ellipsoid (for most purposes, WGS84) is a relatively regular mathematically computed shape. The geoid coincides with the average ocean surface, although the latter shows variations related to tidal and other dynamic oceanographic effects. The geoid surface as shown in Figure 2.9 is more irregular than the ellipsoid.

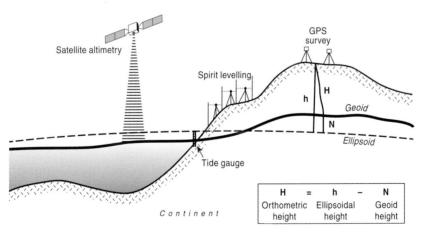

Figure 2.9 Relationship of the geoid to a reference ellipsoid and basis for several survey methods (adapted from Janssen, 2009).

On land, elevations were traditionally derived by survey techniques and the most accurate elevations represent orthometric heights which are reduced to the theoretical geoid. These are usually reported as heights above a datum which approximates mean sea level. Mean sea level itself is not constant in time or space; it is generally determined at particular tide-gauge locations based on a period of observations which would need to span the 18.6-year tidal nodal cycle to account for all variability during that period. Regional sea-level datums have generally been established by adjusting observations taken by spirit-levelling between a series of tide gauges. For example, Australian Height Datum (AHD) was established in 1971 based on 30 tide-gauges linked by 195,000 km of survey; it continues to require refinement (Featherstone, 2006). North American Geodetic Vertical Datum 1929 (NGVD29) was based on 26 tide gauges linked by 105,000 km of survey, but is subject to distortion through vertical land movements, requiring its replacement in 1991 by a new NAVD88 datum tied to a tide gauge at Rimouski in Quebec (Janssen, 2009).

GPS measurements, by contrast, are computed with respect to the geocentric reference ellipsoid (WGS84) and so elevations on the land are ellipsoid heights. The geoid height is the difference between the orthometric and ellipsoid heights, and can be positive or negative. Over the ocean, there will also be discrepancies between ellipsoid and geoidal heights. Even at a tide gauge, it is often the case that mean sea level does not coincide exactly with geoid height. Satellite altimetry now produces previously unavailable data on SSH, which enables refinement of the precision with which the geoid can be computed over the oceans. Repeated measurements of the ocean surface, such as the 10-day repeat orbits of Jason-1 and Jason-2 satellites, provide data on wave and tide conditions around the world, ocean circulation patterns, and the transport of heat, water mass, nutrients, and salt, as well as providing a perspective on the trajectory of sea-level change (see Chapter 8).

The irregular shape of the geoid is a function of the distribution of mass within, and near the surface of the planet, and the Earth's rate of rotation; it is only partly a function of the spatial arrangement of plate-tectonic features. Gaposhkin and Lambeck (1971) published the Smithsonian Standard Earth, a model of the Earth's geoid, based on satellite, laser, and gravity data then available. These early analyses indicated that the South Pole was located approximately 30 m closer to the centre of the Earth and the North Pole approximately 10 m above the reference ellipsoid (King-Hele, 1972). The former was attributed to combined effects of crustal loading of the Antarctic Icesheet and the Earth's overall distortion due to long-term response to axial rotation.

Satellite observations, combined with ongoing gravity measurements, have enabled considerable refinement of the geoid. Satellite altimetry, in particular, provides unprecedented detail over the oceans through repeat altimetric observations with precision of around 2–3 cm. The Earth Geodetic Model (EGM96) was developed in 1996, using altimeter data from TOPEX/Poseidon (launched in 1992). It involved computation of geoid heights (called undulations) to < 1 m, except where

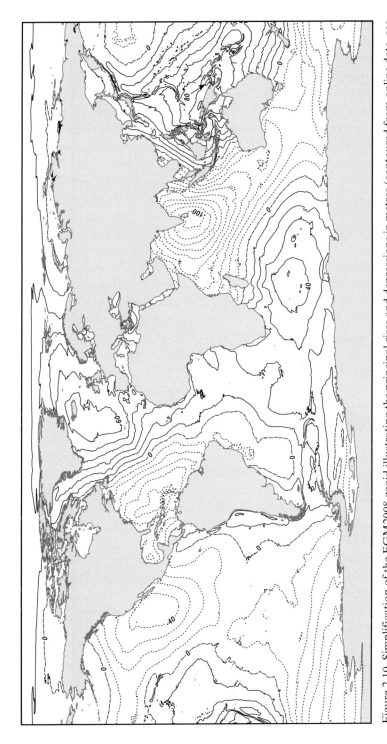

Figure 2.10 Simplification of the EGM2008 geoid illustrating the principal rises and depressions in metres (contoured from online data, see Pavlis *et al.*, 2008).

gravity data were insufficient. This confirmed the broad pattern of geoid variation observed from previous analyses, indicating the lowest point, to the southeast of India, 105 m below the WGS84 ellipsoid, and the maximum point, in eastern Indonesia, 87 m above the ellipsoid. Although the principal causes of variation are anomalies in lithospheric mass, these extremes imply that, at least in these regions, there must also be significant anomalies in the lower mantle.

A more recent, higher precision geoid is the Earth Gravitational Model 2008 (EGM2008), released by the US National Geospatial-Intelligence Agency (NGA), which is complete to spherical harmonic degree and order 2159 (Figure 2.10). The European Space Agency (ESA) launched the Gravity field and Ocean Circulation Explorer (GOCE) gradiometer in 2009. This orbits 260 km above the Earth's surface in a low-altitude orbit to maximise sensitivity to variations in the gravity field. Results from this, and from GRACE (Gravity Recovery And Climate Experiment), will continue to refine the detail with which the shape of the geoid is understood.

Geoidal variations are readily influenced by changes in mass and rotational rate. They are also intimately related to other processes that induce sea-level change such as ice-volume changes and, over longer timescales, plate-tectonic processes that will modify Earth's hypsographic curve (e.g. differential rates of seafloor spreading and resultant changes in ocean volume as well as major changes in the Earth's surface relief, such as the formation of major mountain ranges such as the Himalayas; Flemming and Roberts, 1973; Pitman, 1978; see Figure 2.6).

In the context of the Quaternary record, perhaps the most significant geoidal changes accompany altered loading of Earth's crust due to changes in ice volume. Geomorphological and stratigraphical studies have revealed that regions such as Scandinavia, northern Europe, North America, and Canada were formerly covered by extensive icesheets, which in places attained thicknesses of up to 3,500 m some 21 ka ago (Lowe and Walker, 1997; Ehlers and Gibbard, 2004a,b,c). The relatively rapid deglaciation and withdrawal of continental icesheets resulted in isostatic compensation (uplift) and concomitant changes in the geoid, processes which continue today in view of the delayed crust–mantle response to glacial unloading of icesheets (Johnson and Lambeck, 1999).

The phenomenon of longer-term geoidal changes may be described in terms of changes accompanying glacial cycles, which for the past 1.2 Ma have occurred at a frequency of 100 ka (Raymo, 1992). Glacial cycles are characterised by gradual increase in ice volume over a period of approximately 100 ka and rapid deglaciation in less than 10 ka (Bassinot *et al.*, 1994). During the last glacial cycle, the period of maximum ice accumulation lasted for approximately 8 ka and in the case of the LGM (MIS 2) is assumed to have been of sufficient duration such that isostatic equilibrium was attained before onset of deglaciation (Mitrovica and Peltier, 1991).

Whether changes in distribution of mass within the planet or rotation rate will directly influence sea-level changes at shorter timescales has been the subject of some debate (Mörner, 1976a, 1983). The rate at which the geoid can change is slow, even on geological scales. The more extreme interpretations that high-level fossil evidence is an outcome of migration of geoid anomalies (e.g. Nunn, 1986) can be discounted. Whether more subtle variations in spatial expression of sea-level evidence are related to geoidal variation, such as along-estuary variation of the Senegal River in West Africa (Faure *et al.*, 1980), or variability in Holocene highstand evidence in South America (Martin *et al.*, 1985), has not been resolved. It is known that there are significant geoid anomalies associated with hotspot swells, volcanic plateaus, and seafloor deeps (Cazenave *et al.*, 1986). However, there are also gradients in geoid that manifest in relation to oceanic islands and which appear to exert an influence on the sea surface which need to be taken into account when interpreting sea-level evidence at island sites (Woodroffe *et al.*, 2012b).

2.3.6 *Global variation in geophysical response and equatorial ocean siphoning*

A significant attempt to model spatial variations in response to lithosphere to postglacial sea-level changes was undertaken by Clark *et al.* (1978), building upon insights by Walcott (1972) that postglacial sea-level rise in response to differential ice and water loads on a viscoelastic Earth was not uniform with time over the surface of the Earth. Modelling by Clark *et al.* (1978) suggested that the oceans could be divided into six distinct regions based on their relative sea-level histories since the LGM (Figure 2.11). In formerly glaciated regions (*Zone I*), the relative sea-level curve is dominated by a record of postglacial uplift, represented by isostatic compensation associated with diminishing icesheets. As uplift of land exceeds the rate of sea-level rise, an apparent fall in sea level is registered in these regions. In *Zone II*, the collapsing forebulge, a feature that developed in response to loading by former icesheets, induces widespread submergence of the seafloor, providing additional accommodation space for increased ocean volume due to the meltwater contribution. The results, as documented for many localities along the east coast of the United States such as New Jersey, Virginia, and Georgia (Kraft, 1971), show a progressive, but decelerating, rise in the sea surface over the past 7,000 years, associated with collapse of the glacial forebulge. In *Zone III* limited emergence of less than 0.5 m is recorded with emerged shoreline features likely to be only a few thousand years old. *Zone IV* is represented by oceanic submergence, as recorded in Micronesia (Pirazzoli, 1991). Clark *et al.* (1978) predicted about 0.35 m of emergence in this region since 5,000 years BP followed by a period of submergence of up to 0.15 m between 3 ka and 2 ka BP. *Zone V* is represented by oceanic emergence, with a highstand of 1.5–2 m APSL which, according to Clark *et al.* (1978), commenced some 5 ka ago with the cessation in glacial meltwater contribution to the ocean basins. Finally, *Zone VI* represents many continental shorelines which preserve evidence of a higher relative sea level during the Holocene. These emerged shoreline features are a manifestation of crustal tilting and redistribution of mantle material in response to loading of the ocean floor and increased water load on continental shelves following rapid post-glacial sea-level rise. The hydro-isostatic processes involved are part of the

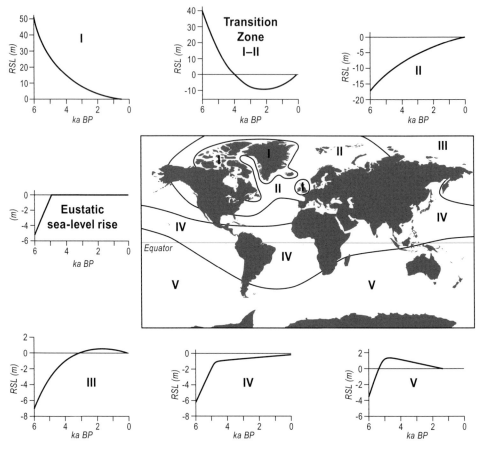

Figure 2.11 The six principal relative sea-level zones identified by modelling by Clark *et al.* (1978) accounting for contrasting relative sea-level histories since the LGM. The boundaries represent broad zones of contrasting relative sea-level behaviour rather than distinct differences. Emergent Holocene shorelines are predicted for Zones I, III, V, and also for most continental shorelines (which was labelled VI by Clark *et al.*, 1978). Continued relative sea-level rise to the present is predicted for Zones II and IV (after Clark and Lingle, 1979, with permission from Elsevier).

overall global, glacio-isostatic adjustment process. The model proposed by Clark *et al.* (1978) is particularly important in the context of demonstrating that there is no need to invoke a net change in ocean volume over the past 5,000 years and, accordingly, represented a benchmark contribution in understanding the basis for the controversy of an apparent higher sea level during the Holocene (e.g. Kidson, 1982), which is discussed in Chapter 7. This is the basis of an equatorial ocean siphoning model proposed by Mitrovica and Peltier (1991), which suggests that significant changes in the geoid accompany growth and decay of major continental icesheets (Figure 2.12). At the time of maximum icesheet development

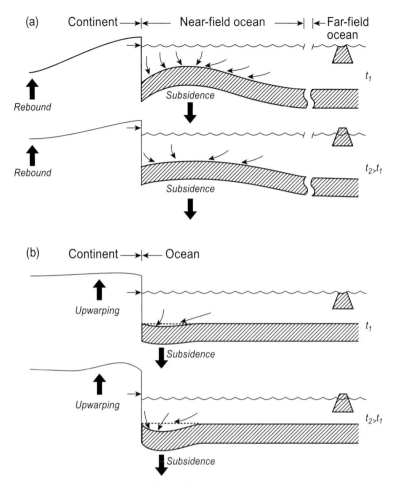

Figure 2.12 A schematic representation of the model of equatorial ocean siphoning as conceptualised by Mitrovica and Peltier (1991), responsible for a late Holocene fall in relative sea level as recorded on ocean islands in far-field localities and remote from continental margins that are otherwise prone to hydro-isostatic changes. During a glacial period continental icesheets cause downward deformation of the lithosphere (crust and upper mantle) promoting sublithospheric flow away from formerly glaciated areas towards mid-latitudes. This results in the formation of a gravity anomaly in lower latitudes and an associated geoid high in the ocean basins. Accompanying and following deglaciation, the formerly glaciated regions undergo glacio-isostatic (viscoelastic) crustal rebound. This results in gradual decay of the gravity anomaly and the oceanic geoid changes configuration, migrating from lower to higher latitudes and toward the formerly glaciated continents (from Mitrovica and Milne, 2002, with permission from Elsevier).

in high latitudes, viscous mantle material was preferentially relocated away from regions directly beneath icesheets towards mid-latitudes, forming a peripheral bulge that encircled the glaciated regions of Antarctica, North America, and northern Europe. These changes are an expression of isostatic compensation due to the stress-induced loading effect of developing icesheets and would have resulted in a fundamentally different global geoid during glaciations. The newly formed geoid highs (forebulges) were primarily located within oceanic regions within mid-latitudes (e.g. in the North Atlantic Ocean in an arcuate zone southeast of Miami, Florida and to the north of the Bahamas). The geographical extent of the forebulge has been determined, in part, based on tidal records from a network of tide gauges along the United States Atlantic Coastal Plain (Chappell, 1974c). A broad domal structure is inferred with apex of the forebulge near Miami, and the trough between the forebulge and zone of uplift (region of the former Laurentide Icesheet) located near Delaware (Chappell, 1974c). The hinge zone separating present-day areas of forebulge decay (subsidence) and isostatic uplift after icesheet decay is located off the coastline of Maine. A recent reconsideration of isostatic adjustment of this coast was undertaken by Kemp *et al.* (2011), as discussed in Chapter 7.

Isostatic crust–mantle compensation (uplift) in high-latitude, formerly glaciated regions accompanied deglaciation, and concomitant collapse of mid-latitude forebulges resulted in mass transfer of ocean water away from the far-field regions (remote from ice loading) and centres of deglaciation now experiencing uplift, towards locations of the collapsing forebulge (i.e. to maintain hydrostatic equilibrium). Present-day modelled rates of isostatic rebound reach up to 22 mm/year near the centre of Hudson Bay (beneath the Laurentide Icesheet), 9 mm/year in Scandinavia (Donner, 1995), and up to 4 mm/year subsidence of the forebulge (Mitrovica and Peltier, 1991). Collapse of the geoid high and mass transfer of water is a latent process that accompanied and followed culmination of deglacial sea-level rise, reflecting rheological properties of the Earth's mantle (inertia in redistribution of mantle material due to its relatively high viscosity). Thus, in order to re-establish hydrostatic equilibrium, ocean water fills the void formerly occupied by the forebulge, and accordingly lowers sea level in the far-field. Accompanying this general process of redistribution of mass within the crust–mantle system, is the process of continental levering, in which ocean basins subside in response to increased load (> 100 m ice-equivalent sea-level rise), and ocean margins experience variable degrees of uplift related to the width of continental shelves. This theory explains widespread evidence for a mid-Holocene highstand in the equatorial Pacific Ocean over the past 4 ka, as discussed in Chapter 7, without detectable variations in global ice volume (Mitrovica and Peltier, 1991; Mitovica and Milne, 2002).

Mitrovica and Peltier (1991) examined relative sea-level histories for the past 5 ka of regions between latitudes 30°N and 30°S to evaluate their model of equatorial ocean siphoning, excluding higher-latitude areas potentially influenced by the peripheral forebulge of icesheets. Sites were also selected remote from continental margins thus avoiding continental flexure (i.e. localised hydro-isostasy). The 18 sites selected are all located in the equatorial Pacific Ocean. Eleven sites revealed good correspondence between theoretical prediction and empirical observations for relative sea-level changes (based on radiocarbon-dated palaeo-sea-level indicators), and 14 of the 18 studied sites showed a sea-level fall that could be accommodated within a model of equatorial ocean siphoning. Plate-tectonic processes appeared to be significant where differences between theoretically predicted and empirically derived relative sea-level histories were apparent; for example, the East Caroline Islands record a relative sea-level rise and are located on the island arc side of the Mariana Trench in a zone of subsidence (Bloom, 1970).

2.4 Tectonism, volcanism, and other processes resulting in relative sea-level changes

A variety of other processes may give rise to a relative change in sea level resulting from subsidence or uplift (see Chapter 5). Mechanisms include vertical tectonic displacements, volcanism, faulting, folding, and epeirogenic movements, the latter two categories requiring more time for their geomorphological effects to be clearly expressed.

2.4.1 Tectonic movements

A variety of neotectonic processes can affect coastlines; these have been reviewed comprehensively by Vita-Finzi (1986) and Stewart and Vita-Finzi (1998), and are described in more detail in Chapter 5. Areas of known tectonic influence on relative sea-level histories include subduction zones where the Earth's lithosphere is reincorporated within underlying mantle, regions of active volcanism, and transcurrent faulting, as well as passive continental margins. In recent years, passive continental margins have been re-evaluated in terms of their apparent tectonic quiescence; several studies have shown that these settings are not as tectonically stable as originally inferred in traditional plate-tectonic models (Summerfield, 1991; Murray-Wallace, 2002; Quigley *et al.*, 2010).

Tectonic processes may compound interpretation of relative sea-level records of a region, particularly where tectonic history is not well understood. No crustal region can be regarded as entirely tectonically stable and, accordingly, crustal movements should be considered even if their rate of process operation is imperceptibly slow. Uplift rates associated with tectonic processes are spatially variable and relate directly to plate-tectonic setting. A logarithmic scale has been used to describe rates of uplift in a global context, namely: quasi-stable, 0–0.005 m/ka; slow, 0.005–0.05 m/ka; moderate, 0.05–0.5 m/ka; rapid, 0.5–5.0 m/ka; and very rapid, 5.0 m/ka (Bowden and Colhoun, 1984).

Development of plate-tectonic theory has provided an important geotectonic framework to explain spatial variations in coastal setting at global scale (Inman and Nordstrom, 1971); Earth lithospheric movements may potentially contribute to relative sea-level histories. For example, crustal uplift on the overthrust plate in subduction-related settings commonly results in development of well-defined flights of marine terraces as seen on Barbados (Schellmann and Radtke, 2004a,b), Huon

Peninsula, Papua New Guinea (Chappell, 1974a), Chile (Ortlieb *et al.*, 1996), and the Ryukyu Islands, Japan (Konishi *et al.*, 1970). These and other situations where vertical land movement is experienced are described in Chapter 5 and their sea-level records are discussed further in Chapter 6.

2.4.2 Volcanism and its link to sea-level changes

The relationship between timing of volcanic episodes and relative sea-level changes and localised ice volumes has been explored by several researchers (Matthews, 1969; Hall, 1982; Wallman *et al.*, 1988; Nakada and Yokose, 1992; McGuire *et al.*, 1997; Church *et al.*, 2005). Numerous interrelated processes may induce individual volcanic episodes in near-coastal settings (e.g. build-up of volatiles, crustal fracturing, seismic events, loading effects of water mass returned to formerly exposed continental shelves). Increased load, such as experienced in the transition from full glacial to interglacial conditions (about > 100 m rise in sea level), may reduce compressive stress within volcanoes by approximately 0.1 MPa (McGuire *et al.*, 1997), in the case of volcanoes that contain differentiated magmas at relatively shallow crustal levels (< 5 km below the ground surface). It has been suggested that mid-ocean volcanoes may have experienced eruptive episodes at the time of major falls in sea level because of lower compressive stresses (McGuire *et al.*, 1997). In a review of the frequency of explosive volcanism in the Mediterranean Basin, based on ages of tephra layers from deep-sea sediments, McGuire *et al.* (1997) found three episodes of increased volcanism, 61–55 ka, 38–34 ka, and 15–8 ka, and noted that the LGM represented the most quiescent phase of regional volcanism during the past 80 ka. They also noted that the most intense phase of volcanism was between 15 and 8 ka, which coincides with the highest amplitude sea-level changes during the past 80 ka. Nakada and Yokose (1992) also speculated that glacio-eustatically induced changing water loads in island-arc settings characterised by a relatively thin lithosphere, such as the circum-Pacific, could potentially induce volcanism.

The effect of volcanic eruptions on sea level is also expressed at a global scale. Large volcanic eruptions can significantly reduce sea-surface temperatures because of aerosols and dust ejected into the stratosphere. The ejecta from volcanoes scatter incoming solar radiation resulting in lower atmospheric temperatures and, consequently, cooling sea-surface waters. The eruption of Mount Pinatubo in 1991 seems to have reduced ocean heat content by 3×10^{22} J, causing a global fall in sea level of 5 mm (Church *et al.*, 2005). Following the eruption of Mount Pinatubo, global sea-level rise from 1993 to 2000 was documented at 3.2 mm/year (Church *et al.*, 2004; Leuliette *et al.*, 2004), in contrast to 1.8 mm/year from 1950 to 2000. The accelerated rate of sea-level rise has been attributed to the combined effects of recovery of sea-surface temperatures following the Mount Pinatubo eruption and recent meltwater contributions from glaciers and icesheets (Church *et al.*, 2005).

2.4.3 Lithospheric flexure

Relative sea-level changes are also associated with lithospheric flexure in response to loading of the plate by accumulations of volcanic rock. McNutt and Menard (1978) realised that emergent reef terraces were commonly found on islands adjacent to relatively young or recently active volcanoes. They proposed that the mass of the extruded volcanic material depressed the plate which resulted in a moat around the island, but rigidity of the lithosphere meant that at a radius of a few hundred kilometres there was an uplifted arch on which emergent islands were situated. This hypothesis appeared to explain islands known as *makatea* islands, which have a volcanic core but are surrounded by emergent reef limestones on their margin (called makatea); examples were known in French Polynesia (uplifted in response to eruption of Tahiti) and the Cook Islands (uplifted in response to eruption of Rarotonga). Models of flexural response based on lithospheric characteristics appear to explain several of these patterns of differential vertical displacement in these archipelagos (Lambeck and Nakiboglu, 1980, 1981), and stratigraphical and geochronological studies of fossil reefs provide further evidence in support of flexure around these recent volcanic loads (Spencer *et al.*, 1987; Woodroffe *et al.*, 1991a).

The island of Hawaii, the largest of the Hawaiian Islands and centre of active volcanism, is considered to have similarly loaded the lithosphere resulting in uplift of an arch in the vicinity of the island of Oahu (Watts and ten Brink, 1989). As Oahu has travelled over the volcanic-induced arch in response to plate-tectonic movements, it has undergone a period of emergence recorded in the relative sea-level record of the island (Grigg and Jones, 1997). This hypothesis has been invoked to explain the contrasting elevations of last interglacial coral reef successions on Oahu compared with the islands of Kauai and Molokai that are more distant from the forebulge (McMurty *et al.*, 2010).

Flexure of the lithosphere is also typical of a plate as it is subducted into an ocean trench. Islands located on a plate that is approaching an ocean trench can also be uplifted as shown by reef limestones at uncharacteristically high elevations on particular islands. The best documented of such situations occurs in the Loyalty Islands, where several uplifted islands represent successive stages of emergence prior to subduction into the New Hebrides Trench (Dubois *et al.*, 1974). Other islands on which a sequence of emergent limestone terraces, of successive interglacials provide evidence of uplift on the arch prior to a plate subducting are Christmas Island in the Indian Ocean adjacent to the Java Trench, and Niue in the South Pacific, adjacent to the Tonga Trench (Woodroffe, 1988b; Kennedy *et al.*, 2012).

2.4.4 Changes in tidal range

Sediment accumulation and associated changes in configuration of the coast may result in changes to tidal amplitude. The clearest evidence for amplification of tidal range is in the Bay of Fundy, which presently experiences the largest tidal range,

> 16 m at equinoctial spring tides, anywhere in the world. As sea level rose during the Holocene and inundated this constricted embayment, tidal range became accentuated (Amos and Zaitlin, 1984). A similar, although lesser, amplification of tidal range is inferred during Holocene evolution of the Wash in eastern England (Hinton, 1995). Alterations in tidal amplitude may not necessarily vary consistently with relative changes in sea level (i.e. in terms of magnitude or direction). Based on tidal modelling using six harmonic tidal constituents, Hinton (1995) concluded that there had been an overall increase in elevation of highest spring tides in the Wash during the Holocene. Coastal changes also resulted in greater reflection of the tidal wave resulting in formation of a standing wave that accentuated the elevation of the highest tides. Human impacts since Roman times, particularly in terms of land reclamation and artificial drainage development, also impacted tidal processes within the region. Ponding of brackish water within coastal estuaries may create the illusion of systematic changes in tidal regimes as has occurred in several Holocene estuaries in southern Australia resulting from closure of the mouths of estuaries due to processes of longshore drift (Sloss *et al.*, 2007).

2.4.5 *Steric changes, meteorological changes, and the role of ENSO events*

Steric changes in sea level can be induced by a variety of meteorological, hydrological, and oceanographic conditions that include changes in atmospheric pressure, current velocity, river discharge, salinity, and water temperature contrasts. Steric effects refer to expansion and contraction in the volume of ocean surface waters accompanying temperature changes; these may be local or regional, giving rise to the 'dynamic sea surface' (Lisitzin, 1974), and can differ by about 2 m from the geoid surface (Mather *et al.*, 1979). Studies of the Gulf Stream reveal a local rise in the sea surface of approximately 3–4 m in response to warmer water temperatures (Menard, 1981).

Short-term (months to several years) climatic variations such as changes in rainfall, atmospheric pressure, and sea-surface temperature, registered across broad regions may also contribute to sea-surface variations. The El Niño–Southern Oscillation (ENSO) phenomenon in the equatorial Pacific Ocean is responsible for changes in mean monthly sea level by as much as 40 cm (Wyrtki, 1985). Tropical air movement in the equatorial Pacific Ocean is dominated by easterly winds which drive surface water away from the eastern Pacific margin and adjacent coastlines of Peru and Chile. Failure of the easterly winds approximately every 2–7 years results in pooling of additional surface water in the eastern Pacific (i.e. development of a Kelvin wave). In the ENSO event of 1982–1983, the sea-surface variation registered across the equatorial Pacific Ocean associated with this event was in the order of 30–40 cm (Harrison and Cane, 1984; Wyrtki, 1985). Continental surface water runoff received during the transition period from a dry ENSO to a wet La Niña phase may also induce local sea-surface variations.

2.5 Geophysical models and the sea-level equation

Geophysical models have been developed to quantify the GIA response to changes in loads of ice and seawater. These can be used to estimate postglacial sea-level history anticipated at any location. The models have been tested and refined through comparison with observational data, primarily since the LGM (Peltier and Andrews, 1976; Clark *et al.*, 1978; Nakada and Lambeck, 1989; Mitrovica and Peltier, 1991; Milne *et al.*, 2002). Recently, there have been attempts to extend application of such models further back in time (Lambeck *et al.*, 2012).

Total change in ice volume equates to *ice-volume equivalent sea level* (ζ_{esl}), which is essentially eustatic sea level as identified by Suess (1888) and described above, and which can be expressed by the following function:

$$\Delta\zeta_{esl}(t) = -\frac{\rho_i}{\rho_o}\int_t \frac{1}{A_o(t)}\frac{dV_i}{dt}\,dt \tag{2.1}$$

where V_i is ice volume at time t, $A_o(t)$ is ocean surface area at time t, and ρ_i and ρ_o are average densities of ice and ocean water, respectively.

GIA models represent Earth's internal structure with simplified, spherically symmetrical, multi-layered internal structures, each having a well-defined viscosity profile with laterally uniform rheological properties. When changes occur in mass distribution, the lithosphere deforms elastically, but viscous flow in the underlying mantle continues for several thousand years. The Earth is parameterised to comprise a lithosphere of known thickness (commonly 65, 80–90, or 120 km) that experiences an elastic response, overlying a two-layered viscoelastic mantle, following concepts proposed by Walcott (1972) and Chappell (1974c). Mantle rheology is modelled using a range of values, increasing viscosity across the seismic discontinuity at depth of ~670 km. Whereas early modelling assumed a largely isoviscous mantle with only a small change from upper to lower mantle, comparison of differential sea-level changes around the coast of Australia led to the suggestion that viscosity might change by up to orders of magnitude (~10^{20} to ~10^{22} Pa s; Nakada and Lambeck, 1989; Lambeck and Nakada, 1990). Further regional studies implied considerable differences between Pacific Island sites and Hudson Bay in North America (Kendall and Mitrovica, 2007). Typically the range for upper mantle viscosity is constrained between 0.3 and 1×10^{21} Pa s, and the lower mantle viscosity ranges between 2 and 10×10^{21} Pa s.

The sea-level equation (SLE) developed by Farrell and Clark (1976) describes relative sea-level variations resulting from transfer of water mass between icesheets and oceans on a viscoelastic Earth. Sea level at any locality (φ) and time (t) can be defined as:

$$\Delta\zeta_{rsl}(\varphi, t) = \Delta\zeta_{esl}(t) + \Delta\zeta_I(\varphi, t) + \Delta\zeta_T(\varphi, t) + \Delta\zeta_{tide}(\varphi, t) + \Delta\zeta_{sed}(\varphi, t) \tag{2.2}$$

where $\Delta\zeta_{rsl}(\varphi, t)$ represents relative sea-level change at location φ and time t compared to its present level. The term $\Delta\zeta_{esl}(t)$ is the global ice-volume equivalent sea-level (eustatic) component, $\Delta\zeta_I(\varphi, t)$ is the isostatic contribution, $\Delta\zeta_T(\varphi, t)$ is the tectonic contribution, $\Delta\zeta_{tide}(\varphi, t)$ represents localised tidal effects, and $\Delta\zeta_{sed}(\varphi, t)$ allows for possible sedimentary processes, such as compaction (Shennan, 2007).

In comparing model simulations with observational field evidence of palaeosea-level indicators, it is preferable to choose sites where local tectonic effects can be excluded, and tidal or sedimentary changes are usually considered negligible. In these circumstances the isostatic component explains geographical variation in elevation of the sea surface, and this is generally divided into glacio-isostasy (ζ_{I-g}) corresponding to deformational and gravitational effects associated with changing ice load, and hydro-isostasy (ζ_{I-h}), which entails response to changes in ocean load (Lambeck *et al.*, 2010), so that the relative sea-level equation then becomes

$$\Delta\zeta_{rsl}(\varphi, t) = \Delta\zeta_{esl}(t) + \Delta\zeta_{I-g}(\varphi, t) + \Delta\zeta_{I-h}(\varphi, t) \tag{2.3}$$

In order to solve the sea-level equation three variables are required: an ice model constrained in space and time, a rheological model of the Earth's lithosphere (depth-averaged effective viscosity of the mantle), and a geometric description of the ocean surface (geoid) to which meltwater is added (Nakada and Lambeck, 1989). The process of modelling relative sea-level histories requires a-priori knowledge of former geographical extent and thickness of Pleistocene icesheets and a chronology of ice-volume change. It is also subject to rotational feedback, as changes in Earth's rotational state that accompany the isostatic adjustment processes also influence sea-level history, particularly in the first few thousand years after onset of deglaciation (Milne and Mitrovica, 1998; Peltier, 2004). Appropriate ice models have been progressively refined as knowledge of past distribution of ice across key sites in the northern hemisphere has developed (Tushingham and Peltier, 1991; Peltier, 1996). ICE-3 represented a major compilation, whereas the more recent, ICE-5G, accounts for ice melt that contributes to an equivalent of about 125 m of sea-level rise since the LGM (updating approximately 120 m of melt implied in the ICE-4G model). Adjustment between these models was made as a result of a larger ice-melt source in the Hudson Bay region (Peltier, 2004), but further evidence of anomalies identified by space geodesy has enabled minor modifications and production of an ICE-6G (Toscano *et al.*, 2011), which also increases the extent of the East Antarctic ice mass to the shelf break. The ice-equivalent sea level for these models, and those for the ice model progressively developed at the Australian National University by Kurt Lambeck and co-workers (Lambeck *et al.*, 1998, subsequently updated in 2005), referred to as KL05, are shown in Figure 2.13. ICE-5G and KL05 apportion greater ice thickness over North America than ICE-3G, and KL05 envisages melting of Antarctic ice continuing into late Holocene.

The sea-level equation is used to delineate a time series of relative sea-level change. The equation can be rearranged so that sea level, present-day three-dimensional motion of the Earth, or present-day changes in geopotential and rotation vectors can be determined (Milne and Shennan, 2007). Figure 2.14 shows schematically how

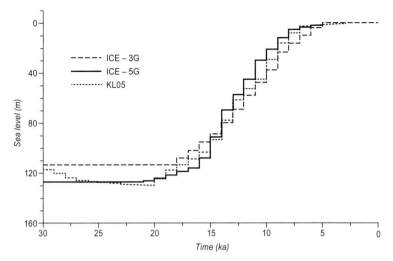

Figure 2.13 Three ice-equivalent sea-level histories (ICE-3G, ICE-5G and KL05) adopted in GIA modelling, shown at 1 ka time steps (after Spada and Galassi, 2012).

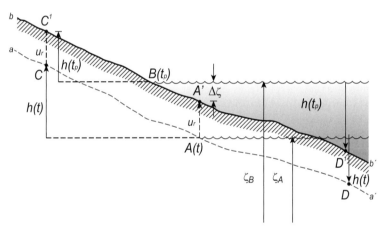

Figure 2.14 Definition of relative sea-level change. At time t the land surface is at a–a', the shoreline is at $A(t)$ and the ocean surface is at a distance ζ_A from the centre of the Earth. In the interval from t to the present t_P the land has been uplifted by u_r and the land surface is now at b–b' and the original shoreline is displaced to A'. In the same interval, the ocean volume has increased so that the ocean surface is now at a distance ζ_B from the Earth's centre of mass and the shoreline is now at $B(t_P)$. The observed relative change in sea level is the position of A' relative to $B(t_P)$ or $\Delta\zeta = u_r - (\zeta_B - \zeta_A)$ (from Lambeck et al., 2010, with permission from Wiley).

change in apparent shoreline position adjusts over a time period, involving both displacement of the land surface and that of the ocean surface.

Accordingly, spatial variation in relative sea-level history, described in Section 2.3.6, is generated. Figure 2.15 shows how this eventuates through incorporation of isostatic and eustatic components. In those near-field areas that were formerly glaciated, crustal rebound exceeds the eustatic component and relative sea level appears to have been falling throughout. In intermediate field sites associated with the peripheral forebulge, subsidence of the crust means that relative sea-level history shows decelerating rise up to present. In far-field sites, particularly on the continental margin, there is much reduced subsidence in response to loading of the ocean floor by meltwater, and some water loading on adjacent continental shelves. In these regions this can result in mid-Holocene palaeo-shoreline evidence occurring above modern sea level. This explains the highstand, associated with ocean siphoning described above. The eustatic sea-level component shown schematically in Figure 2.15 shows the simple case in which melt of icesheets culminated in mid-Holocene. If ice melt continued, as is considered likely in the case of Antarctica (see KL05 in Figure 2.13), then there would be an ongoing eustatic component which would reduce the highstand and affect its timing, as discussed in Chapter 7.

Considerable crustal rebound in the near-field results from glacio-isostatic uplift associated with melt of former icesheets. For sites within the former ice margin, crustal rebound exceeds the eustatic component and the relative sea level is falling throughout. If cessation of global melting is abrupt, this will be recorded as a small cusp in the relative sea-level curve at that time. Rebound becomes less at sites further from former icesheets (dotted line in Figure 2.15). At intermediate sites in the peripheral forebulge zone, the crust experiences subsidence, and the isostatic contribution is of the same sign as the eustatic component; its amplitude depends on distance from the ice margin (two different sites are shown in Figure 2.14). In the far-field, on continental margins, there is subtle subsidence of the seafloor resulting from water loading of the ocean floor (as well as the continental shelf). The relative sea-level curve from many of these sites records a highstand at ~6,000 years ago and a subsequent fall in sea level caused as relaxation of the seafloor dominates once sea-level has stabilised. The highstand will be less apparent if there has been ongoing contribution of meltwater from Antarctic ice over recent millennia (shown schematically by the dotted line in the lower panels of Figure 2.15).

Geophysical modelling of locations near former Pleistocene icesheets has been studied to quantify the internal viscosity profile of the Earth's mantle and lithosphere, and far-field locations have been particularly important in quantifying ice-volume contributions to postglacial relative sea-level histories. Geophysical models of relative sea-level changes are sensitive to both the ice-melt histories of late Pleistocene icesheets and the viscosity structure of the Earth's interior. Close to the icesheets, there will also be an effect associated with gravitational attraction of the ice mass which elevates the sea surface.

The relative importance of these effects varies in time and space. For example, in the case of the Greek mainland and adjacent Aegean Sea, the rigid term that corresponds to the gravitational pull of ice mass, which elevates ocean surface for

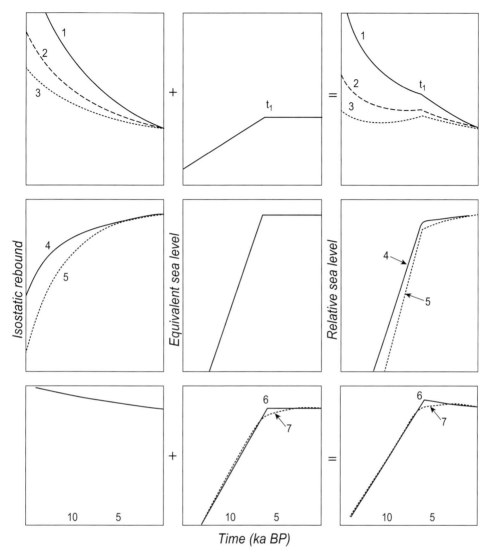

Figure 2.15 Schematic illustration of the modelling process by which crustal rebound is combined with the ice-volume equivalent sea-level (eustatic) component to produce a curve of relative sea-level change which is spatially variable because of the Earth's response to deglaciation. The ice-volume equivalent sea-level (esl) function adopted shows a pattern of ice melt that culminates in mid-Holocene, with no further ice melt, although many recent simulations incorporate an ongoing ice-melt contribution from Antarctica during this period with an appropriately modified eustatic sea-level contribution; gravitational effects are not considered. The vertical scale is not quantified and differs between the three cases: top row represents three sites within the former ice margin, middle row two sites in the peripheral forebulge zone, and bottom row two far-field sites (after Lambeck *et al.*, 2010, with permission from Wiley).

distances of up to 2,500 km from the icesheet, has a more pronounced effect on the sea surface at 18,000 years ago when ice was most extensive than it does at 10,000 years ago. It is partly cancelled by crustal deformation associated with the Fennoscandian ice load; whereas the water load term is positive over land and negative over the sea (Lambeck, 1996b).

The GIA correction, however, is not perfectly determined, but depends on several factors. It depends on the ice-history adopted, both in terms of the ice-equivalent sea-level adopted (Figure 2.13) and the geographical distribution of ice load. The time step used in modelling, whether corrections for the Earth's rotational vector are included, and whether shoreline positions are fixed or adjust as the sea rises, affect the model outcome (Spada and Galassi, 2012). A one-dimensional stratification of the solid Earth is generally adopted, but different values have been used for lithospheric thickness and upper and lower mantle viscosity. Considerable regional and local differences can become apparent between different GIA models (Spada and Galassi, 2012). For example, particularly significant anomalies have been observed between observational evidence along the eastern margin of North America and modelling using the ICE-5G and ICE-6G models, indicating that these regions remain very sensitive to model parameters (Engelhart *et al.*, 2011a,b).

2.6 Synthesis and conclusions

Numerous processes, acting in concert, have been responsible for relative sea-level changes evident along coastlines during the Quaternary. The dominant mechanism for Quaternary sea-level change is glacio-eustasy, resulting from substantial variations in ice volume; ice-equivalent sea-level changes are of ~125 m amplitude from glacials to interglacials. Superimposed on these are other smaller-amplitude changes. Some of these involve feedback effects in response to variable influences of ice and water mass redistributions (i.e. glacio-hydro-isostatic adjustment processes). In the case of hydro-isostasy, there may be evidence in far-field sites (remote from former icesheets) of relative sea-level fall of up to a few metres since attainment of a Holocene sea-level highstand, also referred to as continental levering. In contrast, glacio-isostatic compensation occurs over tens of metres and is most apparent in, and marginal to, areas that were formerly glaciated. Other smaller-amplitude variations in sea level relate to steric effects.

Response of the Earth's lithosphere and mantle to differential ice and water loads has further contributed to the magnitude of observed relative sea-level changes. Geophysical modelling has refined understanding of underlying processes culminating in a more systematic explanation of how different processes contribute to relative sea-level records in space and time. In far-field settings, for example, sea-level records are dominated by postglacial sea-level rise (which overwhelms any other signals inherent within the sea-level record), until the rate of postglacial sea-level rise slows and the effects of hydro-isostasy and equatorial ocean siphoning become evident.

3

Palaeo-sea-level indicators

*Attempting to understand Quaternary sea-level history provides
a vigorous intellectual workout. After negotiating a long path through
data and concepts of mixed quality, one finds global ice volume has
fluctuated by tens of meters sea-level equivalent at rates that are
difficult to resolve.*

(Matthews, *1990, p. 88*)

3.1 Introduction

A diverse range of proxy indicators has been used to identify previous Quaternary
sea-level positions. The most reliable require unambiguous identification of former
sea level, which would appear straightforward, but in practice presents complex
challenges. Only through careful and critical appraisal of reliability and validity
can a precise understanding of the nature and timing of Quaternary sea-level changes
be derived. Reliability refers to the reproducibility of palaeo-sea-level measurements
and validity relates to the degree to which a particular sea-level indicator accurately
quantifies a former sea level without recording effects of other undefined environmental processes.

This chapter critically appraises various lines of evidence, biological, geomorphological, geological, and archaeological, that have traditionally been used to determine
changes in relative sea level from sedimentary successions and landforms of late
Quaternary age (≤ 128 ka). It is demonstrated that, from a range of coastal sedimentary
(depositional) and erosional features used for making inferences about palaeo-sea level,
the most reliable studies are based on investigations undertaken using multiple fixed
sea-level indicators, with detailed geochronological control. Some of the best sites are
those characterised by relatively uniform tectonic behaviour (Chappell, 1987; Siddall
et al., 2006).

Biological indicators of former sea levels comprise well-defined structures
that result from growth of organisms in relation to a distinct tidal datum. Examples
include large bioconstructions such as coral reefs, or individual organisms that
occupy a known position with respect to the sea surface, such as mangroves, and

certain types of molluscs, serpulid worms and other marine invertebrates. Generally, the position of the sea surface, or other physicochemical attributes of water, exerts a control on location of these biota, enabling their use as biological sea-level indicators in space and time. As will be shown, some biological indicators are particularly sensitive to relative sea-level changes. Human settlement patterns, particularly in the archaeological record, can be another form of biological sea-level indicator.

Geomorphological evidence comprises distinctive landforms, and depositional or erosional processes, that relate to tidal datum in some manner. Examples of depositional features include beach ridges and a variety of relict shoreline deposits, whereas examples of erosional features include shoreline notches, shore platforms, and a range of smaller forms, such as honeycomb structures (tafoni), potholes, and solution pans. Geological indicators of former sea level, comprising features such as sedimentary deposits with well-defined *lithologies* (sediment types) or containing sedimentary structures (e.g. planar cross-bedding in foreshore facies) are another type of evidence.

For much of the world's coastlines, Holocene coastal successions display well-defined morphostratigraphy expressed by mutual association of coastal landforms and their preserved stratigraphical successions, providing evidence for former sea levels (e.g. beach ridges, cheniers, foreshore sediments). Over longer periods of time, however, erosional processes destroy many coastal landforms or portions of landforms and their associated sedimentary successions, and only incomplete evidence may remain preserved within thickly stratified deposits, frustrating attempts to unambiguously define palaeo-sea level. With time, distinctive landform features that could otherwise be unambiguously related to former sea levels may be largely destroyed by erosion, making it difficult to accurately quantify former sea levels from fragmentary geological deposits.

3.1.1 Fixed and relational sea-level indicators

Two categories of relative sea-level indicators have been described (Chappell, 1987). *Fixed sea-level indicators* (also called *finite*, *diagnostic*, or *indicative*) represent biological, geological, or geomorphological features that always form within a well-defined elevation with respect to a tidal datum, such as mangroves which are confined to the intertidal zone on tropical coasts. Because fixed sea-level indicators always occur within a restricted elevational range with respect to the sea surface, they represent particularly useful lines of evidence. Intertidal mangroves, subtidal seagrass banks and meadows, and supratidal flats are examples of fixed sea-level indicators with a broad relation to the intertidal zone, whereas specific individual organisms (e.g. barnacles) or growth forms (e.g. coral microatolls) may provide much more precise indications of former sea level in a narrower range within the intertidal. In contrast, *relational sea-level indicators* (also called *directional*) form at a range of elevations above or below the intertidal zone and, accordingly, are of lesser reliability. For example, many species of marine molluscs may potentially occur over a wide range with respect to tidal planes, from inner continental shelf environments to accumulations within high-energy storm or tsunami deposits well above the highest astronomical tides (HATs). In this context, detailed analyses of molluscan or foraminiferal assemblages are essential for refining palaeo-sea-level and palaeoenvironmental interpretations. However, data derived from fixed and relational sea-level indicators, when combined, may potentially yield a very detailed record of relative sea-level changes with respect to time (Sloss *et al.*, 2007; Lewis *et al.*, 2008, 2013).

3.1.2 *Relative sea-level changes, sea-level index points, and indicative meaning*

Methods for systematically reconstructing vertical changes in Holocene sea level were developed as part of the early International Geosciences Programme (IGCP – then International Geological Correlation Programme) and the International Union for Quaternary Science (INQUA, particularly the Shorelines Subcommission) projects (Tooley, 1985). Figure 3.1 illustrates schematically various developments during these projects, from initial attempts to compile a single global sea-level curve using all available data points, to recognition of different sea-level curves for different geographical regions (e.g. Atlantic versus Indo-Pacific sea-level histories), and broader definition of envelopes rather than individual curves. An important consideration has been moving beyond the simple plot of each sample in terms of age and elevation (or depth) to express age estimation errors and other vertical uncertainties (such as survey error). Error terms may be shown as polygons (e.g. Devoy, 1982), horizontal and vertical bars (e.g. Woodroffe *et al.*, 1987), or in more complex ways that depict sea-level tendency (e.g. Shennan *et al.*, 1983).

A *sea-level index* point enables an estimate of the accuracy with which a palaeotidal level, and hence former sea level, can be defined from morphostratigraphical, lithostratigraphical, and biostratigraphical evidence (i.e. position of the palaeo-sea-level indicator within overall tidal range at time of its formation). Sea-level index points take into account four palaeo-sea-level attributes: geographical location, age, altitude, and sea-level tendency. Sea-level index points are established by determining elevation of specific sea-level indicators (e.g. sedimentary structures, landforms, biogenic structures, faunal assemblages) with respect to their tidal datum and radiocarbon or other geochronological age (Shennan *et al.*, 1995; Gehrels *et al.*, 1996; Shennan, 2007). Of particular use in the delineation of sea-level index points are well-defined transitions within sediment profiles such as change up-profile from marine to non-marine sediments recorded within isolation basins (barred basins). In the Arisaig area in north-western Scotland, localised bedrock depressions record a change in deposition from inner continental shelf marine sedimentation to terrestrial sediment deposition (non-marine lakes, bogs, and mires) in response to a relative fall in sea level accompanying regional postglacial isostatic uplift (Shennan *et al.*, 2000a, see

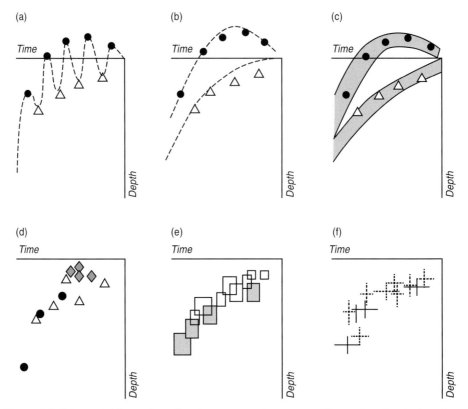

Figure 3.1 Schematic illustration of various approaches to compiling a time-depth plot of sea-level data and interpreting relative sea-level changes: (a) data from disparate sites compiled into one oscillating curve; (b) data from geographically separate locations interpreted in terms of two sea-level curves; (c) similar data interpreted as sea-level envelopes; (d) different data types (symbols) plotted as single points; (e) data plotted with elevation and age uncertainties shown as polygons; and (f) uncertainty terms as horizontal and vertical bars.

Figure 2.7c). The disconformity from marine to non-marine deposition is a critical marker in palaeo-sea-level investigation. Comparison of sea-level index points from different stratigraphical positions or different geomorphological contexts requires a normalisation calculation that accounts for *indicative meaning* of different palaeo-sea-level indicators. Indicative meaning refers to the elevation within the tidal range that the sample is likely to have experienced at time of formation, such as the upper intertidal to supratidal transition to landward of mangrove forests (van de Plassche, 1986). This involves the assumption that tidal range has not changed locally. *Sea-level tendency* refers to whether the sea surface was rising (positive sea-level tendency) or falling (negative sea-level tendency) at a given sea-level index point at time of deposition (Shennan *et al.*, 1983). This analytical approach is significant because few palaeo-sea-level indicators have formed at mean sea level.

3.1.3 Sources of uncertainty in palaeo-sea-level estimation

Several potential sources of uncertainty may contribute to palaeo-sea-level inferences as deduced from single localities and include selection of a specific geodetic reference datum, for example, definition of particular elevation within the tidal range (e.g. Mean Low Water Springs, MLWS), spatial variation in tides within relatively small geographical areas, selection of appropriate palaeo-sea-level indicators, and elevation range in sampling from a single indicator, as well as uncertainties associated with geochronological methods (Kidson, 1986). In this context, individual palaeo-sea-level observations, when expressed in diagrammatic form, should include an age uncertainty term and an elevation uncertainty term (Figure 3.1).

Uncertainties in quantification of former sea-level heights result from several sources, such as the water-depth range in which the sea-level indicator may occur, local water-level anomalies, for example in back-barrier lagoons intermittently closed, and measurement errors during surveying. These can be grouped to derive an overall estimate of height uncertainty according to the following equation (Shennan, 2007):

$$U_{th} = \sqrt{u_1^2 + u_2^2 + u_n^2} \tag{3.1}$$

where U_{th} represents total uncertainty in estimation of a palaeo-sea-level datum, and u_1^2, u_2^2, u_n^2 represent height uncertainty squared of each variable. The square root is then found for the sum of squares of height uncertainties. Other uncertainties in estimation of palaeo-sea level relate to post-depositional changes that modify the original nature of sedimentary deposits. Some of these changes may be subtle and difficult to quantify, such as amount of post-depositional sediment compaction. Sea-level evidence is also dependent on geomorphological setting. For example, morphodynamic changes in the landscape resulting from gradual deposition of coastal landforms can lead to local alteration of tidal prism and, consequently, proxy evidence records water levels locally (Chappell and Thom, 1986).

3.1.4 Palaeo-sea-level curve or envelope?

The way palaeo-sea-level data have been portrayed has changed and graphical forms appear almost cyclic. These reflect implicit assumptions about reliability of particular data sets, statistical validity of data sets, and what constitutes a genuine and resolvable anomaly from a longer-term trend of sea-level behaviour (i.e. sea-level changes compared with noise in the sedimentary record; Bryant, 1988). There has been disagreement as to whether observations, expressed as a series of dots on a diagram, with or without uncertainty terms in defining measurement, should be joined by a line, a curve be interpolated through the data, or whether it is preferable that the data be represented by an envelope (Figure 3.1). The latter approach acknowledges a broader range of potential uncertainty.

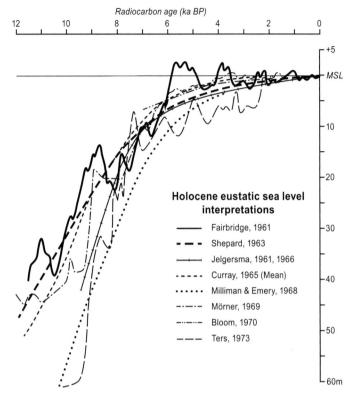

Figure 3.2 A selection of Holocene sea-level curves (after Belknap and Kraft, 1977, with permission him SEPM), showing contrasting reconstructions by Fairbridge and Shepard. See Chapter 7 for further examples and more detail.

Although a seemingly simple issue, perhaps no other issue in palaeo-sea-level studies has been subject to such heated controversy. Estimates of palaeo-sea level at single localities should include acknowledgement of all potential sources of uncertainty and portrayal of any palaeo-sea-level observation by a single, discrete dot on a time–depth plot usually assumes an unrealistic degree of accuracy for most palaeo-sea-level indicators.

Figure 3.2 shows a selection of Holocene sea-level curves compiled from pre-1975 literature. Many indicate a gradual deceleration of sea level as it rose towards its present level. For example, based on a compilation of palaeo-sea-level data from several geographically widely separated regions presumed to have been stable (Texas Shelf, Holland, Argentina, Louisiana, Australia, Florida, Ceylon, and Mexico), the Shepard curve shows a monotonic rise in sea level since the LGM, reaching present sea level only in recent times. If Australian data are excluded from the compilation of Shepard (1963), a sea level higher than present is not evident during the past 6,000 years. In contrast, the Fairbridge curve shows some well-defined anomalies during deglacial sea-level rise and several short-term sea-level stands higher than present during the past 6,000 years. Several of the isolated peaks above present sea level have been defined based on radiocarbon dating of shell beds and other coastal features in Western Australia (Abrolhos Islands, about 500 km north of Perth, as well as Rottnest Island, Point Peron, and Cottesloe Beach near Perth). The question arises as to whether all the isolated sea-level peaks identified by Fairbridge (1961) are real, or an artefact of the graphical depiction of the data.

The Fairbridge and Shepard sea-level curves are both predicated on the concept of a global sea-level curve. In the strictest sense and at the highest spatial and temporal resolution, a truly global sea-level curve is unattainable, as is shown in Chapters 2 and 7. Glacio-hydro-isostatic adjustment processes cause varying responses of coastlines to redistribution of water from icesheets to ocean basins, and result in differing relative sea-level histories accompanying deglaciation. These contrasts are particularly marked for coastlines in near- and far-fields of former continental icesheets. There are also other underlying assumptions. Both assume that vertical and horizontal uncertainties are relatively minor for any individual sample, that ages fall within the uncertainties of counting statistics, which in this case are only at 1σ level (68.26% confident that numerical age falls within associated analytical uncertainty compared with 95.46% at 2σ level; Zar, 1984), and that vertical uncertainties are realistic. Such uncertainties imply that no single point is precise, and it may be unwise to assume palaeo-sea-level data are sampled populations of the real sea-level story, without considering either mathematical modelling or statistical description of the structure of data. These uncertainties mean it may be misleading to extrapolate beyond the data at hand using linear or polynomial curve fitting to interpolate absent data, which may be partially addressed by considering indicative meaning of sea-level index points.

3.1.5 Facies architecture, allostratigraphy, and sea-level changes

The traditional approach of vertical profile analysis, where single sections of sedimentary successions or individual core profiles are measured, does not necessarily provide a representative overview of longer-term sedimentation patterns or depositional processes (Miall, 1990). Consequently, increased emphasis has been placed on *facies architecture*, the three-dimensional geometry of sedimentary facies. Bounding discontinuities of sedimentary facies represent mappable features of particular importance in the study of sea-level changes. *Allostratigraphy* is the branch of stratigraphy that seeks to divide the rock record based on bounding discontinuities rather than using lithological properties which represent the basis of traditional lithostratigraphy. Figure 3.3 illustrates the contrast between a lithostratigraphical and an allostratigraphical approach to mapping genetically related strata (Walker, 1992). Facies architecture and allostratigraphy are relevant in that the genetic significance of strata provides information about sea-level behaviour. Thus, the angle of dip of major bounding discontinuities may indicate

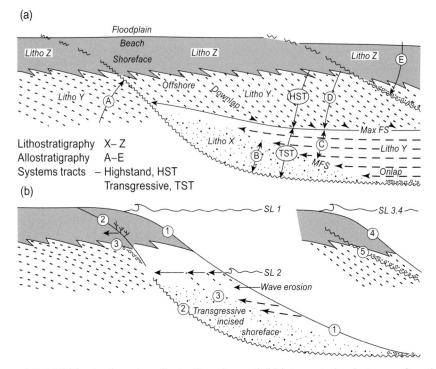

Figure 3.3 (a) Lithostratigraphy, allostratigraphy, and (b) interpretation in terms of sea-level fluctuations (after Walker, 1992). Lithostratigraphic units (litho X = conglomerate; litho Y = shales; litho Z = sandstones) and allostratigraphic units (A–E) are shown in the upper diagram, as also are highstand systems tract (HST) and transgressive systems tract (TST), marine flooding surface (MFS) and maximum flooding surface (MaxFS), and bounding discontinuities (shown by zig-zag between litho Y and Z). This is interpreted in the lower diagram. During time period 1 (SL1), the shoreface built out to profile 1. Sea level then fell to SL2, cutting a new profile (labelled 2). Transgression characterises period SL2 to SL3 with further shoreface truncation (ravinement surface) and deposition of litho X. Sea level stabilised during SL3 and SL4 and the sequence prograded out to form its present morphology.

a relative rise or fall in sea level, such as the erosional contact between coastal sediments lying unconformably on inner continental shelf marine sediments.

3.2 Pleistocene and Holocene palaeo-sea-level indicators compared

Philosophically, there should be no difference in how geological investigations and interpretations of Pleistocene and Holocene sedimentary successions are undertaken. However, in practice, when making inferences about former sea levels, subtle conceptual and methodological contrasts arise, related to age resolving power, sampling constraints, preservation potential, and uniformitarian inferences.

It is significantly harder to resolve time for Pleistocene sedimentary successions. It is still questionable whether time can be validly resolved at millennial and centennial scales within a single late Pleistocene highstand deposit despite recent technological improvements in methods of U-series dating (such as TIMS and MC-ICPMS, see Chapter 4). Whereas uncertainty in physical measurement of radiocarbon for a sample of Holocene age may be ± 80 years at the 95.46% (2σ-level) an uncertainty of $\pm 1,000$ years at the 68.26% (1σ-level) may be obtained on a fossil of last interglacial age (about 128–118 ka) by conventional methods, representing approximately 10% of the duration for this interglaciation. The considerable improvements in precision of latest U-series techniques have reduced this, but still exert constraints on interpretations.

It is necessary to consider which part of a sedimentary deposit should be sampled for dating and how representative that is in relation to inferred palaeo-sea level. These questions highlight the need for a very clear understanding of genetic significance of facies architecture and individual sedimentary structures within deposits, and assessment of dated material as a palaeo-sea-level indicator. For example, the wide range of water depths within which some marine invertebrates live imposes considerable uncertainty as to their palaeo-sea-level significance, and the possibility that fossils have undergone post-mortem transport further complicates interpretations.

Commonly, only thin remnants of former coastal successions are preserved in the stratigraphical record (e.g. only a short portion of an interglacial highstand), making it particularly difficult to quantify duration of coastal deposition. Many primary sedimentary deposits represent an incomplete relative sea-level record in view of erosional processes occurring shortly after deposition and reflect the lower preservation potential of some facies.

Some coastal deposits are more likely to be preserved than others. Coral reefs, for example, have a high preservation potential which explains why fossil reef terraces have been central to understanding Pleistocene sea-level highstands. In contrast, fine-grained sediments, such as mangrove facies, do not preserve well, and although they play an important role in constraining relative sea-level history during the Holocene, equivalent deposits formed during earlier interglacials are scarce. In older deposits it may be more difficult to quantify temporal significance of diastems (i.e. relatively short-term periods of non-deposition). For example, along the embayed bedrock coastline of southern New South Wales, only incomplete remnants of last interglacial (MIS 5e) estuarine sediments are preserved within headward and basal portions of estuarine valley fills (Nichol and Murray-Wallace, 1992), in contrast to northern New South Wales where Pleistocene barriers are commonly preserved (and mapped as the shore-parallel Inner Barrier; Thom *et al.*, 1992).

Modern and Holocene rates of coastal sedimentation are commonly used to infer long-term rates of sediment accumulation and duration of sedimentary events in the earlier geological record. For example, detailed facies architecture and compositional reconstructions of former reefs in the last interglacial reef limestones of Barbados have been interpreted by comparison with modern equivalents (Blanchon and Eisenhauer, 2005). Comparisons of this nature involve uniformitarian assumptions about rates of sediment deposition and biodiversity. However, there are often distinct differences in composition, and perhaps therefore functioning, of reef ecosystems as evident from Pleistocene exposures, compared with their Holocene equivalent (Pandolfi, 1996; Pandolfi *et al.*, 1999).

Irrespective of geological age, some of the most reliable palaeo-sea-level interpretations will be derived from sedimentary successions that preserve an assemblage of sea-level indicators from contrasting water depths. In the following discussion, examples of different types of palaeo-sea-level indicators are discussed. Coral reefs are discussed first as the mapping and dating of fossil reefs has played a central part in reconstruction of Quaternary sea-level history. Other biological indicators are then considered, extending from tropical environments associated with reefs to more temperate shorelines where much early study was concentrated. The potential of geological and geomorphological evidence is outlined, and the value of archaeological evidence is discussed.

3.3 Corals and coral reefs

Coral reefs form large durable structures and have been used extensively in determination of past sea-level changes because corals occur in a narrow vertical depth range, have good geological preservation potential, and can be dated by radiocarbon, U-series, and other dating techniques. Reef-building corals (termed hermatypic, and comprising primarily Scleractinia) grow by calcification as their polyps secrete a carbonate exoskeleton. Although corals are animals, coral polyps support photosynthetic symbionts, called zooxanthellae, which limit them to the *photic zone*, water depths through which light penetrates, typically 0–30 m in clear tropical waters.

Coral reefs result from *in-situ* accumulation of corals and other associated organisms over time, and cover more than $250,000\,km^2$ of the ocean where sea-surface temperatures exceed 18°C (Veron, 1995). Coral growth requires relatively clear and warm waters with minimal suspended sediments, as well as a firm substrate on which corals can establish. They form extensive reef systems in the tropics but reach into subtropical latitudes, extending from 33°N in Japan and 32°N at Bermuda to similar latitudes in the southern hemisphere at Inhaca Island in southern Mozambique and Lord Howe Island in the southern Pacific. The greatest diversity of corals occurs in the Indo-Pacific, with a second region centred on the western Atlantic. Coral reef development is limited in the eastern Atlantic and eastern Pacific Oceans.

Distribution of coral species across an individual reef varies according to water depth, and in relation to several other environmental parameters such as wave energy, light availability, and sedimentation (Chappell, 1980). The growth form of corals varies from delicate branching corals, through encrusting forms, to hemispherical colonies of massive corals and they have discrete tolerances and their distribution appears related to environmental gradients in these stresses. The growth forms are also distributed along these environmental gradients (Hopley, 1986); for example, delicate forms of coral are fractured if wave energy is high, and only more robust forms with greater strength can persist on the reef crest where wave energy is greatest. In the highest wave-energy settings corals will be encrusting, or may be replaced by coralline algal ridges characteristic of Indo-Pacific reefs (Montaggioni and Braithwaite, 2009).

Individual coral colonies can continue to grow for several hundreds of years, and after their death may be incorporated into reef structure, together with other calcareous organisms such as foraminifera, molluscs, and coralline algae. Subaerial exposure imposes a stress on many reef flats and sets an upper growth limit approximately at MLWS. In adjacent backreef environments several coral genera may adopt a microatoll growth morphology (see Section 3.3.4) as a result of exposure at low tide (Hopley *et al.*, 2007).

3.3.1 Reefs and Pleistocene sea levels

Corals and coral reefs are important indicators of palaeo-sea level at several different scales. The significance of prominent reef limestones exposed around tropical shorelines and the corals within them has been recognised for much of the twentieth century (Daly, 1915). Development of radiocarbon and U-series dating enabled discrimination of Holocene reefs from Pleistocene, if corals had not undergone diagenesis, and differentiation of a sequence of successive interglacial sea-level highstands. On uplifted coastlines, such reefs may be preserved as a staircase of terraces that contains not only the principal interglacials but also several interstadials (shorter-lived sea-level highstands related to warm periods within a glaciation, interspersed between full interglacials). An initial estimate of sea level at the peak of the last interglacial was derived from U-series dating corals from reef terraces that were from locations considered tectonically stable. Widespread evidence from across the tropics (from sites such as Bermuda, Bahamas, Oahu in Hawaii, Mangaia in the Cook Islands, the Seychelles, and Western Australia) implied a last interglacial shoreline around 120,000 years old at heights of 2–10 m APSL (averaging 6 m; Veeh, 1966). The evidence in these cases was the elevation of the upper surface of the reefal limestone.

Subsequently, still greater resolution was achieved by extending these radiometric dating programmes to flights of elevated reef terraces on tectonically uplifted coasts. Several such terraces occur above sea level on Barbados, with the older being the more landward. Zonation of coral species and growth forms provided more exact evidence of the relationship between the morphology of the former reef and level of the sea in which it formed (Mesolella, 1967). In the Caribbean, *Acropora palmata* dominates many of the reef crests, and recognition of this species on former reef crests enables inferences to be made about palaeo-sea level (Schellmann and Radtke, 2004a,b).

In contrast, in the Indo-Pacific region, there is not such a clear relationship between water depth and distribution of coral species. Nevertheless, a particularly detailed sea-level reconstruction for the late Quaternary has been derived on the rapidly uplifted Huon Peninsula in Papua New Guinea. More terraces are found on active plate-margin coasts that are uplifting rapidly, such as Sumba in Indonesia, along the Ryukyu Islands in Japan, and on the island of Haiti in the Caribbean (see Chapter 5). With the now increasingly refined sea-level history, it is possible to simulate the morphology that would be anticipated using a one-dimensional forward computer spreadsheet model of reef growth (Koelling *et al.*, 2009). Distribution of species within a reef may provide some evidence for rapid sea-level rise. For example, elevated sea level has been inferred during a short time period within the otherwise prolonged last interglacial stillstand (Blanchon *et al.*, 2009).

Interpretation of fossil reefs, initially at or above sea level, has recently been extended to submerged reefs, providing independent evidence for the pattern of ice melt and concomitant sea-level rise during the postglacial. Submerged reefs on Barbados in the Caribbean were known from bathymetric studies (Macintyre, 1967), but the first radiocarbon reconstruction was based on drilling of those offshore reefs (Fairbanks, 1989, see Figures 2.5 and 7.1). Concurrently, drilling through Holocene reefs on the Huon Peninsula showed a similar pattern of reef growth tracking sea level (Chappell and Polach, 1991). Subsequently, comparable long records of post-glacial sea-level rise have been derived from Mayotte in the Comoros Islands, and Tahiti in the Society Islands (Bard *et al.*, 2010), with only a slightly shorter record from the Abrolhos Islands off the coast of Western Australia (Eisenhauer *et al.*, 1993). Submerged reefs are found at several sites where there is limited modern reef development, for example in the southern Gulf of Carpentaria, Australia (Harris *et al.*, 2008), and on the southernmost Pacific reefs at Lord Howe Island (Woodroffe *et al.*, 2010), as well as along the outer margin of the Great Barrier Reef (Abbey *et al.*, 2011). Following the onset of polar melting around 19 ka, many reefs seem to have been able to keep pace with the rising sea, but gaps in the record are considered to indicate meltwater pulses (Montaggioni and Braithwaite, 2009), discussed in greater detail in Chapter 7.

3.3.2 *Reefs and Holocene sea levels*

Most of the northern hemisphere polar ice appears to have melted by around 7,000 years BP and as the ice-equivalent ocean volume approached its current value, the rate of 'eustatic' sea-level rise slowed, after which regional disparities in the pattern of relative sea-level history, due primarily to hydro-isostasy, have become increasingly evident (Lambeck *et al.*, 2010). For tropical to subtropical regions, corals have played an important role in deciphering these relative sea-level changes. Radiocarbon ages on coral contributed to the sea-level oscillations apparent in the age–depth compilation of Fairbridge (1961), and coral data were central to discrimination between a Pacific sea-level history with a mid-Holocene highstand, and an Atlantic sea-level curve that has been continuously rising until present but at a decelerating rate (Adey, 1978).

The Atlantic sea-level curve has been developed primarily based on ages of *Acropora palmata*, which dominates the reef crest to depths of 5 m on the reef front. It has been widely used to reconstruct sea level because it is considered an indicator of shallow water (Lighty *et al*., 1982; Toscano and Macintyre, 2003). However, fragments of *Acropora* recovered in cores can rarely be unambiguously shown to be in a position of growth. Even where skeletal architecture appears to indicate branches in an upright position, it is difficult to be sure they grew there as opposed to being detrital material that chanced to be deposited in that orientation. Individual colonies of *A. palmata* do occur in water deeper than 5 m, and fragments of this branching coral may have been transported, either to greater depths, or on to land and deposited in storm ridges, implying that uncertainties of sea-level position could range over a vertical range of ~10 m (Blanchon, 2005; Gischler, 2006).

Pacific reefs underwent a period of flourishing reef growth from about 8,000 years ago, after the antecedent platforms (commonly a near-horizontal terrace formed during the last, or penultimate, interglacial) were inundated by the rising sea, and until the sea stabilised close to its present level around 6,000 years ago. Corals grow well in shallow water, and this period of reef re-establishment commenced around 8,000 years BP at depths of 10–20 m below present sea level (depending on antecedent surface topography). In some cases reef growth kept pace with sea level, a growth strategy termed 'keep-up' (Davies and Montaggioni, 1985; Neumann and Macintyre, 1985). In many more cases it lagged behind the sea surface, with reefs adopting a 'catch-up' strategy, or was drowned by rapid sea-level rise, a process called 'give-up' (Figure 3.4).

The classification of reefs in terms of their response to sea level, based on chronostratigraphic studies (Neumann and Macintyre, 1985) can be extended as shown in Figure 3.4 (Hubbard, 1997; Woodroffe and Murray-Wallace, 2012). If the rate of sea-level rise is very rapid, a reef is likely to be drowned. At slightly slower rates of rise, the reef may backstep, with establishment of a discretely different reef with foundations in a more landward location than the earlier reef. In some cases, a reef can keep pace with sea level, but in others it lagged behind the sea surface and only grew to catch up with sea level several thousand years after the sea stabilised close to its present level. During the period over which the sea has been relatively stable there has been little accommodation space for further vertical reef growth and reefs have prograded horizontally. In many parts of the Pacific, there are reefs left stranded (give-up reefs) by a slight fall of sea level from the late Holocene highstand (as a result of ocean siphoning, see Chapter 2, Section 2.3.6).

The pattern of reef growth is not always constrained by sea-level history; this can be effectively demonstrated by a comparison of two particularly comprehensive reconstructions of fringing reef development from contrasting sea-level settings (Stoddart, 1990). Macintyre and Glynn (1976) described how a reef at Galeta Point on the Caribbean coast of Panama had formed during the Holocene based on 13 cores and 32 radiocarbon ages, and Easton and Olson (1976) described a very

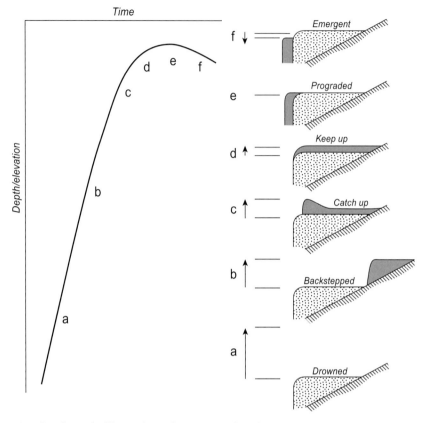

Figure 3.4 A schematic illustration of response of reefs to variations in Holocene sea level (following the keep-up, catch-up and give-up strategies of Neumann and Macintyre, 1985, and its portrayal by Hubbard, 1997; after Woodroffe, 2003).

similar pattern of development for the fringing reef at Hanauma Bay, Oahu in the Hawaiian Islands, based on 10 cores and 63 radiocarbon ages. Although the two studies implied a similar growth history with an early stage of reef initiation around 7,000 years BP followed by a major phase of vertical accumulation to around 3,000 years BP, and relatively little growth over the past 2,000 years, the two sites have actually experienced different sea-level histories. The Panama reef, dominated by *Acropora palmata*, grew as a keep-up reef tracking the decelerating sea level pattern typical of this part of the Caribbean. Easton and Olson dismissed evidence from other researchers for a Holocene highstand in Hawaii and also suggested that the Hanauma Bay reef had adopted a keep-up growth strategy, but this decelerating pattern of sea-level rise is no longer regarded as correct for Hawaii. There is now overwhelming evidence from the Hawaiian Islands that sea level was above present in mid-Holocene (Fletcher and Jones, 1996). An emerged coastal bench and associated beach occurs on Kapapa Island on the eastern side of Oahu from which radiocarbon ages of around 3,500 years BP have been determined (see Figure 7.9a), indicating that sea level was about 2 m higher than present (Grossman and Fletcher, 1998). The Hanauma Bay fringing reef is now considered a catch-up reef (Grigg, 1998).

3.3.3 Conglomerates and recognition of in-situ corals

Debates about the Holocene highstand, or the degree of emergence, have centred on the relationship between coral conglomerates and the sea level at which they formed. A conglomerate platform, described by Charles Darwin from the Cocos (Keeling) Islands (Darwin, 1842) comprises a flat surface that occurs along the seaward margin, and clearly underlies many of the reef islands (Figure 3.5a), containing disoriented coral heads of up to 50 cm diameter. There has been an ongoing controversy as to whether corals preserved in these conglomerate platforms provide evidence of higher sea level or are storm deposits whose elevation is not diagnostic of sea level. Prominent fossil *Porites* corals from Funafuti Atoll in Tuvalu were considered evidence of sea level slightly higher than present (David and Sweet, 1904). Fossil corals in conglomerates in the Maldives were considered by Gardiner (1903) to record a sea level many metres above present. Figure 3.5b shows fossil outcrops of the blue octocoral, *Heliopora*, from Addu Atoll in the southern Maldives but at a level close to, or slightly above, the elevation to which it now grows. Similar *Heliopora* on the reef flat on Onotoa in Kiribati was also considered evidence of emergence (Cloud, 1952; see also Figure 7.10b). An expedition to the Caroline and Marshall Islands in the 1960s was designed to address the sea-level question; however, it did not reach a consensus on whether the coral rubble deposits indicated a higher sea level or were storm deposits. Some researchers felt that, with the exception of tectonically uplifted islands, none of the deposits contained corals that were indisputably in a growth position and that all rubble ridges were storm deposits (Shepard *et al.*, 1967; Curray *et al.*, 1970), whereas other researchers concluded that fossil corals and *Tridacna* clams related to a higher sea level (Newell and Bloom, 1970; Schofield, 1977a).

Subsequently, there have been many dated corals reported from conglomerates on mid-ocean islands, summarised, for example, by Pirazzoli and Montaggioni (1988) for French Polynesia, Grossman *et al.* (1998) for the equatorial Pacific, and Woodroffe (2005) for the eastern Indian Ocean, and discussed further in Chapter 7. A critical question in these studies is whether the corals are *in situ* and have not been reworked since death. The abundant rubble ridges found in high wave-energy reef crest environments, and storm ridges apparent after tropical cyclones, indicate that individual colonies of coral may be transported into reef-flat settings. However, corals can sometimes be shown to be in growth position (Figure 3.5b), and in these cases a comparison between fossil elevation and that of modern equivalents is likely to give an indication of sea-level changes. It is necessary to examine the composition of the conglomerate in detail. Subtle changes in cements enabled changes in sea level

(a)   (b)

(c)   (d)

Figure 3.5 (a) Oblique photograph of reef islands on the Cocos (Keeling) Islands, showing conglomerate outcrops (photograph: Colin Woodroffe). (b) Fossil *in-situ* branches of *Heliopora*, exposed on Addu Atoll in the Maldives, which indicate a sea level close to, and possibly slightly above, modern (photograph: Colin Woodroffe). (c) Large flat-topped microatoll, more than 3 m in diameter (3 m survey staff), Christmas Island, central Pacific. The outer margin is living tissue, growing up to a level coincident with low tide (photograph: Colin Woodroffe). (d) A *Porites* coral that has been uplifted on Simeulue, an island off the west coast of Sumatra, showing evidence of several changes of water level; the entire coral was lifted out of the water and died following the 2004 earthquake (photograph: Kerry Sieh).

to be detected in French Polynesia (Montaggioni and Pirazzoli, 1984). In conglomerates on atolls in the Gilbert chain in Kiribati, there are two units, a lower unit in which *Heliopora* is preserved in its growth position (see Figure 7.10b), and presently above the elevation at which it can grow on the modern reef, indicating a higher sea level. This is overlain by a rubble unit composed of coral boulders thrown by storms onto a former reef flat (Falkland and Woodroffe, 1997; Woodroffe and Morrison, 2001). Even clearer evidence of a higher sea level can be inferred where microatolls, corals that have been constrained in their upward growth by exposure at low tide levels, are found within the conglomerate (Woodroffe *et al.*, 1990a,b). Microatolls provide some of the clearest evidence for past sea level and are described in the next section.

3.3.4 Microatolls

Coral microatolls, discoid corals that have grown laterally, as vertical growth is inhibited by their exposure at lowest tides (Figure 3.5c), provide the clearest evidence of the elevation of sea level in the past. These flat corals have been used as palaeo-sea-level indicators from a number of reefs in the Indo-Pacific because they are commonly well-preserved; they have been used to reconstruct vertical movements on tectonically active coasts where the reef has been elevated by seismic uplift, and it has also been suggested that individual living colonies may have a subtle modulated record of historical water-level changes.

Coral microatolls of up to a few metres in diameter were described in detail from the northern Great Barrier Reef by Scoffin and Stoddart (1978), who recognised their value as sea-level indicators. The principal genus is *Porites*, although intertidal *Goniastrea*, *Platygyra*, and other genera can adopt a microatoll form. Where micro-atolls grow in 'open water' settings they have living rims generally at an elevation around MLWS. However, in habitats not freely connected to open ocean, such as behind storm ridges where the ebbing tide is impeded as it drains off the reef, microatolls are moated, and can be less precisely related to sea level (Hopley and Isdale, 1977). The prime limitation on the majority of microatolls is exposure during lowest tides, although some corals appear to be impeded in their upward growth by accumulation of sediment on their upper surface, and nutrient limitations have also been inferred as a constraint on others (Smithers and Woodroffe, 2000). Never-theless, there can be considerable geographical variation in the elevation that limits that upward growth across or around a reef (Figure 3.6). For example, the tops of microatolls within the same field in areas on the inner Great Barrier Reef with a tidal range of 2–3 m are considered reproducible within ± 10 cm (Chappell *et al.*, 1983). On the Cocos (Keeling) Islands with a tidal range of 1.2 m, the tops of open water microatolls are typically elevated between MLWS and mean low water neaps (MLWN) and accurate surveying of several hundred living specimens showed an overall variation of up to 40 cm, related to subtle hydrodynamic conditions such as wave setup and local tidal conditions (Smithers and Woodroffe, 2000).

Following the pioneering descriptions by Scoffiin and Stoddart (1978), a particu-larly comprehensive study described microatolls from sites along the northern Queensland mainland on the inner margin of the Great Barrier Reef (Chappell, 1983d; Chappell *et al.*, 1983). It demonstrated that microatolls across a reef flat were at progressively lower elevations and of successively younger ages towards the modern reef crest, where living microatolls occur. Microatolls are a particularly valuable palaeo-sea-level indicator because individual corals live for decades to centuries, so contain a filtered record of sea level over a sustained period. For example, a fossil *Porites* microatoll on Pagan in the Mariana Islands has been described exceeding 9 m diameter, and radiocarbon dated 2,195 ± 80 years BP at its centre and 1,535 ± 130 years BP on its outer margin, indicating negligible variation in sea level over this period (Siegrist and Randall, 1989). Similarly, large fossil *Porites*

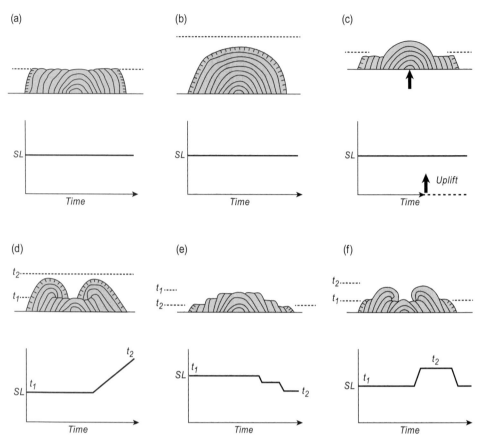

Figure 3.6 Schematic response of microatolls to changes of sea level or uplift of the land, as shown by annual banding within the coral skeleton; illustrating (a) sea level constant from year to year and a microatoll that has grown up to sea level which continues to grow outwards, unlike (b) an unconstrained domed coral. (c) If the coast undergoes uplift, the exposed upper surface dies. (d) When water level increases, a coral can begin to overgrow the formerly dead upper surface. (e) If water level falls episodically terracettes are formed. (f) If there are fluctuations of water level with a periodicity of several years these may be preserved as annuli on the upper surface of the microatoll (after Woodroffe and McLean, 1990, and Lambeck *et al.*, 2010, with permission from Wiley).

microatolls, more than 9 m in diameter, have been reported from the Leizhou Peninsula, in the southernmost part of Guangdong province in southern China (Yu *et al.*, 2009), and from Christmas Island in the Line Islands, central Pacific (Woodroffe *et al.*, 2012b; Figure 3.5c).

At a finer scale, the upper surface microtopography of individual microatolls has been used to reconstruct smaller changes in relative sea level over shorter time periods, determined by slicing the coral to reveal annual growth bands by X-radiography. This approach, known as sclerochronology, has been shown to be very accurate, with some of the largest corals having lived for hundreds of years. Irregularities in surface morphology develop primarily in response to changes in sea level (that is, a change in water depth in which the coral is living altering MLWS). The close correspondence with the height to which living coral can grow within the tidal range is demonstrated in spectacular fashion in those instances in which a reef is rapidly uplifted by an earthquake. Following the Aceh–Andaman earthquake in 2004, the reef was raised by up to a metre and specimens of *Porites* were left protruding above the limit to which polyps could survive (Figure 3.5d), with continued growth only in their lower portions (Meltzner *et al.*, 2006). A record of seismic activity can be deciphered from sequences of fossil microatolls on these tectonically uplifted coasts, using the microatolls as palaeo-sea-level indicators (Zachariasen *et al.*, 2000).

Vertical growth of coral microatolls is constrained by periods of exposure during lowest spring tides. Overall morphology of microatolls in cross-section reflects local changes in sea level that can be detected with an annual resolution (Woodroffe and McLean, 1990). Several distinct growth forms can be seen amongst living microatolls. These were classified by Hopley (1982) as: '*Classical*' microatolls with a relatively horizontal upper surface called the microatoll plane; '*Top hat*' microatolls with an elevated centre and lower outer rims, indicating a lower sea level in the later phase of coral growth; '*Upgrown*' microatolls which have low centres encircled by higher living rims, indicating a rising sea level; and '*Multiple-ringed*' microatolls with concentric annuli across the upper surface of the microatoll. These and similar modes of growth, as indicated by the annual growth banding in the coral skeleton, are shown in Figure 3.6.

If sea level remains constant from year to year, a massive coral that has grown up to sea level continues to grow outwards but with its upper surface constrained at water level (Figure 3.6a). Where the coral does not reach water level, it adopts a domed growth form that is not constrained by water level (Figure 3.6b). If the coast undergoes uplift (or sea level falls), a coral previously not limited by sea level may be raised above its growth limit and the exposed upper surface will die (Figure 3.6c), but with continued lateral growth at a lower elevation. If water level increases, a coral previously constrained by exposure at low tides can resume vertical growth and begin to overgrow the formerly dead upper surface (Figure 3.6d). If water level falls episodically then the microatoll shows a series of terracettes (Figure 3.6e). If there are fluctuations of water level with a periodicity of several years then the upper surface of the microatoll consists of a series of concentric undulations (Figure 3.6f); such a pattern can be seen on microatolls from reef flats in the central Pacific where El Niño results in interannual variations in sea level (Woodroffe and McLean, 1990; Lambeck *et al.*, 2010).

It is important to recognise several limitations to the way in which the coral responds to water-level changes. Coral growth for massive *Porites* is 1–2 cm per year, so the coral cannot track larger changes in water level with annual resolution.

Nevertheless, mean monthly sea-level variations of several tens of centimetres that accompany El-Niño/La Niña episodes in the central Pacific are recorded as minor annuli of a few centimetres on tops of microatolls on several Pacific island reef flats (Woodroffe and McLean, 1990; Spencer *et al.*, 1997).

As with all palaeo-sea-level indicators, caution must be exercised in the use of microatolls in palaeo-sea-level investigations. There is the potential for some micro-atolls to be undercut so it is critical that only microatolls that are *in situ* on firm substrates are analysed. It is also important to note that coral microatolls reflect interannual changes in a low tide level defining the limit of coral growth rather than changes in mean sea level. Recently, it has been shown that there can be significant variations related to geoidal gradients around the coastline of islands, indicating that it is preferable to relate elevations of fossil microatolls to modern, living equivalents in the same geomorphological setting (indicative meaning), rather than to assume horizontal tidal planes over large distances (Woodroffe *et al.*, 2012b).

3.4 Other biological sea-level indicators

Whereas a coral such as *Acropora palmata* occurs in shallow subtidal environments, indicating that the sea surface is slightly above it (Figure 3.7a), several organisms occupy a narrow vertical range relative to the intertidal zone. Mangroves, for example, are almost entirely restricted to the upper intertidal, and commonly show zonation of species (Figure 3.7b). Boring molluscs, such as *Lithophaga lithophaga*, occur predominantly in habitats that are continually wet, providing evidence of submersion, and they have been used to infer that the substrates on which they occur were at some stage under water (Laborel *et al.*, 1994). Those restricted to particular elevations within the intertidal zone are described by the term 'fixed biological indicators' (FBIs). Microatolls are one example of such FBIs, but there are many smaller organisms which also fall into this category.

3.4.1 Fixed biological indicators

On hard substrates, such as cliffs, breakwaters, and seawalls, there are often distinct vertically differentiated zones of intertidal benthic organisms, for example species of encrusting vermetids (*Vermetus triqueter*) and oysters (*Chama* and *Spondylus*). There is an abundant ecological literature on physical parameters determining where particular organisms can survive, but interspecific competition implies they are clearly biologically modulated. Distribution of organisms varies substantially from site to site, but it may be possible to discriminate midlittoral from supralittoral (above) and subtidal (below), as in the Mediterranean (Laborel *et al.*, 1994).

The midlittoral, dominated by barnacles (*Balanus*, *Elminius*, *Tetraclita*, and *Chthamalus*), oysters (*Chama* and *Ostrea*) and mussels (*Mytilus*), is submerged fre-quently as opposed to the supralittoral that is exposed more frequently. A distinct

Figure 3.7 A selection of biological sea-level indicators. (a) The staghorn coral, *Acropora palmata*, which occurs on reef crests in the Caribbean and indicates shallow water (photograph: Colin Wood roffe). (b) Mangrove forests in northern Australia, occupying the upper intertidal. Zonation of species is apparent broadly associated with elevation in the intertidal zone and frequency of inundation (photograph: Andy Short). (c) Oysters forming a prominent visor at Magnetic Island, north Queensland (photograph: Scott Smithers). (d) Molluscs in shell beds, exposed by erosion at North East River, Flinders Island, Bass Strait, Australia. Many of the valves of *Katelysia rhytiphora* are articulate in these sediments, indicated to be last interglacial in age by both amino acid racemisation and electron spin resonance dating, and in the position in which they grew (photograph: Colin Murray-Wallace).

zonation occurs, with elevation depending on tide and wave characteristics. The frondose coralline Rhodophyte *Lithphyllum lichenoides* has been used to reconstruct former sea level along the southern coast of France, and at other sites in the Mediterranean (Laborel and Laborel-Deguen, 1994). The vermetid *Dendropoma petraeum* can form reefs up to 10 m wide with a depth range of over 0.4 m, but in places can be a very accurate sea-level indicator with precision of up to ±0.1 m (Pirazzoli, 2007). Along the Sicilian coast, Antonioli *et al.* (1999) identified platform-type *Dendropoma* reefs which they used to reconstruct relative sea-level changes over the past 700 years.

In intertidal environments in Australia, the polychaete tubeworm, *Galeolaria caespitosa*, secretes durable calcareous tubes, the upper limit of which occurs between mean sea level (MSL) and MHWN, rendering it a useful FBI (Bird, 1988). On sheltered rocky coasts, *Galeolaria* grows over a 0.3–0.5 m vertical range but can grow to higher elevations where there is strong wave action (Baker *et al.*, 2001a). These authors indicate that it can mark former sea level to an accuracy of ± 0.5 m, but this precision may be increased when supplemented using the interface between *Galeolaria* and barnacles, limpets, or other molluscs occupying the upper intertidal (Baker *et al.*, 2001b). Dating of fossil tubeworms provided evidence that Holocene sea level had been higher than present in northern New South Wales (Flood and Frankel, 1989; Baker and Haworth, 1997), and this approach has recently been considerably extended by Baker and Haworth who have undertaken a series of studies using the tubeworms as FBIs (Baker and Haworth, 2000a; Baker *et al.*, 2001a).

In optimum conditions, fossil specimens of sessile intertidal FBIs offer considerable vertical precision. The vertical range is dependent on tidal range and exposure to waves. In microtidal locations, when related to their modern equivalents, it may be possible to achieve vertical comparison between modern and fossil where the uncertainty in palaeo-sea level could be as little as ± 0.05 m (Pirazzoli *et al.*, 1996; Morhange *et al.*, 2001). Precision decreases on coasts with larger tidal range, or with more complex topography or wave conditions, or where preservation is poor. In northeastern Queensland the oysters *Saccostrea cucullata* and *Crassostrea amasa* form conspicuous encrustations from MSL to MHWN, actual elevation varying with exposure (Hopley *et al.*, 2007; Smithers, 2011). *Saccostrea* can form distinct visors protruding > 0.6 m from rock faces (Figure 3.7c), and preservation of these above their modern counterparts offers insights into former sea-level highstands (Beaman *et al.*, 1994; Lewis *et al.*, 2008). Barnacles, such as *Octomeris* and *Tetraclitella*, also provide evidence of former sea levels and have been used in Japan and northern Australia (Pirazzoli *et al.*, 1985; Lewis *et al.*, 2008). However, such evidence is always discontinuous; growth was short-lived, and exposure to erosion means preservation only in very sheltered locations rarely comparable to where living counterparts are observable (Smithers, 2011).

Use of such FBIs to reconstruct sea level has frequently been controversial; for example, higher sea level during mid-Holocene has been inferred from ages on encrustations of *Saccostrea* from Big Wave Bay in Hong Kong (Davis *et al.*, 2000), but this interpretation has been challenged (Yim and Huang, 2002; Baker *et al.*, 2003).

3.4.2 Mangroves

Mangroves are *halophytic* (salt-tolerant) trees that grow in upper intertidal environments in estuaries and on low-wave energy coastlines (Bird, 2000). Mangrove forests are best developed along extensive low-gradient, muddy intertidal zones on tropical shorelines. The greatest diversity of mangroves is found in the Indo-West Pacific, extending from the east coast of Africa to the central Pacific Ocean, with a second province centred on the West Indies, but including the eastern Pacific and extending along the eastern coast of Brazil (Duke *et al.*, 1998). This Atlantic province contains fewer species; much comprises a seaward fringe of red mangrove, *Rhizophora mangle*, backed by black mangrove, *Avicennia germinans*, with *Laguncularia racemosa* and *Conocarpus erectus* on the landward margin. In contrast, mangrove forests in Indonesia and northern Australia contain more than 30 species, including several species of *Rhizophora* and other genera in the Rhizophoraceae, such as *Ceriops* and *Bruguiera*. There is considerably greater diversity of habitats and complex distribution of mangrove species (Figure 3.7b), particularly in areas of high rainfall or abundant river discharge. On open shorelines, a broad zonation can be recognised, generally comprising a seaward zone of *Sonneratia*, then a prominent zone of *Rhizophora*, often with *Ceriops* and *Bruguiera*, and a landward zone of other mangroves such as *Lumnitzera* and *Excoecaria* (Macnae, 1968). *Avicennia* is widespread throughout these mangrove forests, and has the widest salinity tolerance. The latitudinal constraint on mangrove distribution is set by their intolerance of winter frosts (Woodroffe and Grindrod, 1991). In arid and semi-arid settings, they are limited by strongly hypersaline groundwaters (Semeniuk, 1983).

Tidal range determines the extent of mangrove colonisation. The seawardmost mangroves are rooted close to MSL, and zonation in species is broadly related to elevation and frequency of inundation (Watson, 1928). In macrotidal settings it is possible to distinguish elevational ranges at which different mangrove species occur (Baltzer, 1969; Woodroffe, 1995). Mangroves extend several kilometres inland up tidal channels within estuaries, although elevation at which individual species are found becomes more constrained as tidal amplitude decreases and freshwater influences are more apparent (Bunt *et al.*, 1985).

Mangroves have long been regarded as important sea-level indicators in view of their intertidal habitat and their distinctive sedimentary facies in stratigraphical profiles (Thom, 1982). Sediments beneath mangroves vary from organic mangrove-derived peats, typical of microtidal areas with little sediment input, such as on carbonate islands, to terrigenous muds containing woody mangrove-derived fragments. The former are termed *autochthonous*, derived primarily from *in-situ* production of root and other fibrous material, whereas, where there is a supply of sediment from outside, this is termed *allochthonous*. Peat occurs beneath mangroves in the Everglades of Florida, beneath mangrove islands (called mangrove ranges) on the Belize Barrier reef (Macintyre *et al.*, 2004), and on microtidal West Indian islands, such as Grand Cayman, where extensive mangrove forests are underlain by reddish fibrous peat up to 4 m or more deep in which root material from *Rhizophora* is prominent (Woodroffe, 1981). Although most mangrove forests are regularly flushed by tides, peat also forms beneath mangrove forests that are occluded behind beach ridges and inundated primarily by rain, and in these rare situations the peaty substrate accumulates beyond the elevation of the highest tides, reducing the effectiveness with which these deposits can be related to sea level (Woodroffe, 1983).

Mangrove forests on continental margins, such as around Australia, are typically underlain by strongly reduced, bluish-grey clays with abundant woody fragments and fibres. At Port Pirie in northern Spencer Gulf, South Australia, for example, modern mangrove sediments have a vertical range from 1.32 ± 0.2 to 2.2 ± 0.5 m with respect to tidal datum (i.e. upper- and lower-bounding surfaces of the mangrove facies; Barnett et al., 1997). In an undisturbed sedimentary succession within this modern *peritidal* sediment province, mangrove facies is characteristically represented by a wedge-like deposit that interfingers with upper-intertidal sandflat facies (coarse shelly sand or *coquinas*) and is overlain by samphire–algal facies (Barnett et al., 1997; Figure 3.8). Mangrove facies span a broader elevational range in macrotidal estuaries of northern Australia where tidal range exceeds 5 m. Mangroves are rooted at elevations close to MSL at the mouth of the South Alligator River, locally corresponding to the survey datum termed Australian Height Datum (AHD) to 2.8 m above MSL where mangroves are replaced to their rear by samphire. Along the approximately 100 km of that estuary, tidal amplitude is modified, such that mangroves can be rooted as high as 3.7 m above MSL 70 km from the mouth, but with rapid attenuation upstream. The lower limit to mangroves further up the river estuary is difficult to constrain because riverbank mangroves (*Sonneratia*, not found further seaward) frequently undergo erosion, and slumping, which truncates the forest fringe. Thus mangrove substrate could be considered to span an elevational range of –1 to +3.7 m AHD, limiting the precision with which dated mangrove deposits from cores can be related to past sea level (Woodroffe et al., 1987).

In the macrotidal river estuaries of the Northern Territory, extensive coastal and estuarine plains have formed through mid- to late-Holocene accumulation of muds beneath mangrove forests (Figure 3.9). Blue–grey clay with abundant organic fragments has accumulated to depths of 15 m or more, recording inundation of these former river valleys as sea level rose and infilled under tidal conditions since sea level stabilised close to its present level ~7–6 ka. A particularly organic basal layer is found at depths of 6–12 m in which reddish fibrous roots of *Rhizophora* are compressed, marking a lower transgressive unit deposited as the valley was inundated by the rising sea. Stumps of mangroves can be seen in river-bank exposures, with calcareous worm casts, molluscs such as *Geloina*, and calcite nodules around the remains of mud lobsters of the genus *Thalassina*, confirming the mangrove interpretation of this facies in cores (Woodroffe et al., 1989). These record a widespread mangrove forest that occurred around $7,400 \pm 200$ cal. yr BP throughout many estuaries along the coast of northern Australia, termed the 'big swamp' that coincided with stabilisation of sea level at around its present level and which persisted for about 1,000 years until sediment had accreted to the upper limit of tidal inundation and mangroves were replaced by freshwater wetlands (Woodroffe et al., 1985; Lewis et al., 2013).

Identification of mangrove pollen has provided one method to attempt to achieve greater elevational resolution from mangrove sediments. Shore-normal profiles and pollen rain studies through mangroves indicate broad zonation of species (Grindrod,

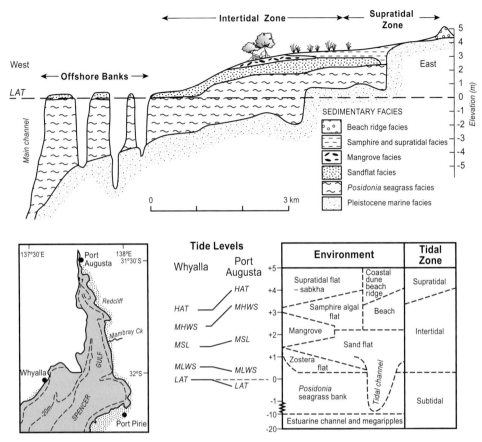

Figure 3.8 Zonation of vegetation along a transect at Port Pirie in relation to tidal datum (LAT) and Australian Height Datum (AHD), and palaeo-sea-level criteria for this coastal region (after Belperio *et al.*, 1984a; Gostin *et al.*, 1984, and Barnett *et al.*, 1997).

1985, 1988). Constraints on using mangrove pollen to reconstruct sea level include variable production and preservation of pollen between species, with *Rhizophora* pollen vastly over-represented compared with *Avicennia*; different ways pollen is dispersed (e.g. some species of *Sonneratia* are primarily bat-pollinated, whereas other genera are wind-pollinated); and differentiation of mangrove pollen is difficult beyond genus level. As a consequence, only broad generalisations can be made in relation to subdividing mangrove facies using pollen. A pollen diagram of sediments below the oxidised upper metre in a core from the Alligator Rivers region (Figure 3.9) recorded a transition from a lower zone dominated by Rhizophoraceae, into one in which *Sonneratia* increased in prominence (interpreted as an increase in the seaward *Sonneratia* fringe), with a return to Rhizophoraceae, then a decrease in mangrove pollen, but with more abundant *Avicennia* pollen, culminating in replacement of mangroves during the late Holocene by grasses and sedges which

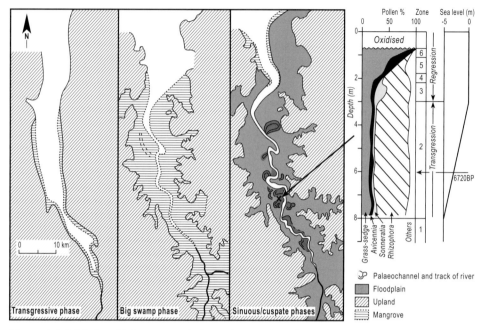

Figure 3.9 Mid–late Holocene evolution of estuarine plains flanking the South Alligator River, northern Australia. The prior valley was inundated around 8,000 years ago during the transgressive phase; extensive mangrove forests flourished across the plains during the 'big swamp' phase, around 6,000–7,000 years ago, as sea level stabilised. Mangroves were replaced once the substrate had accreted beyond the intertidal zone, with deposition of alluvial clays and establishment of freshwater grass–sedge wetland vegetation. This broad long-term change is shown in the pollen diagram of the upper sediments of the plains (after Woodroffe *et al.*, 1985, 1993, with permission from Elsevier).

constitute the principal wetland vegetation of the plains. The *Sonneratia–Rhizophora–Avicennia* transition corresponds with the seaward to landward distribution of these genera implying accretion of sediments under conditions of stable sea level, with the surface building up at a decelerating rate until no longer experiencing tidal inundation and mangrove was replaced by freshwater wetlands (Woodroffe *et al.*, 1985; Figure 3.9). This sequence of pollen transition has now been identified widely, not only in this region (Clark and Guppy, 1988), but across much of northern Australia (Crowley, 1996), and is likely to also be typical across the broader Indo-Pacific region (Engelhart *et al.*, 2007).

Mangrove deposits appear attractive for sea-level reconstruction for several reasons: they are generally deposited intertidally so can be directly related to sea level; they contain organic material suitable for radiocarbon dating which constrains their time of deposition; and it is usually possible to establish that sediments have not been reworked. However, this apparent simplicity hides a number of constraints. First, it is important to establish the elevational range over which mangroves can grow at each site; for example, Scholl and Stuiver (1967) indicated that mangroves grow at 0–12 cm above MSL in Florida, whereas detailed survey of elevational range in the Cayman Islands in a similar microtidal setting indicated that they were growing there at 15–30 cm above MSL, with isolated stands up to 70 cm above MSL where they were occluded behind beach ridges as described by Woodroffe (1981). Subtle distinctions in plant species distribution may be evident in some environments over a narrow vertical range of only a few centimetres, but involving lateral distances over tens of metres, or, in the case of macrotidal estuaries, tens of kilometres. More recent attempts have been made to use foraminifera and in some cases diatoms to reconstruct mangrove environments and their responses to sea-level changes (e.g. Woodroffe, 2009); these are considered below in the section on microfossils.

However, other issues also need consideration. Survey heights record elevation of mangrove substrate, but roots of mangroves penetrate into that substrate. Most species of mangroves are shallow-rooted, but individual roots of some mangroves, such as *Rhizophora*, can penetrate 1 m or more below ground. This not only compounds the relationship of fossil material to observed elevational range, but also means that mangrove sediments are very prone to contamination by younger root material. This yields radiocarbon ages which are younger than expected and implies that mangrove sediments are likely to produce minimum ages that indicate that the indicative level was above the depth sampled, becoming directional rather than fixed indicators.

This is well illustrated by considering the initial reconstructions of sea level in the Caribbean under conditions of a decelerating Holocene sea-level rise. Radiocarbon-dated mangrove peats were used in conjunction with freshwater peats and calcitic muds in developing a sea-level curve from Florida (Scholl, 1964; Scholl and Stuiver, 1967; Scholl *et al.*, 1969), the Cayman Islands (Woodroffe, 1981), and Jamaica (Digerfeldt and Hendry, 1987). Post-depositional root contamination or compaction mean that samples fall below the line interpreted as best representing sea-level rise, derived from radiocarbon ages on coral. Sea level throughout the broader region has been further refined from many more dates on mangrove peat from Belize (Macintyre *et al.*, 1995, 2004). This has continued to be an issue of contention as these authors have interpreted that peat ages that fall above their sea-level interpretation derived from coral data are consistent with that sea-level curve, whereas Gischler (2006) contends that this cannot be correct and that a sea-level higher than that derived by Toscano and Macintyre (2003) would be appropriate for Belize (see Chapter 7).

Mangrove sediments appear to be 'fixed' sea-level indicators where bulk density is high and root penetration and compaction low, for example in the terrigenous clays beneath mangroves on the Australian mainland, but in microtidal areas, root penetration from younger material into mangrove peat renders them directional minimum indicators. The usefulness of mangrove facies as a sea-level indicator appears to be greatest during periods of rapid transgression. During slower rise, as in the Caribbean, the record becomes less clear, and they are not so effective indicators of stillstand (because of continued root penetration) or sea-level fall, because of exposure, oxidation, and degradation of sediments.

Compaction, discussed in more detail below, also needs to be considered as highly organic mangrove sediments undergo post-depositional compaction, apparent by the compressed nature of fibrous roots of *Rhizophora* compressed into ellipsoidal remnants, in some cases appearing simply as fibrous layers. Postglacial mangrove sediments have been identified on the broad Sahul and Sunda shelves, but these have been drowned by sea-level rise (Hanebuth *et al.*, 2000). Back-stepping of mangroves is interpreted from extensive mangrove peat drowned on the floor of the lagoons on both the Great Barrier Reef (Grindrod and Rhodes, 1984), and the Belize Barrier Reef (Halley *et al.*, 1977).

Basal transgressive mangrove facies, beneath coastal plains in northern Australia, contain sequences that are either near-continuous implying that sedimentation kept pace with sea-level rise, or a transgressive–regressive sequence wherein mangrove sediments were initially drowned, but mangrove re-established when sea level stabilised at its present level (Woodroffe *et al.*, 1993). Under conditions of stable sea level, progradation can occur and has given rise to the broad coastal plains around much of southeast Asia and northern Australia. Slight sea-level fall from the Holocene highstand in southeast Asia has left emergent highly oxidised mangrove muds in Malaysia (Geyh *et al.*, 1979).

3.4.3 Salt-marsh sediments and microfossil analysis

Salt marshes are low-lying vegetated areas occupying the upper intertidal zone. Salt marshes occur in low wave-energy environments and are common in estuaries, and in a range of protected coastal settings. Many salt-marsh environments display well-defined, shore-parallel zonation of plant species which relates to degree of tidal inundation. Salt marsh extends from species-poor Arctic ecosystems, often dominated by the grass *Puccinellia phryganodes*, to tropical shores where samphire may be found landward of mangrove forests. The *halophytic* plants on salt marshes, including grasses, herbs, and shrubs, play an important role in further reducing wave energy and promoting deposition of mud. Succulents, such as *Salicornia*, occur to seaward and are frequently inundated, whereas sedges, such as *Eleocharis* and *Carex*, occupy higher, less frequently flooded parts of the marsh. A greater diversity of plants occurs on temperate salt marshes, including grasses such as *Spartina* and *Puccinellia*, and herbs such as *Suaeda maritima*, *Limonium*, and *Aster* (Adam, 1990).

The surface of salt marshes ranges from mid-tide to highest spring tide level (Brown, 1998). Along the Atlantic and Gulf coasts of the United States, marshes are dominated by smooth cordgrass, *Spartina alterniflora*, with its vertical extent varying with tidal range (McKee and Patrick, 1988). Many salt marshes form well-defined, gently seaward-sloping terraces as a result of sediment accretion; elsewhere, the fine-grained sediments may be eroded by wave action with the seaward margin of the marsh truncated by a *saltings* cliff.

The sediments accumulated beneath salt marshes contain a record of accretion that may contain evidence of changes of sea level. Plant remains within the sediments confirm that many marshes have accumulated vertically as sea level has risen (Mudge, 1858; Redfield, 1967), but clearer evidence is derived from microfossils, such as foraminifera and diatoms. However, sea-level reconstruction is complicated as a result of other morphodynamic adjustments that occur in the intertidal zone. For example, detailed morphostratigraphical mapping has revealed complex histories of infill of former tidal creeks within the muddy sediments in the Severn River estuary in western England (Allen, 2000).

Foraminifera, protists that secrete a calcite test (shell), occur ubiquitously in marine and coastal environments. Some species, or assemblages of foraminiferal species, have been found to occur diagnostically within particular sedimentary environments and, accordingly, have been used to infer changes in relative sea level (Scott and Medioli, 1978, 1986; Cann and Gostin, 1985; Cann *et al.*, 2000; Haslett, 2001). Accordingly, one approach has been to define specific sedimentary environments by identification of foraminiferal biofacies. Biofacies refer to packages of sediment that are stratigraphically defined based on their fossil faunas (Raup and Stanley, 1978), and characteristically may be associated with distinct sedimentary environments.

The distribution of foraminifera within salt-marsh environments commonly shows a well-defined zonation, which has been widely used for inferring relative sea-level changes. Scott and Medioli (1978) identified clearly defined vertical zonation in foraminifera across two salt marshes (Tiajuana Lagoon, California, and Chezzetcook Inlet, Nova Scotia in the macrotidal Bay of Fundy region). Their careful analysis of foraminifera within the upper intertidal zone enabled discrimination of palaeo-sea level, with salt-marsh environments divided into two distinct faunal zones. Foraminifera from the upper half of the marsh environment (e.g. *Trochammina inflata, T. inflata macrescens, Tiphotrocha comprimata, Jadammina polystoma*) were only found in that setting. They concluded that the dominant presence of one or all of these species within salt-marsh deposits reflects deposition within the upper quarter of the tidal range and therefore represents a particularly useful palaeo-sea-level indicator (Scott and Medioli, 1978, 1986). This has enabled high-resolution sea-level reconstructions using salt-marsh foraminifera (e.g. Gehrels *et al.*, 1996, 2005), based on the principle that marsh foraminiferal assemblages record the degree to which the salt-marsh surface is able to maintain its position relative to a rising sea level (Figure 3.10).

Higher marsh foraminiferal assemblages (less frequently inundated) occur when marsh accretion exceeds rate of sea-level rise, and lower marsh assemblages occur when marsh sedimentation lags sea-level rise. A 'transfer function' is used to quantify relationships between modern foraminiferal assemblages and marsh-surface elevation. The elevation at which fossil foraminifera lived (their 'indicative meaning') is

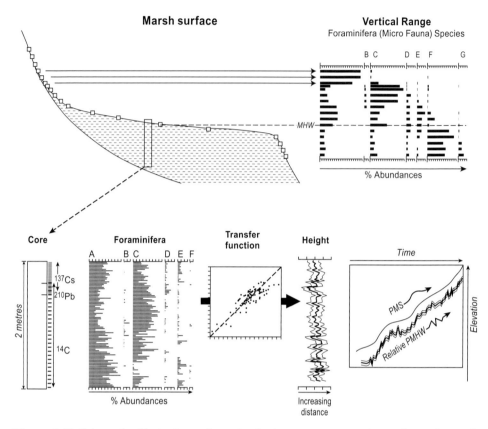

Figure 3.10 Schematic illustration of sea-level change reconstruction using salt-marsh microfossils. The contemporary surface distribution of foraminifera species is related to elevation of marsh surface above a tidal datum (MHW, Mean High Water), and subsequently represented by a transfer function. Palaeo-marsh surface (PMS) indicators are sampled in a sediment core, with dating control of the upper core provided by techniques such as ^{210}Pb or ^{137}Cs and of the lower core by [14]C. The core is also analysed for fossil foraminifera species abundances, which are interpreted in terms of PMS height relative to palaeo-mean high water (PMHW) using the transfer function. These combine to reconstruct the PMS accumulation history and rate of relative PMHW change (after Lambeck *et al.*, 2010, with permission from Wiley).

calculated by applying the transfer function which is based on vertical distribution of modern counterparts (e.g. Horton *et al.*, 1999; Gehrels, 2000). The methods are described in detail by Horton and Edwards (2006) and Engelhart *et al.* (2011a); it is necessary to establish that modern distributions of foraminifera are controlled by environmental factors and that assemblages of dead foraminifera relate to frequency of marsh-surface flooding (Figure 3.10). Modern distributional data provide a 'training set'; the relationship between relative abundances of foraminiferal taxa and environmental variables (elevation) is empirically modelled using regression to derive ecological response functions. This calibration can then be used to estimate palaeo-marsh surface (PMS) elevations from which relative sea level can be determined.

Sea-level index points (SLIPs, m above MTL) can be calculated according to the formula:

$$SLIP = H - D - I + A + C \qquad (3.2)$$

where H is height of the ground surface in metres above or below mean tide level (MTL), D is sample depth below ground surface in metres, I is indicative meaning of the sample in metres relative to MTL as calculated from the transfer function, A is autocompaction of sediment determined by geotechnical correction, and C is core compaction during sampling (Gehrels, 1999; Massey *et al.*, 2008). Optimum SLIPs are derived from samples that have minimal displacement due to sediment compaction; those overlying incompressible Pleistocene basement are termed *base of basal*, whereas those in close proximity to such a basement are termed *basal*; *intercalated* samples are derived from within the Holocene stratigraphical column (Engelhart *et al.*, 2011a; see also Figure 7.15).

This approach, first developed in Europe, has been widely applied along the east coast of North America (van de Plassche *et al.*, 1998; Gehrels, 1999; Donnelly *et al.*, 2004; Gehrels *et al.*, 2005; Horton *et al.*, 2006; Kemp *et al.*, 2012). Age constraints on the lower sections of cores are generally by radiocarbon dating, extended into historical times with shorter-lived isotopes, such as ^{210}Pb, or marker layers, such as ^{137}Cs, pollution horizons, and pollen. It has been suggested that individual palaeo-sea-level estimates can be derived with a vertical precision of ± 5–$20\,$cm and age resolution of $\pm 10\,$years in the nineteenth century and $\pm 5\,$years in the twentieth century. Some validation of recent sea-level reconstructions can be achieved through comparison with tide-gauge records (Donelly *et al.*, 2004; Gehrels *et al.*, 2005). It is important that there are suitable modern analogues for fossil assemblages; there has been debate as to whether to include living foraminifera (in total assemblage) or include only those that are dead, whether estuarine fringing marshes provide analogues for back-barrier marsh settings, the bias that post-mortem dissolution of calcareous species may have, and how to interpret fossil assemblages for which no modern analogues are available (Massey *et al.*, 2006; Wright *et al.*, 2011; Barlow *et al.*, 2013).

Recently, this approach has been extended to the muddy estuarine environments in Australia. Foraminifera had been used in studies of relative sea-level changes during late Pleistocene interstadials (MIS 3) in Gulf St Vincent (Cann *et al.*, 1988, 1993, 2006) and Spencer Gulf, southern Australia (Cann *et al.*, 2000). These studies are based on the ratio of *Elphidium crispum*, which favours shallow water, to *E. macelliforme*, which occurs in deeper water. Distributions of foraminifera in Tasmania indicate that the technique can be used there (Callard *et al.*, 2011). There is less confidence in its application in tropical Queensland. Haslett (2001) examined distribution of foraminifera within intertidal environments of a microtidal, distributary channel of the Barron River Delta in northern Queensland. Three foraminiferal biozones were identified based on appearance of species in distinct sedimentary environments located within different portions of the tidal prism. Salt-marsh environments were represented by a monospecific assemblage of *Trochammina inflata* (Biozone 1). In contrast, tidal flat settings were dominated by *Ammonia beccarii* and two biozones were identified within this environment. In the upper portions of the intertidal flats, foraminiferal assemblages were dominated by *A. beccarii* (> 70% of the total assemblage) with remaining assemblages showing low species diversity (Biozone 2a). In contrast, in the lower intertidal zone characterised by regular inundation, although the foraminiferal assemblage was still dominated by *A. beccarii* (55–65%), the subordinate population showed greater species diversity (Biozone 2b), attributed to *allochthonous* input to the estuary from adjacent open-marine environments (Haslett, 2001). Biozone 1 was found between High Water and Mean High Water (MHW), Biozone 2a between MHW and MTL, and Biozone 2b between MTL and Low Water (Haslett, 2001). It is not clear whether these techniques can be as effectively applied to tropical coastlines. For example, applying a transfer function in Cleveland Bay in Queensland, Woodroffe (2009) inferred a sea-level highstand of 2.8 m, considerably higher than evident from other proxy sea-level indicators in the region. However, detailed taphonomic studies show complex mixing and degradation processes that affect preservation and modify assemblages, compounding relationships between surficial foraminiferal assemblages and their inferred intertidal elevation and those assemblages observed down core (Berkeley *et al.*, 2008, 2009).

3.4.4 Seagrass

Seagrasses are marine flowering plants that have adapted to live under water in fine-grained sediments in sheltered settings. Seagrass plays an important role in promoting sedimentation by baffling currents. Seagrasses are found in both temperate and tropical settings. In the Caribbean, turtle grass, *Thalassia testudinum*, grows on shallow carbonate banks and in lagoons and also preserves as a peaty facies that can be diagnostic of shallow water. In temperate latitudes, eel grass, *Zostera marina*, offers some potential but has not been widely used as a sea-level indicator. The occurrence of seagrasses preserved within sediments has been used in some studies of sea-level changes. The bladed seagrass *Posidonia australis*, for example, has been examined in studies of Holocene relative sea-level changes in northern Spencer Gulf, South Australia (Belperio *et al.*, 1984a). Although occurrences of *P. australis* have been documented within the intertidal zone (Shepherd, 1983), the species flourishes in the shallow subtidal zone between spring low water level (about –2 m MSL) to a water depth of about 4–6 m (Belperio *et al.*, 1984a). In their modern environmental setting, *Posidonia* (*P. australis* and *P. sinuosa*) form extensive meadows.

The coastal environments of northern Spencer Gulf in South Australia show an ordered plant biofacies zonation that is directly related to tidal range (Belperio *et al.*, 1988). *Posidonia* seagrass facies is represented by skeletal carbonate sands bound by rhizome and leaf and sheath fibres. These deposits have been interpreted as representing a minimum (low-water) sea-level indicator (Belperio *et al.*, 1984a; Barnett *et al.*, 1997). The *Posidonia australis* seagrass facies forms a distinctive Holocene carbonate bank deposit along the eastern coastline of northern Spencer Gulf, South Australia. The sediment record in this region indicates a 2.5 m higher relative sea level from about 6 ka until 1.7 ka (Belperio *et al.*, 1984a). Radiocarbon dating of *Posidonia* seagrass and bioclastic sand post-date and pre-date the depositional event, respectively (Belperio *et al.*, 1984b). By using a combined sampling strategy, it is possible to more clearly define timing of deposition within seagrass depositional environments.

A further approach to defining the position of relative sea level is to carefully trace the facies boundary between uppermost *Posidonia* seagrass facies and overlying intertidal sandflat facies (Belperio *et al.*, 1984a; Barnett *et al.*, 1997). As peritidal depositional environments characteristically give rise to shallowing-upward sedimentary successions (i.e. an upwards transition from subtidal, through intertidal to supratidal facies; Pratt *et al.*, 1992), superposition of intertidal sandflat sediments occurs over subtidal seagrass facies. In northern Spencer Gulf, this boundary is represented in modern sediments by a sharp change upwards from poorly sorted, fibrous shelly sand and mud into cleaner, better sorted shelly sand and at Port Pirie was defined at 0.25 ± 0.25 m with respect to modern tidal datum (Barnett *et al.*, 1997; see Figure 3.8).

3.4.5 Marine molluscs

Marine molluscs have been used extensively in coastal geomorphological research because they provide information about former environments of deposition, and other palaeoenvironmental attributes, such as palaeowater temperatures (Ludbrook, 1978, 1984; Murray-Wallace *et al.*, 2000), as well as their relevance to coastal archaeological studies (Cann *et al.*, 1991; Bateman *et al.*, 2008). Approximately 10,000 species of bivalve molluscs have been formally described. The majority have an *infaunal* life habit, living mainly within sediments in near-coastal environments (Petersen, 1986).

Molluscs represent particularly useful palaeoenvironmental indicators because they are amenable to dating by radiocarbon and other geochronological methods. Despite these advantages, molluscs should be used with caution when making inferences about palaeo-sea level, as many species can tolerate wide water-depth ranges. For example, although generally frequenting shallow water (≤ 5 m), occurrences of the scallop *Pecten fumatus* alive at the time of collection have been found in extreme instances in up to 80 m water depth in southeastern Australia (Lamprell and

Whitehead, 1992; Murray-Wallace *et al.*, 1996b). At the most general level, it is possible to distinguish intertidal, shallow- and deeper-water assemblages of molluscs (Petersen, 1986). Inferences about palaeo-sea level may be more reliably determined based on molluscan faunal assemblages rather than the presence of a single species which could be reworked from older sediments.

Many shells do not appear to have been transported far from their place of death (Figure 3.7d); however, robust specimens are able to survive post-mortem transport and individual valves of the intertidal cockle *Cerastoderma edule* (previously known as *Cardium edule*, and found in intertidal to shallow-subtidal coastal settings, living in sand or mud and capable of tolerating brackish water) were transported considerable distances from their place of origin during the postglacial marine transgression (Flessa, 1998). Some older shells in the southern North Sea had been transported landwards and some younger shells had been transported seawards, with reference to the postglacial sea-level curve of Jelgersma (1979), during the past 10,000 years. Landward transport of living and recently deceased shells was attributed to storms moving shells above high tide level. Movement of intertidal shells to deeper-water settings was attributed to strong, seaward bottom water flows following storms. In the most extreme case, Flessa (1998) found a specimen of *C. edule* dated at $7,355 \pm 80$ years BP and now found in the intertidal zone, which had been vertically displaced by about 18 m and transported landwards by up to 10 km following its death. Near Darwin in northern Australia, extensive beds of sub-fossil *Austriella sordida* are being eroded out of muds underlying sandy chenier ridges and incorporated into the ridges, implying that ages of shells in the ridge give a biased older age than the time of ridge deposition (Woodroffe and Grime, 1999). Similarly, in a study of molluscs washed up on modern beaches along the United States Atlantic Coastal Plain, Wehmiller *et al.* (1995) found that many well-preserved shell specimens were in fact of last interglacial age, although superficially resembling modern shells. This reinforces the point that single mollusc specimens should not be relied on for inferring palaeo-sea level.

Some marine mollusc species have been studied extensively in view of their significance in understanding long-term environmental and sea-level changes. In Australia and New Zealand, for example, the arcoid bivalve *Anadara trapezia* (Sydney Blood cockle) is restricted to the intertidal zone (Murray-Wallace *et al.*, 2000). It is an estuarine, filter-feeding, shallow-burrowing bivalve of semi-infaunal habit. It is commonly found within stable, soft, fine-textured tidal-flat sediments with the posterior part of the shell commonly exposed above the substrate. The species had an extensive distribution during the last interglacial maximum (MIS 5e; 125 ka) in both Australia and New Zealand and appears to have had a slightly wider geographical range in southeastern Australia in mid-Holocene (Murray-Wallace *et al.*, 2000). It became extinct in New Zealand sometime after MIS 5e. The significantly wider geographical range of *A. trapezia* in southern Australia during the last interglacial appears to be associated with enhanced ocean boundary currents (e.g. Leeuwin Current of western and southern Australia) at that time, coinciding with higher, less seasonally concentrated precipitation and river discharge.

The Goolwa cockle *Donax deltoides* occurs prolifically within the lower swash zone on beaches (Ludbrook, 1984) and represents a particularly useful palaeo-sea-level indicator. The species was harvested extensively by Aboriginal people living near the mouth of the River Murray in southern Australia during Holocene times. Large middens occur along much of this coast and have been dated to between 4,000 and 300 years BP (Bourman and Murray-Wallace, 1991). The species also occurs within last interglacial (125 ka) shoreline facies on the landward side of the Coorong Lagoon and indicates a relative sea level of about 1 m above the present level in this region.

Uncertainty about the degree of post-mortem transport of bivalves can be reduced where they are found articulated. Particularly useful are shell beds where it can be seen that bivalves are all in their position of growth. As with other biological indicators it is important to compare elevation of fossil bivalves with modern living equivalents, and it is preferable that these are known to be in comparable environments with similar tidal range. In the southern Gulf of Carpentaria, winnowing of fine sediments from the surface of the McArthur River delta has exposed articulated bivalves in beds preserved in their position of growth. The cockle *Anadara granosa* forms a widespread shell bed, and *Anodontia edentula* frequently occurs slightly above the cockle beds. Shell beds of mid-Holocene age further inland on the delta surface are found at a higher elevation and support occurrence of a Holocene highstand at this site (Woodroffe and Chappell, 1993).

Pelagic gastropods have been of considerable interest in view of their potential for biostratigraphical correlation of coastal sedimentary deposits over very wide geographical areas. In southern Australia and New Zealand, for example, the janthinid gastropod *Hartungia dennati chavani* has been commonly used in biostratigraphical correlation of Pliocene and early Pleistocene deposits (Ludbrook, 1983; Kendrick *et al.*, 1991). The significance of the altered distribution of *Strombus bubonius* in Mediterranean raised beach deposits since the last interglacial is discussed elsewhere (see Chapter 5).

3.4.6 Submerged forests

As the sea rises across the land surface, it inundates former terrestrial environments. It has long been recognised that there are many places where remains of stumps and trunks of trees record submerged forests. Evidence of submerged forests from the Thames estuary was described by Thomas Huxley who used it to surmise that these remnants, together with similar sites from Cornwall, Devon, and Wales, recorded the submergence of southern Britain (Huxley, 1878). A more detailed account was given by Clement Reid, who described such forests from the docks in Tilbury, as well as along much of the southern coasts of Britain (Reid, 1913). Similar stumps, overlain by salt-marsh muds, were considered evidence of the submergence of the North American coast from New England to Nova Scotia by Douglas Johnson (1911). An early attempt was made to understand the age of these forests using tree-ring analysis (Lyon and Goldthwait, 1934). Radiocarbon dating of well-preserved trunks has shown that many of those in Britain are mid-Holocene in age (6.5–3 ka), representing late Mesolithic to mid-Bronze Age. Transition from forest to salt marsh is recorded where blue–grey clay overlies peat layers with tree stumps, often with prominent tree trunks, generally with a dominant alignment as if the trees were felled by a single event (Allen, 1993). These organic-rich layers form a series of interbedded strata in the Severn Estuary; detailed studies have been undertaken, for example, on Caldicot Level in southern Wales (Allen and Haslett, 2002).

3.5 Geomorphological and geological sea-level indicators

Near-horizontal terraces, sometimes with distinct notches on uplifted coasts provide evidence of past sea-level changes, similar to, but rarely as convenient to date, as flights of uplifted reef terraces on some tropical shorelines around plate margins. On more stable shorelines, shore platforms at the foot of active clifflines may contain some palaeo-sea-level information, but the problem of confidently delineating the age of platform formation makes it difficult to use in sea-level reconstruction. Isolation basins have provided valuable information on uplifted coasts, whereas beach ridges and cheniers are more useful on coasts that have prograded because of abundant sediment supply. Geological evidence includes aeolianite, calcretes, and beachrock.

3.5.1 Marine terraces and shore platforms

The relentless action of the sea on the shoreline gradually cuts back poorly con-solidated lithologies. Marine abrasion is very effective at eroding even the most resistant lithologies given sufficient time, and past episodes of erosion are preserved particularly strikingly on uplifted coasts where successive sea-level highstands may be preserved as flights of marine-cut terraces. For example, terraces along the coast of the Santa Monica mountains in California were interpreted by William Morris Davis to result from rapid cliff erosion, stranded by the gradual uplift (Davis, 1933). Suites of terraces can be seen on uplifted coasts around the world (Bloom and Yonekura, 1985; Ota and Kaizuka, 1991), and terrace formation can be simulated by modelling where direct dating is not possible (Anderson *et al.*, 1999).

   Shore platforms are the near-horizontal rock platforms at the foot of cliffs. They are erosional landforms formed where the cliff face has been cut back by marine processes. Recession of coastal cliffs results in platforms that characteristically develop in the intertidal zone, although some are supratidal, being above regular inundation of all but the roughest seas. These features are clearly related to the level of the sea, but the actual processes that erode them and the time at which they were cut remain a subject of debate. In the past, the platforms have been referred to as 'wave-cut platforms', but this term is now largely replaced by 'shore platform', as waves are not the sole agent, nor necessarily the prime agent, responsible for their formation (Twidale *et al.*, 1977; Pethick, 1984). There remains contention about whether such platforms are eroded directly by wave action, or have been bevelled by subaerial physiochemical processes acting on their upper surfaces (Stephenson and Kirk, 2000; Trenhaile, 2002). Wave erosion has been inferred as the key process by some researchers (Edwards, 1941). Subaerial weathering has been suggested to be a more significant process than wave exposure controlling platform width, but influenced by inhomogeneities in the rock associated with joints and fissures (Abrahams and Oak, 1975). An alternative interpretation, that platforms result from disintegration of rock above the water table, was proposed for '*old hat*' islands that occur in the Bay of Islands, northern New Zealand, where platforms occur on both exposed and sheltered sides of islands (Bartrum, 1916; Kennedy *et al.*, 2011).

Striking examples of shore platforms occur along the south coast of New South Wales. These platforms were described by James Dana when he visited Australia on the US Exploring Expedition (Dana, 1849). They are particularly horizontal (Sunamura, 1992; Trenhaile, 1997) with a distinctive low-tide scarp at their seaward margin, in contrast to more sloping platforms (called ramps) in other parts of the world. It is clear that dip of bedding planes in the Triassic Hawkesbury Sandstone exerts a control on some platforms exposed around Sydney in southeastern Australia (Kennedy, 2010). Many researchers infer a variety of processes in platform erosion (Jutson, 1939; Hills, 1949). Structural features such as bedding, frequency of principal bedding surfaces of strata, distribution of joint sets, and relative contrasts in resistance of different rock types to weathering (e.g. differential cementation) appear to have influenced formation of shore platforms. In one of the first systematic descriptions of these benches south of Sydney, Bird and Dent (1966) recognised that their characteristics, whether sloping or horizontal, were related to lithology and structure of the coast. The relative importance of these various processes is likely to be expressed differently on contrasting lithologies, with higher platforms on more resistant rock (Thornton and Stephenson, 2006).

Although this 'waves versus weathering' debate remains a focus of discussion, it seems clear that platforms are formed in a narrow elevational range that is closely associated with the level of the sea. However, the processes of formation are important if the elevation of the platform is to be used as a sea-level indicator. In an early overview of the morphology of platforms in the UK, Trenhaile (1974) proposed that formation of platforms in macrotidal settings occurred at an elevation that was related to tidal variation; in extending this concept to other platforms around the world, it was proposed that platforms occurred at mid-tide because that was the elevation at which the water surface occurred for the greatest duration (Trenhaile and Layzell, 1981). However, duration of tidal immersion is not greatest at mid-tide level; the water surface has been shown to spend longer durations at low and high tide levels than close to mean sea level (Carr and Graff, 1982). Moreover, retention of water on emerged surfaces at low tide, termed 'water layer levelling', may be more important, enabling significant downwearing by physiochemical processes (Wentworth, 1938). Until factors responsible for shore platform development are fully elucidated, it will remain difficult to unambiguously relate many platform surfaces to a precise sea level.

In view of uncertainty about formative processes, it is perhaps not surprising that there is also contention over when such platforms formed. Uplifted shore platforms provide evidence about past land movements. For example, there was much debate about the significance of erosional shoreline features in Scotland, particularly where there seemed to be associated raised beaches (Jamieson, 1908; Sissons *et al.*, 1966). Cosmogenic dating, providing an age based on exposure, has now confirmed that these are postglacial (Stone *et al.*, 1996), as discussed further in Chapters 4 and 7.

Wide shore platforms developed on resistant lithologies may have formed during more than one interglacial sea-level highstand. Rates of modern weathering processes, measured by micro-erosion meter for more than 30 years, indicate that shore platforms on the Great Ocean Road in Victoria have been lowering at an average rate of ~0.3 mm/year (Stephenson *et al.*, 2012). Particularly wide platforms seem unlikely to have formed solely during the present Holocene interglacial when rates of process operation are extrapolated; those along the Illawarra coast, south of Sydney, appear too wide to have been truncated entirely within the Holocene and their morphology seems inherited from previous Pleistocene episodes of marine planation (Brooke *et al.*, 1994), or partially inherited from one or more earlier stillstands (Young and Bryant, 1993).

3.5.2 Shoreline notches and visors

Shoreline notches and their associated visors (overhanging rock) are common features of many of the world's coastlines (Pirazzoli, 1996, 2007). Notches represent erosional depressions typically within vertical rock faces formed by various processes acting on different rock types. Structural notches result from differential weathering on lithologies of contrasting resistance. Notches are most apparent at or close to sea level, but they can be formed at one of several different elevations, sometimes excavated at the highest point that waves reach, elsewhere at other elevations in the intertidal range, and occasionally subtidal (Laborel *et al.*, 1999). Many are presumably still actively being formed and provide the datum against which to compare elevation of notches that are no longer in the zone where they formed. Emergent notches have received the most attention; however, there are submerged erosional notches that can also provide evidence of former lower sea levels (Evelpidou *et al.*, 2012a).

Tidal notches are particularly distinctive mid-littoral erosional features on limestone coasts, with the deepest point generally coinciding with mean sea level

(Spencer, 1988). Studies aimed at addressing processes that form such notches indicate that abrasion by sediment accelerates retreat, but that much of the erosion occurs through the action of organisms, rather than through direct solution. Bioerosion by boring organisms, such as sponges and bivalve molluscs (e.g. *Lithophaga*), is accentuated by grazing invertebrates, such as chitons and gastropods. Eventually, undercutting in the intertidal zone will produce a distinct 'visor', and ultimately the cliff above the notch will collapse (initially with the rock fall persisting as talus, but subsequently removed by further erosion), changing the profile from being notched into that of a bench.

Figure 3.11 shows a schematic representation of how notch morphology might be related to past sea level, and subsidence, assuming gradual bioerosion that deepens the notch over several centuries of relative sea-level stability. Such notches become abandoned when they are no longer within the tidal range. Subsidence, as has characterised several shorelines in Greece, can abruptly submerge a notch,

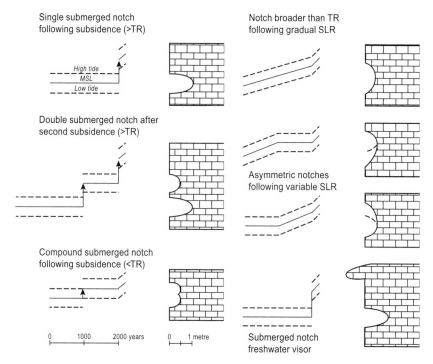

Figure 3.11 Schematic representation of mid-littoral bioerosional tidal notches on limestone coasts. In these examples an average rate of notch erosion of 0.5 mm/year for ~2,000 years is presumed, with a tidal range (TR) of 0.4 m, typical of the Mediterranean Sea. Abrupt periods of sea-level change represent episodes of subsidence, whereas each scenario is shown culminating in gradual sea-level rise (SLR), as is occurring this century, and which appears to leave no notch because rate of rise is faster than rate of bioerosion (after Evelpidou *et al.*, 2012b, with permission from Elsevier).

preserving it beneath the water line if the displacement exceeds the tidal range (Evelpidou *et al.*, 2012b). In the case where there is a gradual rise of sea level, a notch forms that is broader than the tidal range, and which may be asymmetrical if the rate of rise is irregular. Evelpidou *et al.* (2012b) suggest that the current rate of sea-level rise is too fast for bioerosion to erode deep notches.

Erosion is imperceptibly slow, but rates can be inferred at between 0.2 and 5 mm/year, averaging 1–1.5 mm/year on tropical limestones and 1 mm/year in the Mediterranean (Pirazzoli, 2007; Evelpidou *et al.*, 2012b; Moses, 2012). Morphology of cliffs varies in relation to exposure to wave energy. On the Caribbean Island of Curaçao, subtidal undercutting occurs in sheltered areas where there is minimal wave action; prominent intertidal notches occur around much of the island, and supratidal benches are found in the most exposed sites, where waves ensure that there is a spray zone that keeps much of the rockface wet (Focke, 1978).

Notches serve to mark former shoreline position and are particularly prominent evidence of former highstands, sometimes associated with coastlines that have been uplifted, as in the case of the MIS 5e terrace on Barbados (Figure 3.12a), but at other times considered stable and indicating that the sea was higher than present, as with the notch seen in the Cayman Islands (Figure 3.12b), which is inferred to have been eroded during the last interglacial (Woodroffe *et al.*, 1983b; see also Chapters 5 and 6). Supratidal surf benches may form in the more exposed settings on many coasts, particularly on limestone shores (Figure 3.12c), termed *trottoirs* in the Mediterranean. They are erosional in origin, but generally have an accretionary cover of vermetid gastropods or coralline algae. Reliability of erosional notches and benches as indicators of palaeo-sea level can therefore be seen to depend on identification and understanding of processes that were active in their formation, together with associated features including potholes and honeycombing (also called *tafoni* for larger forms), which are commonly observed in intertidal and supratidal zones but are not themselves diagnostic of a particular elevation (Pirazzoli, 2007).

In areas of low wave action, where a distinct intertidal notch is evident, Pirazzoli (2007) has suggested that elevated or submerged notches can be used to specify palaeo-sea-level position with confidence of ± 0.1 m. Abrasion notches are commonly variable in depth and elevation of visor, deepest point of retreat and floor, and uncertainties of local tidal amplitude and exposure mean that their precision as palaeo-sea-level indicators is low (± 1 or more metres). Notches are better preserved where relative sea-level changes have been rapid, particularly through abrupt tectonic movements; if uplift exceeds tidal range, an existing notch may be removed from the zone in which it could be modified, and a new one initiated, providing a convenient modern reference. Such situations are rare, and the complexity of processes as implied in Figure 3.11 indicates a confused record.

Once elevated above the active zone of erosion, notches (and sometimes sea caves which were formed at a similar sea level) can persist for a long time (Figure 3.12d), but there is often no indication of the time at which they were formed. There have

Figure 3.12 (a) A distinct notch cut in an uplifted terrace on the island of Barbados (photograph: Colin Murray-Wallace); (b) a notch at 6.4 m elevation above present sea level cut into Tertiary limestone on Cayman Brac, in the West Indies. Last interglacial reef limestone, termed the Ironshore Formation, is seen in the background (photograph: David Stoddart); (c) a surf bench at Vaikona on the high-energy coast of Niue, south Pacific (photograph: David Kennedy); (d) isolated rock with tidal notch indicating emergence offshore of the town of Poros on the Greek Ionian Island of Kefalonia. The uplift may have occurred during the 1953 earthquake (photograph: Colin Woodroffe).

been situations where it cannot be established whether a particular notch is Holocene or Pleistocene in age. Secondary features, such as flowstones or speleothems (particularly stalactites hanging from the roof of the notch) may be dated and provide minimum ages.

3.5.3 Isolation basins

Isolation basins represent broad bedrock depressions that have partially infilled with marine sediments and have been progressively uplifted above the upper marine limit (HATs). The basins become stranded from the sea when the bedrock sill, which separates open marine from brackish waters, is no longer breached by the sea. Isolation basins occur in areas of well-defined uplift such as the Arisaig area in northwestern Scotland (Shennan *et al.*, 1999, 2000a; Lloyd, 2000; see Figure 2.7c). Here, glacio-isostatic crustal rebound in response to the retreat of the Last Glacial Scottish Icesheet is the primary mechanism of uplift, which has served to raise basins beyond the level they can be inundated. Based on the radiocarbon ages of sediment deposited during marine incursions of isolation basins, a comprehensive record of relative sea-level changes may be constructed. For example, in western Greenland, initial glacio-isostatic rebound means that a basin isolated 8 ka is now 41 m APSL. Below this one there is a staircase of isolation basins, indicating relative sea level falling below present around 4.5 ka with a lowstand of about –5 m around 3 ka, before rising to present in the late Holocene (Long *et al.*, 2006).

3.5.4 Beach ridges

Beach ridges are laterally persistent, shore-parallel convex ridges formed by swash-related processes during high- or low-wave energy conditions, with an upper surface that may be shaped by aeolian processes (Hesp, 1984; Taylor and Stone, 1996; Otvos, 2000; Tamura, 2012). They may be composed of sand or gravel, or mixtures of these particle sizes. Where aeolian processes are particularly significant in concentrating large quantities of sand, these features have been termed relict foredunes (Hesp, 1984). Beach ridges characteristically occur in areas of high sediment supply, resulting in formation of a series of shore-parallel ridges (a strandplain or beach-ridge plain). Many beach-ridge plains show subparallel sets of ridges which may have resulted from localised coastal erosion during storms. At Guichen Bay, in South Australia, up to 80 well-defined beach ridges occur (Sprigg, 1952), and have formed across a 4 km embayment within the past 5,000 years, as determined by OSL dating of quartz grains from mixed skeletal carbonate–quartzose sand (Murray-Wallace *et al.*, 2002).

The characteristic product of beach-ridge deposition and progradation is a relatively thin (≤ 10 m), but laterally persistent regressive shoreface sandbody that overlies a thinner succession of transgressive littoral sands (Walker and Plint, 1992). Regressive beach-ridge plains developed on the microtidal Nayarit coast of Mexico reveal up to 280 subparallel beach ridges that have formed as a result of coastal progradation during the past 4,500 years (Curray *et al.*, 1969).

Although beach ridges represent relational sea-level indicators, they have been used to infer relative sea-level changes (Davies, 1961; Heteren *et al.*, 2000). In Tasmania, Davies (1961) inferred a relative sea-level fall since attainment of the Holocene highstand based on linear regression analysis, showing progressive seaward lowering in elevation of beach-ridge crests and swales. Although low correlation coefficients were obtained, each transect studied revealed a general seaward lowering in elevation across the Holocene beach-ridge plains with a relative sea-level fall of 0.6–0.9 m in southeastern Tasmania, and 1.5–1.8 m in eastern and northern Tasmania (Davies, 1961).

A more reliable approach for inferring relative sea-level changes from beach-ridge successions is to map the facies contact between the base of the beach-ridge aeolian sediment component and the underlying backshore or berm sands (Heteren *et al.*, 2000). Along many coastlines, backshore sands are represented by low-angle, landward dipping, planar cross-stratified sand and overlying dune facies show multiple sets of trough cross-stratification of varying orientations with reactivation surfaces. In a study of the beach-ridge succession of Sandy Neck, Massachusetts, USA, Heteren *et al.* (2000) found that the lithological boundary between beach-ridge dune facies and the underlying beach sands and gravels represented a well-defined, laterally persistent surface. Mapping of this surface by augering and GPR, together with OSL dating of the basal aeolian sands, indicated a relative sea-level rise during the past 4,000 years.

The origin of beach ridges has been the subject of debate, and summaries of the principal models of beach-ridge formation are provided by Hesp (1984), Taylor and Stone (1996), Tamura (2012), and Otvos (2012). Unambiguous models of beach-ridge formation are clearly needed to infer palaeo-sea level. Several nearshore processes can give rise to beach-ridge formation, and the relative importance of different processes depends on conditions such as coastal configuration, wave energy, sediment supply, sediment composition, and grain size. Foredune development has been observed by repeated surveys of the foreshore at Bengello Beach, Moruya, southern New South Wales, over more than 40 years (McLean and Shen, 2006), but even here not all stages of formation of successive ridges are fully understood.

Fairbridge and Hillaire-Marcel (1977) examined a succession of glacio-isostatically emerged Holocene, mixed sand and gravel beach ridges in Richmond Gulf on the eastern shore of Hudson Bay in Canada. The role of storms is therefore important when attempting to resolve palaeo-sea-level histories from beach-ridge successions, as the higher sea surface indicated by the storm deposits is not necessarily a true reflection of long-term mean sea level. A series of 185 sand-gravel beach ridges formed during the past 8.3 ka at a presumed average rate of 1 ridge every 45 ± 5 years. The ridges were attributed to storm events and 'hightide crescendos' (Fairbridge and Hillaire-Marcel, 1977, p. 413). Fairbridge and Hillaire-Marcel (1977) proposed two mechanisms to explain their origin. In situations where glacio-isostatic uplift and eustatic sea-level rise were in phase and coincident in rate, they suggested that the shoreline will remain sufficiently static to foster development of a well-defined beach ridge. During storms, strong onshore winds mimic the effect of a relative sea-level rise.

3.5.5 Cheniers

Cheniers are ridges of sand, gravel, or shell that overlie finer-grained muddy coastal sediments. The term derives from the Cajun word 'chene' for oak trees that grow on a series of parallel ridges in southern Louisiana, and this classic chenier plain, west of the Mississippi River delta, was first described and dated using radiocarbon (Russell and Howe, 1935; Gould and McFarlan, 1959). Although one of the earliest applications of radiocarbon dating, the wide spread of ages from any one ridge implies that shells can be reworked and re-incorporated into subsequent ridges. Chenier ridges are found in at least two principal settings. They form extensive plains along the coast of many open 'bights', such as on the broad mudflats that have accreted along the northern coast of South America associated with the Amazon River, and on the Texas–Louisiana coast of the Gulf of Mexico, where mud accumulates west of the mouth of the Mississippi River. They also occur in 'bayhead' or 'bayside' locations, such as in the Gulf of California, Broad Sound in Queensland, Australia, and the Firth of Thames in New Zealand (Otvos and Price, 1979; Short, 1989). In these locations, fine sediment is supplied from rivers, or by reworking of mud deposits in the nearshore during the postglacial transgression. The episodic deposition of coarse sand, shell, or gravel ridges implies a periodic driving mechanism; in the case of the Mississippi River this has been linked to distributary switching (although further dating indicates delta dynamics and ridge formation are rarely synchronous; Penland and Suter, 1989; McBride *et al.*, 2007), but it is also possible that there could be internal mechanisms associated with shell bed production, or external factors such as storms, that trigger ridge formation (Chappell and Grindrod, 1984; Woodroffe and Grime, 1999).

Chenier ridges are narrow in shore-normal cross-section ($\leq 2\,m$ or several tens of metres), and low in elevation. Their value in sea-level reconstruction is that they are wave-formed landforms that accumulate in low-energy settings, resting directly on tidal mudflat facies. In a pioneering use of chenier ridges, Schofield (1960) inferred a series of sea-level oscillations based on constructional features of the ridges themselves. He described 13 ridge systems, 20–200 m wide, south of Miranda on the western margin of the Firth of Thames in New Zealand. Radiocarbon ages on the cockle, *Austrovenus stutchburyi* (then known as *Chione stutchburyi*), which is the principal component of the ridges, provided age control, and each ridge comprised a lower wave-constructed unit with abundant shell hash, and an upper storm unit and associated washover fans composed almost entirely of cockle valves. Schofield used subtle surface features (the storm-ridge elevation, high spring-tide benches, and tidal flat surfaces between ridges) to reconstruct sea level and considered that erosional phases between ridge deposition resulted from periods of rising sea level. However, a laterally continuous shell bed of *Mactra ovata* beneath much of the plain suggests that there have not been rises and falls of sea level as implied by Schofield, but that the plain has accreted during gradual fall of sea level (Woodroffe *et al.*, 1983a; Dougherty and Dickson, 2012). Detailed studies on similar ridges composed of this cockle elsewhere in New Zealand demonstrate that ridges can be quite mobile and are reworked across whatever substrate lies behind them, whether that be mudflat at a range of elevations in the intertidal zone, with overwash that can move the ridge into backing mangrove woodlands (Ward, 1967).

Sequences of chenier ridges have been used to constrain palaeo-sea-level history elsewhere, particularly in Australia (Cook and Polach, 1973; Clarke *et al.*, 1979; Rhodes, 1982). The feature which has been used in relative sea-level studies in the case of extensive chenier plains along the southern shore of the Gulf of Carpentaria is the contact between the tidal mudflat sediments and the overlying chenier (Chappell *et al.*, 1983). The base of chenier reconstruction indicates 1–2 m of Holocene emergence which appears consistent with hydro-isostatic modelling for this region; however, given observations elsewhere of mobility of chenier ridges into and across upper intertidal environments at a range of elevations (Woodroffe and Grime, 1999), it is doubtful if this geomorphological evidence can be regarded as having sufficient elevational accuracy.

3.5.6 Aeolianites

Aeolianites (eolianites in American spelling) are geographically extensive, wind-blown deposits of sand-sized (0.063–2 mm) skeletal carbonate fragments derived from the physical breakdown of marine invertebrates such as molluscs, echinoids, bryozoans, and corals. The term was adopted by Sayles (1931) in his study of Bermuda. These sediments have also been described using the lithological term *calcarenite* to avoid genetic definition, as many facies within these successions were deposited by subaqueous processes. Although by definition, calcarenites contain $\geq$ 50% calcium carbonate as framework grains and cement, some aeolianites have been noted to have as little as 10% by mass of calcium carbonate (Murray-Wallace *et al.*, 2001). Aeolianite deposits represent elongate, composite ancient coastal dunes (McKee and Ward, 1983; Fairbridge, 1995; Vacher and Quinn, 1997; Brooke, 2001; Bateman *et al.*, 2004). Particularly well-described successions occur in the Bahamas, Bermuda, Hawaii, western India, the Mediterranean, South Africa, and southern Australia (McKee and Ward, 1983; Brooke, 2001). The aeolianite successions of southern Australia extend from Shark Bay, Western Australia, to western Victoria, including the Bass Strait Islands which represents the world's largest aeolianite province.

Since the early observations of aeolianites by Charles Darwin during his voyage on HMS. *Beagle* and subsequent pioneering study of carbonate dune and *palaeosol* (ancient soil) successions of Bermuda by Sayles (1931), the origin of aeolianites has been debated (Fairbridge, 1995). Aeolianites preserved along some coastlines have clearly formed at times of interglacial highstands analogous to the Holocene (Sprigg, 1952; Murray-Wallace *et al.*, 1998). This has been confirmed by dating using a variety of geochronological methods, the results of which imply correlation with particular sea-level highstands during the past 300 ka (Vacher and Quinn, 1997). In contrast, other coastlines provide evidence suggesting aeolianite formation at times of lower sea level during glacial events (e.g. parts of the Hawaiian Islands: Stearns, 1985). In these situations, it is argued that aeolianites formed in response to stronger onshore winds carrying sand-sized skeletal carbonate fragments landwards from partially or totally exposed continental shelves. The dunes were then covered by vegetation and soil developed. Further evidence for the glacial age (*sensu lato*) of some aeolianite successions is that basal units occur below present sea level, implying a lower sea level at time of their formation, and scuba diving has identified erosional shoreline notches within dune limestones that formed at times of lower sea level (Stearns, 1985).

Numerous sedimentary structures are preserved within aeolianites from which inferences about relative sea level may be derived (McKee and Ward, 1983). Aeolian trough cross-bedding and other dune bedforms (e.g. topset, bottomset, grain ava-lanche structures) represent relational sea-level indicators, having clearly formed above tidal datum. In situations where active aeolian dunes have migrated into standing bodies of water such as coastal lagoons, however, it may be possible to infer relative sea level, albeit crudely, based on presence of distinctive fossils and interfingering relationships with lagoonal muds. In the modern Coorong Lagoon in South Australia, for example, specimens of the intertidal bivalve mollusc *Katelysia* have been found within tabular cross-beds in the lee of landward migrating parabolic dunes. The molluscs moved up the sediment profile but were unable to keep pace with the influx of sediment and were asphyxiated within the advancing dune sediment. Based on analogous relationships of sedimentary facies, a similar pattern of sedimentation is evident for the last interglacial (125 ka) Woakwine Range, a prominent coastal barrier within the same region (Murray-Wallace *et al.*, 1996a, 1999; see Chapter 6).

Rhizocretions (plant root and trunk moulds) represent relational sea-level indi-cators, clearly indicating a lower relative sea level at time of formation, whereas beach-related facies formed close to sea level. However, confident discrimination of aeolian and subaqueously deposited calcarenites can be a frustratingly difficult task in some aeolianite field contexts, particularly if distinctive fossils are absent. This problem is exacerbated by the presence of coarse-grained marine shell fragments up to 10 mm in diameter evident on surfaces of modern (active) aeolian dunes implying that coarser-grained bioclasts are not necessarily *prima facie* evidence of subaqueous deposition. Some foreshore deposits have been identified based on presence of large, reworked coral clasts that can only have been deposited by subaqueous processes (Woodroffe *et al.*, 1995). Beach facies may also be recognised based on the presence of coarser bioclastic sand and associated fossils, but noting the caveat discussed above about coarse-grained stringers within aeolian dune deposits.

3.5.7 Calcretes

A variety of weathering, pedogenic, and diagenetic features that relate to subaerial exposure of carbonate sediments may be of use for inferring relative changes in sea level, but are only of limited value as palaeo-sea-level indicators. Calcretes and

caliche horizons (B-horizon of carbonate soil profiles) result from pedogenic processes and are characteristically represented by fine-grained, strongly indurated
chalky deposits of low magnesian calcite (Esteban and Klappa, 1983).

Karst and caliche facies represent distinctive products of subaerial exposure
(Esteban and Klappa, 1983). Karst-related features include a complex association
of subterranean landforms such as caves with speleothems (stalactites, stalagmites,
flowstones) and landsurface features directly attributable to dissolution (e.g. sinkholes,
limestone pavements, Rillenkarren). The dominating characteristic of karst landscapes
is absence of well-defined surface drainage (Jennings, 1985). Karst-related features
will form in areas of moderate to high rainfall and, accordingly, may span a wide range
of climate zones.

Caliche facies, in contrast, are restricted to semi-arid climates and form within
near-surface environments within the vadose zone. Their presence in marginal
marine stratigraphical records can be directly related to relative changes in sea level.
In northern Spencer Gulf, South Australia, for example, presence of caliche-related
features on the upper-bounding surfaces of late Pleistocene marine sediments, and
stratigraphically beneath Holocene marine sediments, indicates subaerial exposure of
this region at times of lower sea level during the last glacial cycle (Billing, 1984).
Laminated subaerial soilstone crusts have been described from Florida (Multer and
Hoffmeister, 1968), and are a prominent feature over the surface of Pleistocene
carbonate deposits throughout much of the Caribbean, but they appear to contain
little evidence specifically related to past sea level.

3.5.8 Beachrock

Beachrock results from lithification of unconsolidated beach sediments by calcium
carbonate cements in the intertidal zone of primarily tropical, but also subtropical,
beaches (Hopley, 1986a). The term has, however, been used in a range of contexts
and to denote cementing agents of different composition, such as iron and silica.
Although beachrock formation, *sensu stricto*, is essentially confined to the intertidal
zone, caution should be applied in using such features for making inferences
about palaeo-sea level because calcium carbonate is highly mobile in many coastal
environments, and the exact elevation and time at which lithification occurred can
scarcely ever be determined. Diagenetic recrystallisation and secondary cementation
processes are pervasive in carbonate depositional environments whether tropical
or temperate.

Beachrock can form relatively quickly in geological terms as attested to by
the presence of fragments of deeply embedded beer bottle glass and other human
artefacts firmly cemented within these sediments. Outcrops of beachrock are a
common and conspicuous feature of the shorelines of many calcareous reef islands.
Beachrock is prominently bedded at angles that correspond with modern beach
gradients, and preservation of bedding is strikingly apparent, whereas there is

commonly little to reveal that modern beaches are so clearly stratified. Chemical processes that lead to cementation have been subject of considerable debate and remain contentious (Vousdoukas *et al.*, 2007), but it is clear that outcrops of beachrock are remnants of former beach sediments marking a formerly active beach. Often there are extensive lineations of beachrock across a reef flat; for beachrock to become exposed, it requires erosion of unconsolidated sand, and the occurrence of beachrock where there is now no longer a shoreline is compelling evidence that former shorelines have been eroded. Beachrock, therefore, clearly marks past shoreline positions, but it is less certain whether it can be used to detect changes in sea level.

Outcrops of beachrock are seldom flat on top; indeed, the laminated and dipping nature of the successive layers of lithified beach sand give it a particularly undulating upper surface. Where a near-horizontal surface does occur in such deposits, it more often seems to be a feature of post-lithification truncation than a marker of the upper limit to which the beach has cemented. Lack of a distinct upper limit to cementation makes use of beachrock as a sea-level indicator rather imprecise. On the Great Barrier Reef, it has been suggested that beachrock can be formed up to HAT (Hopley, 1986a). Cementation may occur above the highest tide level elsewhere; a view propounded by Kelletat (2006) but disputed by Knight (2007). Further confusion is likely to exist where cemented beach sediments intergrade into other cemented deposits, such as cay sandstone that has been cemented by carbonate cements associated with the water table within reef islands, or phosphatic limestones that occur under forests of the tree *Pisonia grandis* and are inferred to form through the accumulation of bird droppings (guano).

A further constraint on use of beachrock as an indicator of palaeo-sea level is the challenge of dating time of sand deposition and differentiating this from time of cementation. Radiocarbon dating of components within the sand will record cessation of ^{14}C uptake by organisms which broadly corresponds with time of death of contributing organisms. However, there may have been considerable time since death of a coral and deposition of fragments of its skeleton on a beach. Coral can also be eroded and reworked. Sand-sized grains of coral may have been through several stages of destruction and even pieces of coral rubble or shingle can have been through several phases of transport. Attempts to date cement within beachrock are more likely to be of value in determining the time at which beachrock was formed, but there still remain problems. Preliminary attempts to date beachrock on siliciclastic beaches using luminescence dating techniques appear more likely to provide ages that represent deposition of components (Tatumi *et al.*, 2003), but it is not clear that convincing sea-level chronologies will be resolved from these deposits. There is a wide range of deposits that are covered by the term 'beachrock' and it is doubtful if generalisations from carbonate islands on coral reefs can be extrapolated to other types of beachrock in different lithological and climatic settings.

3.6 Geoarchaeology and sea-level changes

Geoarchaeology, the subdiscipline of archaeology that integrates observations from geology and geomorphology to contextualise human–landscape interactions, can contribute particularly valuable information about relative sea-level changes. Preservation of various built structures in near-coastal, and today, inner continental shelf environments may provide a chronicle of coastal change in the absence of other natural proxies. Historical coastal settlements and ports provide important insights into relative sea-level changes during past millennia, particularly in the Mediterranean Sea. Pioneering geophysical interpretations from archaeological indicators enabled sea-level change estimation (Flemming, 1969; Caputo and Pieri, 1976; Pirazzoli, 1976).

Although there are many known archaeological remains in coastal or near-coastal locations, only a small fraction can be used to obtain precise information on their former relationship to sea level. One of the best-known examples is the so-called Temple of Serapis (Figure 3.13a) that was described by Lyell (see Chapter 1). There are several limitations which include uncertainty about the use of a particular site – poor preservation, or ambiguity in geologically unstable areas as to whether elevation may have been altered as a consequence of tectonic activity, subsidence (for example, in soft sediments at river deltas), or extreme events such as tsunami.

The relationship of evidence to former sea level can be of two types; first, those structures or artefacts which must have been terrestrial but have now been flooded; and second, those coastal structures that were built with a precise relation to water level and which no longer function because of a change in sea level (Lambeck *et al.*, 2010). Palaeolithic rock art in the Cosquer cave at a depth of 37 m BPSL in Calanque de Morgliou in southern France, and which has been dated to at least 27 ka, represents an example of the first and provides an archaeological constraint on sea-level change in the Mediterranean (Lambeck and Bard, 2000). An Early Neolithic burial site at –8.5 m in northern Sardinia with an age of ~7,600 ka BP implies that sea level was more than 8.5 m lower than present at that time (Antonioli *et al.*, 1996). Submerged mosaics associated with the sunken city of Baia and a submerged breakwater at Capo Malfatano in Sardinia, together with submerged wells off the coast of Israel, also indicate gradual sea-level rise since 8,000 years ago (Sivan *et al.*, 2004; Lambeck *et al.*, 2010). The origin of many 'flood' myths and legends illustrates the human capacity to reason and, accordingly, to respond to sea-level changes, implying an awareness of landscape alteration associated with sea-level changes (Oppenheimer, 1998).

In the more recent archaeological record various human built structures have provided evidence for relative sea-level changes, such as inundated Roman age dwellings, tombs, drainage canals, cisterns, piers, fish tanks, and quarries, constructions generally occupied after 2.5 ka BP (Mastronuzzi and Sanso, 2003; Anzidei *et al.*, 2011; Florido *et al.*, 2011). Although these forms of palaeo-sea-level indicators lack

Figure 3.13 Geoarchaeological sea-level evidence. (a) Marble pillars of a market place popularly known as the Temple of Serapis as observed by Charles Lyell in 1836 and featured in the frontispiece of the ninth edition of *Principles of Geology* (1853). The marble pillars indicate a recent relative change in sea level as attested by borings made by the marine mollusc *Lithodomus* on the lower portion of the pillars (photograph: Marco Anzidei). (b) Roman fish tanks connected by tidally controlled canals cut in sandstone at Briatico, Calabria, southern Italy (photograph: Marco Anzidei). (c) Partially submerged Roman ruins in southern Italy (photograph: Colin Murray-Wallace). (d) Wooden structures, from the eighteenth or nineteenth century, now located in the intertidal zone, Hayling Island (photograph: copyright Hampshire & Wight Trust for Maritime Archaeology).

the resolution of many of the indicators considered earlier in this chapter, they do nonetheless provide compelling evidence for sea-level changes, most notably a rise in relative sea level. Variations of sea level have been reconstructed from structures in the Vieux Port of Marseilles (Morhange *et al.*, 2001).

The remains of the Roman market at Pozzuoli in the Phlaegrean Fields, described by Charles Lyell, as summarised in Chapter 1, provides an interesting case study of relative sea-level change at one site. The remaining standing columns have marine borings that can be seen up to 7 m APSL, indicating submergence followed by uplift. Recent radiocarbon dating of these boring organisms implies sea-level highstands at three periods: during the fifth century AD, the early Middle Ages, and before the 1538 eruption of Monte Nuovo (Morhange *et al.*, 2006).

Roman fish tanks, *piscinae*, were constructed as holding tanks for fish culture with a direct correspondence to sea level to enable tidal flushing (Figure 3.13b). Their use, particularly from between the first century BC and the first century AD, is recorded by contemporary authors such as Plinius, Varro, and Columella. They were generally carved into rock to a depth of 2.7 m or less, with various installations, such as sluice gates with posts, sliding grooves, thresholds, and footwalks (Lambeck *et al.*, 2004; Antionioli *et al.*, 2011). There are many places around the Italian coast where Roman remains provide indicative meaning in terms of past sea level (Figure 3.13c).

Geoarchaeological and palaeoenvironmental reconstructions have, for example, explained the presence of former sea corridors based on the spatial distribution of *middens* (shell mounds or accumulations resulting from harvesting of marine shells) in regions that had not previously been considered to have experienced marine inundation (Cann *et al.*, 1991; Cann and Murray-Wallace, 1999). For example, the location of shell middens in northern Australia needs to be considered in the context of geomorphological changes in coastal configuration that may explain apparent anomalous or isolated examples (Beaton, 1985; Woodroffe *et al.*, 1988). Similarly, palaeoenvironmental and geomorphological analysis of Port Phillip Bay in Victoria, Australia, indicates that it is likely to have largely dried up in the late Holocene between 2.8 and 1.0 ka due to barrier closure which could be mistakenly inferred as a relative fall in sea level (Holdgate *et al.*, 2011). The modern Port Phillip Bay is approximately 1,950 km^2 and illustrates the magnitude of this late Holocene environmental change.

Geoarchaeological remains are also important in northern Europe. On the north-western coastline of Germany in the Ostfriesland region, an area characterised by extensive coastal marshes and back-swamp areas of raised bogs of Sphagnum peat, people responded to relative sea-level rise by constructing a network of dykes for the preservation of farming land and constructing dwelling mounds (*Wurten*) (Petzelberger, 2000). There are many sites in southern Britain where archaeological remains have been found within intertidal settings, indicating submergence (Figure 3.13d).

3.7 Synthesis and conclusions

Palaeo-sea-level indicators are needed in order to reconstruct past sea-level positions. However, it is clear that many types of evidence should be treated with caution. First it is necessary to establish what the relationship was between the particular proxy and sea level at time of formation. Then there are concerns about how that relationship might have been obscured by alteration since time of formation. For example, compaction or bioturbation of coastal sediments means that fossils may no longer be at the elevation at which they lived (Horton and Shennan, 2009). This is still more likely where microfossils are easily transported. Weathering and erosion reduce the extent to which evidence is preserved. Subtle morphodynamic adjustments in response to evolving coastal environments can distort tidal planes and compound interpretation of former elevations.

In order to use fossil evidence as sea-level index points, it is necessary to establish the relationship of the local environment in which the biological, morphological, or lithostratigraphical evidence accumulated to a contemporaneous reference tide level. This provides the indicative meaning and its associated uncertainty (van de Plassche, 1986). The indicative meaning varies with different types of evidence and is commonly expressed as a vertical range from a reference water or tidal level.

A range of biological indicators have been described in this chapter, as well as other geomorphological and sedimentological lines of evidence, including both depositional and erosional features. There is a rich history of settlement around some coasts, for example, in the Mediterranean Sea; constructions, such as Roman fish tanks, provide a further line of evidence as to past sea level. The sea-level researcher needs to know the age of the sample or its time of deposition. These are not always apparent or easily distinguished. The next chapter examines methods which can be used to determine age, and discusses issues such as time-averaging, reworking, and post-depositional diagenesis. In general it is easier to determine the age of younger material than older material. In some cases, there may be evidence of human activities which can provide a proxy for sea level.

4

Methods of dating Quaternary sea-level changes

Like a piece of music, Pleistocene history has its themes, rhythms, harmonies, disharmonies and varying tempos. To follow that history one must be like the young pianist, who learns his notes and practices scales. Vision and memory of principal events are necessary, and some numerical figures should be steadily in mind.

(Daly, *1934, p. 50*)

4.1 Introduction

Geochronological methods have been instrumental in deriving a record of relative sea-level changes for the Quaternary Period. The development of geochronological methods, particularly during the latter twentieth century, revolutionised understanding of the nature and rates of Quaternary sea-level changes. In particular, radiocarbon and uranium-series disequilibrium dating (hereafter U-series dating) represent important methods for assigning numerical ages to geological and archaeological deposits of Holocene and late Pleistocene age. The potential of radiocarbon dating was soon recognised by the Earth Science community and applied to a diverse range of carbon-bearing materials of organic and inorganic origin, from a variety of geological contexts. Radiocarbon dating has provided timeframes for the complex postglacial relative sea-level histories evident from around the world (Pirazzoli, 1991, 1996; Smith *et al.*, 2011). Similarly, U-series dating enabled assignment of ages to interglacial and interstadial coastal successions beyond the range of radiocarbon, and, with recent refinements such as thermal ionisation mass-spectrometry (TIMS), provided support for the Croll–Milankovitch Hypothesis concerning the 'orbital pacemaker' (Hays *et al.*, 1976) of long-term climate change and glacio-eustasy. Radiocarbon and U-series dating remain the most commonly applied methods for assigning numerical ages to coastal deposits, particularly in carbonate depositional environments (Vita-Finzi, 1973), and luminescence methods are increasingly being applied particularly to aeolian successions (Jacobs, 2008).

Additional geochronological methods were developed in the late twentieth century and are now extensively used in studies of coastal evolution and Quaternary sea-level changes (Rutter and Catto, 1995; Noller *et al.*, 2000; Walker, 2005). These include luminescence-based methods (thermoluminescence, OSL, and infrared-stimulated luminescence), electron spin resonance, and amino acid racemisation. Cosmogenic dating is increasingly being used to determine the duration of subaerial exposure of bedrock surfaces, and offers potential to assess the age of coastal landforms, such as shore platforms. Other techniques have been applied to older sedimentary successions such as palaeomagnetism to locate the Brunhes–Matuyama geomagnetic polarity reversal (780 ka) and are applicable in particular circumstances. This chapter reviews the methods that have commonly been applied in palaeo-sea-level investigations, focusing on the fundamental principles of each method, their specific requirements, relative advantages, and limitations and difficulties of applying these methods in particular field contexts.

4.1.1 *Terminology*

Geochronology refers to the study of the timing of geological events in Earth history, especially numerical and relative age assessments. Geochronological methods have been classified using the physico-chemical basis of the method or the type of chronological information each method may yield (e.g. relative or numerical ages; Colman *et al.*, 1987; Noller *et al.*, 2000). Radiocarbon and U-series dating methods are based on the radioactive decay of radioisotopes. Radioactive decay occurs at a uniform rate and is not influenced by external environmental conditions. In contrast, methods such as luminescence and electron spin resonance determine the age of materials based on the time-dependent, cumulative effects of natural radioactive decay on discrete mineralogical components of sediments during their burial history; the mineral grains act as natural dosimeters. These methods are sensitive to changes in environmental conditions over time, particularly those variables that influence radiation dose rates. Finally, chemical methods such as amino acid racemisation measure the rate of change of a specific chemical reaction, and are sensitive to long-term changes in environmental conditions, particularly diagenetic temperature.

Geochronological results may be expressed in four ways: numerical, relative, calibrated, and correlated ages (Colman *et al.*, 1987; Noller *et al.*, 2000). As the name implies, numerical ages refer to ages expressed in number of years (termed sidereal when referring to calendar years). The highest degree of age resolution is derived from annual records such as tree rings, coral banding, and *varves* (annually deposited couplets of sediment represented by a coarser-grained lamina of sand and silt deposited in summer and spring, and a finer-grained lamina of silt and clay deposited during winter). Relative ages refer to the assignment of ages within a time series from oldest to youngest, for example, deeper sedimentary units are considered older than near-surface ones, or topographically higher shorelines showing greater erosion might be regarded as older than lower, less dissected ones. Calibrated ages are those in which a relative sequence is tied to independently derived ages; for example, amino acid racemisation ages may be calibrated by linking the rate of the racemisation reaction to an independent method such as radiocarbon or U-series dating. Correlated ages refer to the designation of time-equivalence of disjunct sedimentary deposits, for example where these are directly compared with fossils from different sedimentary successions.

The *precision* and *accuracy* of numerical ages is an important consideration in geochronology (Figure 4.1). *Precision* refers to reproducibility of measurement; it can be thought of as the number of decimal places to which a measurement can realistically be made. For example, the results of successive replicate radiocarbon measurements on a single sample are shown in Figure 4.2, together with a measure of uncertainty represented by the 1-σ uncertainty terms reported with conventional radiocarbon ages derived on the basis of liquid scintillation counting (Barker, 1970). In this case, the uncertainty refers solely to the physical measurement of residual radiocarbon within the sample and should account for all the sources of uncertainty in the laboratory measurement. It would be misleading to quote a radiocarbon age for the sample down to an individual year, as that level of precision cannot be reproduced in counting processes. The 1-σ uncertainty term indicates that there is a 68.26% level of confidence that the derived age falls within the limits of

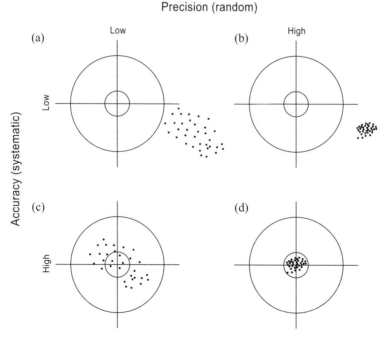

Figure 4.1 Precision and accuracy in geochronological methods; (a) low precision and accuracy; (b) high precision but low accuracy; (c) low precision but higher accuracy; and (d) high precision and high accuracy (modified after Wagner, 1998, with kind permission of Springer Science + Business Media).

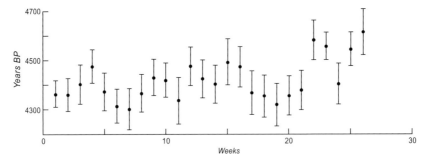

Figure 4.2 Results of replicate radiocarbon measurements by the liquid scintillation method for a single sample, repeatedly counted over a period of 6 months. The derived radiocarbon ages range from 4,300 to 4,650 years BP and the variation observed is consistent with that expected for counting statistics in replicate analyses (from Barker, 1970, by permission of the Royal Society).

uncertainty. At the 2-σ level the confidence level rises to 95.46% and to 99.73% at the 3-σ level (Zar, 1984). In contrast, *accuracy* refers to the deviation of determined age from the 'real' age of a geological deposit. This is a far more difficult issue to address and requires several independent lines of evidence to assess the integrity and validity of the results (e.g. the stratigraphical and geomorphological context of the dated materials and ideally, an independent geochronological method to assess the concordance of results). An age might be reported to the nearest 10 years because measurements can be made to that precision, but if the sample was contaminated, the age would be completely inappropriate because the determination would not be accurate (it would be off-target as indicated in Figure 4.1). Both precision and accuracy are usually required, although there may be situations where high precision is not required to address fundamental research questions, such as whether an abraded shore platform is Holocene or Pleistocene in age (Vita-Finzi, 1987; Stone *et al.*, 1996).

4.1.2 Historical approaches used for evaluating geological age of coastal deposits

Before the discovery of radioactivity and subsequent development of numerical dating methods, other approaches were commonly used to infer age that relied on assumptions about the rates of geological or geomorphological processes. For example, age–height relationships of marine terraces, thicknesses of sedimentary successions, relative stratigraphical relationships, or presence of distinctive fossils were used to assign notional ages to deposits. *Index fossils* were particularly important, represented by morphologically distinctive organisms of wide geographical range, but which had rapidly evolved and therefore were of restricted age-range. In many instances biostratigraphical studies, particularly of foraminifera, have shown that species changes may reflect a distinct life-habitat preference (i.e. biofacies) rather than representing an evolutionary appearance, and many species are *diachronous*, cutting across time at a scale of several tens of thousands of years (McGowran, 2005).

Fossils have also been used to make climatic inferences based on presence of faunal assemblages, reflecting ingressions of warm- or cold-water related organisms. The distinctive large gastropod *Strombus bubonius*, for example, has enabled correlations of emergent shoreline sequences in the Mediterranean (Gignoux, 1913; Issel, 1914; Hearty *et al.*, 1986). Now confined to the west African coast from Western Sahara to Angola and found on the Cape Verde Islands, it has been regarded as representative of a transient, late Pleistocene, 'Senegalese', warm-water fauna and a reliable biostratigraphical marker species for the last interglacial *sensu lato* (i.e. all of MIS 5; ~130–80 ka) in the Mediterranean region. Its disappearance from the Mediterranean is attributed to global cooling associated with the LGM (Hillaire-Marcel *et al.*, 1986). Similarly, the distinctive arcoid bivalve, *Anadara trapezia* (Sydney Blood Cockle), represents an important biostratigraphical marker species for the last and penultimate interglaciations in Australia and the last interglacial maximum (~130–118 ka) in New Zealand (Murray-Wallace *et al.*, 2000).

Charles Lyell championed the notion of 'statistical palaeontology' (Rudwick, 1978), whereby ages were assigned to deposits based on percentages of molluscan fossil faunal assemblages represented by extant forms. On this basis the Quaternary was defined by 70% of fossil faunas being represented by modern equivalents (Lyell, 1853). Rates of erosion and the degree of weathering have been used to assign relative ages to glacial deposits (e.g. weathering rinds on drop stones and various moraines), as well as assessing the potential number of sea-level highstands based on the cross-sectional width of shore platforms.

Detailed records with annual resolution of deglacial sedimentation in Scandinavia were established by Gerard de Geer (1888) based on the analysis of varves. The Swedish varve record shows an annual resolution back to approximately 13,300 varve years and provides insights into the history of ice margin recession and associated estuarine and deltaic sedimentation (Walker, 2005).

Since the early work of Emiliani (1955), the potential to reconstruct the Quaternary using biota, particularly foraminifera, from deep-sea sediment cores has been realised. This developed a new rigour when Shackleton and Opdyke (1973) revealed the value of oxygen-isotope records from deep-sea sediments to constrain ice and ocean volumes. The mutually consistent record of ice-volume change derived from hundreds of deep-sea cores from all of the ocean basins, and their correlation with radiocarbon- and U-series-dated palaeo-shorelines, such as those from the Huon Peninsula, Papua New Guinea, and Barbados, has established a coherent global palaeo-sea-level framework.

4.2 Radiocarbon dating

Radiocarbon dating is the most commonly used method of determining the age of geological and archaeological deposits of latest late Pleistocene and Holocene age. Its widespread use reflects its suitability for a diverse range of carbon-bearing materials, and a half-life that permits a relatively high degree of age resolution for Holocene time.

The potential of radiocarbon (^{14}C) as a geochronological method was realised by Willard Libby and co-workers in the mid-1940s in a series of interrelated experiments (Arnold, 1992). A comparison of the radiocarbon activity of 'modern' methane produced from a Baltimore sewage waste plant with petroleum methane (background) provided a first-order test of the method (Anderson *et al.*, 1947). Modern sewage waste yielded a count rate of 10.5 disintegrations per minute, whereas petroleum methane recorded no measurable radiocarbon activity. Analyses of wood of known age from archaeological and geological contexts provided further validation of the method (Arnold and Libby, 1949, 1951).

4.2.1 *Underlying principles of the radiocarbon method*

Radiocarbon is produced in the upper atmosphere (around 15 km above the Earth's surface) as a result of cosmic ray neutrons bombarding atmospheric ^{14}N. In the reaction, ^{14}N transmutates to ^{14}C with the addition of a neutron, and the eviction of a proton (Faure, 1986). The ^{14}C rapidly combines with oxygen to form ^{14}CO (which has a residence time of several months) and subsequently undergoes further oxidation to form ^{14}CO$_2$ (Trumbore, 2000). Approximately 7.5 kg of ^{14}C is produced in the upper atmosphere each year, of which approximately 75% is produced in the stratosphere and 25% in the upper troposphere (Taylor, 1987; Trumbore, 2000). During photosynthesis and cellular respiration, ^{14}CO$_2$ is continuously taken up by organisms and subsequently incorporated through the food chain.

The event dated by the radiocarbon method is the cessation of radiocarbon uptake, which for short-lived organisms broadly coincides with death. For longer-lived organisms, such as trees, corals, or some molluscs (e.g. the giant clam, *Tridacna gigas*), radioactive decay of ^{14}C incorporated in tissue during life can be resolved in the older portions of the organism that are no longer biologically active (e.g. successive tree rings and growth bands in molluscs). Following the cessation of ^{14}C uptake, radiocarbon is progressively lost from the host matrix due to radioactive decay, by emitting an electron (beta particle) which results in conversion of a neutron to a proton, reverting to ^{14}N. Half the original mass of ^{14}C remains after one half-life has expired (5,730 years: Stuiver and Polach, 1977; Figure 4.3). The half-life of ^{14}C was originally determined to be 5,568 ± 30 years (Engelkeimeir *et al.*, 1949); this value is still used by international convention in the calculation of uncorrected radiocarbon ages, although the half-life has been recalculated to 5,730 ± 40 years (Godwin, 1962). Uncorrected refers to radiocarbon ages as reported by laboratories without consideration of variables such as the marine-reservoir effect or potential discordances between ^{14}C and calendar years; ages corrected to calendar years are referred to as 'calibrated' ages. More recent comparisons of ages based on radiocarbon and U-series analyses suggest that a further assessment of the ^{14}C half-life may be warranted (Chiu *et al.*, 2007).

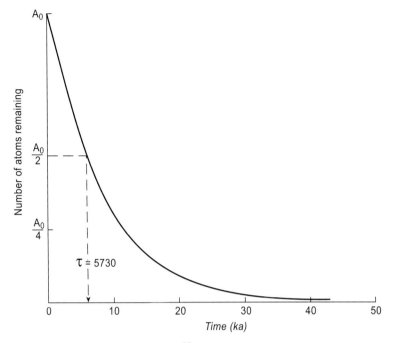

Figure 4.3 The exponential decay curve for ^{14}C based on the revised half-life (τ) of 5,730 years. After eight half-lives, equal to the elapse of 45,840 years, it becomes particularly difficult to distinguish residual radiocarbon from background levels and, accordingly, defines the age range of the method. A_0 represents the residual radiocarbon activity at the time of cessation of ^{14}C uptake and $A_0/2$ the expiration of 1 half-life (τ).

The radiocarbon age is defined by the relation:

$$A_{SN} = A_{ON}e^{(-\lambda t)} \tag{4.1}$$

where A_{SN} refers to the radiocarbon activity of the sample of unknown age; A_{ON} is the activity expressed as the net count rate (counts per gram of carbon per minute) of the modern reference standard (95% NBS oxalic acid standard), corresponding to wood grown in 1890 AD, and therefore not subject to the fossil fuel effect, but corrected for the ^{14}C decay over that period to 1950 AD; t is time in years since the cessation of ^{14}C uptake; $-\lambda$ is the radioactive decay constant as defined by the Libby half-life of $\tau = 5,568$ years, where $\tau/\ln2 = 8,033$ years. Accordingly, by rearranging Equation 4.1, the radiocarbon age (t) is defined by:

$$t = -8033 \ln(A_{SN}/A_{ON}) \tag{4.2}$$

where A_{SN}/A_{ON} is equivalent to the *fraction modern*, and thus the age is a ratio of the residual ^{14}C activity of the sample and the modern reference standard, and normalised to a δ^{13}C value of –25‰ Peedee Belemnite (PDB) (see Section 4.2.4 on isotopic fractionation) (Gupta and Polach, 1985; Hua, 2009).

The above relations can be re-expressed in terms of the *fraction modern*, which accounts for the degree of ^{14}C enrichment or depletion within a sample. This is of interest for very young materials, such as those regarded as 'modern'. The fraction modern is defined by the millesimal difference (d) of ^{14}C activity of the sample, compared with the modern reference standard, expressed as:

$$d^{14}C = (A_{SN}/A_{ON} - 1) \times 1000‰ \qquad (4.3)$$

Accordingly, a positive value will represent an age in the future, and therefore reflect the incorporation of ^{14}C derived from atmospheric weapons testing post-1950 AD or other sources of nuclear enrichment (e.g. labelled ^{14}C). Radiocarbon ages pre-1950 AD, however, will have a negative value. A $d^{14}C = 0‰$ represents an activity equivalent to 95% of the activity of the oxalic acid standard and a $d^{14}C = -1000‰$ is equivalent to background and, accordingly, *finite radiocarbon ages* (ages that can be statistically resolved from background) span this range of $d^{14}C$ values.

4.2.2 Age range

The exponential decay of ^{14}C sets the practical limit of the radiocarbon method which is between 50 and 60 ka, and corresponds with approximately 10 half-lives of radioactive decay (Figure 4.3). Technological refinements in instrumentation and sample pre-treatment strategies have been critical when analysing samples whose ages are close to background. After about eight half-lives of ^{14}C decay it is difficult to measure the residual radiocarbon and distinguish it statistically from natural background levels (i.e. in materials not normally regarded as containing radiocarbon, such as Permian coal). In this context, when measuring very low levels of residual ^{14}C activity, pre-treatment chemistry procedures become critical in removing potential modern contaminants from samples of unknown age. Similarly, it is imperative that the chemicals used in sample pre-treatment do not contain traces of modern ^{14}C.

The older limit of the radiocarbon method is obtained when the residual radiocarbon activity (A) is not significantly different to twice its standard deviation ($A = 2\sigma_A$). The standard deviation,

$$\sigma_A = \sqrt{\sigma_u^2 + \sigma_b^2}$$

where σ_u refers to the residual radiocarbon activity of the sample of unknown age and σ_b is the background as defined by a laboratory background standard (Arnold, 1995).

The dating of younger materials is complicated by several issues and it is difficult to define a specific upper limit of dating. By international convention, materials younger than 200 years (before 1950 AD) are regarded as modern (Gupta and Polach, 1985). As examined below, further issues arise in the calibration of radiocarbon ages to sidereal (calendar) years. There may potentially be more than one solution, a particularly frustrating problem for studies that require high temporal resolution over the past 1,000 years. This phenomenon, termed the *de Vries Effect*, relates to short-term, variable rates of ^{14}C production in the upper atmosphere. For example, radiocarbon activities for the years 1660 AD and 1790 AD are not significantly different (Stuiver, 1982; Arnold, 1995).

Additional complications in the dating of young materials from the mid-nineteenth century onwards, following the Industrial Revolution, are associated with combustion of fossil fuels which do not contain ^{14}C. As a consequence, atmospheric levels of ^{14}C have lowered relative to ^{12}C, known as the *Suess Effect*. A further complication arose in the early 1950s with the detonation of high-yield atom bombs which nearly doubled the atmospheric level of ^{14}C (Gupta and Polach, 1985; Hua, 2009), producing a well-defined bomb spike registered in both the northern and southern hemispheres, although with different intensities. This can produce radiocarbon ages with activities suggestive of an age in the future (i.e. higher than pre-1950s atmospheric ^{14}C levels) due to the higher ^{14}C concentration in modern materials.

4.2.3 Measurement techniques

Three approaches have been used for measuring the radiocarbon activity of geological and archaeological materials: gas proportional counting, liquid scintillation counting, and accelerator mass spectrometry (AMS). In the early history of the radiocarbon method, gas proportional and liquid scintillation counting methods were commonly used. Gas proportional counting involves the detection of ^{14}C in CO_2, while liquid scintillation measures the count rate per minute as represented by the emission of β^- particles within benzene, produced following a trimerisation reaction of CO_2 using a catalytic convertor (Gupta and Polach, 1985). The practice of deriving a ^{14}C age based on β^- particle emissions has been likened to measuring a population based on its death rate (Barker, 1970). Since the 1980s, direct atom counting of ^{14}C by AMS has been in the ascendancy. AMS directly measures ^{14}C atoms relative to ^{13}C and ^{12}C. Besides the direct quantification of ^{14}C atoms, AMS also has the advantage of permitting the analysis of samples of significantly smaller mass by a factor of at least a thousand. Recent measurements have been undertaken on 10–20 µg of carbon (Hua, 2009). The great improvements in measurement capability have more recently provided the opportunity to undertake compound-specific radiocarbon dating. Several issues affect the application of radiocarbon dating in studies of sea-level change, depending on environmental setting, and it is imperative that these are considered in palaeo-sea-level investigations, particularly where high temporal resolution is required.

4.2.4 Isotopic fractionation

In nature, ~98.9% of carbon is ^{12}C, ~1% is ^{13}C, and ^{14}C accounts for less than one in a billion atoms. During life, the rate of ^{14}C uptake relative to ^{13}C and ^{12}C within different groups of organisms does not reflect this ratio (Gupta and Polach, 1985). Accordingly, at the time of cessation of radiocarbon uptake, different biological materials can have contrasting initial radiocarbon activities, which potentially differ significantly from atmospheric levels due to the depletion of ^{14}C, relative to ^{13}C and ^{12}C. This phenomenon, termed *isotopic fractionation*, is particularly significant for studies that require a high degree of temporal resolution, and where the ages of contrasting sample types are compared. Differential rates of ^{14}C assimilation within plants appear to relate to contrasting photosynthetic pathways. For many species of plants, the modern activity of ^{14}C is some 3–4% lower than atmospheric levels, corresponding with an apparently older age of 240–320 years (Aitken, 1990).

The depletion of ^{14}C due to isotopic fractionation can be determined based on analysis of the ratio of ^{13}C to ^{12}C within samples. This is because of the well-established relation that the amount of depletion of an isotope of one mass unit difference such as ^{14}C will be twice that of ^{13}C relative to ^{12}C (Craig, 1953). The ratio of ^{13}C to ^{12}C, expressed as $\delta^{13}C$‰ (parts per thousand – per mil) is routinely determined by mass spectrometry and is calculated based on the following equation:

$$\delta^{13}C = \left[\left({}^{13}C/{}^{12}C \right)_{sample} / \left({}^{13}C/{}^{12}C \right)_{PDB} - 1 \right] \times 1000 \qquad (4.4)$$

where PDB refers to the Peedee Belemnite, *Belemnita americana* from the Cretaceous Peedee Formation in South Carolina. As the PDB standard is no longer available, the primary modern reference standard is a 1950 sample of oxalic acid from the United States National Bureau of Standards, and the radiocarbon activity is based on 95% of the count rate. According to international agreement, the $\delta^{13}C$ of the oxalic acid standard is −19‰ in relation to PDB (Gupta and Polach, 1985). A range of representative $\delta^{13}C$ values for different organic materials is shown in Figure 4.4 illustrating the effect of isotopic fractionation on determined age, expressed as a departure from −25‰.

Radiocarbon ages were originally calculated based on an assumed $\delta^{13}C$ value of −25‰, which corresponds with an average value for wood (Gupta and Polach, 1985). In marine-sourced carbonate, such as shells with a $\delta^{13}C$ value commonly in the range of 1 ± 1‰, the departure from $\delta^{13}C = -25$‰ is equivalent to an apparent age excess of 430 ± 50 years (Gupta and Polach, 1985). Variable sources of ^{14}C uptake such as within estuarine systems where freshwater input may contribute to the growth of marine invertebrates is a potential complication, and highlights the necessity of measuring the $\delta^{13}C$ value for each sample, rather than using assumed $\delta^{13}C$ values in radiocarbon age calculations. Studies have shown that $\delta^{13}C$ values on marine carbonate in estuarine environments may vary significantly depending on the relative inputs of ^{14}C from terrestrial and marine systems (Ulm, 2002; Petchey and Clark, 2011).

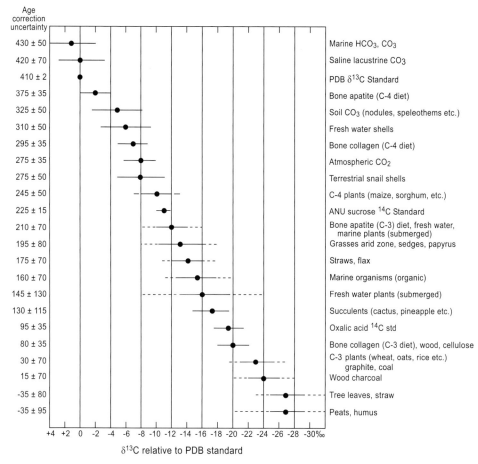

Figure 4.4 Representative $\delta^{13}C$ values for a range of natural materials, and the effect of isotopic fractionation from a value of −25‰ expressed in years on the y-axis (i.e. age correction and associated uncertainty term). The isotopic fractionation age correction is usually undertaken in the initial radiocarbon age calculation (Gupta and Polach, 1985, p. 114).

4.2.5 Marine reservoir and hard-water effects

Ocean surface waters are not in isotopic equilibrium with the atmosphere, in terms of the concentration of radiocarbon. They differ from atmospheric $^{14}C/^{13}C$ values and the difference is influenced by the degree to which ocean waters are stratified, time during which ocean surface water is in contact with the atmosphere, and rates of exchange through the entire water column. The marine carbon reservoir is about 60 times larger than the atmospheric reservoir, and residence time of ^{14}C in the ocean is longer (Gordon and Harkness, 1992). Ocean currents and upwelling processes bring ^{14}C-depleted waters, due to radioactive decay of carbonate in solution and other carbon-bearing compounds, into inner continental shelf environments. Accordingly, aragonite or calcite of marine invertebrates, such as molluscs, living in such radiocarbon-depleted seawater have an 'apparent age' reflecting the lower ^{14}C concentrations in the marine environment; this is termed the marine reservoir effect (Ascough et al., 2005).

There are significant geographical variations in the magnitude of the marine reservoir effect (Bard, 1988; Figure 4.5). It is also likely to have varied through time, although insufficient studies have been undertaken to accurately determine the magnitude of these variations. Short-term fluctuations in atmospheric ^{14}C production are less accurately recorded by marine fossil biota in view of changing ocean surface current activity and inherent complexities of post-depositional, taphonomic processes.

The marine reservoir approximates 400 radiocarbon years ($R(t) = 400\ ^{14}C$ years) in the open Atlantic, Pacific, and Indian Oceans between the latitudes of 40°N and 40°S (Bard, 1988; Ascough et al., 2005; Figure 4.5). This correction is automatically applied when a conventional radiocarbon age is calibrated using a modelled marine ^{14}C calibration curve (http://calib.org/marine; http://radiocarbon.ldeo.columbia.edu/research/resage.htm). A full marine calibration curve is also available (Marine 09) to calibrate marine radiocarbon ages, calculated using an ocean–atmosphere box diffusion model for the period 0–10.5 ka and foraminifera and coral data 10.5–12.5 ka (Hughen et al., 2004). However, there are significant local and regional variations in the actual magnitude of the marine reservoir at a site, and these are represented by a value termed ΔR, the deviation from this marine average, which has to be subtracted from the conventional radiocarbon age in order to compare with results for materials from terrestrial environments, or with ages determined by other methods such as U-series dating.

A marine reservoir value of $450 \pm 35\ ^{14}C$ years ($\Delta R = 50\ ^{14}C$ years) has been widely used. During the second half of the twentieth century, ^{14}C levels increased as a consequence of nuclear weapons-testing, so it has been necessary to directly date shells collected alive at known times before 1950 AD (Ascough et al., 2005). The 450-year value was derived using samples of marine molluscs alive at the time of their collection (before 1950) from southern Australia (Narooma in NSW, Adelaide in SA, and Garden Island in WA) and Torres Strait in northern Australia (Gillespie and Polach, 1979; Gillespie, 1990). A similar reservoir value of 460 ± 40 years ($\Delta R = 60\ ^{14}C$ years) was reported by Bowman and Harvey (1983) for several species of marine molluscs, alive when collected in 1934 AD and 1938 AD from Spencer Gulf and Gulf St Vincent in southern Australia. Other values reported include $365 \pm 20\ ^{14}C$ years for Iceland (Håkansson, 1983), $405 \pm 40\ ^{14}C$ years for the United Kingdom (Harkness, 1983) and $336\ ^{14}C$ years for New Zealand (Jansen, 1984).

It is important to establish the ΔR and apply the appropriate correction for different sites in order to compare ages on marine samples with other evidence, and in the calibration of radiocarbon ages. Other attempts to quantify the marine reservoir effect have involved comparative analyses of pairs of ages on marine shell and charcoal from middens (Gillespie and Temple, 1977; Ortlieb et al., 2011), comparison of radiocarbon and high precision U-series ages on a series of samples,

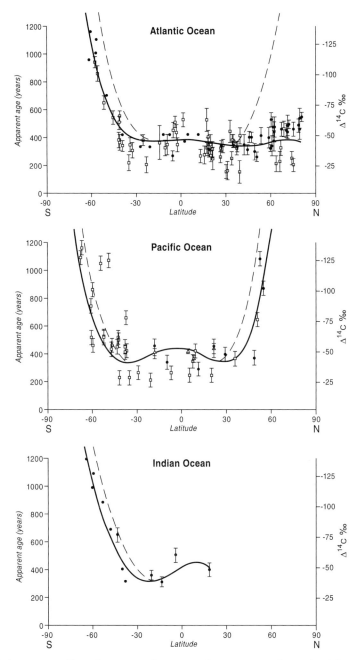

Figure 4.5 Apparent radiocarbon ages of ocean surface waters as determined from analysis of corals (open triangles), molluscs (closed circles), and seawater CO_2 samples (open squares). Closed squares denote pre-atom bomb $\Delta^{14}C$ concentrations (Broecker *et al.*, 1985). The solid lines represent polynomial curve fits, while the dashed lines depict a hypothetical profile of the world oceans during the Last Glacial Maximum, assuming that North Atlantic Deep Water did not form during that interval (Bard, 1988).

or dating of annual band sequences in long-lived corals (Fairbanks *et al.*, 2005; Guilderson *et al.*, 2009), and analyses of planktonic foraminifera from marine sediments and terrestrial peats both covered by coeval tephra (Ikehara *et al.*, 2011). Paired charcoal and shell ages can be unreliable because of the possibility of site disturbance and assumptions that shells and wood were contemporaneous (i.e. that the wood was not old at the time it was burnt – the *old wood effect*). In locations where people have been collecting marine molluscs as a food source over lengthy periods of time, further problems arise from *palimpsest sediments* – condensed sedimentary deposits that reveal the prolonged effects of accumulation and reworking.

The numerical value for the marine reservoir correction can range from $\Delta R = 0$ ^{14}C years in equatorial regions to 800 ^{14}C years or higher for Antarctic waters (Negrete *et al.*, 2011). Some of the highest values have been reported on materials from Antarctica, South Georgia, and the South Shetland Islands, based on analysis of a range of materials including marine shells, seaweeds, and bones. The upper limit to the marine reservoir has been determined as approximately 1,100 years by Broecker *et al.* (1985), was shown to be up to 1,300 years on seal bone (Negrete *et al.*, 2011), and may potentially exceed 1,300 years (Gordon and Harkness, 1992). The deep ocean can have an apparent age of several thousand years. Younger values have also been reported for marine molluscs from different regions of the Antarctic coastline; for example, Negrete *et al.* (2011) report a value of 750 ± 60 ^{14}C years on a modern specimen of the mollusc *Nacella concinna*. Collectively these high values have been attributed to the intermixing of deep and intermediate water with ocean surface waters and the role of sea ice in inhibiting carbon dioxide exchange between the atmosphere and the sea surface (Bard, 1988; Gordon and Harkness, 1992). In Scandinavia the marine reservoir is approximately 400 years (Olsson, 1974).

As more determinations of ΔR become available, systematic variations can be seen across the major oceans. For example, in the Indian Ocean there is evidence of strong depletion of ^{14}C on the western margin (with ΔR exceeding 200) due to upwelling in the Arabian Sea, compared with much lower values in the South China Sea, many of which are negative (Southon *et al.*, 2002). A suite of analyses from sites in the South Pacific indicate less variability (Phelan, 1999; Petchey *et al.*, 2008), and Pacific waters influence the eastern Indian Ocean as a consequence of the Indonesian throughflow (Hua *et al.*, 2004). Higher values are found on the eastern and northeastern Pacific coasts (Ingram and Southon, 1996; Kovanen and Easterbrook, 2002).

Open marine corrections may not be appropriate for coastal environments, particularly estuarine settings where seawater is diluted. A relatively high reservoir age (1,200–1,300 years) is suggested for the Lagoon of Venice (Zoppi *et al.*, 2001). Systematic analyses of shells from estuaries along the east Australian coast indicate differences from open water; for example, estuary-specific values of up to -155 ± 55 years have been found (Ulm, 2002). The periodic release of compounds such as methane from the shallow subsurface in marine environments also contributes to local marine reservoir effects. In the Black Sea region, for example, ^{14}C-CH_4 releases from cold-water seeps, mud volcanoes, and methane clathrates have remained within the water column and contributed 'aged' ^{14}C to marine molluscs, such as *Mytilus galloprovincialis*, living within the region. The radiocarbon activity of the seeps has yielded radiocarbon ages ranging between 18.2 and 24 ka ^{14}C years BP (Guilin *et al.*, 2003; Kessler *et al.*, 2006).

The modern value for the marine reservoir has commonly been applied to radiocarbon ages on Holocene and late Pleistocene materials. However, the magnitude of the marine reservoir effect has almost certainly changed over the course of the late Quaternary and requires further investigation to increase temporal resolution and for accurate comparisons of different geochronological methods. Radiocarbon analyses of shell and charcoal from coastal deposits between 14 and 24°S in Chile and Peru have shown changes in the magnitude of the marine reservoir during the Holocene (Ortlieb *et al.*, 2011). For the interval 10,400–6,840 years BP, a reservoir value of 511 ± 278 years was determined, whereas between 5,180 and 1,160 years BP the average was 226 ± 98 years. In the past 1,000 years, the marine reservoir fluctuated from 355 ± 105 years to a late twentieth century value of 253 ± 207 years. Changes in the intensity of upwelling events, possibly related to different intensities in the ENSO, may be responsible for changing the influx of ^{14}C-depleted marine waters to the inner shelf environment of northern Chile and southern Peru. Similarly, paired ^{230}Th/^{234}U and radiocarbon analyses on fossil corals from Koil and Muschu Islands, northern Papua New Guinea, also reveal variability in the marine reservoir effect during the Holocene (McGregor *et al.*, 2008). For the period 7,220–5,880 years BP, the marine reservoir value of 185 ± 30 ^{14}C years is evident compared with the modern value of 420 ^{14}C years. The increase in the magnitude of the marine reservoir occurred from 5880 to 5420 years BP and is attributed to the increasing influence of the South Equatorial Current in the southwestern Pacific Ocean possibly associated with the onset of the modern ENSO regime.

Contrasting ocean current patterns may be initiated by different coastal palaeogeographies, a function of sea-level changes, particularly at times of glacial maxima (e.g. the closure of gateways). Estimates suggest that during the LGM, the world's ocean surface area was reduced by up to 10–20% due to the abstraction of water in continental icesheets and subaerial exposure of continental shelves (CLIMAP Project Members, 1976). This would naturally result in a reduced ocean surface area for CO_2 exchange, and an associated increase of Δ^{14}C between atmosphere and oceans of between +45 and +100 years has been suggested (Bard, 1988). General circulation modelling of the marine realm has shown that a reduction by 30% in Atlantic meridional overturning circulation may result in an increase by up to approximately 500 years in the value of the marine reservoir in high-latitude regions, 300 years in parts of the subtropics, and as much as 1,000 years in the Southern Ocean (Franke *et al.*, 2008). Large variations in the magnitude of the marine reservoir are further illustrated by the fact that during Heinrich Event I (~14,300 years ago) the marine reservoir in the North Atlantic Ocean may have been as much as 2,000 years

(Sarnthein *et al.*, 2007). Changes in the intensity of the Earth's magnetic field have also been implicated in influencing the magnitude of the marine reservoir, particularly the Laschamp geomagnetic event, which has been dated to 40.4 ± 2 ka based on analysis of lava flows (Guillou *et al.*, 2004). The U-series dated coral record of Chiu *et al.* (2007) reveals a progressive increase in $\Delta^{14}C$ for the period 50–40 ka ago, and corresponds with a global increase in ^{14}C production culminating in the Laschamp geomagnetic event.

Organisms can also be depleted in ^{14}C if they have grown close to water with depleted solute load. Waters in estuarine and fluvial systems that traverse large tracts of limestones and other carbonate lithologies are likely to transport significant quantities of calcium biocarbonate as the solute load. In such circumstances, the source rocks and the waters in contact with them, invariably have lower $\Delta^{14}C$ values, resulting from the complete loss of ^{14}C from the bedrock through radioactive decay. Freshwater molluscs and aquatic plants that frequent these environments will assimilate a proportion of their total carbon uptake from such waters and, accordingly, will have an initial radiocarbon activity at the time of death (or cessation of ^{14}C uptake) that is lower than would otherwise be expected for modern materials. This phenomenon, termed the *hard-water effect*, may potentially yield ages that are apparently older by up to 2,000 years (Arnold, 1995). The magnitude of the hard-water effect varies with geographical location, and so no universal value can be assumed. On Tonga in the South Pacific, molluscs of the genus *Gafrarium* from lagoonal waters on Tongatapu show the hard-water effect which is equivalent to a ΔR value of 273 ± 34 ^{14}C years, in contrast to the surrounding open-ocean waters which have a ΔR of 11 ± 83 ^{14}C years (Petchey and Clark, 2011). The hard-water effect can be significant in coastal studies, especially where radiocarbon ages on materials from different sedimentary subenvironments, such as open-marine, estuarine, and freshwater, are being compared.

4.2.6 Secular $^{14}C/^{12}C$ variation and the calibration of radiocarbon ages to sidereal years

One of the fundamental assumptions originally made in the radiocarbon method is the constancy of radiocarbon production in the upper atmosphere (Libby, 1955). However, *dendrochronology* (tree-ring studies), dating of varves, and comparisons of ages determined by radiocarbon and U-series dating of annual bands in fossil corals, have shown that ages expressed in radiocarbon years do not conform to sidereal years on a one-to-one basis (Klein *et al.*, 1982; Stuiver *et al.*, 1998; Chiu *et al.*, 2007; Reimer *et al.*, 2009). Variations in past ^{14}C concentrations, expressed as $\Delta^{14}C$, have been shown to occur at high frequency, revealing century to decadal variability (Hua, 2009). There is a tendency for the radiocarbon method to overestimate the age of materials younger than approximately 1,000 years, while closer correspondence between radiocarbon and sidereal age is evident for materials around 3,000 years

old. Radiocarbon ages underestimate the 'true' age of fossils from the early Holocene. For example, an uncorrected radiocarbon age of 6,500 years represents an age
underestimate of about 800 years. The magnitude of this offset is even greater for
materials that formed during the LGM (~21 ka ago) as revealed by the paired
radiocarbon and U-series dating of corals (Bard *et al.*, 1990b). Radiocarbon ages
underestimate sidereal ages by approximately 3,000 years for a sample with a radiocarbon age of 15,000 years BP and up to 4,000 years for materials determined to be
20,000 years BP (Stuiver *et al.*, 1998). Between 30 and 40 ka (sidereal years) the offset
in radiocarbon years exceeds 4,000 years (Fallon, 2011; Figure 4.6). The Younger
Dryas cold event (12.9–11.6 ka) also represents a complex part of the radiocarbon
record (Fiedel, 2011). The onset of the Younger Dryas is associated with a sudden
increase in Δ^{14}C such that radiocarbon ages fall from 11 to 10.6 ka (^{14}C years) within
an interval of approximately 100 calendar years. Conversely, a radiocarbon plateau
characterises the interval of 10.6–10.3 ka (^{14}C years), corresponding with the calendar
period of 12.7–11.9 ka (Reimer *et al.*, 2009).

Calibration of radiocarbon ages to sidereal years is complex, especially for some
parts of the record where more than one calibrated age is possible for a single
radiocarbon age (Bowman, 1990; Hua, 2009). A further problem relates to the slope
of the calibration curve and that, in some instances, a larger age-range may result in
the calibrated age when transformed from ^{14}C years (Figure 4.7).

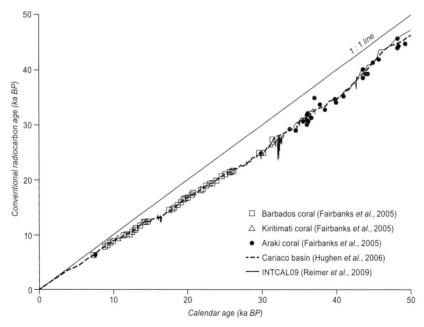

Figure 4.6 Offset in radiocarbon years from calendar for the past 50 ka (after Fallon, 2011,
with kind permission from Springer Science + Business Media B.V.).

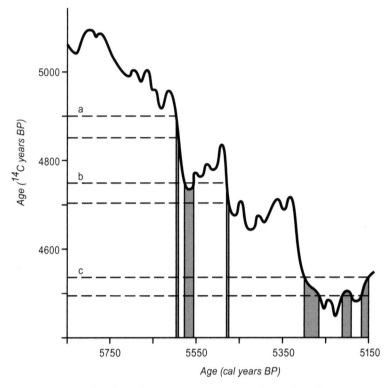

Figure 4.7 Conversion of radiocarbon ages as expressed in radiocarbon years to sidereal (calendar) years, illustrating the potential complexity in calibration. In example (a) the steep slope of the calibration curve results in a lower uncertainty in calibrated age. In example (b), the noise in the record due to cosmic ray flux results in two possible calendar ages of contrasting probability, based on the intersection of the radiocarbon ages with the calibration curve, and in (c), the low slope of the calibration curve results in three possible calendar ages.

Several processes have been identified to explain the discordance between radiocarbon ages and calendar (sidereal) ages. Changes in the Earth's geomagnetic field, which in turn may modulate the amount of cosmic ray generation of ^{14}C in the upper atmosphere, have been identified as one mechanism, as well as periodic changes in the Sun's solar activity. Variations in solar activity involving fluctuating emissions of protons and electrons from the Sun, termed the *solar wind*, may be involved in changes in ^{14}C production with time. An additional mechanism related to long-term climate change involves changes in *thermohaline circulation*, deep-ocean circulation of cold, dense waters that serve as a major driving mechanism for heat exchange around the planet. Thermohaline circulation may be involved in the periodic ventilation of CO_2 containing ^{14}C that has a lower activity (because of radioactive decay) being introduced to the Earth's near-surface environments. *Heinrich events*, episodic iceberg-rafting events, recorded within the latitudes

40–60°N by deposition of terrigenous–clastic sediment at times of dramatic ocean surface cooling in the North Atlantic Ocean (Heinrich, 1988; see Section 6.12), have also been implicated as a mechanism for $\Delta^{14}C$ variations.

Radiocarbon calibration curves and associated programs provide a means of expressing radiocarbon ages in calendar (sidereal) years. The INTACL09 and MARINE09 calibration curves extend the record back to 50 ka, and are based on cross-comparisons of radiocarbon and U-series ages on corals (Reimer *et al.*, 2009). In terms of age resolution, the INTACL09 record was determined at 10-year intervals for the age range 12–15 ka (calendar years), 20-year intervals for 15–25 ka, 50-year intervals for 25–40 ka and 100-year intervals for 40–50 ka. A constant marine reservoir of 405 ^{14}C years was applied to the MARINE09 record. A variety of calibration programs are now available (the four most widely used are available online: Calib, http://intcal.qub.ac.uk/calib/; CalPal, http://www.calpal.de/; OxCal, http://c14.arch.ox.ac.uk/embed.php?File = oxcal.html; Fairbanks, http://radiocarbon. ldeo.columbia.edu/research/radcarbcal.htm). However, given subsequent improvements in the calibration programs as more data become available, it is imperative that uncorrected (conventional) radiocarbon ages are published with the calibrated ages (as well as the radiocarbon age laboratory code), as this will enable future researchers to evaluate primary ^{14}C data.

Calibrated ages are reported as an average age with a 2σ-uncertainty (95% confidence interval). When reporting radiocarbon ages these are usually rounded off appropriately: ages less than 1,000 years rounded to the nearest 10 years with the uncertainty term rounded to the nearest five years; ages between 1,000 and 10,000 years rounded to the nearest 10 years, with the uncertainty to the nearest 10 years; ages falling between 10,000 and 25,000 years rounded to the nearest 50 years with the uncertainty to the nearest 10 years; and ages > 25,000 years rounded to the nearest 100 years and the uncertainties to the nearest 50 years (Gupta and Polach, 1985). Unrounded ages imply an unrealistic precision, not upheld in the palaeoenvironmental context of most dated samples or in the physical measurement of residual radiocarbon.

4.2.7 Contamination and sample pre-treatment strategies

Contamination within sedimentary deposits during diagenesis can be a potentially serious problem for radiocarbon dating. It may be difficult to assess which processes have occurred, or to recognise potential contaminants. For example, relative sea-level changes may influence the position of the water table in near coastal landscapes, which in turn may enhance diagenesis in the *vadose zone* (region of percolating rainwater above the fluctuating water table). This may result in precipitation of authigenic calcium carbonate with a contrasting radiocarbon activity as a pore-filling cement between carbonate grains, and render sample pre-treatment more complex. There are no universally applicable criteria to reliably assess the likelihood of contamination and, accordingly, samples need to be assessed on a case-by-case basis.

In most geological environments there is a greater likelihood of contamination from younger carbon sources, particularly introduced from soil-forming processes. The influence of modern ^{14}C on a derived age is far greater than a comparable level of contamination by 'aged' ^{14}C that has commenced radioactive decay. For example, a sample with a known age of 10,000 years (^{14}C years) and contaminated by 10% radiocarbon with a modern activity, will yield an apparent age of 9,100 years (Olsson, 1974). In contrast, a sample with a 'true' age of 20,000 years but contaminated by 20% older (background) radiocarbon will yield an older apparent age by only about 1,600 years (Figure 4.8). More dramatically, a sample 'infinitely' old in terms of the radiocarbon method (i.e. not containing radiocarbon, such as Permian coal), when contaminated by 1% ^{14}C with a modern activity, will yield a finite, apparent age of 37 ka (Gupta and Polach, 1985). Figure 4.8, which illustrates the effects of contamination, is based on the assumption of 'instantaneous' contamination, rather than a prolonged period of contamination involving differential ^{14}C activities.

Strategies for assessing potential contaminants vary according to the sample matrix. Effects of contamination can be difficult to model and require a thorough understanding of the stratigraphical context of the dated materials. For example, volcanism introduces CO_2 into the atmosphere that does not contain ^{14}C and consequently plants growing close to volcanic centres may have lower initial radiocarbon activities, resulting in older apparent ages. In ideal situations, independent methods of dating may establish if contamination has occurred, given the assumption that any discordance in age is not a function of selective reworking of older materials from a different source. In order to remove contaminants from a sample, it is necessary to consider their potential sources (e.g. soil humic acids and other organic molecules such as the unstable amino acids serine and threonine, younger pedogenic calcium carbonate, bioturbation effects, intrusive modern plant roots, or bacterial action). Rigorous pretreatment strategies are needed to remove these potential contaminants from a sample using physical and chemical strategies, which may significantly reduce sample mass (or in exceptional circumstances totally destroy samples of low mass), and could result in a larger uncertainty in the derived age. Accordingly, a balance must be sought between removal of potential contaminants and retaining sufficient sample for radiocarbon assay, predicated on the palaeoenvironmental question under consideration, and the age-resolution required. In the case of AMS, where very small samples are analysed, the issue must be addressed of how representative a selected portion of a sample is for analysis, in relation to the larger deposit, both in terms of potential contamination and whether a single fossil is *in situ* and reflects the age of an entire faunal assemblage.

Advances in pre-treatment strategies and laboratory sample preparation procedures have been developed to accommodate samples of smaller mass for AMS analysis, as well as addressing the issue of potential contamination of older materials close to the practical limits of radiocarbon dating. Particularly useful summaries of

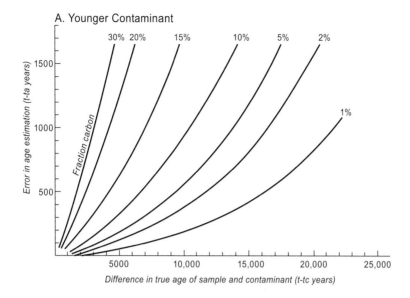

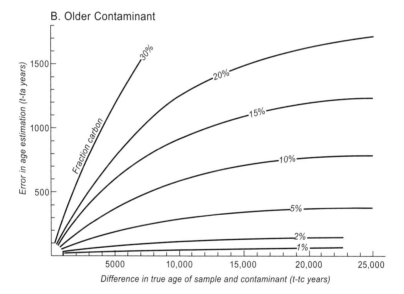

Figure 4.8 Graphs illustrating the effect of variable contributions of contaminants, either containing modern or 'aged' ^{14}C on resultant radiocarbon ages (Gupta and Polach, 1985). In example A, the influence of younger contaminants in the form of a higher radiocarbon activity is shown, while in B, the influence of older contaminants and lower radiocarbon activities are shown. In A, a sample with a 'true' age of 20,000 years, but contaminated by 2% ^{14}C with a modern activity, will result in a younger radiocarbon age by 1,600 years. In contrast, in graph B a fossil of equivalent age but contaminated by 2% older carbon with no measurable ^{14}C (sometimes described as 'dead carbon') will yield a radiocarbon age of only 100 years younger, yielding an apparent age of 19,900 years.

pretreatment strategies for a range of materials are given by Gupta and Polach (1985), Hedges (1992), and Trumbone (2000). One of the challenges has been preventing adsorption of modern $^{14}CO_2$ on the surfaces of pretreated samples, particularly for materials close to the practical limit of radiocarbon. The acid–base–wet oxidation (ABOX) procedure, for example, is an effective means of overcoming this problem for charcoal by undertaking combustion and graphitisation steps in a vacuum line isolated from the atmosphere (Bird *et al.*, 1999).

4.2.8 *Statistical considerations: comparisons of radiocarbon age and pooling of results*

In palaeo-sea-level investigations involving the compilation of numerous radiocarbon ages, issues commonly arise concerning the equivalence of ages and whether ages differ significantly between different data sets. When assessing whether ages are significantly different, they must first be corrected for isotopic fractionation, any reservoir effects, and ideally calibrated to sidereal years, particularly for comparison of charcoal and marine shell ages. In cases where the analytical uncertainties are relatively low and *Gaussian* in nature (normal distribution), the Z-score may be calculated to determine if the derived ages are significantly different (Gupta and Polach, 1985). For example, Belperio *et al.* (1984b) determined Z-scores to assess if radiocarbon ages on seagrass facies, skeletal carbonate sand, and entire shells were significantly different in the Holocene coastal successions of Spencer Gulf, southern Australia. A Z-score is calculated based on the following equation:

$$Z = |(t_1 - t_2)/\sqrt{[(\sigma_1)^2 + (\sigma_2)^2]}| \tag{4.5}$$

where t_1 and t_2 are the corrected radiocarbon ages, and σ_1 and σ_2 are the associated uncertainty terms and the result of the calculation is expressed as an absolute value. Given that 95% of the results will fall in the range of $t_m \pm 1.96 \, \sigma$ [$t_m = (t_1 - t_2)/2$], at the 5% level, the hypothesis is rejected that results are equal if $Z > 1.96$. The above equation is inappropriate for finite radiocarbon ages close to the practical limits of the method, those that have non-Gaussian uncertainties, or where the uncertainties are large compared with the derived age. In these situations, it is more appropriate to calculate a Z-score based on the $d^{14}C$, the residual radiocarbon activity of the sample, normalised for the degree of isotopic fractionation as determined in wood ($\delta^{13}C$ of $-25‰$).

Calculation of the t-statistic can be used in instances where several ages are to be compared. This involves the initial calculation of a pooled mean age (t_p) and its associated uncertainty (σt_p):

$$t_p = \Sigma[t_i/(\sigma t_i)^2]/\Sigma \, 1/(\sigma t_i)^2 \tag{4.6}$$

and

$$\sigma t_p = \sqrt{[1/\Sigma(1/\sigma t_i)^2]} \tag{4.7}$$

where t_i, σt_i, t_p and σt_p are as defined in Equation 4.5. The t-statistic (T) then may be determined according to:

$$T = \Sigma[(t_i - t_p)^2/(\sigma t_1)^2] \tag{4.8}$$

T conforms to a chi-squared distribution with $n-1$ degrees of freedom. Accordingly, if the value of T is less than the tabulated chi-square value at a selected level of significance, the hypothesis is accepted that the results are not significantly different [i.e. the pooled mean age (t_p) and its associated uncertainty (σt_p) can be regarded as statistically valid]. An example of a pooled mean age calculation presented by Gupta and Polach (1985) involved the pooling of the following four radiocarbon ages and their uncertainties [$4{,}900 \pm 100$, $4{,}700 \pm 90$, $4{,}800 \pm 70$, and $5{,}100 \pm 200$ years BP].

$$\text{Pooled age}(t_p) = \frac{[4900/100^2 + 4700/90^2 + 4800/70^2 + 5100/200^2]}{[1/100^2 + 1/90^2 + 1/70^2 + 1/200^2]}$$

$$\text{Uncertainty term}(\sigma_t) = \frac{1}{\sqrt{[1/100^2 + 1/90^2 + 1/70^2 + 1/200^2]}}$$

Thus, the unrounded result for the pooled mean age and uncertainty is $4{,}811 \pm 47$ years BP. In this example there are 3 degrees of freedom (DF). Accordingly, the value of T is calculated thus:

$$T = \frac{(4900 - 4811)^2}{100^2} + \frac{(4700 - 4811)^2}{90^2} + \frac{(4800 - 4811)^2}{70^2} + \frac{(5100 - 4811)^2}{200^2}$$

$$T = 4.43$$

The determined value of $T = 4.43$ is less than the chi-square value, which at the 5% confidence level is $(0.05, 3\ DF) = 7.82$. Thus, the results are not significantly different at the 95% confidence level and the determined pooled mean age is statistically valid. The pooled mean age and its uncertainty would normally be rounded to the nearest 10 years and therefore be $4{,}810 \pm 50$ years BP.

4.3 Uranium-series disequilibrium dating

U-series disequilibrium dating has played a central role in delineating the timing of sea-level highstands such as the last interglacial maximum (130–118 ka; Edwards *et al.*, 1987a; Shackleton *et al.*, 2002), and previous interglacials (Edwards *et al.*, 1987b; Stirling *et al.*, 2001), as well as late Pleistocene interstadials (Stearns, 1984; Cutler *et al.*, 2003). The potential of a method that involved the uranium-series decay chain was first recognised by Bateman (1910), and α-counting of the daughter isotopes was first applied to date corals by Barnes *et al.* (1956). With the development of more sophisticated techniques, it remains widely used in studies of sea-level changes with the dating of corals from tectonically uplifted fringing reef complexes (Chappell, 2002; Muhs *et al.*, 2002) and from 'stable' continental margins (Muhs, 2002).

4.3.1 Underlying principles of U-series disequilibrium dating

U-series dating methods involve the application of a range of daughter isotopes derived from the radioactive decay of the parent nuclides ^{238}U, ^{235}U, and ^{232}Th, through three complex decay series, culminating in stable ^{206}Pb, ^{207}Pb, and ^{208}Pb (Faure, 1986; Ku, 2000). These decay chains involve a total of 10 elements (U, Pr, Th, Ac, Ra, Rn, Po, Bi, Pb, and Tl) and 39 isotopes, and various α, β, and γ decay reactions. Uranium and thorium have long half-lives, but several of the daughter isotopes are short-lived. The half-lives of ^{234}U and ^{230}Th are 246 and 75.4 ka, respectively (Wagner, 1998), and with recent developments in the mass-spectrometric analysis of the U-series, the age range of ^{230}Th/^{234}U is approximately 500 ka under ideal conditions (Figure 4.9). The dating range of the U-series method is governed by the point at which the activity ratio approaches a state characterising secular equilibrium, which in the case of mass-spectrometric methods is comparable to approximately seven half-lives of ^{230}Th ($7 \times \tau = 528$ ka).

Given sufficient geological time, a mineral phase containing uranium will achieve secular equilibrium and the number of radioactive decay events for each member in the decay series will equal that of the long-lived uranium parent isotope. A variety of geological events may cause disequilibrium, and the tendency for a system to return

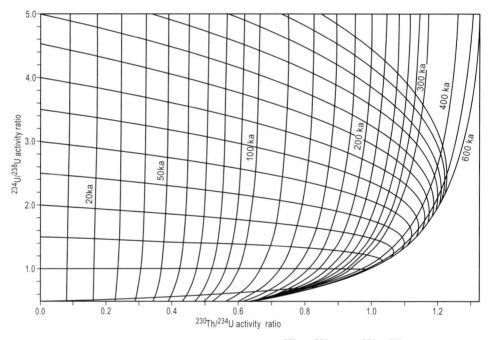

Figure 4.9 Concordia diagram for the activity ratios of ^{230}Th/^{234}U and ^{234}U/^{238}U showing the dating range of the U-series method and applicable to closed systems and without ^{230}Th at t_o. The subhorizontal lines denote the decay paths for materials with an initial activity ratio as defined on the y-axis (^{234}U/^{238}U). The near-vertical lines represent isochrons in ka.

to secular equilibrium provides the basis for this method of dating. Events which may disturb secular equilibrium include partial melting through igneous and metamorphic processes, prolonged weathering, sedimentation, and the biological precipitation of aragonite and calcite in marine invertebrates such as corals and molluscs. Such events interrupt the decay chain and mean that the ratio of one isotope to another does not accord with secular equilibrium; two groups of U-series methods have been identified based on the initial absence or presence (excess) of the daughter nuclide (Blackwell and Schwarcz, 1995; Ku, 2000).

The daughter-deficient method is based on the absence of the daughter isotope at the time of mineral formation. It is this method which has been used to date corals, and some other organisms, that incorporate uranium but not thorium into the carbonate skeleton. This is the basis for $^{230}Th/^{234}U$ dating of corals, also referred to as $^{238}U-^{234}U-^{230}Th$, U/Th, or Thorium-230 dating in the literature. Uranium is found in trace amounts in water (usually 0.01–100 ppm) and can be incorporated into marine carbonates, but as thorium is highly insoluble it is not precipitated, and the ^{230}Th present within a coral is assumed to be due solely to the radioactive decay of ^{234}U.

In contrast, in the daughter excess method, the daughter isotope is initially present in a greater concentration than its parent nuclide, and the method involves quantifying the rate of loss of the daughter isotope from the system, assuming the system is closed. Examples of these methods include ^{210}Pb dating of lacustrine sediments of recent age (the past ~120 years) and excess ionium (^{230}Th; ionium was an early name for this isotope of thorium) and protactinium (^{231}Pa) for the dating of marine sediments (the past ~300 and ~150 ka, respectively; Wagner, 1998).

In the early applications of U-series dating, α-spectrometry was used to measure the radioactive decay of ^{234}U and ^{230}Th, involving counting for periods of up to 3 days. Analytical uncertainties associated with these analyses were commonly up to 5% at the 2σ-level on corals of last interglacial age (125 ka) (Neumann and Moore, 1975). In the 1980s, developments in mass spectrometry enabled atom counting; TIMS enabled greater analytical precision with 2σ uncertainties on ages of 125 ka of ± 1,000 years. The significant improvements in accurately measuring the U-series, which resulted in derivation of ages with high precision, led to debates on the duration of the last interglacial maximum as defined by palaeo-sea levels inferred from fossil corals (e.g. 130–122 ka: Edwards *et al.*, 1987a; 132–120 ka: Chen *et al.*, 1991; 132–120 ka: Stein *et al.*, 1993; 127–122 ka: Stirling *et al.*, 1995). In the wider literature, this discussion was based on highly precise ages on poorly preserved fossils which would have been excluded from analysis if appropriate screening criteria were invoked, such as $\delta^{234}U$ values for fossil corals in the range of modern seawater and live corals of approximately 138–150‰, uranium concentrations of between 2 and 4 ppm and only trace quantities of ^{232}Th (which is considered an indication of groundwater contamination) in corals (< several ppb) (Chiu *et al.*, 2007).

The age (t) of a fossil is given by the equation:

$$t = \frac{2.303 \log \left[1 - \left(^{230}\mathrm{Th}/^{234}\mathrm{U} \right)_A \right]}{\lambda_{230}} \tag{4.9}$$

where $\lambda_{230} = 9.217 \times 10^{-6}$ years.

Before analysis, a sample is dissolved and spiked with a tracer consisting of a mixture of artificially enriched $^{229}\mathrm{Th}$ (or $^{228}\mathrm{Th}$) and $^{233}\mathrm{U}$ (or $^{236}\mathrm{U}$) of known concentrations. U and Th in the sample are pre-concentrated, separated, and purified by ion-exchange chromatography, and then the isotope ratios are determined. Mass spectrometry, initially by inductively coupled plasma ionisation using a sector magnet and a single collector system, provides considerable improvements on α-counting. More recently, multi-collector, inductively coupled plasma mass spectrometry (MC-ICP-MS), and particularly using laser ablation (LA-ICP-MS), methods have been used to isolate the U-series (Eggins *et al.*, 2005). MC-ICP-MS offers many advantages over former methods, including high sample throughput, greater sensitivity, and predictable mass biases which may be more easily corrected (McCulloch and Mortimer, 2008; Zhao *et al.*, 2009). This has also presented an elegant approach for multiple $^{230}\mathrm{Th}/^{234}\mathrm{U}$ analyses of discrete sections (100 μm in diameter) within shells or coral, on samples with as little as 1 ppm U. Although this latter approach involves larger analytical uncertainties, it provides the opportunity to derive multiple ages from profiles scanned across individual fossils, permitting assessments of the geochemical integrity of fossils and their potential post-mortem uptake of uranium. This method has also permitted a high sample throughput and the opportunity to scan samples to assess relative ages before undertaking more detailed numerical dating (McGregor *et al.*, 2011). Inductively coupled plasma–quadrupole mass spectrometry has also recently been developed permitting derivation of analytical uncertainties of 4–5% at the 2σ-level on 500 ng of $^{238}\mathrm{U}$ within a dating range of approximately 1–400 ka (Hernández-Mendiola *et al.*, 2011).

A range of materials has been dated in studies of Quaternary sea-level changes using the U-series method. Corals have provided particular insights into former interglacials, whereas there has been much attention also focused on speleothems. The method has been extended to date marine molluscs and pedogenic deposits with varied success.

4.3.2 U-series dating of marine carbonates

The incorporation of uranium into coral and the formation of $^{230}\mathrm{Th}$ from the radioactive decay of $^{234}\mathrm{U}$ represents the most widely used system for numerical dating of marine carbonates in studies of late Quaternary sea-level changes (e.g. see references in Blackwell and Schwarcz, 1995 and Hernández-Mendiola *et al.*, 2011). The concentration of uranium in biogenic marine carbonate is commonly 0.1–5 ppm.

Under oxidising conditions in near-surface seawater, uranium is present in the form of the uranyl ion (UO_2^{2+}) which is transported in solution and incorporated into calcium carbonate, whereas thorium is largely insoluble in seawater due to its high ionic potential (Th^{4+}) and thus is readily hydrolysed and removed from solutions by precipitation or scavenging processes, meaning ^{230}Th will be present only in negligible quantities, if at all, in corals and other marine organisms. Assumptions implicit in $^{230}Th/^{234}U$ dating include: (1) the presumed absence of ^{230}Th in the sample at time of death (t_0), and (2) that the system remained closed from the time of death to the time of analysis.

Corals are composed of aragonite and are particularly suitable for U-series dating analysis. Corals typically contain 2–3 µg/g of uranium but no thorium or protactinium at time of death, and the $^{230}Th/^{234}U$ activity ratio in modern samples has been determined at 10^{-6}, indicating an apparent age of 8 years (Eisenhauer *et al.*, 1993). Coral matrices appear to be relatively stable under a range of diagenetic conditions, and accordingly do not appear to readily take up thorium from the surrounding environment, in contrast to other carbonate fossils. However, fossil corals in humid, tropical environments are less likely to behave as closed systems and may exchange isotopes with the surrounding environment. The recrystallisation of metastable aragonite to its polymorph, calcite, is also likely to violate closed-system behaviour.

Since the early work of Veeh (1966) and Mesollela *et al.* (1969), a huge literature has emerged on the dating of corals using the U-series method for studies of late Pleistocene sea-level changes and particularly detailed reviews are given by Blackwell and Schwarcz (1995) and Muhs *et al.* (2002). Key assumptions of U-series dating are that the initial $^{230}Th/^{234}U$ activity ratio is close to zero (Th is absent), that the fossil represents a closed system during its diagenetic history without uptake of U-nuclides in the chain from ^{238}U to ^{230}Th, and that the initial activity ratio of $^{234}U/^{238}U$ is known. In many field contexts these assumptions are not upheld and during diagenesis, uranium progressively diffuses into the aragonitic matrix of corals-thus violating the assumption of closed-system behaviour. Potential mechanisms for the enrichment of ^{234}U and ^{230}Th within fossils include uranium mobilisation by dissolution, producing aqueous ^{234}Th and ^{230}Th and thorium mobilisation by α-recoil, ejecting ^{234}Th and ^{230}Th from crystal lattices (Thompson *et al.*, 2003). In view of these difficulties, different analytical approaches have been invoked to model the open-system behaviour of the uranium series (Villemant and Feuillet, 2003; Thompson and Goldstein, 2005). The ultimate rate of these processes is governed by the rate of radioactive decay. Accordingly, excess ^{234}U within fossils may indicate open-system behaviour and a corresponding increase in ^{230}Th during diagenesis.

Considerable controversy surrounds the dating of marine molluscs by the U-series method, in view of the widely recognised open-system behaviour of these matrices (Kaufman *et al.*, 1971; McLaren and Rowe, 1996). Modern molluscs typically have low concentrations of U and Th (0.1 ppm), but higher concentrations are evident in equivalent fossil species (≥ 4 ppm), implying post-mortem U-uptake (Kaufman *et al.*,

1971). U-series dating of molluscs has involved assumptions about the rate at which uranium is taken up by the aragonite matrix during diagenesis (whether this is a linear rate of uptake or an early, more rapid uptake). Hille (1979) proposed an open-system model in a reassessment of ^{230}Th/^{234}U ages previously reported by Szabo and Rosholt (1969). Hille's model assumed that all the *in-situ* decay products within the shells remained chemically stable, and that the uptake of uranium within the sample conformed to an exponential rate. Accordingly, the recalculated ages for the lowest marine terraces at San Pedro (78 ka) and Santa Monica (80 ka), in southern California, are correlative features to the interstadial marine terrace of southern Barbados.

An important check for the validity of ^{230}Th/^{234}U ages is the activity ratio of ^{234}U/^{238}U (δ^{238}U). Departures from values representative of modern seawater (δ^{238}U value of 1.144 ± 0.002; Chen *et al.*, 1986) may indicate diagenetic contamination from meteoric groundwaters, implying that U-series ages on molluscs are highly dependent on the nature of the host sediments. Hillaire-Marcel *et al.* (1986) have shown that apparently reliable ages may be obtained where the host sediments are represented by relatively impermeable clays, reducing the likely migration of U or Th throughout the sediment profile.

4.3.3 U-series dating of other materials

Speleothems are reprecipitated deposits of calcium carbonate that form in terrestrial environments such as limestone caves. They include a variety of forms such as *stalagmites* (vertical, well-laminated accumulations of carbonate that project up from cave floors), *stalactites* (similar to stalagmites but hanging from the roof of caves) and *flowstones* (tabular sheets of limestone formed on cave floors or parallel with cave floors). Speleothems have been used extensively in palaeoclimatological studies in view of their suitable geochemical characteristics, as well as the lengthy and relatively reliable records they may provide. Speleothems generally behave as closed systems to uranium uptake after calcite precipitation. A potential problem, however, is the uptake of ^{230}Th during calcite formation represented as a *detrital* component (i.e. included within pre-existing rock fragments).

The study of sea-level changes has benefited from the dating of speleothems, such as those formed in aeolian calcarenites (also known as aeolianite or dune limestone) at times of glacial low sea-level stands (Brooke, 2001; see Section 3.5.6). The dating of speleothems helps constrain the timing of subaerial cave formation, marine flooding events and the 'forbidden zone' in which sea level has not risen (Richards *et al.*, 1994). In Bermuda, the Bahamas and Lord Howe Island (southwestern Pacific Ocean), speleothem development within calcarenites has coincided with periods of low sea level, during the last glacial cycle (Harmon *et al.*, 1978; Richards *et al.*, 1994). Following early studies in the West Indies (Li *et al.*, 1989; Lundberg *et al.*, 1990), speleothems have been used in the Mediterranean to reconstruct aspects of sea-level history over the past 250,000 years (Dutton *et al.*, 2009). The utility of speleothems for inferring palaeo-sea level is illustrated by their age–height relationships in the Bahamas for the sea-level highstand of MIS 5a. Speleothems constrain palaeo-sea-level between –15 m and –18 m for MIS 5a at about 79.7 ± 1.8 ka, in excellent agreement with the uplift-corrected sea level inferred from the Worthing Terrace of southern Barbados (Richards *et al.*, 1994; Schellman and Radtke, 2004a,b). U-series and radiocarbon dating methods have been used to establish the age of these deposits and stable isotope measurements on the same materials corroborate the palaeoclimatic inference of speleothem development during glaciations.

There may be other sedimentary settings within which the relative behaviour of uranium and its daughter isotopes provide a perspective on the age of coastal landforms. Concretions and crusts incorporate isotopes, and their suitability for dating has been examined (Short *et al.*, 1989). Ferruginous deposits on the surface of shore platforms provide rather limited age constraints on the time of formation of these surfaces, and apart from some initial attempts to incorporate such analyses with other techniques around the coast of Australia (Bryant *et al.*, 1990; Woodroffe *et al.*, 1992), have not been widely adopted. There have also been efforts to date calcrete, indurated accumulations of calcium carbonate formed by pedogenic processes, in arid and semi-arid landscapes in which annual evaporation significantly exceeds rainfall. In view of the lengthy history of their formation, the open-system behaviour and the detrital contribution of uranium within the soil-forming environment, calcrete and related deposits are particularly challenging for U-series dating, as with the radiocarbon method. Calcrete profiles represent the B-horizons of calcareous soils, although other modes of formation have been noted (Summerfield, 1991). In near-coastal environments, they have commonly formed through subaerial exposure of carbonate sediments following a relative fall in sea level (Esteban and Klappa, 1983). The onset and cessation of calcrete development may potentially be determined by analysing the central and outer portions of calcrete nodules, respectively, although the ages derived cannot be considered very accurate.

4.4 Oxygen-isotope stratigraphy

Although oxygen-isotope analysis does not represent a geochronological method per se, its application to time-series analysis of stratigraphical information is invaluable for inferring long-term palaeoclimatic and sea-level events and, accordingly, the broad principles of its application are outlined here. The importance of oxygen-isotope analysis of deep-sea and ice cores in studies of long-term climate and glacio-eustatic sea-level changes was introduced in Chapter 1 in the context of emerging ideas on Quaternary sea-level changes. In the following discussion, the application of oxygen isotopes for inferring relative changes in global sea level is examined in greater detail.

Oxygen occurs as three *stable* (non-radioactive) isotopes, ^{16}O, ^{17}O, and ^{18}O, with relative abundances of 99.76%, 0.04%, and 0.2%, respectively (Bradley, 1999).

Oxygen-isotope analysis is based on quantifying the relative abundances of ^{16}O and ^{18}O by mass spectrometry, expressed as $\delta^{18}O$ according to the following relation:

$$\delta^{18}O_{fs} = (^{18}O/^{16}O)_{fs} - (^{18}O/^{16}O)_s/(^{18}O/^{16}O)_s \times 1000 \qquad (4.10)$$

where $\delta^{18}O_{fs}$ is the $^{18}O/^{16}O$ of the fossil sample and $^{18}O/^{16}O_s$ is the $\delta^{18}O$ of a standard such as the PDB or Standard Mean Ocean Water (SMOW). The results are multiplied by 1,000 to express as parts per thousand (‰). Accordingly, lower $\delta^{18}O$ values indicate that the fossil sample is depleted in ^{18}O, and conversely, positive values indicate enrichment in ^{18}O relative to ^{16}O.

The isotopic composition of calcite secreted by organisms, measured as the ratio of ^{18}O to ^{16}O ($\delta^{18}O$) is related to temperature; there is a biological fractionation of oxygen isotopes between water and carbonate, which is a function of temperature. The remains of organisms on the ocean floor preserve a record of the former environmental conditions in the water column during their life. Emiliani (1955) pioneered the application of oxygen-isotope variations in fossil foraminifera as a surrogate of environmental change. At first it was considered that temperature alone was the prime cause of $\delta^{18}O$ variations. However, recognition that deep-ocean waters in the equatorial realm had not changed significantly in their mean temperature between glacial and interglacial times, meant that the oxygen-isotope signal had to be recording another factor besides temperature. Shackleton and Opdyke (1973) showed that up to 70% of the fractionation signal was due to ice-volume changes. As seawater evaporates and is precipitated in polar regions to form continental icesheets, the lighter ^{16}O isotope is preferentially removed and, as a consequence, the remaining water is enriched in heavier isotopes. The stable isotope signature of the frozen water is depleted in ^{18}O. The stable isotope chemistry of marine organisms, such as foraminifera, records the chemistry of the ocean waters. Foraminifera die and settle on the seafloor, and if above the *carbonate compensation depth* (CCD), that is not in deep water in which calcium carbonate undergoes dissolution, the microfossils record sequential changes in the stable isotope geochemistry of the world's oceans.

The ocean floors preserve a significantly longer and more complete record of global ice-volume changes as revealed in deep-sea cores than can be found anywhere on land. The terrestrial record is very patchily preserved because it has been reworked by successive glacial advances, whereas the record from selected deep-sea cores is one of almost continuous accumulation (Bowen, 1985; Bradley, 1999). Oxygen isotopes record the variations of ice volume, which is a function of ocean volume, and hence sea level. A 0.1‰ change in $\delta^{18}O$ is equivalent to about a 10 m change in eustatic sea level with due allowance for local variations in isotopic signals in response to water temperatures and salinity (Chappell and Shackleton, 1986).

Complications in the application of oxygen isotopes for inferring climate and sea-level changes include contrasting resistance of foraminiferal species to dissolution, the possibility that a sample includes microfossils from contrasting water depths and thus significantly different temperatures and *vital effects*. Vital effects refer to differential rates of isotopic fractionation that occur in some genera of foraminifera that are not in equilibrium with their host waters. The metabolic production of CO_2 within organisms may also enhance this contrast in isotopic ratios. Similarly, some *planktonic foraminifera* may undergo optimum calcification at a specific water depth and therefore not reflect sea-surface temperatures. Significantly, different rates of sedimentation in different sectors of ocean basins may also complicate correlation of oxygen-isotope data.

Oxygen-isotope results as revealed from studies of deep-sea cores and trapped gas within ice cores provide invaluable time series of ice-volume changes, and as a corollary, glacio-eustatic sea-level changes (see Figure 1.7). The application of oxygen isotopes in this manner is commonly referred to as *oxygen-isotope stratigraphy* (Grossman, 2012). The data represent a ratio scale form of geochemical information and geochronological methods are required to calibrate the time series. Time-series oxygen-isotope data are also available from speleothems, but the length of these records commonly does not extend over multiple glacial cycles for individual speleothems.

Many studies have shown that the structure of long-term (several hundred thousand years) oxygen-isotope time series are reproducible from all of the world's oceans, as well as in the Greenland and Antarctic ice cores (see Figure 1.7). For example, the last glacial cycle consistently shows a gradual increase in ice volume following a rapid fall in sea level at the end of the last interglacial maximum and their correlation with raised reef records has been established (Steinen *et al.*, 1973; Chappell, 1974a; Chappell and Shackleton, 1986). Similarly, all records reveal a rapid deglaciation before the onset of the Holocene (Bintanja *et al.*, 2005). The oxygen-isotope record therefore represents an essentially continuous time series of ice-volume changes, and in instances where there do not appear to be *hiatuses* (depositional breaks), provides detailed records which may also be used to infer first-order patterns of glacio-eustatic sea-level changes. The oxygen-isotope record is divided into a series of stages, termed Marine Isotope Stages (MIS; Shackleton, 2006), in which warm intervals such as interglacials and interstadials are assigned odd numbers (e.g. MIS 5e for the last interglacial maximum and MIS 1 for the current, Holocene interglacial) and cold stages such as stadials and glacials are assigned even numbers (e.g. MIS 2 for the LGM). Marine isotope stages represent events that may be correlated at a global scale and their identification in the first instance may be achieved based on the matching of spectral signatures of oxygen-isotope data. Geochronological information is required from the longer isotopic record to determine the validity of such correlations, and commonly the Brunhes–Matuyama geomagnetic reversal at 780 ka represents an important datum for correlation (Shackleton and Opdyke, 1973).

The meticulous investigations by many researchers have resulted in a mutually consistent set of ice-volume records that permit relatively reliable inferences about the nature of glacio-eustatic sea-level changes for the entire Quaternary. In general, the greatest temporal resolution of sea level and climatic events is evident since the beginning of the middle Pleistocene (780 ka) through to the present. This record is dominated by 100-ka cycles of sea-level changes. Caution should be exercised when inferring glacio-eustatic sea level as the magnitude of some isotopic signals appears to vary in different ocean basins (e.g. the magnitude of inferred sea level for the high-stand of MIS 11 at 420 ka ago). Notwithstanding this issue, some general comments can confidently be made about the nature of Quaternary sea-level changes from oxygen-isotope records. The European Project for Ice Coring in Antarctica (EPICA) Dome C record for example, reveals that MIS 5e and 11 represent the warmest interglacials during the past 800 ka and MIS 2, 12, and 16 represent the most pronounced glacials (Masson-Delmotte *et al.*, 2010). Each glacial cycle during the past 800 ka is initiated with a rapid change in ice volume within a few thousand years, followed by a sawtooth pattern of sequential decline.

The sophistication of palaeoenvironmental inferences derived from oxygen-isotope records is dependent on the type of sediment (i.e. marine sediment, ice or speleothem calcite) and the environmental parameters that control sedimentation. Bioturbation by marine organisms significantly reduces the resolution of marine-based oxygen isotope records, together with the requirement of numerous foramini-fera to attain sufficient sample mass for analysis. Thus a single sample of foraminifera (~50 individuals) may potentially contain *remanié* fossils of significantly differing age, and hence isotopic chemistry. The duration and temporal resolution of isotopic records from deep-sea cores is in part controlled by sedimentation rates within ocean basins. Ice-core records have consistently shown a higher level of event resolution than deep-sea cores, providing more detailed records of short-term ice-volume and sea-level changes such as those associated with *Dansgaard–Oeschger Events* (~500–2,500-year events involving atmospheric temperature changes of up to 15°C as identified in Greenland ice cores; Dansgaard *et al.*, 1984, 1993; Oeschger *et al.*, 1984). A more detailed discussion of relative sea-level changes related to these events, as well as their correlation with the emergent coral reef record of the Huon Peninsula, Papua New Guinea is given in Chapter 6.

4.5 Luminescence dating methods

Luminescence dating methods potentially offer the most accurate means of deter-mining the age of coastal depositional landforms, although with lower precision than radiocarbon or U-series dating. The greater accuracy is because the method deter-mines the age of depositional events (Liam and Roberts, 2006). Luminescence methods record the last exposure to light (also termed the *bleaching event*) sediment-ary particles have experienced. Luminescence dating methods represent electron capture techniques (also termed *trapped charge dating*) and are based on the principle that minerals such as quartz and feldspar act as natural dosimeters, and record the amount of exposure to ionising radiation derived from the radioactive decay of a range of isotopes of the U-series, as well as isotopes such as ^{40}K and ^{87}Rb and the influence of cosmic rays (Prescott and Hutton, 1994).

Following burial, natural environmental radiation creates free electrons which are subsequently trapped at lattice defects and electron traps which act as negative ion vacancies within minerals. The number of trapped electrons is directly proportional to the length of time that mineral grains were subjected to ionising radiation during their burial history, assuming they were adequately *zeroed* immediately before deposition (i.e. eviction of all formerly trapped free electrons). This assumption is valid for baked materials such as pottery, but varies in different depositional environments. The residual signature of any remaining free electrons, after exposure to sunlight is significantly less for optical dating than for thermoluminescence (TL) and, accordingly, represents a significant advantage in the application of optical dating (Aitken, 1998). In TL analysis, upon laboratory heating a sample, free electrons that were progressively trapped during diagenesis are evicted from electron traps. The light signal registered during heating is in the form of a glow curve (Figure 4.10), the intensity of which is directly proportional to the number of trapped electrons, and accordingly the total radiation dose the sediment grains experienced during burial (Forman *et al.*, 2000). Eviction of free electrons is achieved through the use of lasers or light-emitting diodes in the case of optical dating and generates a 'shine-down' curve of optically stimulated luminescence (OSL) or infrared-stimulated luminescence (IRSL) intensity against duration of stimulation. These forms of analysis mean the same sediment grains cannot be analysed more than once due to eviction of free electrons.

Optical methods were introduced by Huntley *et al.* (1985) to improve the sensitivity and efficiency of luminescence methods. Optical dating (also termed OSL) involves shining a beam of light (photons) emitted by lasers or light-emitting diodes on a sediment sample or near-infrared radiation (IRSL) to quantify free electrons accumulated in negative ion vacancies, rather than the incremental heating of samples as performed in TL (Price, 1994; Aitken, 1998). IRSL has been successfully applied in the dating of K-feldspar from raised beach deposits in Norway (Fuchs *et al.*, 2012).

4.5.1 Quantifying the cumulative effects of environmental radiation dose

In principle, luminescence ages are determined based on the following equation:

$$\text{Age} = \text{Equivalent dose(De)}/\text{Dose rate} \qquad (4.11)$$

The equivalent dose (De) represents the total amount of laboratory-induced radiation required to derive a luminescence signal of equivalent intensity to that of the natural sample (palaeodose) since the last bleaching event at deposition and the onset of its burial history. The De is, therefore, the best approximation of the palaeodose as

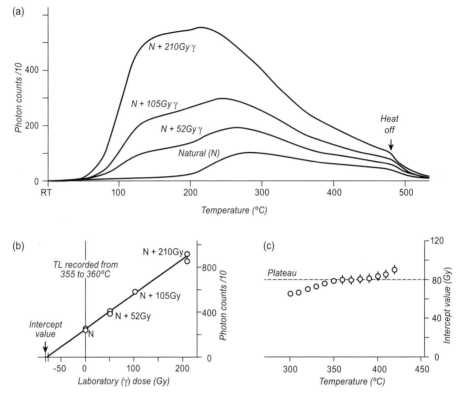

Figure 4.10 TL dating using the additive dose method for finding De of a sediment sample. (a) Glow curves as aliquots are heated; N is natural, whereas the other curves represent successive added doses of radiation from a ^{40}Co source. Deeper or more thermally stable traps are being emptied as the temperature increases; (b) dose–response at 355–360°C, showing the natural and added doses and extrapolation back to determine the intercept value (~80 grays); (c) plot of dose intercept as function of glow curve temperature showing the plateau (this example is a fluvial silt, modified from Lian and Huntley, 2001, with kind permission from Springer Science + Business Media B.V.).

only beta or gamma radiation is applied to a sample in laboratory analyses, whereas in the natural field setting, the sediment is also exposed to alpha radiation and cosmic rays. The dose rate, or rate of energy absorption from environmental radiation per annum, can be determined from dosimetry measurements undertaken on subsamples of the dated sediment either in the laboratory or during fieldwork using scintillometers.

Contrasting approaches have been used to determine De (Lian and Roberts, 2006). The *additive method* involves subjecting successive subsamples (aliquots) of the same sediment to increasing doses of radiation and documenting the resultant luminescence signals in a dose–response curve. Characteristically, this involves several subsamples for each dose increment so that an average value may be determined in view of non-systematic responses of individual mineral grains to irradiation. In addition, a subsample is analysed that is not irradiated so that the natural luminescence signal (OSL or TL) can be quantified as sampled from the sedimentary deposit. The *x*-axis intercept derived from linear regression of the luminescence intensities compared with the laboratory-induced dose rates permit the estimation of the De (Aitken, 1998; Forman *et al.*, 2000; Figure 4.10). A potential limitation of this analytical protocol is the possibility that for older samples, involving larger doses, the growth–response is non-linear (sublinearity) such that the increase in luminescence signal does not continue to increase at the same rate that characterised lower doses.

An alternative approach for quantifying De is termed the *regenerative method* and involves bleaching several subsamples of the sediment to near zero to remove as many of the trapped electrons as possible. Samples are then subjected to increasing radiation doses in the laboratory and then compared with the natural OSL or TL for the same sediment. Although this approach does not involve linear interpolation of De, it may be subject to problems of sensitivity between determination of the natural palaeodose, and the luminescence responses induced by laboratory-induced radiation of a different intensity.

A *combined additive–regenerative procedure* has also been widely adopted. In this approach a subsample of aliquots is subjected to additive-dose treatment, while a second subsample is bleached to empty the light-sensitive electron traps, and the regenerative dose–response for the same series of laboratory doses is determined (Lian and Roberts, 2006). The shape of the dose response curves should be similar, and the De is determined by the 'shift' that is necessary to overlay the additive on the regenerative (Figure 4.11). This approach has been termed the 'Australian slide' method in view of the translation of data along the dose axis and its application to a series of coastal barriers in southern Australia (Prescott *et al.*, 1993).

Estimation of palaeodose is complicated by fluctuating water content within sediment profiles, which may absorb some of the radiation that would have otherwise been absorbed by sedimentary particles. Several variables contribute to the palaeodose and include radioactive contributions from α-, β-, γ-radiation and cosmic rays, and their attenuation by fluctuating diagenetic conditions, determined by factors such as changes in the Earth's surface environment due to fluctuating water content within the vadose zone, as well as the interplay of erosional and depositional processes affecting burial depth of a sediment horizon of interest (Prescott and Hutton, 1994). The effect of water within sediment profiles is more of a problem for sediments that have experienced significant fluctuations in moisture content over prolonged periods (e.g. continental shelf sediments affected by sea-level hemicycles). These issues are potentially less difficult to quantify in arid environments or permanently saturated sediments.

A significant advance in luminescence dating is the analysis of single grains (Murray and Roberts, 1997). Single grain dating enables an assessment of sediment

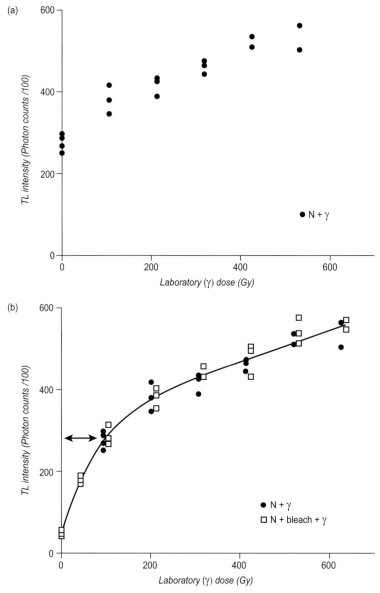

Figure 4.11 The Australian slide method of combined additive- and regenerative-dose technique TL dating. (a) A series of aliquots are given various laboratory irradiation doses to determine their dose–response (additive, $N + \gamma$); (b) other aliquots are 'bleached' to empty their light-sensitive electron traps (exposed to light). These are then subjected to laboratory irradiation to regenerate the dose–response (regenerative, $N + \text{bleach} + \gamma$). The set of additive results is then shifted along the dose axis until it coincides with the regenerative results and the shift is an estimate of the palaeodose with allowance for incomplete laboratory bleaching (deep traps). (This example is for sand-sized quartz extracted from a relict dune at Policeman's Point in South Australia with a TL age of ~95 ka, redrawn from Lian and Roberts, 2006, with permission from Elsevier.)

dynamics and an improved understanding of whether sediment grain populations were adequately bleached before deposition. Radial plots are commonly used to depict single grain analyses as well as results for aliquots of multiple grains (Figure 4.12). Such an approach assists in evaluating sediment reworking and post-depositional pedogenic processes as well as the effects of *bioturbation* (sediment disturbance by organisms). The development of radial plots has also significantly improved understanding of the statistical attributes of sediment grain populations and led to development of different models for age calculations (e.g. central age model, minimum age model; Lian and Roberts, 2006).

Sediment samples for luminescence dating need to be screened from natural light during and after collection to avoid the release of trapped electrons from the more light-sensitive traps. Sampling at night, or by driving opaque cylinders into the sediment and ensuring they are sealed to minimise exposure to light overcome this problem. In the laboratory, sediment needs to be prepared under filtered light (dim red or orange lighting). Aliquots of suitable quartz or feldspar of an appropriate particle size (commonly 90–125 µm as this is generally abundant) are commonly analysed.

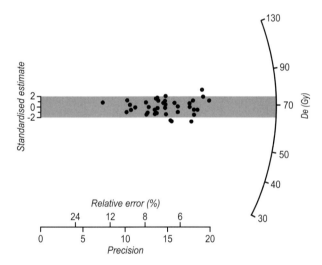

Figure 4.12 Paleodose (De) distribution for a coastal aeolianite sample from East Beach, East London, South Africa (sample NHN2 of Jacobs and Roberts, 2009). Each point represents a single aliquot for which the De can be read by extending a line from the 'standardised estimate' axis on the left-hand side, through the point of interest, to intersect the radial axis on the right; the point of intersection is the De. The uncertainty on this estimate can be read by extending a line vertically from the data point to intersect the horizontal axis running along the bottom of the plot. This axis shows the relative standard error in % (i.e. the standard error divided by the De estimate, multiplied by 100) and its reciprocal (the precision). In such plots, the most precise De estimates fall farthest to the right. The shaded band is centred on the weighted mean De value of the sample (figure courtesy of Zenobia Jacobs).

4.5.2 *Age range of luminescence methods*

It is difficult to generalise about the age range of luminescence methods in view of widely variable dose rates experienced by sediments in different environments. The age limit is determined by the point at which additional environmental radiation doses do not result in the retention of additional 'free' electrons or the discernible increase in luminescence signal, a point termed *saturation*. Numerous workers have suggested that TL dating is applicable for only the past 400 ka and in exceptional situations may be extended past this limit (Aitken, 1998). Huntley *et al.* (1993, 1994), however, reported TL ages up to 800 ka for quartz sand derived from Pleistocene aeolianite (West Naracoorte Range) in southern Australia. In this region the mixed quartz–skeletal carbonate sediments show a consistently low dose of 0.5 Gy/ka (Huntley *et al.*, 1993, 1994; Murray-Wallace *et al.*, 2002). The potential age range of conventional analytical protocols in OSL dating is approximately 150–200 ka (Wagner, 1998), but significant variations to the age range are site-specific and directly a function of the dose rate. The upper limit of luminescence dating has been extended by development of thermally transferred optically stimulated luminescence (TT-OSL), also termed recuperated OSL. TT-OSL saturates at far higher radiation doses than does the fast component of OSL signal (at 325°C) enabling determination of ages back to 1 Ma in some circumstances (Duller and Wintle, 2012). Jacobs *et al.* (2011) used TT-OSL to determine earlier interglacial ages for coastal sediments along the Cape coast of South Africa than would have been possible with conventional means, identifying sediments from MIS 11 (about 400 ka). OSL has the potential to date geologically recent sedimentary events. Murray-Wallace *et al.* (2002) reported an age of 51 ± 5 years on a recently formed coastal foredune.

4.5.3 *Anomalous fading and partial bleaching*

The integrity of luminescence ages is in part based on the ability of electron traps and lattice defects to retain free electrons over geologically significant periods of several Ma. Some minerals, however, particularly feldspar and zircon, experience the progressive loss of electrons from these centres, a phenomenon termed *anomalous fading* (Aitken, 1998). The result of anomalous fading is to yield apparently younger numerical ages. Anomalous fading involves the leakage of electrons from traps without having entered the conduction band, and occurs at a rate that is faster than otherwise predicted based on the depth, and therefore the apparent stability of electron traps (Aitken, 1998).

Partial bleaching refers to instances during sedimentation where sediment grains were not exposed to sunlight sufficiently long for all the formerly trapped electrons to have been evicted. Given the higher sensitivity of OSL dating, partial bleaching is less of a problem than for conventional TL. The likelihood that sediments have been adequately bleached also varies with sedimentary environment. Sediments that have not been subaerially exposed or were deposited in shallow subtidal or fluvial environments are less likely to have been adequately bleached before deposition. In contrast, sediments within aeolian environments are more likely to have been well-bleached and accordingly have been commonly analysed in studies of sea-level changes (Huntley *et al.*, 1993, 1994; Murray-Wallace *et al.*, 1999; Bateman *et al.*, 2004; Jacobs, 2008; Jacobs and Roberts, 2009; Carr *et al.*, 2010). Sediments that have only been partially bleached at the time of deposition will result in older apparent ages.

4.6 Electron spin resonance dating

Electron spin resonance (ESR) dating is a radiogenic method that, as with luminescence methods, quantifies the number of electrons captured in electron traps and lattice defects within minerals due to radiation damage, in particular, the detection of paramagnetic centres and radicals which are created by irradiation in the natural environment (α-, β-, γ-particles and cosmic rays). The number of paramagnetic centres (free electrons in the conduction band) is proportional to the strength of the radioactive field (dose rate) and the duration of exposure to the radiation sources (i.e. time). In the context of Quaternary sea-level changes, ESR may be applied to the dating of fossil molluscs, corals, and foraminifera. It has also been applied to the dating of teeth, bone, flints, travertine, and terrigenous clastic sediments (Blackwell, 1995; Wagner, 1998).

ESR dating is based on the principle that the self-rotation of electrons, which gives rise to a magnetic moment, can be measured within mineral matrices such as aragonitic fossil molluscs. During analysis, when a fossil sample is exposed to a strong magnetic field, the free electrons trapped within paramagnetic centres experience an adjustment such that their spin axes undergo a precession movement around the direction of the applied magnetic field. The precession frequency is proportional to the strength of the applied magnetic field and the g-value (Figure 4.13). The g-value is a dimensionless number that differs according to mineral phase and represents a ratio between magnetic moment and angular momentum. The g-value ranges around the value for a free electron ($g = 2.00232$). During analysis, a microwave is applied to the sample at right angles to the magnetic field and induces ESR, at which point the electron spinning direction will reverse. The magnitude of the resonance is directly proportional to the number of trapped electrons and hence the age of the fossil. In contrast to luminescence-based methods of dating, electrons are not evicted from minerals during ESR analysis, permitting replicate analyses on the same samples.

Given that ESR is an electron-capture method, many of the parameters that influence the uncertainty in ESR-derived ages are similar to those considered in luminescence dating. This includes environmental factors influencing accumulated radiation dose during burial history, that is contributions from U, Th, K, and Rb as well as cosmic rays, which are affected by moisture in the sediments. These and other factors have a variable influence on the overall radiation dose at different field sites,

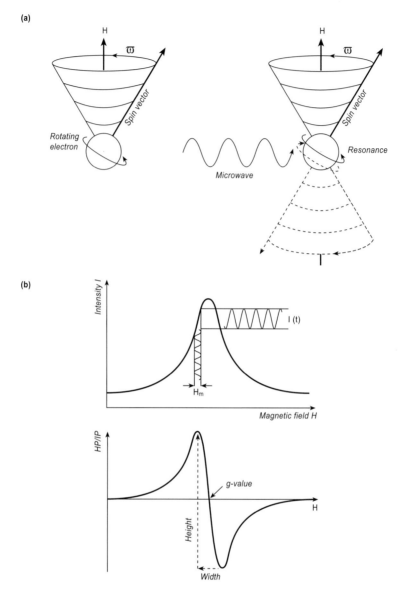

Figure 4.13 Diagrammatic representation of ESR and the application of g-values in geochronology. (a) A free electron in a negative ion vacancy is subjected to a strong magnetic field, and the spin axis of the electron defines a precession encircling the direction of the applied magnetic field defined by vector H. The contemporaneous application of a microwave at right angles to the magnetic field induces resonance behaviour of the electron, resulting in a change in its direction of spin. (b) The ESR spectrum and defined g-value are derived from a comparison of the microwave absorption signal compared with magnetic field intensity (H), applied in the form of an oscillating magnetic field (H_m) (after Wagner, 1998, with kind permission of Springer Science + Business Media).

so it is important that detailed dosimetry studies are undertaken in the laboratory and in the field.

ESR is a useful method for dating Pleistocene corals and molluscs, but has not been as widely applied in studies of Quaternary sea-level changes as the preceding methods (Grün *et al.*, 1992). The method has recently been applied in the dating of quartz grains from interglacial coastal barriers (Rittner, 2013). The potential age range extends from several hundred years before present to beyond 1 Ma, although few ages have been reported towards this upper limit because diagenetic processes affect ESR signal growth (Blackwell, 1995). Numerical ages on Pleistocene fossil corals from Barbados, with 2σ uncertainties between 5% and 9%, reveal evidence for sea-level highstands at the sub-Milankovitch scale, indicating the potential age resolution and utility of the method in Quaternary sea-level investigations. Based on meticulous field mapping, Schellman and Radtke (2004a,b) identified discrete episodes of reef growth as defined by the emergent reef morphostratigraphies. Three distinct events were identified in the last interglacial MIS 5e reef sequences in the Maxwell (118 ± 9 ka) and Rendezvous Hill (128 ± 11 ka and 132 ± 13 ka) Terraces, and three distinct events were also identified within the terraces correlating with MIS 7, 9, and 11. They identified two events, dated at 74 ± 5 and 85 ± 7 ka, respectively, in the late Pleistocene interstadial MIS 5a (Worthing Terrace) as originally defined by Bender *et al.* (1979).

ESR-based chronostratigraphies require the analysis of multiple replicates from individual sample sites, numerous sample sites from the same emergent shoreline succession and the use of only the oldest ESR-derived ages from the successions. The latter point, as revealed in comparisons of U-series and ESR ages, is based on a tendency to display abrupt increases in radiation sensitivity as seen in the use of the dating signal $g = 2.0006$, observed particularly in corals of middle Pleistocene age (Schellman and Radtke, 2004a,b). Although the physical basis for the change in sensitivity in the dose–response curves using the additive method is not clearly understood, it is manifested by inflexion points in dose–response curves such that lower than predicted ESR intensities correspond with a given value of additive radiation dose to a sample. This in turn means that a single exponential saturation function cannot be used to adequately define the growth–response curve.

4.7 Amino acid racemisation dating

Amino acid racemisation (AAR) has been widely applied in studies of Pleistocene stratigraphy, sea-level changes, and coastal evolution. Applications have included relative and numerical age assessments of geological and archaeological deposits, stratigraphical correlation of widely separated (> 1 km) deposits, palaeotemperature studies, identification of reworked fossils, and the screening of samples before undertaking radiocarbon or U-series dating. The method has also been applied in dating materials younger than 1,000 years (Goodfriend *et al.*, 1992, 1995; Sloss *et al.*, 2007), a problematic interval for radiocarbon, and has been applied to 'whole-rock' dating of sedimentary successions (Hearty, 1998; Murray-Wallace *et al.*, 2001).

The potential of the time-dependent AAR reaction as a method of dating was first reported by Hare and Abelson (1968), who suggested that the extent of amino acid racemisation (epimerisation) may be used to assess the age of materials within, and beyond, the range of radiocarbon dating. They noted greater concentrations of amino acids in the D-configuration with increasing fossil age in specimens of the marine bivalve mollusc *Mercenaria* sp., but negligible racemisation in modern specimens. The method has been refined and reviewed extensively (Schroeder and Bada, 1976; Davies and Treloar, 1977; Williams and Smith, 1977; Wehmiller, 1982, 1984; Rutter and Blackwell, 1995; Wehmiller and Miller, 2000), including discussions of sources of uncertainty, precision, and age-resolving power (Miller and Hare, 1980; McCoy, 1987; Miller and Brigham-Grette, 1989; Murray-Wallace, 1995, 2000).

4.7.1 The amino acid racemisation reaction

Amino acids are present in all living organisms as polymers linked together through covalent peptide bonds to form enzymes and proteins (Schroeder and Bada, 1976). The α-amino carboxylic acids are non-volatile, organic molecules of low molecular mass (about 100–200 Daltons) within proteins. Amino acids are characterised by a centre of asymmetry at the α-carbon position, to which is attached an amino group (-NH$_2$), a carboxyl group (-COOH), a hydrogen atom (-H), and a hydrocarbon group (-R). Structurally, all amino acids except glycine contain at least one asymmetric central tetrahedral carbon atom.

Amino acids and their precursor peptides undergo numerous diagenetic reactions that include oxidation, decarboxylation, deamination, hydrolysis, and racemisation (epimerisation) (Figure 4.14). In biologically active tissue during life, amino acids are exclusively left-handed molecules (L-amino acid: *levorotatory*), due to enzymic reactions (Williams and Smith, 1977). Following the cessation of protein formation or the death of an organism (the two events are not necessarily contemporaneous, as reflected in teeth and eye lenses where racemisation is known to occur during life and in the older growth bands of long-lived molluscs such as *Tridacna*), the enzymic reactions that formerly maintained the disequilibrium condition cease (i.e. exclusively L-amino acids corresponding with a D/L = 0). Amino acids then slowly interconvert from a left-handed molecule to a right-handed counterpart (D-amino acid: *dextrorotatory*). This process is termed amino acid racemisation for amino acids with one chiral carbon centre (i.e. enantiomers such as aspartic acid) and epimerisation for amino acids with two carbon centres, such as isoleucine. The interconversion of L- to D-amino acids continues until equilibrium (D/L = 1 for enantiomers and approximately 1.3 for diastereoisomers; Williams and Smith, 1977). Epimerisation results in a chemically, not just optically, different molecule, and is therefore readily distinguished on conventional amino acid analysers (Rutter and Blackwell, 1995).

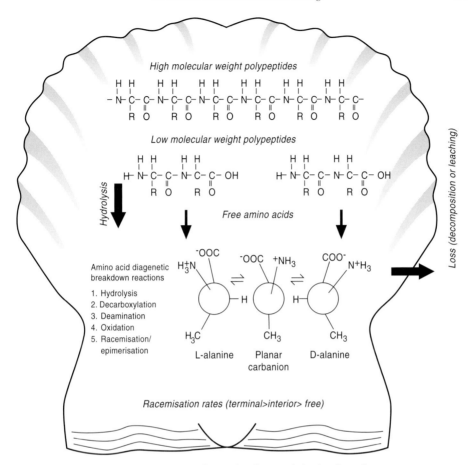

Figure 4.14 Schematic representation of protein diagenesis in fossil molluscs.

Advantages of AAR dating include the wide range of organic material that can be analysed, the small sample requirement (μg of calcium carbonate before sample pre-treatment permitting the analysis of single foraminifera or ostracods; Hearty *et al.*, 2004), speed of analysis, and the potential to derive ages beyond the range of U-series dating (> 500 ka). The majority of studies have, accordingly, involved dating marine and terrestrial molluscs (Wehmiller, 1982, 1993; Miller and Mangerud, 1985; Murray-Wallace, 1995, 2000; Bowen *et al.*, 1998; Brooke *et al.*, 2003a), because the residual protein is well protected from external environmental influences and unambiguously related to the dated event, despite uncertainties associated with diagenetic temperature changes.

Numerical ages are less commonly derived by AAR because of the complex nature of racemisation kinetics in natural systems, which for most materials do not conform to extended-linear first-order kinetics from the onset of the reaction to the attainment of equilibrium (Wehmiller, 1984; Clarke and Murray-Wallace, 2006). However, numerical ages have been derived based on a model of apparent parabolic kinetics (Mitterer and Kriausakul, 1989), and power curve fitting routines (Clarke and Murray-Wallace, 2006). The parabolic model involves plotting the extent of racemisation against the square root of age, and requires age calibration by another geochronological method. Ages are calculated using the equation:

$$t = [(D/L)_s / M_c]^2 \qquad\qquad (4.12)$$

where t is fossil age, $(D/L)_s$ is the extent of racemisation (epimerisation) in a fossil of unknown age and Mc is the slope of the line defined as $[= (D/L)/t^{1/2}]$. Fossils of last interglacial age (~125 ka) represent a particularly useful calibration as they will have experienced temperature changes throughout a full glacial cycle. This is particularly important for the dating of materials from older interglacial deposits as they are likely to have been subjected to a similar range of long-term temperature changes as indicated by oxygen-isotope records (Bassinot *et al.*, 1994; Masson-Delmotte *et al.*, 2010). Numerical ages determined using the apparent parabolic model in several geological studies appear consistent with independent assessments by other geochronological methods (Hearty, 1998; Murray-Wallace *et al.*, 2001) or geomorphological or stratigraphical evidence (Wehmiller *et al.*, 1995).

4.7.2 Environmental factors that influence racemisation

Racemisation in natural systems may be influenced by (1) the diagenetic temperature history; (2) fossil genus and matrix; (3) moisture regime; (4) hydrolysis state of peptide residues; and (5) pH and clay mineral catalysis (Pillans, 1982). Both the *rate* and *extent* of AAR may vary. For example, water within the micro-environment of protein residues accelerates racemisation (Bada, 1985), but can also result in diffusion of protein residues from poorly preserved fossils. These may then yield younger apparent ages, showing poor reproducibility because they are degraded or recrystallised. The racemisation reaction is also sensitive to the matrix that hosts the residual protein, as well as the sedimentary environment in which the fossil is buried. Although pH and clay mineral catalysis are known to influence racemisation, molluscan carbonate generally protects protein residues from these influences.

The most significant parameter influencing AAR in natural systems is the diagenetic temperature history, the integrated effect of all temperatures a fossil has experienced since the onset of racemisation to the time of analysis (Williams and Smith, 1977; Rutter and Blackwell, 1995). Temperature changes have an exponential influence on racemisation rate, such that for every 1°C increase in temperature, the racemisation rate will increase by approximately 18%, approximately doubling for every 5°C increase in temperature (McCoy, 1987). Aspartic acid, for example, will undergo 10 times the extent of racemisation in 1,000 years at 25°C than in 2,000 years at 12.5°C (Davies and Treloar, 1977).

The diagenetic temperature history directly determines the potential age range of the AAR method, which will vary according to geographical location. Results from Arctic environments with current mean annual air temperatures (CMATs) of –12°C to –7°C, indicate that racemic equilibrium is generally obtained within 10 Ma from the onset of racemisation (Williams and Smith, 1977). Miller and Mangerud (1985) have suggested that 20 Ma or longer may be required for isoleucine to attain equilibrium values for Arctic sites (CMAT below –10°C). In contrast, equilibrium is obtained significantly faster in temperate and tropical climates. In mid-latitude regions (e.g. CMAT ~10°C), it takes approximately 2 Ma for isoleucine to attain equilibrium (Miller and Brigham-Grette, 1989). Bowen *et al.* (1998) report near-equilibrium values for isoleucine (D/L of 1.22 ± 0.08) in specimens of the fossil mollusc *Tawera spissa* from the Butlers Shell Conglomerate, Castlecliff Beach, Wanganui Basin, New Zealand (CMAT 13.5°C) independently assigned to MIS 31 (1.06–1.08 Ma) of the chronology of Bassinot *et al.* (1994). In tropical environments such as Papua New Guinea (CMAT 28°C), near-equilibrium values for isoleucine have been reported for marine molluscs of last interglacial age (125 ka; Hearty and Aharon, 1988).

The *effective diagenetic temperature* (EDT) represents the integrated kinetic effect of all temperatures a fossil has experienced during diagenesis. The EDT will always be higher than the average temperature a fossil has experienced during diagenesis, because higher summer temperatures disproportionately influence the extent of racemisation. Collecting samples from deeply buried contexts (i.e. ≥ 1 m) will significantly reduce the influence of seasonal temperature extremes, and is more likely to reflect temperatures associated with long-term climate change. If a sample has been exposed subaerially for a lengthy part of its diagenesis, high summer temperatures will disproportionately influence the rate of racemisation, and yield a higher D/L value than if the material had been continuously buried. Rates of racemisation have also been found to vary with fossil genus and within different materials (Miller and Brigham-Grette, 1989) and are controlled by the original protein configuration and the ease with which hydrolysis reactions can break down proteins.

4.7.3 Sources of uncertainty in AAR dating

Numerous studies have examined potential sources of uncertainty in AAR geochronology (Miller and Hare, 1980; Brigham, 1983; McCoy, 1987; Miller and Brigham-Grette, 1989). Much uncertainty in amino acid D/L values relates to sample preparation and analysis, but there are also differences between samples and differences within, and between, deposits (Figure 4.15). Analytical precision related to within-species intra-fossil D/L variations can vary by $\leq 8\%$, and within-species inter-fossil D/L variations can vary by $\leq 12\%$, whereas sample pre-treatment, preparation, and chromatography may only contribute $\leq 3\%$ uncertainty (Murray-Wallace, 1995). D/L variation within a deposit and between-site differences at the regional geographical level are potentially more difficult to quantify, but can be significantly reduced by

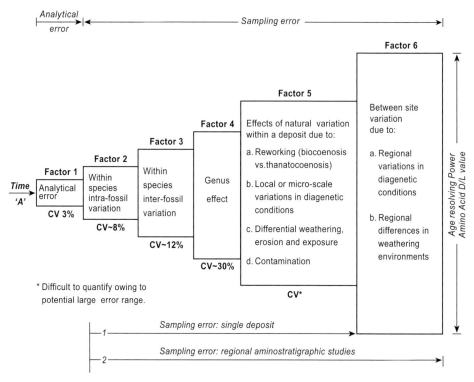

Figure 4.15 Sources of uncertainty in AAR dating. See text for discussion.

undertaking careful field sampling and systematic laboratory work. The most reliable results come from routinely analysing the same anatomical region of a single genus (e.g. the hinge region in bivalve molluscs or the columella or operculum of gastropods) from deeply buried (i.e. > 1 m) and well-preserved fossils. Ideally, a minimum of five different fossil specimens should be analysed from each site. The relationship between the number of fossils analysed and the precision with which a stratigraphical unit may be characterised has been summarised mathematically by Miller and Hare (1980) as:

$$N = [(Z_1 - 1/2\alpha)\sigma/d]^2 \qquad (4.13)$$

where N is the number of sample replicates required for analysis at a given level of precision, $(Z_1 - 1/2\alpha)$, which at the 2σ-level (95% confidence level) is approximately equal to 2, σ is the standard deviation of the population, and d is the desired precision of the mean.

Potential sources of contamination include: (1) removal or addition of amino acids by groundwater, (2) diagenetic formation of amino acids, (3) additions from bacteria and other microorganisms, and (4) human-induced contamination during sample collection and analysis (Pillans, 1982). Recent attempts to improve the accuracy of AAR dating have focused on analysing the intracrystalline fraction of amino acids which are less susceptible to contamination or diffusive loss from fossils (Penkman *et al.*, 2008).

Wehmiller (1984) proposed several acceptability criteria for assessing the suitability of fossil taxa for AAR dating. These include: (1) reproducible analytical results on replicate fossils of the same genus from a single deposit, (2) good mineralogical and structural preservation of the fossil material, (3) the potential of each genus to demonstrate increasing extents of racemisation with age, (4) the ability of each genus to achieve, and to retain racemic equilibrium in very old samples (e.g. shells of early Pleistocene age), (5) consistent relative rates of racemisation of different amino acids in replicate fossil samples, and (6) progressively increasing D/L values towards lower latitudes for coeval fossils. More recently, the identification of outliers in amino acid data sets and their significance in numerical age assessments has been examined in the context of age calibration curves and data screening (Kosnik *et al.*, 2008; Kosnik and Kaufman, 2008).

4.7.4 Application of AAR to dating coastal successions

The most common application of AAR has been in the aminostratigraphical (chronostratigraphical) subdivision of sedimentary successions, particularly in palaeo-sea-level investigations (e.g. Miller *et al.*, 1979; Kennedy *et al.*, 1982; Miller and Mangerud, 1985; Wehmiller *et al.*, 1988; Hsu *et al.*, 1989; Murray-Wallace, 1995; Bowen *et al.*, 1998; Hearty, 1998; Murray-Wallace *et al.*, 2010). *Aminostratigraphy* involves the classification of strata into chronostratigraphical units or *aminozones* that correspond with intervals of time, termed geochronological units (Salvador, 1994). Litho-, bio-, and morphostratigraphical evidence is generally used to support these chronostratigraphically defined units, as well as calibration by other geochronological methods. An aminozone refers to a single time-stratigraphical unit representing a well-defined interval of time (e.g. single marine oxygen isotope stages) and is delineated by the clustering of amino acid D/L values on replicate fossils from a single, mappable lithostratigraphical unit (Miller and Hare, 1980).

Kennedy *et al.* (1982) first applied racemisation data to the regional correlation of coastal sedimentary successions over distances of hundreds of kilometres, where regional temperature differences yield contrasting extents of racemisation for fossils of equivalent age. They plotted the extent of leucine racemisation in fossil molluscs against latitude in a transect from the State of Washington to northern California. This empirical approach overcomes many of the assumptions associated with kinetic model ages. Implicit is the assumption that diagenetic temperatures will always have been higher in lower latitudes, and that differences in long-term temperature histories would be of comparable magnitude to present day temperature differences between sample sites (Wehmiller, 1984). The approach to plotting D/L values against latitude (or now more commonly, CMAT), has been used in many field settings including the United States Atlantic Coastal Plain (Wehmiller, 1993), the Mediterranean (Hearty *et al.*, 1986), northern Europe (Miller and Mangerud, 1985), and southern Australia (Murray-Wallace, 2000). Although latitudinal temperature gradients generally represent a linear function, the D/L values for fossils of the same genus and of a common age should define an exponential trend with latitude in view of the exponential effect of contrasting temperatures on diagenetic racemisation (Figure 4.16).

A particularly effective application of AAR has been in taphonomic studies to establish if fossils have been reworked (Murray-Wallace and Belperio, 1994; Wehmiller *et al.*, 1995). This is important in many field situations as reworked fossils may not be immediately evident on the basis of their preservation. In a comprehensive study, Wehmiller *et al.* (1995) examined the extent of isoleucine epimerisation in molluscs (predominantly *Mercenaria*) from 21 beaches between New Jersey and Florida to determine the frequency of reworking of marine shells along the coastal plain and quantify residence time for shells in these littoral environments. AAR is unable to resolve time within Pleistocene interglacials but can detect reworking of shells from older interglacial deposits into younger sedimentary successions.

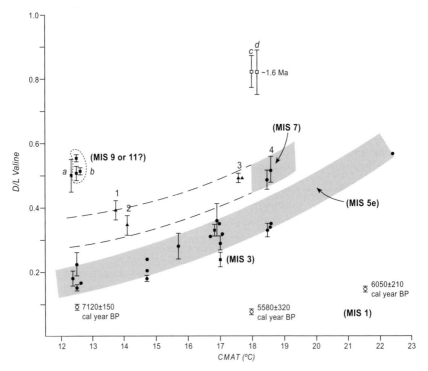

Figure 4.16 Plot of the extent of valine racemisation in Quaternary molluscs from southern Australia against CMAT. The exponential effect of temperature on diagenetic racemisation is evident in the plot of data for the last interglacial MIS 5e coastal deposits.

Wehmiller *et al.* (1995) showed a positive correlation between the blackening of shells and their older inferred ages and found that approximately 60% of the shells collected from modern beach surfaces of the United States Atlantic Coastal Plain were derived from a Pleistocene source. AAR has also been used to evaluate degree of reworking within glacial-age, lowstand continental shelf sediments (Murray-Wallace *et al.*, 1996b).

AAR has been applied in the dating of sediments (whole-rock method), particularly where macro-fossils and other mineralogical constituents that would otherwise be suitable for dating by other methods are absent (Kindler and Hearty, 1997; Hearty *et al.*, 1992; Hearty, 1998; Murray-Wallace *et al.*, 2000). This is the most complicated application of AAR, as a particularly detailed understanding is required of sedimentation dynamics. Coarse-grained (1–2 mm) skeletal carbonate sand (bioclasts) from *aeolianite* (coastal aeolian dunes) is most suited to the whole-rock method (Hearty, 1998; Murray-Wallace *et al.*, 2000), as pre-treatment strategies are easier with larger bioclasts.

A further application of AAR has been to derive ages within the past few hundred years, which has proved difficult with radiocarbon due to complexities in conversion of radiocarbon to calendar years. The fast-racemising amino acid, aspartic acid, has been used to assign ages to young materials (Goodfriend *et al.*, 1992, 1995). In a pilot study, Goodfriend *et al.* (1992) examined the extent of aspartic acid racemisation in a 350-year-old living coral, *Porites australiensis*, from the Great Barrier Reef, Queensland. They showed that aspartic acid racemisation in *Porites* is the fastest racemising of any amino acid in any biogenic carbonate, and that the initial rate of racemisation was equivalent to 0.6% per year but slowed to 0.04% per year in older growth bands of the corals. Aspartic acid D/L values ranged from 0.05 for AD1985 to 0.26 for AD1632. The mean annual sea-surface temperature for the sample site is 24.3°C. Goodfriend *et al.* (1992) suggested that extent of aspartic acid racemisation could be used as a method to assess whether corals have missing bands. This approach has potential for application to other materials and to offer the opportunity to verify recent ages where other techniques are ambiguous.

4.8 Cosmogenic dating

Cosmogenic dating involves measurement of nuclides that have accumulated in rock surfaces as a result of interactions of cosmic rays with particular minerals. Concentrations of these nuclides provide quantitative estimates of the time the surface has been exposed, or the rate at which it has been eroded. Where a rock surface has been formed, or exposed, by an event, accumulation of nuclides provides the basis for estimating the time at which the event occurred. If the rock surface undergoes surface lowering, then the nuclide profile provides insights into the rate of erosion. Depending on continuity of surface exposure and rate of denudation, cosmogenic dating can provide information on age or rates of process operation from thousands to millions of years (Cockburn and Summerfield, 2004).

The Earth is continually bombarded by high-energy cosmic rays originating from supernova explosions in our galaxy. These primary rays (fast protons and alpha particles) interact in the atmosphere to create secondary cosmic rays, particularly neutrons, but also muons. These particles can penetrate about 40 cm, but sometimes a metre or more, into rock or sediment and can result in nuclear interactions with certain target materials, which produce long-lived radionuclides (Granger, 2007). Attenuation with depth is a function of rock density. Several cosmogenically derived isotopes have been employed in dating, including ^{3}He, ^{21}Ne, ^{36}Cl, ^{10}Be, ^{26}Al, ^{41}Ca, and ^{53}Mn (Catto, 1995). Whereas the stable isotopes ^{3}He and ^{21}Ne can be measured using noble gas mass spectrometry (NGMS), diffusion out of quartz limits their application to many rocks. Concentrations of the radionuclides, which are not stable but undergo decay, are very low, perhaps a few thousand atoms per gram, requiring complex sample preparation and measurement by accelerator mass spectrometery (AMS).

These isotopes form in rocks as a result of cosmic rays and their rate of production on the Earth's surface varies with latitude due to the variable influence of the Earth's geomagnetic field in modulating incident cosmic ray flux, and the altitude of the sample site, as well as the mineralogical composition of the rocks (Stone, 2000; Gosse and Phillips, 2001). It is also important to correct for shielding of the surface materials by regolith and as a function of the surrounding topography which means that the site may have experienced reduced cosmic ray flux. This is generally corrected by measuring vertical angles to the horizon and incorporating this into geometric shielding models.

For a stable nuclide, such as ^{3}He, concentration increases linearly through time, but for radionuclides, the concentration is also influenced by decay, and can be expressed by the following equation:

$$N = P/\lambda(1 - e^{-\lambda T}) \tag{4.14}$$

where N is concentration of the radionuclide (atoms/g); P is production rate of the nuclide (atoms/g/year); λ is the decay constant of the nuclide; and T is the exposure time. More complex formulae are required where pre-exposure concentrations of nuclides occurred, or where exposure has been variable over time.

Glacial retreat has been widely constrained using cosmogenic dating, since the first attempts to use low-level decay counting to measure ^{36}Cl in samples from the Rocky Mountains (Davis and Schaeffer, 1955). The exposure of fresh rock surfaces previously buried beneath ice enables the calculation of timing of that exposure on the basis of concentration of radionuclides since exposure occurred. Exposure ages can be estimated using the equation above if the rate of production of nuclides is known, the surface has remained exposed since the event, and the surface sampled is that which was originally exposed with no subsequent denudation.

However, more often than deriving the exposure age of an event, the nuclide concentrations reflect incremental rates of denudation, which is examined in terms of changes in nuclide production and retention with depth. Eventually a rock surface exposed for long enough will become saturated, and will come into secular equilibrium, as the rate of production and that of radioactive decay reach a constant level, which occurs after about 4–5 half-lives for each isotope. Rates of denudation can be modelled assuming a 'steady-state erosion model'. If saturation has not been reached, this will be an overestimate of the rate, and if denudation has been episodic, or the history has been complex, then further complications will arise.

In some cases the ratio of two radionuclides provides a test of the denudation modelling. Most commonly this has involved ^{10}Be and ^{26}Al, with half-lives of 1.5 and 0.72 Ma, respectively. In quartz, ^{26}Al is produced approximately six times faster than ^{10}Be, regardless of the absolute production rate, but ^{26}Al decays more than twice as fast as ^{10}Be. This implies that the ^{26}Al : ^{10}Be ratio will be 6 : 1; however, if the surface remains exposed this decreases the ^{26}Al : ^{10}Be ratio until secular equilibrium is produced, and steady-state erosion can be calculated for denudation rates of > 10 m/Ma (Nishiizumi *et al.*, 1989).

Cosmogenic nuclides have considerable potential for the dating of past shorelines, being particularly appropriate where rocks have been exposed for the first time through marine planation, and were previously shielded from cosmic rays. They have been used to date palaeo-lake shorelines, for example, those of Lake Bonneville in the western US (Cerling and Craig, 1994), Lake Mannix in the deserts of California (Meek, 1989), and Tibetan glacial lakes (Kong *et al.*, 2007). There have been some initial results from marine terraces; for example, those at Punta Caballos in southern Peru, where cobbles were collected from a desert pavement, but the ages derived were ambiguous (Nishiizumi *et al.*, 1993). In terms of former sea levels, the most successful use of cosmogenic dating was the determination of an age for the Main Rock Platform in western Scotland (see Section 7.7.1), which is a prominent terrace along the eastern margin of the Firth of Lorn (Stone *et al.*, 1996). In this case, ^{36}Cl, the product of spallation of ^{39}K and ^{40}Ca in calcite in calcareous schists, was used. In the context of ^{36}Cl analysis, the duration of subaerial exposure is defined by the relation;

$$t = -1/\lambda_{36} \ln\left[1 - \lambda_{36}\, N_{Cl}(R_{36} - R_o)/P\right] \qquad (4.15)$$

where t is time (duration of subaerial exposure), λ_{36} is the decay constant for ^{36}Cl (2.303×10^{-6} years), R_{36} is the atomic ratio of ^{36}Cl to stable ^{35}Cl, R_o refers to the ratio of ^{36}Cl/^{35}Cl resulting from U- and Th-derived neutrons, N_{Cl} refers to the number of Cl atoms per mass of rock, and P is the production rate of ^{36}Cl. The cosmogenically formed isotopes can be measured accurately by AMS, permitting detection of one atom of ^{36}Cl in 10^{15} atoms of ^{35}Cl (Phillips, 1995). The times required for the observed ^{36}Cl concentrations to be produced, assuming continuous exposure, indicate that the platform, up to ~100 m wide in places, originated between 8.9 and 10.4 ka. These exposure ages imply a young age as opposed to a Pleistocene age for the feature. They fall slightly short of the Loch Lomond stadial, considered equivalent to the Younger Dryas, during which it was suggested the platform was cut (Sissons, 1974), but they support the contention that it records a late glacial sea-level event rather than an older period of erosion, in which case it would have been subject to ice-scouring during the last glacial advance.

There have been relatively few other attempts to provide age control of rocky coast erosion using cosmogenic dating, despite the apparent potential to date either exposed platform surfaces or boulders on these (Naylor *et al.*, 2010). Shore platforms are subject to complex shielding issues as well as immersion by seawater, particularly if there have been periods of higher sea level during which they may have been submerged. It is also necessary that rocks of suitable lithology are sampled. Minimum ages of 1–2 Ma were obtained using ^{10}Be, ^{26}Al, and ^{21}Ne for an uplifted shore platform that is up to 4 km wide in the foothills of the Cantabrian Mountains in northern Spain (Alvarez-Marrión *et al.*, 2007); on the basis of these ages and the likely shielding and past sediment cover, this was considered Pliocene in age. A range of younger ages have recently been derived for shore platforms in Korea (Choi *et al.*, 2012).

4.9 Other dating techniques

There is a range of other techniques which have been used for dating, with some such as potassium/argon or fission-track dating appropriate, and extending into, the pre-Quaternary, and others such as ^{210}Pb dating being used over a shorter, historical period. Many fall beyond the scope of this chapter, but a short summary of several key techniques follows.

4.9.1 Event markers

Sometimes the record of a particular event provides a chronological marker. This is clearest where there is physical, chemical, or biological alteration or deposition that is preserved in the sedimentary record (Gale, 2009). In contrast, erosional events are less commonly used as markers. Layers of volcanic ash are one of the clearest such markers, and tephrochronology represents the relative dating of events constrained by ash (tephra) layers which record discrete eruptions. For example, in New Zealand the stratigraphy of tephra and ignimbrite layers with distinctive geochemical compositions provides a framework for the regional correlation of landforms (Froggatt, 1983; Pillans *et al.*, 1994). The clearest tephras are seen in areas that are known to be volcanically active, but regional and even global recognition of some eruptions may be possible through microscopic techniques. Specific episodes of coastal deposition may be inferred where accumulation of sea-rafted pumice is identified; even on the east coast of Australia which does not experience volcanism, the recognition of particular sources for pumice can provide important markers (Ward and Little, 2001).

Similarly, a major storm or tsunami can deposit a distinct sediment layer that can provide a chronological constraint on other coastal sediment sequences. A well-researched example is the sandy horizon first identified in the carselands of the estuaries of eastern Scotland and attributed to a tsunami event that resulted from the Storegga slides on the continental shelf off Norway (Dawson *et al.*, 1988). This event has now been identified extensively around parts of the Scottish coast and elsewhere in the British Isles (Smith *et al.*, 2004), and criteria have been described to differentiate such deposits from those that occur during sea-level transgression (Bondevik *et al.*, 1998).

Markers tend to be more common and better preserved in the recent sedimentary record, and many are the outcome of anthropogenic impacts. For example, people have often transported items to locations where they would not have occurred naturally, and their identification in the stratigraphical record can then be assumed to represent a point after the time of introduction. Introduced plants contribute to the pollen record. Pollen analysis provides many opportunities for discriminating the timing of sedimentary deposits; in the British Isles, the composition of plants differs between successive interglacials, and the relative appearance of taxa is often time-specific. The postglacial colonisation of the British Isles by the land snail *Discus rotundatus* is a further example of a well-dated event that provides a clear time horizon if appropriate material is preserved (Gale, 2009). In contrast, the introduction of invasive plants also hints at the age of sediments. Genera such as *Ambrosia* and *Rumex*, introduced into North America, and *Pinus*, introduced into Australia, provide distinct time horizons, indicating that where this pollen type is found it follows European settlement, although there can be considerable variation in when the pollen is first found at any site. The occurrence of charcoal is often associated with human burning; however, charcoal can also result from natural burning, such as lightning strikes, and is not always an unambiguous marker.

Pollutants provide further markers (Chenhall *et al.*, 1995). Coal dust and lead from leaded fuel are examples (Allen, 1987). Recognition of metal pollution in the sediments of the Severn estuary has been used to indicate timing of depositional events (Allen and Rae, 1986). A particularly widely used marker is the appearance of anthropogenically derived isotopes, such as caesium that resulted from nuclear weapons-testing. Peaks in ^{137}Cs in cores, corresponding with the periods around 1952–1954 and 1962–1965, have provided insights into rates of coastal sedimentation and relative sea-level change, such as in the rapidly subsiding marshes of the Mississippi delta (Delaune *et al.*, 1978).

4.9.2 Palaeomagnetism

Although not strictly a geochronological method in the sense of providing numerical ages by independent means, palaeomagnetism has been invaluable in Quaternary studies and particularly in investigations of deep-sea cores. *Palaeomagnetism* (ancient magnetism or remanent magnetisation) is based on the premise that magnetically susceptible minerals adopt a preferred orientation within sediments at ambient temperatures or within igneous rocks as their magmas or lavas cool below the Curie temperature (approximately 575°C and 675°C for the iron oxides magnetite and hematite, respectively; Hailwood, 1989). The orientation of the minerals will statistically reflect the direction of the Earth's magnetic field at time of formation of the minerals in igneous lithologies (thermal remanent magnetisation), or following sediment deposition, particularly fine-grained sediments (depositional remanent magnetisation). The magnetic reversal at the end of the reversely magnetised Matuyama chron, marking the beginning of the normally magnetised Brunhes chron at 780 ka ago, corresponds with MIS 19, and represents a critical time marker for stratigraphical correlation of early to middle Pleistocene marine and coastal sedimentary successions (Shackleton and Opdyke, 1973; Shackleton *et al.*, 1990). Palaeomagnetic signals revealing polar wander of the Earth's magnetic field have been detected in artefacts such as baked pottery and bricks. The palaeomagnetic signal in the form of polarity reversals within *palaeosols* (relict soil profiles) may have been acquired a considerable time after sediment deposition and the onset of soil formation and, accordingly, will register a post-depositional remanent magnetisation, perhaps by as much as 60 ka (Pillans and Bourman, 1997).

Several attributes of the Earth's geomagnetic field are considered in palaeomagnetism and may include the magnetic polarity (normal or reserved magnetisation), field declination and inclination, secular variation, and magnetic susceptibility of sediments (Barendregt, 1995). In addition, palaeointensity variations in the Earth's magnetic field are increasingly being examined in stratigraphy (Roberts *et al.*, 2013). Magnetic polarity refers to the direction of the Earth's magnetic field which appears to have oscillated approximately every 1–1.5 Ma (the Matuyama chron, for example, extended from 2.581 Ma to 780 ka), such that the magnetic north appears in the southern hemisphere during intervals of reversed magnetic polarity. The cause of reversals in magnetic polarity is not clearly understood, but may relate to periodic changes in the motion of free electrons within the Earth's core. The duration of the transition period may range from 10 to 100 ka (Hailwood, 1989). As the transition from normal to reversed polarity (and vice versa) occurs relatively rapidly in a geological sense, the transition represents a basis for establishing the time equivalence of strata at a global scale.

Coastal sedimentary successions suitable for palaeomagnetism studies should be dominantly fine-grained (silt to clay particle size), strongly magnetised with only limited authigenic minerals, and not show signs of deep weathering and secondary iron mineralisation. It is critical that the preferred orientation of magnetically susceptible minerals is the result of their realignment within the Earth's magnetic field, rather than reflecting a former, dominant direction of sediment transport within the sedimentary environment.

Magnetostratigraphy involves the subdivision of the stratigraphical record based on the application of palaeomagnetic data. Palaeomagnetic studies of K/Ar-dated lava flows and deep-sea sediments, constrained by oxygen-isotope (climate-based) event stratigraphies provided the original means of constructing the geomagnetic timescale (see Figure 1.1). In the context of palaeomagnetism, *geomagnetic polarity chrons* represent the fundamental time intervals of Earth history, defined by the timing of polarity reversals such as the Matuyama–Brunhes boundary at 780 ka. Shorter intervals within chrons, termed subchrons, represent discrete events lasting several thousand years such as the Jaramillo polarity subchron from 940 to 890 ka and the Olduvai from 1.95 to 1.77 Ma. Short-term geomagnetic excursions have also been identified, but many may only have local stratigraphical application. The Mono Lake (28–27 ka), Laschamp (~42 ka), Blake (112–108 ka), Pringle Falls (218 ka), and Big Lost (~565 ka) excursions have been identified within the Brunhes chron and appear to have wide geographical applicability (Bradley, 1999).

The practical application of palaeomagnetism in Quaternary stratigraphy and relative sea-level changes needs to be considered on a site-by-site basis. Factors that may limit the applicability of palaeomagnetism include post-depositional compaction modifying the preferred orientation of magnetically susceptible minerals, the influence of palaeocurrents in affecting sediment fabric, issues relating to sediment particle size, bioturbation and other forms of post-depositional sediment disturbance, the possibility of self-reversal, diagenetic formation of magnetically susceptible minerals during an interval of contrasting polarity, as well as complications associated with collecting reliably oriented sediment samples (Barendregt, 1995).

4.10 Synthesis and conclusions

Several geochronological methods are commonly used in dating Quaternary sea-level events, and their applicability to specific field settings is constrained by their physical and geochemical requirements, their age range, and the nature of the research questions under investigation. The age resolution required by specific questions such as relative sea-level changes during the Holocene clearly favours the application of radiocarbon and U-series dating (Pirazzoli, 1991; Collins *et al.*, 2006; Smith *et al.*, 2011). Commonly, preference is given to methods that provide numerical ages based on well-established rates of radioactive decay, which has also favoured radiocarbon and U-series methods.

Luminescence dating, electron spin resonance and amino acid racemisation have been used in a range of field contexts, particularly for Pleistocene sedimentary successions beyond the range of radiocarbon dating, and in contexts where materials otherwise suitable for U-series are not available for analysis. It is important to recognise that no geochronological method should be regarded as routine in nature, and numerical ages that at face value appear consistent with their morphostratigraphical context should be accepted cautiously and, where possible, independently verified.

The most credible studies are those involving multiple geochronological methods, permitting rigorous assessment of the reliability of results and more critically constraining the timing of palaeo-sea-level events. Ongoing efforts are needed to quantify the marine reservoir effect in space and time, and resolve discordances between ^{14}C and sidereal years for late Pleistocene time to improve the temporal resolution of relative sea-level changes in interstadial MIS 3, the LGM, and the transition to the Holocene. Similarly, continued efforts in improving pre-treatment strategies for the removal of potential contaminants will further improve radiocarbon and other methods, and in the dating of hitherto intractable sample types.

There are several challenges related to producing higher-resolution records for palaeo-sea-level investigations. These include increasing the age-resolving power and reducing analytical uncertainties, reducing sample mass requirements for several methods, and extending the age range of some geochronological methods. Significant advances accompanied the adoption of accelerator mass spectrometry for radiocarbon and various mass-spectrometric techniques in U-series dating, but these have raised new questions about sampling protocols to ensure that representative samples are analysed, in terms of age and general preservation state. Reliable dating of early Pleistocene records also remains a major challenge in palaeo-sea-level investigations. Oxygen-isotope records from deep-sea cores have identified successive sea-level events associated with the 41-ka cycles in the early Pleistocene, but there remain few options to date these apart from a 'count down from the top of core' approach to age assessments.

5

Vertical displacement of shorelines

Geologic history discloses the fact that some great areas of the Earth's surface which were in former ages below sea level are now thousands of feet above it. It also gives us reason to believe that other areas now submerged were in other ages terra firma.

(Dutton, *1889, p. 63*)

5.1 Introduction

Up to half of the world's coastlines are situated within geotectonically active settings and consideration of tectonism is an essential step in the investigation of Quaternary sea-level changes. A rigorous understanding of geologically long-term crustal behaviour (i.e. over timescales of thousands to millions of years) is essential for inferring glacio-eustatic (ice-equivalent) sea-level changes. The perceived tectonic stability of a coastal landscape is strongly constrained by the spatial and temporal scales of observation. The presence of former shorelines provides important evidence to evaluate the long-term behaviour of tectonically active regions. It is important to discriminate patterns of vertical displacement from broader sea-level signatures at such sites.

This chapter demonstrates the wealth of information that can be deduced about tectonic behaviour of coasts from palaeo-sea-level investigations. It examines the effectiveness with which shorelines preserved from the last interglacial maximum can be used as a benchmark for quantifying vertical crustal movements during the past 125 ka. The chapter begins by examining the broad plate-tectonic framework, contrasting tectonic domains that include so-called 'stable' trailing-edge coastlines (passive margins), subduction settings, coastlines associated with transform faults, and those associated with intra-plate volcanism.

Much of the chapter considers neotectonism, and shows how an understanding of neotectonic characteristics of coastlines can help derive detailed reconstructions of Quaternary sea-level changes (Fairbridge, 1981). Neotectonics refers to recent movements within the Earth's crust that may be vertical, horizontal, or a combination. Bates and Jackson (1987, p. 445) defined neotectonics as the '... post-Miocene structures and structural history of the Earth's crust'. In the context of this book, neotectonism relates to tectonic movements that have occurred over the course of the entire Quaternary Period (past 2.59 Ma), although there remains debate as to exactly what constitutes recent tectonism, and where the dividing line should be drawn between it and older tectonism. A detailed knowledge of neotectonism is also essential for understanding future relative sea-level changes (i.e. provides essential baseline information to evaluate inferences about recent relative sea-level changes derived from tide-gauge records, for example, Ballu *et al.*, 2011), and this is examined in more detail in Chapter 8.

Vertical displacement is also evident as a consequence of isostatic and flexural adjustments of the Earth's crust. These processes, which were introduced in Chapter 2, can result in adjustment that affects shorelines globally, or can account for local anomalies in elevation of former shorelines, apparent, for example, from the occurrence of contemporaneous shorelines at different heights on adjacent islands, or at different places along a coastline.

5.2 Plate tectonics and implications for coastlines globally

The concept of plate tectonics has provided an integrated and geohistorically coherent framework for explaining spatial and temporal distribution of the Earth's major surface relief, and the basis for long-term, tectonically driven geological changes in the Earth's lithosphere. The theory also cogently explains the 'non-fixist' nature of continents in space and time. It represents a major paradigm shift, based on recognition of geomagnetic reversals, mid-ocean ridge spreading centres, transform faults, and seafloor spreading. It is able to explain not only relative motions between lithospheric plates, based on Euler's theorem for rotations of spheres, but also absolute motion with respect to fixed features in the underlying mantle.

Although foreshadowed by Arthur Holmes in his classic *Principles of Physical Geology* (1944), plate-tectonic theory emerged as an extension of the hypothesis of seafloor spreading (Dietz, 1961; Vine and Matthews, 1963). It provides viable explanations for long-term horizontal movement ('drifting') of continents, origin of fold belts (mountain ranges), disjunct distributions of species (living and extinct) between continents, similarities in bedrock geology and inferred palaeoclimates from now widely separated continents (e.g. Permo-Carboniferous glacigene deposits of the former Gondwana continents), as well as a model for the 'rock cycle' (i.e. formation of igneous, sedimentary, and metamorphic rocks) and myriad large-scale geomorphological features such as tectonic landforms and Earth's macrotopography (Summerfield, 2000; Kearey *et al.*, 2009). Part of the elegance of the plate-tectonic paradigm is its capacity to integrate seemingly disparate data and observations from so many fields of science. The evolution of ideas leading to acceptance of plate-tectonic theory is reviewed extensively elsewhere (Cox, 1973; Hallam, 1983; Le Grand, 1985; Menard, 1986a; Rogers and Santosh, 2004; Nield, 2007; Kearey *et al.*, 2009).

Central to the theory of plate tectonics is discrimination of the Earth's lithosphere into oceanic crust that is formed by passive upwelling of basaltic magma (with an average density of 2.9–3.0) from continental crust that underlies landmasses and is composed of less dense rocks such as granites with densities of 2.7–2.8. The crust and upper mantle, which together in vertical profile represent approximately the outer 100 km of the Earth, form a series of surface plates. Approximately 7 major plates and 12 minor plates have been identified (Summerfield, 1991; Kearey *et al.*, 2009; Figure 5.1).

5.2.1 Lithospheric plate domains

The Earth's surface can be divided into a series of plate-tectonic domains within which to view relative sea-level histories of coastlines, at the mega-scale ($> 10^6 \, \mathrm{km}^2$) and over the longer term ($> 200 \, \mathrm{ka}$). The present spatial configuration of continents and oceans has resulted from fragmentation of an older continental mass, Pangaea, which commenced break-up approximately 250 Ma ago. During the past 2.59 Ma, the broad outline of modern coastlines was established by lithospheric plate movements.

The pattern of ocean-floor development and demise can be seen most clearly for the Pacific Ocean, which is dominated by the Pacific Plate, but with several other smaller plates. The Pacific Plate, which represents approximately two-thirds of the Pacific Ocean, is being created in its southeastern sectors along mid-ocean ridges and is undergoing subduction dominantly on its northern and western margins. These subduction zones, together with those of the smaller plates to the east, are marked by what is known as the Pacific Rim, also popularly called the 'ring of fire' for its volcanic activity (Figure 5.1).

The northernmost part of the Pacific Plate is bounded by subduction into the Aleutian Trench along the Alaskan Peninsula. The Aleutian Islands, an island arc formed in response to this subduction, broadly demarcate the southern extent of the Bering Sea. The westernmost portion of the Aleutian trench joins with a major transform, which in turn joins with the subduction zone defining the northwestern margin of the Pacific Plate. The latter zone extends from southern Kamchatka Peninsula, along the southern side of Japan and the Ryukyu Islands. The western subduction zone of the Pacific Plate bifurcates, extending along the Mariana Trench to the south and the southwest of Japan, along the eastern coastline of Taiwan, along the eastern Philippines (represented by the Mindanao Trench), and south across the northern coastal sector of Papua New Guinea. The southwestern subduction setting of the Pacific Plate is defined by the Tonga and Kermadec Trenches, which collectively extend from slightly south of Samoa to the southern North Island of New Zealand.

In the eastern Pacific there are several smaller plates spreading from mid-ocean ridges to collide with the Americas. The South American continent, on the overriding

South American Plate, has an ocean trench along almost its entire western margin. The Nazca Plate collides with this continent and descends into the Peru–Chile Trench. North of Peru, the northeastward trending Cocos Plate is being subducted beneath Nicaragua, El Salvador, Guatemala, and southern Mexico. Farther south, the Antarctic Plate descends into an extension of this subduction zone. Sectors that are not represented by subduction are characterised by major transform faults such as along the Californian coast. A restricted zone of subduction is located in western Canada south of Vancouver Island.

The Atlantic Ocean province is dominated by passive continental margins with the exceptions of the Caribbean Sea region in which the North American Plate is being subducted beneath the Caribbean Plate, and the subduction zone trending broadly east–west along the southern Mediterranean Sea resulting from collision of the African and Eurasian Plates. In the Caribbean region, islands of the Lesser Antilles represent an island-arc system, whereas Barbados has developed on the accretionary prism within the fore-arc basin defined by the North American Plate as it descends beneath the Caribbean Plate (Schellman and Radtke, 2004a).

The Indo-Australian Plate, encompassing Peninsular India in the northwest, and in the east, continental Australia, Papua New Guinea, and the North Island of New Zealand, is trending in a northerly direction. The Indian subcontinent initially collided with the Eurasian Plate approximately 55–50 Ma ago leading to formation of the Himalayas (Kearey *et al.*, 2009). To the north of Australia, the northern margin of the Indo-Australian Plate is represented by the subduction zone defined by the Java Trench, Timor Trough, and Banda Arc.

The Antarctic continent is located within the central Antarctic Plate, which is bounded by the southernmost portions of the Indo-Australian, Pacific, Nazca, South American, Scotia, and African Plates. The plate boundaries are dominated by mid-ocean ridges and associated transform faults with limited subduction being restricted to southern Chile and the eastern sector of the Scotia Plate. The Scotia Plate is a longitudinally elongate plate extending from Tierra del Fuego to South Georgia and its northern and southern boundaries are defined by major transform faults.

Current rates of seafloor spreading range between 4–6 mm/year for the Antarctic Plate and 100 mm/year for the central Pacific Plate (Summerfield, 1991). Rates of plate movement have not remained uniform throughout geological history, as illustrated by the northward movement of the Australian continent. In the early post-rift phase, Australia moved northward at a rate of 8 mm/year, by 43 Ma ago began to move at 30 mm/year, and is currently moving slightly in excess of 60 mm/year (Veevers, 2000). Not only have rates of seafloor spreading altered through time, but also the directions of plate motion have changed. One notable example is the apparent change in Pacific Plate movement around 47–50 Ma marked by a sharp inflection (a 60° change in direction) in the Emperor seamount chain northwest of the Hawaiian Islands.

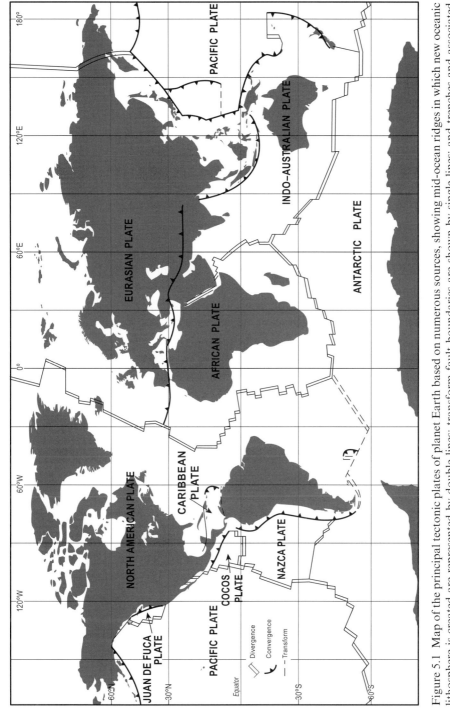

Figure 5.1 Map of the principal tectonic plates of planet Earth based on numerous sources, showing mid-ocean ridges in which new oceanic lithosphere is created are represented by double lines; transform fault boundaries are shown by single lines; and trenches and associated subduction complexes are depicted by lines with solid triangles, the apexes pointing to the direction of the descending lithospheric slabs.

5.2.2 *Plate margins*

Oceanic crust is formed at mid-ocean ridges, and as it moves away from these spreading centres it progressively cools and contracts as it changes in density, resulting in an increase in depth of the seafloor. Formation of new lithosphere and its destruction at subduction zones have the effect of inducing horizontal plate motion, a process also thought to be aided by large-scale convective cells within the Earth's mantle. Newly formed oceanic lithosphere initiates ridge-push due to the contrasting elevation of lithosphere at mid-ocean ridges, compared with lithosphere in abyssal contexts. This is recorded as a series of parallel zones of alternate reversed and normal polarity which enable age of the seafloor to be determined. There is a well-established relationship between ocean depth and time, with depth increasing with the square root of age, relative to height of the mid-ocean ridge at which it formed. Oceanic crust is subducted into ocean trenches when it reaches the far side of the lithospheric plate.

In contrast to oceanic crust which is ultimately subducted back into the mantle, and therefore is nowhere older than a few hundred million years, continental crust is too buoyant to be subducted and persists as the landmasses we see as continents. As a consequence coastlines of continents frequently coincide with plate margins and it is important to recognise vertical movements associated with plate boundaries. Three principal types of plate boundaries have been identified: divergent boundaries from which plates spread, convergent boundaries where they meet, and transform boundaries where they slide past each other (Le Pichon *et al.*, 1973; Summerfield, 1991; Figure 5.2). Triple junctions are important where three plates interact and may comprise elements of spreading centres, transform faults, or subduction zones.

Divergent boundaries, which primarily occur at depth within ocean basins along mid-ocean ridges, are where new lithosphere is formed as a result of ascent of magma to near the seafloor. Iceland is one of the few regions where a mid-ocean ridge is exposed above present sea level. The active Reykjanes Rift Zone, trending NE–SW in central Iceland, has produced extensive lava flows which have led to progressive southeastern development of the island over the past 24 Ma. Iceland remains volcanically active, with the abrupt appearance of the small island of Surtsey in 1963 and recent eruption of Eyjafjallajokull and other volcanoes on the island itself. St Paul and Amsterdam Islands in the southeast Indian Ocean are also associated with mid-ocean ridges (Nunn, 1994).

The relative sea-level record from Iceland is primarily a geologically recent record represented by emergent *Nucella* shell beds and sequences of isolation basins of Holocene age that are confined relatively close to the modern coastline (Símonarson and Leifsdóttir, 2002; Lloyd *et al.*, 2009). The general paucity of coastal successions within the region is likely the result of formerly more extensive glaciation, episodic *jökulhlaups* (glacier outburst floods), and periods of more pronounced subaerial volcanism which have destroyed any long-term record of relative sea-level changes.

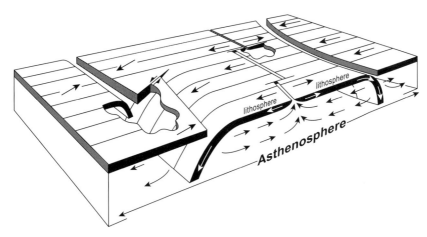

Figure 5.2 Schematic representation of plate-tectonic processes showing spreading centres, subduction zones, and transform faults (redrawn from Isacks *et al.*, 1968).

The neotectonic history of Iceland is complicated by episodes of glacio-isostatic uplift in response to deglaciation, as well as periods of localised uplift associated with onset of significant volcanism, followed in turn by subsidence due to deposition of thick piles of lava and volcanic ejecta, resulting in differential neotectonic histories around the island. In the lowland coastal area of Flói in southwestern Iceland, the Thjórsá lava covers an area of 270 km^2 and is between 15 and 20 m thick and is one of the largest subaerial Holocene lava flows on Earth (Símonarson and Leifs-dóttir, 2002). Crustal subsidence in the past 8 ka, since extrusion of the Thjórsá lava flow, is approximately 1 m with much subsidence presumed to have occurred after the first few hundred years following its formation (Símonarson and Leifsdóttir, 2002). In northwestern Iceland, sedimentary records of near-coastal isolation basins indicate that uplift of up to 56 mm/year has occurred since late glacial time resulting from glacio-isostatic compensation, and the upper marine limit reaches up to 80 m APSL (Lloyd *et al.*, 2009). Le Breton *et al.* (2010) report uplift rates of 21–92 mm/ year from 10 ± 0.3 to 8.15 ± 0.35 ka in response to glacio-isostatic compensation following the withdrawal of the Last Glacial (Weichselian) icesheet. They showed that relaxation time in lithospheric adjustment to deglaciation has varied spatially around the island and ranged between 2 and 4 ka.

Convergent boundaries are locations at which plates meet. When continental crust on one plate collides with another continent, spectacular mountain ranges are built, as has occurred with the Himalayas as the Indian landmass has met Eurasia. However, fossil shoreline evidence is found only in coastal settings where oceanic crust on one plate meets either another largely oceanic plate, or converges with continental crust on another plate. In these cases oceanic lithosphere is re-incorporated into the mantle by a process termed *subduction*. Subduction occurs at ocean trenches; the physical basis for its initiation is not entirely understood, and has remained a subject of controversy since the early arguments for an expanding Earth (Carey, 1976). The oceanic lithosphere appears to initiate a slab-pull force, by which negatively buoyant lithosphere is reincorporated within the mantle (asthenosphere). The descending slab is characterised by a dipping succession of earthquake foci, termed the Benioff Zone (Benioff, 1949).

Vertical displacement can occur on either the descending (underthrust) lithospheric slab, or on the plate which is not subducted, termed the overthrust side of the subduction zone (Pirazzoli, 1994). The descending lithospheric slab generally contains a zone of flexure as relatively rigid plate undergoes a transition from broadly horizontal ocean floor to a subducting slab at an angle of 30–45° as it is re-incorporated within the mantle. Such flexure will affect the few islands that may be on this ocean crust with uplift on the arch prior to subsidence on the descending plate. More significant vertical displacement is typical of the overthrust plate margin, whether that is oceanic or continental crust. Subduction-related settings are characterised by active volcanism, high frequency of earthquakes, and significant metamorphism at depth, each of which may result in uplift. Regional uplift in these settings is enhanced by development of an accretionary prism, comprising marine sediments that are not easily subducted with the descending lithosphere (characterised in vertical profile by a series of arcuate-shaped reverse faults). As the descending lithosphere is re-incorporated within the asthenosphere, the position of the trench relative to a fixed reference point within the mantle, migrates in the opposite direction to the subducting plate, a phenomenon termed *rollback*.

In the context of Quaternary sea-level changes, classical and well-documented examples of uplifted shorelines from convergent plate-margin settings include the west coast of South America (Chile–Peru) (Leonard and Wehmiller, 1992; Ortlieb *et al.*, 1996), Barbados (Schellmann and Radtke, 2004a,b), Cyprus (Poole *et al.*, 1990), the Huon Peninsula, Papua New Guinea (Bloom *et al.*, 1974; Chappell, 1974a; Chappell *et al.*, 1996a,b), the Japanese islands (Konishi *et al.*, 1970; Berryman *et al.*, 1992; Ota, 1992), New Zealand (Ota *et al.*, 1989, 1991), Sumba (Pirazzoli *et al.*, 1993b), Taiwan (Liew *et al.*, 1993; Pirazzoli *et al.*, 1993a), and Timor (Chappell and Veeh, 1978). Several of these are discussed in the following sections with particular reference to vertical displacement of the last interglacial shoreline (MIS 5e – 125 ka).

The third form of plate margin comprises *transform boundaries*. The rate of plate formation and subduction are not equal for individual plates, but are balanced when all the Earth's plates are considered. Differential rates of plate movements therefore require transform faults to accommodate complex interactions of continental-scale lateral displacement of lithosphere (Kearey *et al.*, 2009). These plate margins do not generate crustal assemblages, but are characterised by considerable lateral displacement (strike-slip motion) of bedrock and significantly lower levels of vertical crustal displacement at a regional scale. At a more restricted scale of only a few tens of kilometres, however, significant uplift or subsidence may occur in areas of *transpression* (crustal compression) or *transtension* (crustal extension), respectively. The San Andreas Fault in California is the pre-eminent example of a continental transform plate boundary.

Transform boundaries are commonly characterised by multiple, near-parallel transform (strike-slip) faults which may extend along their strike-length for hundreds of kilometres. Crustal deformation associated with transform faults may extend over a narrow zone of 20–40 km in regions of relatively cool and rigid lithosphere, and several hundred kilometres in areas such as the San Andreas Fault where there are significant contrasts in the strength of the lithosphere (Kearey *et al.*, 2009). Several large-scale geomorphological features characterise continental transform faults including linear fault scarps and laterally offset features such as offset streams. Localised areas where transform faults adopt a curved form and release stress, producing pull-apart basins due to transtension or areas of uplift resulting from thrust faulting and folding associated with transpression, are typical of transform settings.

The continental transform boundary between the Pacific and North American Plates, represented by the dextral San Andreas, San Jacinto, and associated faults extends for over 1,500 km. The plate margin was formerly a subduction setting that was active until 12.5 Ma ago. Active spreading within the Gulf of California commenced approximately 10 Ma ago and to the north, the San Andreas fault system migrated eastward across the offshore continental borderland with formation of deep transtensional depressions, such as the Ventura Basin that preserves up to 5,000 m of Plio-Pleistocene sediments, representing one of the thickest recorded successions of strata of this age on Earth.

Mapping of the upper limit of Pleistocene coastal sedimentation and erosion along the transform margin of California has revealed contrasting rates of uplift and subsidence along this plate boundary. The highest recorded level of marine influence, as revealed by the junction of former shore platforms and their backing cliffs, is called the *shoreline angle*. The shoreline angle represents the maximum height attained by a given sea-level highstand, and these features have commonly been mapped to define relative uplift rates on coastlines of emergence. It is up to 700 m APSL on the northern margin of the Ventura Basin on the western Transverse Ranges (Orme, 1998; Figure 5.3). The modern shoreline angle is at mean high water at about 1 m APSL. The considerable extent of uplift in the Ventura region is also shown by MIS 3 (60–40 ka) and Holocene shorelines. MIS 3 coastal sediments have been uplifted to 100 m APSL at Carpinteria and up to 360 m APSL immediately inland from Pintas Point. Similarly, Holocene coastal deposits in this area occur up to 37 m APSL. To the south, on the seaward side of the Peninsular Ranges, Baja California, the upper marine limit ranges from 100 to 357 m APSL (Orme, 1998).

The last interglacial shoreline (MIS 5e) represents the best preserved relict shoreline along the California coastline. On the Coastal Range coastline seaward of the San Andreas Fault, the MIS 5e shoreline ranges from 6 to 90 m APSL. On the Palos

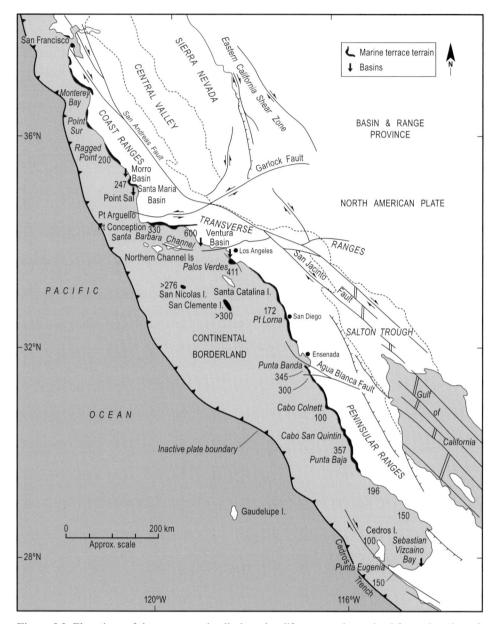

Figure 5.3 Elevations of the upper marine limit and uplift rates as determined from elevation of last interglacial (MIS 5e) marine terraces along coastal California (modified after Orme, 1998).

Verdes Peninsula, a doubly plunging anticline on which 13 marine terraces have formed, the MIS 5e shoreline occurs up to 37 m APSL. Rates of uplift deduced from the MIS 5e shoreline on the Pacific coast side of Baja California range from 0.03 to 0.29 m/ka.

5.2.3 Plate tectonics and coastal classification

Plate tectonics provides a conceptual framework for classifying coastlines which is particularly relevant in terms of their vertical stability. The contrast of the volcanically and seismically active mountainous coasts of North and South America with their more subdued, sedimentary east coasts had already been described by Suess (1888), discriminating them as Pacific- and Atlantic-type coasts, respectively. However, plate tectonics provided a framework that more clearly explained these differences. Adopting the plate-tectonic paradigm, Inman and Nordstrom (1971) presented a tectonic and morphological classification of the world's coastlines to describe what they termed first-order features, large-scale coastal landscapes that are directly related to a particular style of plate-tectonic behaviour, such as convergent boundaries. They recognised that Pacific-type coasts were at convergent plate boundaries, which they termed collision coasts, whereas Atlantic-type coasts were not at the margin of plates, which they termed trailing-edge coastal systems. Collision coasts are associated with subduction zones and can be divided into *continental collision coasts* and *island-arc collision coasts*. In the former, oceanic crust is subducted beneath continental crust (e.g. Peru–Chile Trench), while in the latter, oceanic crust is subducted beneath oceanic crust, such as the Banda and Aleutian island arcs. The collision coasts are all characterised by uplift; they contain linear mountain ranges, coastal cliffs and narrow continental shelves with a deep trench offshore into which oceanic crust is subducted. *Trailing-edge coasts* represent coastlines in mid-plate positions commonly with extensive sedimentary basins, in what have been considered passive tectonic settings. Inman and Nordstrom (1971) identified three principal types of trailing-edge coastlines that included *Neo-trailing-edge coasts* such as the Red Sea and Gulf of California coasts, *Afro-trailing-edge coasts* such as Africa in which the opposing coastline is also a trailing edge, and *Amero-trailing-edge coasts*, in which the trailing edge accumulates sediment brought by rivers from active erosion on the mountainous collision side of the continent. Historically, such trailing-edge coasts have been considered as passive margins, not generally characterised by significant neotectonism, although this has now been questioned (Murray-Wallace, 2002; Sandiford, 2007; Quigley *et al.*, 2010; Pedoja *et al.*, 2011). Finally, Inman and Nordstrom identified *marginal sea coasts* in which the coasts are protected from the open ocean due to the presence of an adjacent *island arc*; examples include the coasts of Vietnam, southern China, and Korea.

5.2.4 Ocean plate dynamics and island types

Significant intra-plate variations can be observed on oceanic crust. As plates gradually move away from mid-ocean ridge spreading centres, basaltic volcanoes can be extruded at hotspots, weak points in the crust. Many of these volcanoes do not build sufficiently to reach the sea surface, but remain as seamounts on the ocean floor.

However, larger volcanoes, particularly those associated with superswells, do form islands (Adam and Bonneville, 2005). The Big Island of Hawaii rises from water depths of around 4,000 m, to heights in excess of 9,000 m from the seafloor, and is presently active at Kilauea. It is a shield volcano composed of basalt which has upwelled, and has two prominent surface expressions, low-angle fluid lavas that form billowy 'pahoehoe', and angular, clinker-like 'aa' (Moore and Clague, 1992).

Such islands undergo vertical displacement as a result of isostatic adjustments of the plate onto which they have been extruded. Where the crust is not sufficiently rigid to support the island, it will be locally supported, and tends to undergo rapid subsidence following eruption. If the crust is rigid enough, the island will be regionally supported. Rigidity of crust generally increases with age (Menard, 1986b). Young volcanic islands, such as Iceland or the Galapagos, are locally supported because the crust is not rigid enough to support the mass.

Volcanic islands are frequently found in linear chains, particularly in the Pacific Ocean (McDougall and Duncan, 1980). The Hawaiian chain of islands is the classic example of this, and James Dana observed that those islands to the northwest of the active volcanic island, with its peaks of Mauna Loa and Mauna Kea, are progressively more dissected by erosion, which he correctly interpreted as indicative of their increasing age (Dana, 1890; Wilson, 1963). In the case of the Hawaiian Island chain, which contains volcanic islands at the younger southeastern end of the chain and coral atolls, such as Midway Atoll, at the northwestern end, the chain continues as a series of seamounts, known as the Emperor seamount chain. It was initially considered that the source of magma had migrated along the island chain, but with recognition of lithospheric plate motion this has now been re-evaluated and it is inferred that the plate has moved across a 'hotspot' in the mantle (Clouard and Bonneville, 2001). Recent re-evaluation of the Hawaiian–Emperor chain questions whether the hotpsot has remained stationary; an interpretation that it has moved south since the earlier Emperor seamounts were formed appears to fit better with the location of seamounts beyond the change of direction at longitude 172°E.

As an oceanic plate migrates, islands in these chains are carried into progressively deeper water. Plate tectonics thus provides a mechanism that can explain the subsidence that Darwin inferred in his far-reaching coral reef theory (Scott and Redondo, 1983). Darwin envisaged that fringing reefs, barrier reefs, and atolls represented successive stages on subsiding mid-ocean volcanoes as a result of vertical growth of reefs (see Figure 1.4). Subsequent drilling has provided support for this idea that the three types of reefs are stages in an evolutionary sequence, driven by gradual subsidence of the volcanic island around which the reef had initially formed. Subsidence occurs, first rapidly through isostatic adjustment of ocean crust to the mass of the volcanic island, and later at slower rates because of continued cooling and deepening of the plate as it migrates further from the mid-ocean spreading centre. Islands in the Society Island chain show the early stages, with reefs fringing much of the perimeter of the youngest islands, Tahiti and Moorea, and with barrier reef development on the remnant volcanic islands to the northwest, such as Bora Bora and Huahine. These volcanic remnants with a barrier reef around them, called *almost-atolls*, would be anticipated to be 'converted into atolls, the instant the last pinnacle of land sinks beneath the surface of the ocean' (Darwin, 1842, p. 109). Although atolls are scarce at the northwestern, the oldest, end of the Society chain, they do occur along the northwestern Hawaiian Islands, such as Midway Atoll, terminating at Kure Atoll, at 29°N, where coral growth rates and reef accretion rates are slow and subsidence appears likely to overwhelm reef growth, a threshold called the 'Darwin point' (Grigg, 1982).

5.3 Styles of tectonic deformation and rates of uplift or subsidence

In this section a range of neotectonic behaviours is examined that are significant for the interpretation of relative sea-level histories. Long-term and large-scale geological processes may be superimposed on plate-tectonic processes, and in combination these have a further legacy, creating antecedent conditions for processes associated with glacio-hydro-isostasy. For example, many passive margins have resulted in development of wide continental shelves (e.g. southern Africa, southern Australia, and southern Argentina) which are influenced by hydro-isostasy. In contrast, convergent margins, such as the subduction of the Juan de Fuca Plate beneath western Canada, the Nazca Plate beneath Chile and Peru, and the Indo-Australian Plate beneath the southern South Island of New Zealand, have led to the formation of fold mountain belts favouring development of an icesheet, ice caps, and valley glaciers, which in turn are associated with local glacio-isostatic adjustments.

Three main considerations in the study of neotectonism include the nature of deformation processes, their timing and duration, and mechanisms responsible for deformation (Vita-Finzi, 1986). Deformation of strata by warping or folding is likely to be more difficult to observe than offset beds or fault scarps in view of slower process rates associated with ductile deformation. In contrast, the effects of brittle fracture, such as faulting, generally result from episodic release of stress within rocks and, accordingly, may be readily recognised through discontinuities of structural features within rocks such as distinctive marker beds.

Rates of uplift vary significantly with contrasting tectonic setting. A logarithmic scale (log base 10) has been used to assist qualitative description of uplift rates in a global context (Bowden and Colhoun, 1984), namely, quasi-stable (0–0.005 mm/ka), slow (0.006–0.05 mm/ka), moderate (0.06–0.5 m/ka), rapid (0.6–5 m/ka), and very rapid (> 5 m/ka).

5.3.1 Coseismic uplift

Coseismic uplift refers to localised uplift associated with earthquakes. Earthquakes can cause uplift of the ground surface by tens of centimetres to several metres across a horizontal distance of several tens of kilometres. Characteristically uplift is delimited by a discrete fault plain, or fault plains, which may be exposed at the ground surface (or the seafloor), although this does not happen in all earthquakes. An earthquake represents the abrupt release of elastic strain that has built up in the lithosphere. A range of geometric styles of faults may be related to coseismic uplift that include normal, reverse, and thrust faults. In some regions, older normal faults have been reported as having been reactivated by a reverse displacement (Stiros et al., 1994). Such events may be followed by further post-seismic slip. However, further interseismic adjustments can also be detected, often in the opposite direction to the coseismic displacement; for example, abrupt coseismic events result in uplift of Sumatran reefs during earthquakes, but gradual subsidence occurs between these events. It has been possible to reconstruct a general history of these changes from the constraints these changes of level have on corals that have grown up to the intertidal level at which they can grow no higher (Zachariasen et al., 1999, 2000; see Figure 3.5d).

Darwin witnessed the after-effects of a major earthquake on the western coast of South America when he visited Concepcion in Chile shortly after the earthquake that devastated it on 20 February 1835. He wrote that the town 'is now nothing more than piles and lines of bricks, tiles and timbers', but this experience, and his observation of marine shells at elevations of more than 2,000 m along the Andean cordillera, convinced him of the significance of vertical movements of land. Concepcion was devastated again by the Maule earthquake (M = 8.8) in 2010, and the largest magnitude on record, M = 9.5, was for the 1960 earthquake in Chile. Similarly, the coastal city of Antofagasta in northern Chile, experienced a high-magnitude earthquake (M = 7.3) in July 1995, which resulted in uplift of the local coastline by variable amounts (5–30 cm) with the greatest uplift ($\geq$ 40 cm) centred on south-western Mejillones Peninsula, 20 km north of Antofagasta (Ortlieb, 1995). Uplift associated with the event was clearly defined by bleached *Lithothamnium* (coralline algae).

The well-documented Prince William Sound megathrust earthquake of M = 9.2 that occurred on 27 March 1964 in Alaska resulted in up to 8 m of uplift of the shoreline in a restricted area on Montague Island southeast of Anchorage. Elsewhere widespread uplift of 1–3 m was experienced, although with localised subsidence in other localities (Plafker, 1965; Shennan, 2009). There was as much as 40 m of vertical displacement on the seafloor associated with the Aceh–Andaman earthquake (M = 9.2) which occurred on Boxing Day in 2004. Up to 1.5 m of uplift occurred on the island of Simuele, off the west coast of Sumatra, in association with earthquake activity during and following this event, as seen from emergence of corals on the island (Meltzner et al., 2006, 2010; see Figure 3.5d). The 1953 magnitude (M = 7.2) earthquake off the Greek island of Kefalonia (Cephalonia) resulted in regional differential uplift of the island ranging between 0.3 and 0.7 m over 40 km (Stiros et al., 1994; see Figure 3.12d). The devastating Tohoku–Oki earthquake (M = 9) on

11 March 2011, which resulted in the tsunami that inundated Sendai and the surrounding region in northern Japan, actually resulted in coseismic subsidence of ~1.2 m along the coast (Ozawa *et al.*, 2011).

Repeated coseismic uplift has been one of the most significant features of convergent plate-margin coasts around the world which has resulted in the emergence of suites of coastal terraces, either depositional reef units on tropical plate margins or erosional platforms on non-reef coasts. Averaged rates of late Pleistocene coseismic uplift have been significantly faster in subduction-related settings than for almost any other mechanism that can cause uplift, such as isostasy or epeirogeny. Although fast rates of uplift occur in the initial stages of glacio-isostatic compensation, as have been registered in Scandinavia, the uplift rate has exponentially decayed (parabolic function) with time (Mörner, 1979). In contrast, in subduction-related settings, the long-term averaged rates of uplift indicate sustained high rates of coastal emergence.

Rates of uplift differ between subduction settings as attested by elevation of the last interglacial shoreline. At its maximum elevation, the fringing reef complex (Reef VII) of the last interglacial (MIS 5e) occurs up to 400 m APSL in southeastern Huon Peninsula in Papua New Guinea where rates of coseismic uplift are in the order of 3–3.5 m/ka (Chappell, 1974a; Chappell *et al.*, 1996b), representing one of the world's highest recorded elevations of last interglacial coastal strata. In Japan, the highest reported coastal successions of last interglacial age (MIS 5e) occur between 186 and 190 m APSL on the northeastern coastline of Shikoku (Ota, 1986, 1992). In the southwestern South Island of New Zealand, the highest evidence of the MIS 5e shoreline is in the Preservation Inlet region, where terraces up to 370 m APSL have been recorded (Ota and Kaizuka, 1991). For much of the South Island, however, MIS 5e terraces range from 10 to 55 m APSL. On the North Island of New Zealand, MIS 5e terraces commonly range between 30 and 190 m APSL, with an exceptionally high terrace at East Cape some 300 m APSL.

The rapid rate of tectonic uplift within some subduction settings makes possible the identification of sedimentary successions deposited during sea-level lowstands. Marine sediments deposited some 14,000 years ago (^{14}C years) in a former water depth of between 80 and 100 m, now crop out at 17 m APSL along the eastern coastline of Taiwan (Pirazzoli *et al.*, 1993a). Rates of uplift in excess of 5 mm/year have been reported for parts of the Coastal Range of eastern Taiwan (Lundberg and Dorsey, 1990).

5.3.2 Epeirogenic uplift

Epeirogenic uplift refers to regional uplift affecting areas of hundreds of square kilometres without demonstrable deformation of the Earth's crust (i.e. minimal folding associated with uplift). Epeirogenic uplift has been known to affect the interiors of continents (King, 1962), but its influence on coastal landscapes and relative sea-level histories has been less appreciated. The Coorong Coastal Plain in southern Australia has undergone epeirogenic uplift throughout much of the Pleistocene (Murray-Wallace *et al.*, 1998; Figure 5.4). Individual interglacial barrier shoreline structures can be traced from the northwestern extremity of the coastal plain in a southeasterly direction towards the locus of maximum uplift. The last interglacial shoreline as a distinct coastal landform, for example, can be traced over 300 km from close to present sea level in the Murray River mouth region up to 18 m APSL near Mount Schank, a mid-Holocene volcano (Murray-Wallace *et al.*, 1996a). The area of maximum uplift is associated with a Quaternary volcanic complex and the deep crustal emplacement of magma chambers (Sheard, 1990). The uplift is expressed by regional doming of the Oligo-Miocene Gambier Limestone as seen in the subcrop map illustrating the folded upper surface of the formation which is directly associated with known centres of Quaternary volcanism (Figure 5.4). The volcanism is in the form of a hot region rather than a single mantle hotspot, and evidence for genetically related volcanism extends to western Victoria (Joyce, 1975; Price *et al.*, 2003).

The inversion of sedimentary basins may also relate to epeirogenic uplift. Uplift of portions of the southern Australian continental margin (e.g. Nullarbor and Roe Plains; James *et al.*, 2006) has been related to changes in the dynamic topography of the entire continent in response to subduction-related processes at its northern margin (Sandiford, 2007).

Hypotheses to explain the mechanical basis for epeirogeny relevant to the Quaternary Period include thermal anomalies within the mantle (e.g. upwelling of hot asthenosphere), relief amplification resulting from within-plate stresses, flexural loading on the margin of the lithosphere adjacent orogenic belts, epeirogeny related to subduction processes, and comparative changes of thickness in the ratio of crust to mantle–lithosphere (Van der Pluijm and Marshak, 2004).

5.3.3 Folding and warping

In view of the slower rates of ductile deformation compared with brittle fracture, folding and warping processes cannot be directly observed within a human timeframe, although subtle changes in elevation may be quantified by long-term GPS (Milne *et al.*, 2004). However, some localities show evidence for ductile deformation during the Pleistocene, illustrating that folding processes do not necessarily require tens of millions of years to occur. At Vrica in Calabria, southern Italy, for example, early Pleistocene (< 2.59 Ma) marine sequences reveal a prominent regional dip in response to folding and crustal deformation.

Folding involves significant crustal deformation and potentially a range of fold styles may develop depending on the magnitude of strain (e.g. gentle open folds, tight folds and isoclinal folds where fold limbs are essentially parallel). In sedimentary successions where relative ages of strata are known and when viewed in two dimensions, folds with a hinge area at the crest of the structure are termed anticlines and

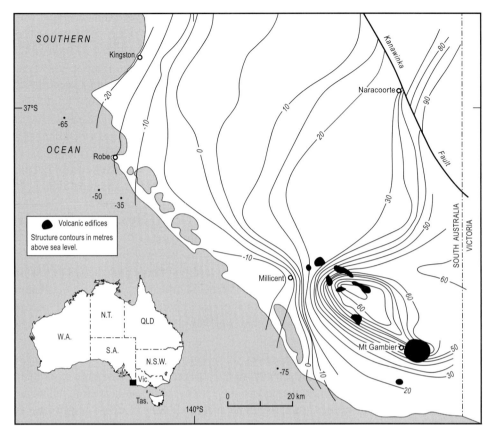

Figure 5.4 Subcrop map of the upper surface of the Oligo-Miocene Gambier Limestone, Otway Basin, southern South Australia. A well-defined dome structure is evident near the known centres of Quaternary volcanism suggestive of a genetic relationship between uplift and volcanism.

those with the hinge area within a trough are termed synclines. In a three-dimensional context, the structures are part of dome and basin features and the two are commonly associated in deformed terrains. A fold axis representing an imaginary mirror plain separates one limb of a fold from the other and transects the fold structure through its hinge area. The line representing the intersection of a fold structure with its axial plain is termed an axial trace. When the axial trace departs from the horizontal (i.e. has a discernable angle of dip) the fold is said to have a plunge. Thus, a doubly plunging syncline defines a basin structure. In the most general sense, very gentle, open folds may occur over larger regional landscape areas than more tightly formed fold structures. A subtle fold-like feature with only one limb is termed a monocline.

Warping resembles gentle open folds and may be associated with epeirogenic and subduction-related processes. The emergent marine terraces of the South Taranaki region, Wanganui Basin in the southern North Island of New Zealand, show effects of regional warping (see Figure 6.18; Pillans, 1983). The terraces extend parallel with the modern coastline for more than 100 km, up to 350 m APSL and some 20 km inland, with remarkably uniform surfaces to the bevelled terraces. Domal uplift has occurred in response to deformation of the Indo-Australian Plate immediately above the descending southwestern Pacific Plate resulting in preservation of 47 Plio-Pleistocene cyclothems (Carter and Naish, 1998).

5.3.4 Isostasy

The principle of isostasy was introduced in Chapter 2. The tendency of the Earth's crust–mantle system to attain a state of equilibrium with respect to its mass, thickness, and surface relief continues to have a profound effect on the Earth's surface. This is most clearly expressed by glacio-hydro-isostatic adjustment processes that continue today in response to deglaciation following the LGM (e.g. glacio-isostatic uplift associated with melt of icesheets and hydro-isostasy inducing continental levering due to loading of water in ocean basins and on continental shelves following return of meltwater to the global ocean; Mitrovica and Milne, 2002). Quaternary isostatic processes represent a dominantly passive response to ice- and ocean-volume changes which induce crustal loading or unloading (i.e. with lower levels of crustal loading, isostatic compensation will occur without necessarily inducing faulting). Accordingly, isostatic processes differ from tectonically induced changes associated with plate-tectonic movements (e.g. ridge push associated with mid-ocean ridges and draw-down of subducting lithospheric slabs). As a qualification, it is noted that some isostatic responses, such as uplift accompanying erosional unloading of mountainous terrains, are in part related to active tectonism (i.e. active tectonism creating surface relief through subduction processes, and denudation promoting isostatic compensation). Isostatic compensation is also associated with volcanoes and their thick piles of volcanic ejecta and is understood in the context of lithospheric flexure.

5.3.5 Lithospheric flexure

Volcanic islands that are regionally supported result in flexure of the lithosphere which forms a moat near the island, and produces an arch 150–330 km from the load, depending on the age and rigidity of the crust (McNutt and Menard, 1978). Within the immediate vicinity of the volcano, the lithosphere is depressed, forming a moat, which becomes filled with sediment to varying degrees. However, at a critical distance (of the order of 200 km but depending on rigidity of the crust) there is slight upward movement, which will be recorded by emergence of shoreline deposits on any islands that are located such a distance away from the volcanic load.

Emergent reef, coralline beach deposits, and shoreline notches and platforms have been described on the islands of Oahu, Molokai, and Lanai (Stearns, 1978). These have been attributed to lithospheric flexure and development of an arch resulting from progressive and relatively constant subsidence of the Big Island of Hawaii at an average rate of 2.6 mm/year, and of neighbouring Lanai at 1.9 mm/year, in response to continued hotspot volcanism (Moore, 1970; Campbell, 1986; Ludwig *et al.*, 1991; Grigg and Jones, 1997). The islands of Oahu, Molokai, and Lanai have migrated across the zone of flexure as the Pacific Plate has moved northwest, and they have been progressively uplifted, resulting in a late Quaternary record of coastal emergence. Shoreline features radiometrically dated to the last interglacial (MIS 5e) from a unit known as the Waimanalo Limestone have been examined in detail on the island of Oahu (Ku *et al.*, 1974; Muhs and Szabo, 1994). The last interglacial shoreline occurs at a maximum elevation of 8.2 m at Waimanalo, and as much as 12.5 m at Kahe Point. Dating of this fossil reef and several older interglacial reefs indicates that Oahu has experienced uplift at a constant rate of 0.06 mm/year for at least the past 500 ka (McMurty *et al.*, 2010).

Similar lithospheric flexure has been recorded in response to volcanic loading by Tahiti in the Society Islands and by Rarotonga in the Cook Islands (Pirazzoli, 1994; Dickinson, 1998). One of the advantages of the latter is the presence of reefal limestones on islands on the arch which provide the opportunity to assess the extent of differential movement experienced (Woodroffe *et al.*, 1991a).

Lithospheric flexure is also evident in relation to raised limestone islands associated with bending of the plate prior to subduction into an ocean trench. Niue in the Pacific is approaching the Tonga Trench, and appears to have been uplifted with a series of emergent terraces (Schofield, 1959); however, the last interglacial seems elusive on this island, with a prominent terrace at 23–28 m elevation containing corals of penultimate interglacial age (Kennedy *et al.*, 2012). Christmas Island in the Indian Ocean, approaching the Java Trench, is a similar isolated island on which the prominent Lower Terrace (MIS 5e) occurs from an elevation of 12 m to around 20 m (Woodroffe, 1988b). A more complete sequence can be seen in the Loyalty Islands as they approach the Vanuatu Trench, with Ouvéa showing the first stages of uplift (which is confined to the eastern part of the island), and Lifou and Maré showing successive stages (Dubois *et al.*, 1974).

5.3.6 Mantle plumes

Increasing attention has been paid in recent years to the role of isolated mantle plumes as a mechanism for uplift. Mantle plumes represent localised vertical sources of heat flow within the mantle that may directly affect the dynamic topography of the Earth's surface. Rather than very broad convection cells within the mantle at the scale of thousands of kilometres, as originally conceived in plate-tectonic theory, mantle plumes represent spatially more discrete structures, potentially in the order of 100–200 km thick. In the North Atlantic Ocean, V-shaped ridges that formed within oceanic lithosphere, between Greenland and Iceland, have been attributed to local influence of mantle plumes (Poore *et al.*, 2011). Similarly, episodic 'blobs' of ascending mantle material may explain distinct pulses of localised uplift in intra-plate settings. The area to the immediate northwest of the Orkney Islands in the North Atlantic Ocean is considered to have been uplifted by mantle plume processes during the Palaeocene-Eocene thermal maximum, resulting in development of a subaerially influenced landscape with a well-established drainage system (Hartley *et al.*, 2011). Following a period of subaerial exposure possibly lasting 1 Ma, the landscape subsided and was reburied. Mantle plumes may remain active for tens to hundreds of millions of years.

5.3.7 Subsidence and submerged shorelines

Whereas most of the processes considered above relate to uplift, there are many regions of the world that have experienced subsidence, and several that are continuing to subside. The lithosphere subsides when a large mass is emplaced upon it. An example of subsidence in response to extrusive volcanism has already been described, in relation to the volcanically active island of Hawaii, and subsidence that can be deduced from the occurrence of reef terraces of last interglacial age at depths of tens to hundreds of metres around its margin. This gradual deepening associated with cooling and contraction of ocean lithosphere explains why atolls in Micronesia comprise a kilometre or more of carbonate sediments above a volcanic basement (Guilcher, 1988).

Subsidence can also be associated with other loading on the lithosphere. Subsidence occurred beneath the major icesheets, although now those areas are undergoing the late stages of rebound following melting of most glacial ice. As a component of that conspicuous rebound, more subtle subsidence is now being experienced in fore-bulge regions, as outlined in Chapter 2.

Subsidence is also associated with accumulation of thick sediment piles within deltas, particularly megadeltas, such as those of the Mississippi, Ganges–Brahmaputra, Mekong, and Yangtze (Changjiang) Rivers. Some subsidence in such cases may result from compaction of Holocene deltaic sediments, or dewatering, but another component represents an imperceptible form of neotectonism (Törnqvist *et al.*, 2008). Progressively with time, sediment wedges up to hundreds of metres in thickness may develop within these settings. *Listric normal faults* (arcuate-shaped, concave downward faults that develop whilst sediments are accumulating within the receiving basin) are common in these settings. Graben structures may also be characterised by periods of subsidence that may continue long after the original initiation of the structural feature.

5.4 The last interglacial shoreline: a reference for quantifying vertical displacement

Shorelines associated with the last interglacial maximum (MIS 5e; 130–118 ka) are widespread around the world. They provide a pragmatic method of estimating vertical displacement over the past 125 ka, by measuring elevation differences between observed shorelines and the global average for MIS 5e. Last interglacial deposits are sufficiently old to quantify even modest rates of vertical movement and, as a consequence, elevation of last interglacial coastal deposits has been widely used as a benchmark to delineate recent and ongoing tectonic behaviour at continental scales (Ota, 1986, 1994; Murray-Wallace and Belperio, 1991; Bordoni and Valensise, 1998; Zazo *et al.*, 1999; Ferranti *et al.*, 2006; Pedoja *et al.*, 2011; Muhs *et al.*, 2012). The sea level is generally considered to have been at about 6 m above present at the peak of the interglacial (Veeh, 1966), although a broader range of 2–10 m is also widely adopted. Although much of the ocean floor has been gradually subsiding, the Seychelles in the central western Indian Ocean represent an anomalous outcrop of granite and are considered stable. Last interglacial reef units occur at up to 8 m on several of the islands in the Seychelles (Israelson and Wohlfarth, 1999). However, the diverse range of observations in elevation of MIS 5e shoreline deposits relates at the global scale to the combined effects of differential rates of tectonism, and the fact that the crust–mantle system during MIS 5e was not necessarily in isostatic equilibrium (i.e. still responding to deglaciation following MIS 6; Dutton and Lambeck, 2012). A range of sedimentary facies and palaeo-sea-level indicators constrained by several geochronological methods have been used in examining the last interglacial shoreline and an extensive literature has emerged (see Chapter 4).

5.4.1 Terrace age and elevation

Coastlines undergoing tectonic uplift generally preserve evidence for multiple sea-level highstands. The number of highstand events preserved within a tectonically active setting is a function of both factors that operate at time of formation of the shoreline features and subsequent processes that affect their preservation. A former shoreline, evident as a discernible terrace in the landscape, may have been formed as a result either of erosion of the underlying substrate, for example planation of rock platforms, or through deposition. In many tropical locations distinct reef terraces have formed, each recording successive periods of fringing reef growth. Where former shorelines comprise sedimentary deposits, their formation has been a function of several depositional parameters, such as sediment type and rates of sediment supply, and their maturity depends on duration of the highstand event. The degree of sophistication of palaeo-sea-level inferences derived from tectonically active settings is in part a function of preservation potential of the sedimentary facies, which is likely to be climatically dependent (e.g. water availability for fluvial erosion at times of sea-level lowstands and post-depositional *in-situ* weathering).

Initially, correlation of fossil shorelines was achieved solely on the basis of their elevation. This concept of *thalassostatic terraces* was applied around the Mediterranean by Zeuner (1945), who adopted the ideas developed by Milankovitch and identified a sequence of terraces: Sicilian, Milazzian, Tyrrhenian, Monastirian, and Nissian (see Figure 1.6). Zeuner incorrectly assumed that the correlations could

be extrapolated to eustatic terraces around the world based on height alone. However, since development of a robust suite of geochronological techniques, recognition
of shoreline features dating to the last interglacial is generally based on relative, or in
many cases numerical, age assessments.

Rates of vertical displacement calculated on the basis of observed elevation of
features of last interglacial (MIS 5e) age can, in some circumstances, be substantiated
where elevation of features of Holocene age is consistent with rates of uplift of the
last interglacial. Radiocarbon dating of palaeo-sea-level indicators provides greater
age resolution for quantifying rates of vertical movement during Holocene time than
can be achieved when dating Pleistocene successions. This offers a greater opportunity for determining frequency of episodic coseismic uplift events than is possible
in Pleistocene time, but is much more sensitive to the accuracy with which elevations
are constrained. One of the challenges in comparing inferred rates of vertical
displacement from the Holocene with late Pleistocene successions, however, relates
to the time averaging effect that longer records have in terms of rates of change.
A high-magnitude uplift event in the Holocene, for example, may have a disproportionate effect on calculation of long-term uplift rates, compared with the same event
in a longer record. Extrapolating rates of uplift in the late Pleistocene from
radiocarbon-dated Holocene records is likely to result in erroneous conclusions,
unless uplift is very rapid. The example of coseismic uplift on the Huon Peninsula is
examined below, and then in subsequent sections, several case studies are selected for
further discussion.

There is usually greater uncertainty about time of formation of shorelines older
than last interglacial in view of inherent difficulties in dating early Pleistocene and
early–middle Pleistocene sedimentary successions (e.g. finding geochronological
methods appropriate for the dating of specific sedimentary features). The majority
of uplifted settings, however, preserve shoreline structures in an ordered time series
extending back only three to four interglacials. The Huon Peninsula record, for
example, with reefs upto 1,000 m APSL, extends back to before 250 ka (Bloom
et al., 1974; Chappell, 1974a; Chappell *et al.*, 1996a,b). However, there are several
locations where sequences of shorelines offer potential for longer reconstructions.
For example, Sumba Island in Indonesia has a record extending back approximately
1 Ma with the highest and oldest reef terrace up to 475 m APSL (Pirazzoli *et al.*,
1993b; see Figure 6.19). The Wanganui Basin in the southwest of the North Island,
New Zealand, has a similarly long record preserving 12 marine terraces that extend
up to over 350 m APSL (Pillans, 1983, 1990; Bowen *et al.*, 1998; see Figure 6.18).
The Coorong Coastal Plain in southern Australia has a dated record of at least 1 Ma
(Huntley *et al.*, 1993, 1994; Murray-Wallace *et al.*, 2001), and genetically similar
coastal barriers extend farther inland within the Murray Basin. These, as yet, have
not been reliably dated, but may be up to 6 Ma (Kotsonis, 1999; see Figure 6.14).
Some coastlines are characterised by a more highly punctuated depositional record
such that early Pleistocene successions are preserved in settings that would intuitively

have been considered to have long been eroded (e.g. the early Pleistocene Point Ellen Formation preserved in eroded granite depressions close to present sea level on Kangaroo Island in southern Australia (Ludbrook, 1983; Milnes *et al.*, 1983)).

Evidence for significantly lower sea-level stands, such as those formed during glacial maxima or stadials of glacial cycles, tends to be less commonly preserved on tectonically emerging coastlines. During these time intervals, the sea surface was significantly below the modal zone of coastal sedimentation of glacial cycles (~60 m BPSL). Erosion is likely to have been actively incising into emerged terraces (Chappell, 1974b). Accordingly, the magnitude of sea-level rise accompanying degla-ciations far exceeded that of tectonic uplift, such that lowstand shoreline successions generally remain below sea level during succeeding interglacials and interstadials. A further complexity in some of these settings is that deposits associated with older highstand successions are inaccessible in outcrop because they are mantled by younger highstand successions.

The nature of the stratigraphical record within marine terraces on emergent coastlines varies depending on uplift rate, duration of sea-level highstand, and type of sedimentary facies deposited. Commonly, marine terraces are preserved in a stepped sequence. The backing cliff or escarpment represents the marine limit at the time of terrace formation and the associated deposition of coastal sediment. The planar surfaces of terraces typically slope in a seaward direction and represent an unconformity surface formed by marine erosion. On the seaward margin of marine planated terraces there is an abrupt break of slope forming the backing escarpment to the lower and more recently formed terrace, a feature termed the *shoreline angle* (Lajoie, 1987; Pillans, 1990; Orme, 1998; Figure 5.5).

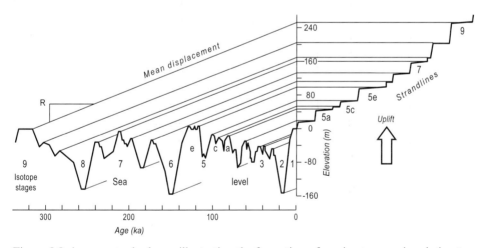

Figure 5.5 A conceptual schema illustrating the formation of marine terraces in relation to a late middle Pleistocene to Holocene glacio-eustatic sea-level history (from Lajoie, 1987, with permission from the National Academy of Sciences, courtesy of the National Academies Press, Washington, DC).

Many emergent coastlines record the late stages of relative sea-level rise in the form of a transgressive facies association, commonly represented by a thin lag deposit. These successions may in turn be overlain by a highstand facies association, typically comprising a series of seaward-dipping clinoforms deposited in response to coastal progradation (regressive facies). This latter association might not involve a detectable change in relative sea level, but commonly involves continued sediment supply during the highstand. The coral reef terraces of the Huon Peninsula have been termed *transgression limited* (Chappell, 1983b), in the sense that their growth appears to have ceased at the culmination of transgressions when the rate of tectonic uplift has exceeded the rate of relative sea-level rise, as illustrated by development of reef VIIa at 135 ka ago. Following a period of relative sea-level fall, a subsequent sea-level rise at 120 ka promoted a later phase of reef development with the marginal reef terrace (VIIb) with a landward limit some 5 m lower elevation in the landscape than terrace VIIa. In a similar way, Ota *et al.* (1988) found that the culmination of postglacial sea-level rise in sedimentary deposits in New Zealand was older at those sites that had the most rapid rate of uplift. In some situations, the early portion of a regressive facies may also be represented in the stratigraphical record (i.e. a fall in relative sea level and hence a *forced regression*).

5.4.2 Constraints on using the last interglacial shoreline as a benchmark

The last interglacial represents a peak in the marine oxygen-isotope record at around 125 ka, and is generally considered a time when sea level was significantly above its present level. Numerous high-precision U-series ages, of varying reliability, provide a range of estimates of the duration of the highstand, and imply that it was of comparable length to, or longer than, the period of relative sea-level stability experienced during the Holocene. As indicated above, Broecker *et al.* (1968), Bloom *et al.* (1974), Chappell (1974a), and many subsequent workers, have since used a 6 m reference level as a *de facto* global 'eustatic' sea level for the last interglacial maximum, based on the study by Veeh (1966), from which corrections for local tectonic deformation have been derived. Much of the evidence for the elevation of the MIS 5e shoreline falls in a broader range of 2–10 m. In a recent statistical compilation of observed elevations of last interglacial shorelines, Kopp *et al.* (2009) have suggested that the sea was as much as 9 m above present. However, such a statistical treatment of elevations from selected locations where the 5e shoreline is observed above present is likely to be inappropriate because it does not adequately allow for those areas where there is no evidence, or where the MIS 5e shoreline is not found above present sea level, as discussed further in this chapter. It is also important to recognise that the Earth will have experienced isostatic adjustments during the last interglacial in a similar manner to those observed during the Holocene, and that the peak of the sea-level highstand will not have been experienced at the same time in both the Atlantic province (focused on the Caribbean) and the Pacific region.

If sea level was above its present level at the peak of the last interglacial, this implies that ocean volume was greater and that polar icesheets were less extensive during MIS 5e than they are today. There is some evidence to support the belief that it was warmer; for example, tropical fauna, such as the hippopotamus, occurred in Europe at that time, and coral reefs appear to have extended beyond the latitudinal limits to which they are presently constrained (Marshall and Thom, 1976; Greenstein and Pandolfi, 2008; McCulloch and Mortimer, 2008).

Veeh (1966) showed that there was widespread evidence throughout the tropics for a last interglacial shoreline, around 120 ka determined by U-series dating, at heights averaging 6 m APSL. He used islands and continental shorelines, such as the Seychelles, Pacific islands, and Western Australia, which he considered stable. Several of these island sites, such as Oahu in the Hawaiian Islands, are now considered to have undergone flexural adjustments in response to loading of the ocean floor by adjacent volcanic islands, and may not have been stable (see Section 5.3.5). Mangaia in the Cook Islands, for example, is on the flexural arch surrounding the younger volcanic island of Rarotonga (Spencer *et al.*, 1987; Woodroffe *et al.*, 1991a).

There are some other important caveats that should be considered in the application of last interglacial shoreline successions to defining recent crustal movements. First, at the time of their deposition they will have been subject to a similar global variability in sea level as evident for Holocene successions due to glacio-hydro-isostatic effects. Lambeck and Nakada (1992) address this issue and indicate that geophysical modelling may explain discrepancies in ages and duration of the high-stand between those studies that date fossil reefs in the Atlantic, and those from the Pacific region. They indicate that isostatic adjustments around the globe might have been sufficient to explain many of the emerged MIS 5e shorelines without invoking a greater ocean volume. Second, the timing of the highstand will differ to the timing of MIS 5e as defined in deep-sea cores (i.e. attainment of the highstand will post-date the beginning of the marine-based isotope stage).

There are other significant assumptions implicit in the study of emergent coast-lines and in quantifying long-term rates of uplift, related to the value inferred for glacio-eustatic (ice-equivalent) sea level at the last interglacial maximum, the notion of a constant (linear) rate of uplift (Bloom *et al.*, 1974) of emerged coastlines since emplacement of the last interglacial shoreline, and the confidence with which the last interglacial shoreline can be identified, particularly in the absence of rigorous geochronological evidence. For example, it seems likely that uplift rates at Huon have varied through time (Kikuchi, 1990).

Numerous studies have indicated that sea level was not stable throughout the last interglacial highstand, but experienced one or more peaks. An early inference of such fluctuations came from the Huon Peninsula, with the concept of a double peak shown in the Reef VII complex. However, other researchers have indicated a rapid rise of sea level towards the end of the interglacial (Hearty *et al.*, 2007; Blanchon *et al.*, 2009). Even in early studies such as those of Neumann and Moore (1975) on the Bahamas, there was a discrepancy between evidence that sea level reached an elevation of about 6 m above present recorded by a prominent shoreline notch, and its presumed association with fossil corals that reach elevations of 2–3 m on Grand Bahama and Abaco, northern Bahamas.

While there is uncertainty as to what elevation sea level reached, from 2 m to as much as 9 m above present during the last interglacial, the level of uncertainty is relatively small compared with the magnitude of uplift commonly experienced on emergent coastlines such as Huon Peninsula, Barbados, or other locations where uplift has been rapid. With the long-term rate of uplift determined from elevation of MIS 5e shoreline successions, relative sea level at the time of deposition of younger coastal successions may be interpolated with a relatively high degree of certainty. Palaeo-sea-level estimations for coastal successions older than the last interglacial maximum, however, represent extrapolations beyond the temporal and spatial range of empirical observations and therefore may represent generalised inferences of potentially lower reliability, particularly in the context of longer-term uplift rates.

5.5 Coastlines in tectonically 'stable' cratonic regions

Coastlines located in intra-plate settings and remote from former icesheets, such as continental Africa, Madagascar, India, and Australia, appear relatively stable compared with the formerly glaciated areas or plate margins (see Figure 5.1). Many of these represent passive trailing-edge margins. The tropical to mid-latitude portions of such continental coastlines were not directly glaciated during the Quaternary, and are in the far-field; they may have experienced local neotectonic effects or hydro-isostasy. In contrast, continental shorelines in higher latitudes are more likely to have been subject to the effects of glacio-isostasy associated with collapse of forebulges that developed in the periphery of icesheets, such as along portions of the United States coastline (New Jersey, Virginia, and Georgia). In regions characterised by forebulge development, postglacial relative sea-level records are represented by a progressively rising sea surface with attainment of the Holocene highstand only in the past 2,000 years.

The southern Cape coastline of South Africa is an example of a passive margin located in the far-field of former icesheets. The region has been considered to show a high degree of tectonic stability. Coastal successions of last interglacial age (MIS 5e) in the southern Cape region occur between 3.8 and 8.5 m APSL (Carr *et al.*, 2010). The higher limit attained by the MIS 5e shoreline in this region may relate to hydro-isostatic processes given the wide continental shelf, rather than an expression of tectonic uplift. Terraces occur along the east coast of Africa (Alexander, 1968); the last interglacial has been reported at elevations of 5–6 m in Tanzania and as high as 10–12 m in Kenya (Ase, 1981; Braithwaite, 1984).

Australia is similarly remote from former icesheets; nevertheless, subtle vertical movement can be inferred by comparison of the evidence for elevation of last interglacial shorelines around its margin. These trends are examined below and, in the sections that follow, the emerged shorelines around the Mediterranean will be considered, and then those of the Caribbean region. These are both closer to the influence of former icesheets, and also have regions within them that have experienced a more complex tectonic history. They will be described as a prelude to examining plate-margin areas which have been even more tectonically active.

5.5.1 Australia

Australia, the flattest of all continents, is composed of extensive stable cratonic regions and experienced minimal ice accumulation during Quaternary glaciations. The southern Australian passive margin began to form with onset of rifting and separation from Antarctica during the final phase of fragmentation of the super-continent Gondwana, which commenced approximately 160 Ma (Veevers and Eittreim, 1988; Veevers *et al.*, 1991). The rifting regime changed to a drifting phase approximately 45 Ma (Veevers *et al.*, 1991). The last interglacial shoreline is relatively well preserved, partly as a result of aridity of much of the continent, which has also limited the supply of sediment to many coasts. The age and elevation of the last interglacial shoreline have been studied around the margin of the mainland, and on many adjacent islands, revealing evidence of subtle vertical displacements over the last glacial cycle (Murray-Wallace and Belperio, 1991; Murray-Wallace, 2002; Figure 5.6).

Cratons are represented by extensive areas ($> 1000 \, km^2$) of crystalline rock of igneous and metamorphic origin which, although showing evidence for crustal deformation, have not experienced significant recent crustal disturbances. The Yilgarn (Western Australia) and Gawler Craton (South Australia) comprise granites–greenstones and granites–gneisses, respectively. They were last deformed between 2.6 and 1.4 Ga and have since remained tectonically stable. Evidence for tectonic stability may take a variety of forms that include preservation of relict landforms, extensive landscapes with low surface relief with deep weathering profiles reflecting prolonged periods of continental-scale denudation, general absence of recently formed fault scarps and limited present-day earthquake activity. Similarly, recently deposited sediments show an absence of soft sediment deformation structures that could be attributed to seismic activity.

Geotectonically stable areas are less likely to preserve a lengthy record of Quaternary sea-level highstand successions due to erosion of older shoreline sequences during younger highstands, given that in the most general sense, interglacial sea levels repeatedly returned to a similar elevation during the middle to late Pleistocene. Thus many regions considered to be stable preserve evidence for only the last interglacial (MIS 5e) and Holocene highstands. The last interglacial shoreline maintains an elevation of around 2 m APSL for some 500 km along western Eyre Peninsula and adjacent regions (Murray-Wallace and Belperio, 1991). Although

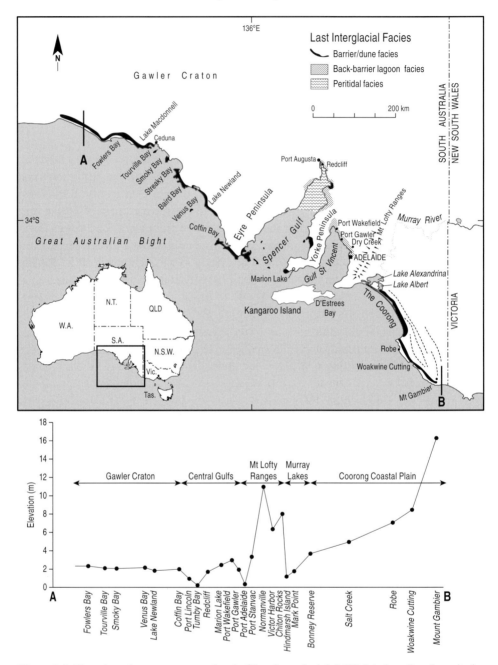

Figure 5.6 Elevation above present sea level of last interglacial (MIS 5e) shoreline deposits in southern Australia (after Murray-Wallace, 2000).

cratonic regions around the world display a high degree of tectonic stability, they are not immune to glacio-hydro-isostatic adjustments. As an example, the southern tip of Eyre Peninsula in southern Australia registered 0.5 m of flexure in response to hydro-isostasy following culmination of the postglacial marine transgression approximately 7 ka ago (Belperio *et al.*, 2002).

The western margin of Australia is a high-energy coastline composed of carbonate sediments which have formed a sequence of prograding beach ridges or massive dunes. In the Perth Basin a sedimentary record of siliciclastic shorelines, inner shelf and regressive-dune facies extends back from the early Pleistocene into the Pliocene, whereas subsequent middle and late Pleistocene sequences comprise marine bioclastic sediments (Fairbridge, 1954; Kendrick *et al.*, 1991).

Ningaloo Reef fringes 260 km of the Cape Range coast. Holocene reef has formed over a last interglacial reef limestone which is encountered at a depth of about 18 m, but a prominent last interglacial terrace, the Tantabiddi terrace, occurs above sea level on the adjacent mainland. Reefs were also extensive along much of this western coast, on prominent headlands or offshore islands, during the last interglacial. They extended as far south as Margaret River, where isolated corals have been dated to MIS 5e, implying that conditions were warmer than now as this is several hundred kilometres south of the present limit to corals on this coast (McCulloch and Mortimer, 2008). High-precision U-series ages on corals in their growth position from other sites along this coast provide important constraints on timing and duration of the last interglacial highstand, indicating that it extended from 128 to 116 ka (Stirling *et al.*, 1995, 1998). The corals show that sea level was at least 3 m APSL. A key site is Fairbridge Bluff on Rottnest Island, named after Rhodes Fairbridge who made a major contribution to reconstructions of sea level and particularly recognised the significance of morphological and faunal evidence on this coast. The site on Rottnest Island was described by Fairbridge and dated by Veeh (1966) and Szabo (1979). It comprises the Rottnest Limestone which contains platy species of *Acropora* corals, as well as other genera such as *Goniastrea*, up to 3 m APSL. The assemblage resembles that now found on tropical reefs many hundreds of kilometres to the north (Geenstein and Pandolfi, 2008).

Around Geraldton, MIS 5e reef has been found on the mainland at Leander Point and Burney Point, and at several outcrops on the Abrolhos Islands, 60 km offshore. On Wallabi Island, reef framework is reported up to an elevation of 4 m APSL; drilling has revealed a substantial period of reef growth in the last interglacial (Collins *et al.*, 1993; Zhu *et al.*, 1993). Fossil reef evidence at 16 separate sites along this coast indicates the last interglacial shoreline at about 3 m APSL. The only significant exception is at Cape Cuvier, where it has been reported that there are two morphologically distinct units: a lower, well-developed accretional reef between 3 and 5.5 m elevation, and an upper erosional terrace and incipient coralgal rim between 8.5 and 10.5 m elevation. Poorly constrained U-series ages were obtained on corals up to 5 m elevation, and although the site has been cited as evidence of a period of higher sea level on this coast (O'Leary *et al.*, 2008), it seems likely that it was subject to neotectonic activity, perhaps associated with the Cuvier Dome.

The last interglacial shoreline reveals subtle neotectonic changes in the southern Australian continental margin over the course of the last glacial cycle (Murray-Wallace and Belperio, 1991; Figure 5.6). Elevation of the MIS 5e shoreline is defined by the upper marine limit and the upper bounding surfaces of stratal units of last interglacial deposits. In this context, relative sea level at time of deposition may have been potentially higher for some deposits, but not hugely so, as the successions identified were predominantly deposited in intertidal to shallow subtidal environments as indicated by associated molluscan faunas. Notwithstanding this limitation, the data set provides a first-order approximation of differential neotectonism along the southern Australian margin since the last interglacial maximum. The different shoreline elevations directly relate to contrasting morphotectonic provinces.

Several distinct tectonic domains of pre-Quaternary origin occur along the southern Australian continental margin and have directly influenced Quaternary coastal evolution within these regions (Murray-Wallace, 2002). The contrasting records of coastal evolution within the different morphotectonic regions are supported by differences in broad-scale sediment geometries of last interglacial coastal facies and their vertical disposition with respect to present sea level (Figure 5.6). Approximately two-thirds of the southern margin coincides with the Yilgarn and Gawler Cratons (Preiss, 1993). The most consistent datum for the MIS 5e shoreline in southern Australia, mapped on Eyre Peninsula (Gawler Craton) as the Glanville Formation (Belperio, 1995), occurs for a distance of some 500 km of coastline, with the upper bounding surface of the formation at a consistent datum of 2 m APSL. Many of the coastal successions in this region are represented by prolific intertidal shell beds dominated by the estuarine bivalve mollusc *Anadara trapezia* that is presently found on the surface of mudflats (Murray-Wallace *et al.*, 2000).

Spencer Gulf, Yorke Peninsula, Gulf St Vincent and the Adelaide Foldbelt (Mount Lofty Ranges-Fleurieu Peninsula), to the east of the Gawler Craton, delimit another tectonic domain previously termed the South Australian Shatter Belt (Fairbridge, 1954; Figure 5.6). The gulfs and peninsulas represent graben and horst structures, respectively. A record of subsidence since the Tertiary is evident for the gulfs, with up to 90 m of displacement of the early Pleistocene Burnham Limestone in Gulf St Vincent (Belperio, 1995). The gulfs region also reveals greater variability in elevation of last interglacial coastal strata. The higher elevation of the last interglacial coastal deposits adjacent to the southern Mount Lofty Ranges reflect ongoing tectonic uplift.

The Murray Basin and Coorong Coastal Plain defines another distinct morphotectonic region to the east of the Adelaide Foldbelt (Figure 5.6). The Murray Basin was a large epicratonic marine basin during the Mesozoic and Tertiary. Immediately to the east of the foldbelt, the Murray Lakes region (Lake Alexandrina and Lake Albert) occurs within an area of known subsidence (Bourman *et al.*, 2000). This area grades into a broad region of epeirogenic uplift to the southeast (Sprigg, 1952). The record of uplift of the Murray Basin, evident since early Neogene time, is also evident for the Coorong Coastal Plain as shown by a series of emergent coastal barriers. In the latter, epeirogenic uplift appears in part to be associated with Quaternary volcanism (Murray-Wallace *et al.*, 1998). The Quaternary volcanic provinces of Victoria and southern South Australia represent an extensive region of over 400 volcanic centres (Matcham and Phillips, 2011) and thus the predicted position of the hot region as the Australian continent has moved north is now in the general region of northern Tasmania.

The last interglacial shoreline has been elusive across much of northern Australia. Initial sedimentological work in the macrotidal King Sound in Western Australia suggested that a 4–6 m-thick mangrove mud facies, termed the Christine Point Clay, was late Pleistocene in age (Semeniuk, 1980). However, mangrove stumps from a similar unit were radiocarbon dated by Jennings (1975) and shown to be Holocene, and the Christine Point Clay is now acknowledged as Holocene (Semeniuk, 2008). It is equivalent to a Holocene mangrove mud, which formed in a 'big swamp' across much of macrotidal northern Australia as postglacial sea level decelerated and stabilised close to its present level (Woodroffe *et al.*, 1985, see Figure 3.9).

Coral reefs are relatively poorly developed along this macrotidal coast which is highly turbid, but there is prolific reef growth on several shelf reefs on the northwest shelf. Coring has identified Pleistocene reefs beneath a Holocene reef margin. Detailed stratigraphy and age analysis have been undertaken for Scott Reef and several of the reefs that comprise Rowley Shoals. These offshore reefs formed on a former Miocene continental margin and have persisted there despite rapid subsidence of the shelf edge since the mid-Miocene. Seismic reflectors attributed to the last interglacial occur at 36 m below sea level in Mermaid Reef and 18 m below sea level in Imperieuse Reef, implying differential subsidence rates between these adjacent reefs. It is found at even greater depth (around 25 m) beneath Scott Reef (Collins, 2010).

Elevation of the last interglacial shoreline is poorly known across the mainland of northern Australia. Shore platforms on Cobourg Peninsula have been described at modern sea level. Their surface comprises a ferricrete in which there are well-preserved ripple marks; a number of U-series ages on the Fe/Mn accumulations implied a late Quaternary age (Woodroffe *et al.*, 1992). Several phases of ferricrete formation appear likely, and both last and penultimate interglacials may have contributed to the platforms that are several hundred metres wide in the area around Port Essington. Lithified barriers occur around the southern margin of the Gulf of Carpentaria with ridges up to 5 m elevation (Smart, 1976). Outcrops of calcarenite, rising to 3 m on Rose Reef in the southern Gulf, were inferred to be late Pleistocene (Nott, 1996).

Last interglacial foundations provide antecedent topography over which Holocene reefs of the Great Barrier Reef have formed. Although there are Pleistocene sedimentary deposits along the mainland coast at up to ~5 m elevation, similar to those across northern Australia, last interglacial reef limestones are almost all below sea level on the reef itself. A limestone of unknown age, inferred to be last interglacial, has been reported outcropping on the island of Saibai in Torres Strait but, under reefs in central Torres Strait, a Pleistocene foundation, again inferred to be of last interglacial age, occurs at 5–7 m depth (Woodroffe *et al.*, 2000). South of Mackay, last interglacial limestone has been found on Digby Island, but occurs at successively deeper depths with distance across the reef (Hopley *et al.*, 2007). There are few locations at which the reef limestone underlying the modern Holocene reefs has been U-series dated to confirm it is of last interglacial age (Marshall and Davies, 1984). The contact with the Pleistocene appears to be as shallow as 4 m on Bewick Island (Thom *et al.*, 1978), but occurs much deeper beneath other islands (11 m on Raine Island; 10–23 m on One Tree Reef; 13 m on Heron Island; 15 m on Hayman Island; 17 m on Redbill Reef; 15–22 m on Stanley Reef; and as deep as 28 m on Myrmidon Reef), as outlined by Hopley *et al.* (2007). Subsidence can be inferred for parts of the Great Barrier Reef, although the irregular topography of the underlying Pleistocene surface, as revealed by both seismic profiling and coring, indicates that this may have been accentuated by solution and weathering during the lowstand. These earlier reefs may not all have grown to the sea level that was characteristic of the peak of the last interglacial.

Barriers of Pleistocene age are known along much of the coast of New South Wales and southern Queensland (Murray-Wallace and Belperio, 1991). These lithified barriers are known as the Inner Barrier, and are exposed above the surface north of Newcastle (Chapman *et al.*, 1982; Thom, 1984). Corals from beneath these barrier sands have been dated indicating last interglacial ages from Grahamstown, Evans Head, as well as on North Stradbroke Island (Marshall and Thom, 1976; Pickett *et al.*, 1989). Those at Evans Head indicate a sea level ~4–6 m APSL. Similarly, ages on the arcoid bivalve, *Anadara trapezia*, exposed in estuarine sediments at Largs in the Hunter Valley, indicate last interglacial sea level ~4–5 m APSL (Thom and Murray-Wallace, 1988). A last interglacial sea level has been inferred to be 3 m APSL at Tathra in southern NSW (Young *et al.*, 1993a).

Tasmania has a distinct neotectonic history. The island is characterised by emergent last interglacial coastal landforms and sedimentary successions of shallow-marine origin (Bowden and Colhoun, 1984; Murray-Wallace and Goede, 1991, 1995). The last interglacial successions represent the highest topographic occurrences of coastal strata of this age on the Australian continent (11–32 m APSL) and thus tectonic processes must be considered to explain their anomalously high elevation. Glacio-isostatic and hydro-isostatic crustal readjustments cannot explain the magnitude of uplift in view of the narrow shelf surrounding Tasmania together with the limited extent and thickness of ice in the central highlands of the island during the LGM (Colhoun *et al.*, 2010). Shallow subtidal, shelly sands at Mary Ann Bay near Hobart, extend up to 24.6 m APSL (Murray-Wallace and Goede, 1991). Other high level deposits of the last interglacial have been described from northeastern and northwestern Tasmania, and collectively their high elevation has been attributed to regional crustal doming associated with an emerging hot region, or crustal under-plating during northward movement of the Indo-Australian Plate (Bowden and Colhoun, 1984; Murray-Wallace and Goede, 1991, 1995).

5.5.2 *Southern Africa*

The southern continental margin of Africa preserves a rich array of Pleistocene coastal landforms and relict shoreline deposits. The early mapping by Krige (1927) identified a prominent 'shoreline of emergence' at 6 m APSL along 1,500 km of the coastline of South Africa extending from Durban on the eastern coast to Port Nolloth on the west coast. Extensive coastal barrier systems comprising dune lime-stone (aeolianite) have been correlated to the last two glacial cycles based on OSL dating (Bateman *et al.*, 2004, 2011). The coastal barriers represent major repositories of skeletal carbonate sand generated from the comminution of marine invertebrates on the 100 km-wide Agulhas Shelf. Relict gravel beach, sandy shoreface, and estuar-ine facies at Cape Agulhas point to a relative sea level of approximately 7.5 m APSL during the last interglacial maximum and dated to 118 ± 7 ka based on OSL dating of quartz sand (Carr *et al.*, 2010). At Swartvlei and Groot Brak estuaries, regressive successions of tidal inlet facies indicate a relative sea level of up to 5.6 m APSL at approximately 122 ± 7 ka ago based on the OSL dating of superposed aeolian dune facies.

5.6 Coastlines of emergence

In the following discussion, selected examples are examined of Quaternary neotec-tonism at plate margins that include convergent continental margins and island arcs. Uplift is a characteristic feature of coastal regions located near subduction complexes. In these settings, marine terraces are generally very well developed in view of the mutual interplay of coastal processes operative during discrete sea-level highstands, and the relatively rapid rates of uplift of these coastal landscapes. Emergent marine terraces, for example, are particularly well developed along large sectors of the Peruvian–Chilean coastline (Ortlieb *et al.*, 1996) and have long been the subject of study.

Island-arc systems have developed where oceanic lithosphere is subducted beneath adjoining oceanic lithosphere on convergent plate margins at which rapid uplift is commonly evident (Hamilton, 1988). Rapid uplift ensures that marine abrasion platforms and coral reef successions are removed from the effects of younger sea-level highstands, resulting in development of a succession of emergent marine terraces preserved in a 'staircase' fashion. Average long-term rates of tectonic uplift in subduction settings have been documented up to 3.5 m/ka on the Huon Peninsula, Papua New Guinea (Chappell *et al.*, 1996a,b) and up to 5.5 m/ka in eastern Taiwan (Petley and Reid, 1999).

5.6.1 Huon Peninsula

The Huon Peninsula, part of the northern cordillera of Papua New Guinea, is the foremost example of an emergent coastal complex in a subduction zone (Figures 5.7 and 5.8a). It is situated in a particularly complex geotectonic setting on the South Bismarck Sea Plate, a microplate between the northward trending Indo-Australian Plate and the southwesterly trending Pacific Plate in a zone of one of the fastest rates of plate convergence in the world (~120 km/Ma). Uplift of the peninsula relates to its proximity to the subduction zone at the New Britain Trench and a series of thrust complexes to the southwest of the peninsula. The significance of the terraces was first recognised by Fairbridge (1960) based on interpretation of reconnaissance oblique aerial photographs taken during the Second World War. The region has been the focus of meticulous studies undertaken by John Chappell and colleagues since the early 1970s which have provided a particularly detailed record of eustatic sea-level changes for the late Quaternary (Veeh and Chappell, 1970; Bloom *et al.*, 1974; Chappell, 1974a,b; Chappell and Polach, 1976, 1991; Aharon and Chappell, 1986; Chappell *et al.*, 1996a,b). Collectively, they conclusively demonstrated that the marine terraces provide a record of glacio-eustatic (ice-equivalent) sea-level change. In particular, the similarity in relative order of age–height relationships of different interstadial and interglacial terraces of the Huon Peninsula with the reef succession on Barbados (Mesolella *et al.*, 1969; Steinen *et al.*, 1973) provided compelling evidence that the late Pleistocene glacio-eustatic sea-level record was fundamentally different to that previously inferred from the Mediterranean Basin (i.e. that the record was primarily one of sea-level oscillations in which successive interglacial highstands tended to attain a common datum, rather than representing a continuum of progressively falling sea levels).

A staircase of more than 20 former fringing coral reefs is preserved along an 80-km sector of the Huon Peninsula. At their maximum, they extend up to 1,000 m APSL (Chappell *et al.*, 1996b). The sequence comprises offlapping fringing and barrier reefs, with lagoonal successions and associated uplifted gravel fan-delta facies. Barrier reefs with lagoon facies characterise the terraces along the north-western sector of the peninsula, while fringing reefs and gravel terraces are more common in the southeast. The lagoons are relatively narrow, elongate features and seldom exceed depths of 10 m. Preservation of the entire reef complex is due to limited presence of rivers along the coastline and rapid uplift. The sedimentary successions provide a detailed archive of sea-level highstands during the past 340 ka. Based on U-series dating, early research on the peninsula identified sea-level highstands at 30, 40, 50, 60, 80, 105, 120, 140, 185, and 220 ka (Chappell, 1974a). Bloom *et al.* (1974) also reported U-series ages from several of the Huon

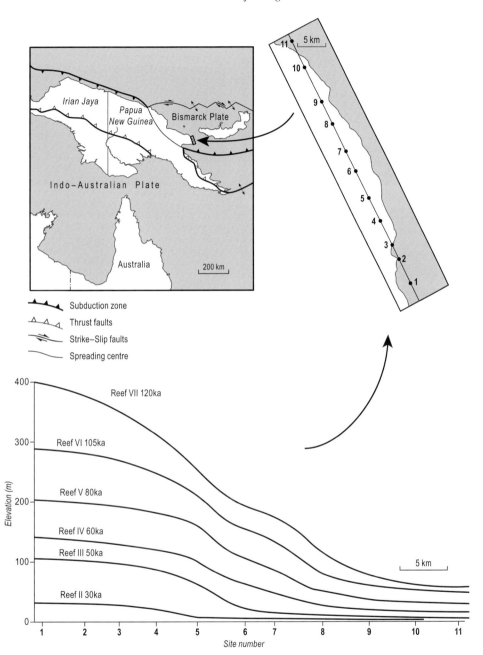

Figure 5.7 Location map of the Huon Peninsula, Papua New Guinea, showing the general geographical setting of the emergent coral reef complex within a plate-tectonic framework and variation in elevation of principal reef terraces from south to north (Chappell, 1974a).

(a)   (b)

(c)   (d)

Figure 5.8 Photographs of emergent coastal landforms; (a) oblique aerial photograph of the Huon Peninsula, Papua New Guinea. The arrow points to the reef terrace associated with the last interglacial maximum (MIS 5e – 125 ka; photograph: Sandy Tudhope); (b) Robe Range, Coorong Coastal Plain, southern Australia, an emergent coastal barrier aeolianite complex associated with MIS 5c (105 ka; photograph: Colin Murray-Wallace); (c) MIS 5c marine terrace fronting the Coastal Cordillera, Hornitos region, northern Chile (photograph: Colin Murray-Wallace); and (d) marine terraces on the west coast of San Clemente Island, California. The MIS 5e terrace is the second emergent terrace from the modern shoreline (photograph: Dan Muhs).

Peninsula terraces. Based on a calibration of the last interglacial terrace at 125 ka with an assumed sea surface of 6 m APSL at the time of its formation, Bloom *et al.* (1974) concluded that late MIS 5 interstadial sea levels were at –15 m at 103 ka (MIS 5c), and –13 m at 82 ka (MIS 5a). In addition they concluded that sea level was at 28 m BPSL at 60 ka, about 38 m BPSL at 42 ka and 41 m BPSL at 28 ka, but noted in view of the uplift history of the peninsula that if the 28 ka reef is in fact older, then the palaeo-sea level would actually be lower. The Huon Peninsula sea-level record also reveals evidence for short-term sea-level events associated with Dansgaard–Oeschger and Heinrich Events (Chappell, 2002; see Section 6.1.2) and, accordingly, represents one of the world's most detailed archives of eustatic sea-level change in the later Quaternary.

The Huon Peninsula has been a favoured site for studies of late Quaternary sea-level changes as there is a clear physical separation between highstand events as represented by morphostratigraphically distinct regressive coral terraces. Late Pleistocene and Holocene records of tectonism reveal broadly comparable rates of uplift as shown from radiocarbon dating of emergent *in-situ* corals (Chappell *et al.*, 1996b). The Holocene reef succession reveals up to six well-defined steps represented by relatively thin regressive terraces formed upon an underlying transgressive reef complex. The latest Pleistocene record (MIS 3; 52–33 ka) reveals up to 15 stepped terrace features. The magnitude of uplift associated with earthquake events varies along the length of the peninsula, with the highest rates of 3.3–3.5 m/ka confined to the southeastern portion closest to the subduction zone and a lower rate of 0.5–0.7 m/ka occurring in the northwestern region.

The reef terraces preserve facies associations typical of emergent reef complexes with shallow-water reef, fore-reef talus, deep fore-reef, back-reef, and lagoon facies (Figure 5.9). The generalised reef facies structure includes the corals *Acropora humilis* and *A. cuneata* in the reef crests, resting stratigraphically above *A. palifera*. Restricted patch reefs occur in the back-reef lagoonal environments. Considering the last interglacial reefs as a representative example of other reef terraces, individual reef terraces record evidence of rising sea level (transgression), and the ensuing highstand, the latter represented by regressive lithofacies associations. Contrasts in facies architecture of the reefal units are largely a function of differential rates of eustatic sea-level changes and tectonic uplift (Chappell, 1983b). As the time-averaged rate of tectonic uplift along the peninsula is not uniform, differences in internal geometry of reefs and their elevation are evident along their strike. At times of relative sea-level rise (submergence), individual reefs build upwards and seawards above their fore-reef facies resulting in a landward dip in facies boundaries. In this context, the horizontal distance between the reef crest and the land increases, resulting in a wider back-reef lagoon. In contrast, during intervals of relative sea-level fall (emergence), shoreline notches are formed on the seaward side of the reef crests.

The last interglacial (125 ka) reefs are regarded as transgression-limited in the sense that growth of the main structure (reef VIIa in Figure 5.9) ceased when sea level stopped rising at about 130 ka (Chappell, 1983b). A subsequent relative rise in sea level at ~120 ka resulted in reef VIIb forming as a veneer on the principal reef structure. In the contrasting examples illustrated, the reef from northwestern Huon Peninsula with the lower uplift rate (0.5 m/ka) is wider in shore-normal profile with a well-developed lagoon and shallow-water reef facies association (Figure 5.9). This is a function of the higher relative rate of eustatic sea-level rise in relation to the rate of tectonic uplift, as well as the ability of the reef to keep up with relative sea-level rise. An additional factor influencing these contrasting reef geometries is the slope of the substrate on which the reefs have developed. As the substrate slope is twice as steep in the northwestern reef, there is less accommodation space for development of a lagoon landward of the reef crest.

In addition to depositional evidence for relative sea-level changes, erosional shoreline notches are also developed on the reefal limestones, attesting to former

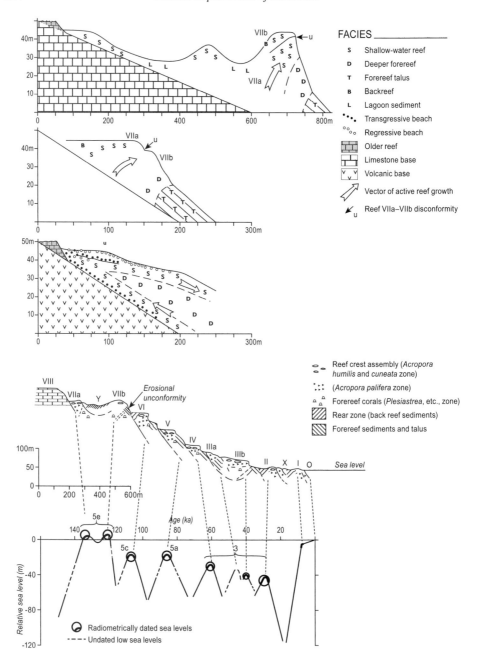

Figure 5.9 Facies architecture of the Huon Peninsula reef complexes in response to the combined effects of vertical crustal movements (uplift) and glacio-eustatic sea-level changes (modified after Chappell, 1983b).

sea-level highstands. For reef terraces I to VIIb the notches are still relatively well preserved and provide an opportunity to obtain samples close to the culmination of transgressive events (Chappell, 1974a). The reef terraces older than the last interglacial are more degraded, however, and have only poorly preserved wave-cut cliffs and shoreline notches.

At Kwambu on the northwestern sector of the Huon Peninsula, the Holocene record from this reef complex is based on radiocarbon dating of 54 coral samples (Chappell and Polach, 1991; Chappell *et al.*, 1998). Rates of coseismic uplift deduced from radiocarbon dating of corals younger than 7 ka, and from corals of last interglacial age (MIS 5e), are broadly similar. The long-term rate of uplift as deduced from the last interglacial reef complex at Kwambu, based on U-series dating is 1.9 m/ka (Stein *et al.*, 1993). The upper surface of the Holocene barrier and fringing reefs at Kwambu have been uplifted to approximately 13 m APSL and ceased growth at 7 ka when coastal emergence raised the reef above its contemporary sea level. Coseismic uplift events involving approximately 1 m of uplift, identified within the Holocene coral record, occurred at 10.5, 9.6, 8.5, 6.9, 6.2, and 2.5 ka BP (Chappell *et al.*, 1998). In contrast, observed coseismic events at Huon, such as that of 15 May 1992, resulted in only limited uplift and temporary disruptions to coral communities (Pandolfi *et al.*, 1994, 2006). Although uplift appears to be coseismic in nature (Ota, 1996), it is now apparent that the major terrace features formed during MIS 3 and 4 owe their origin to fluctuations in sea level that have been superimposed on the seismic record. For example, the sequence of six terraces, each of 10–20 m thickness, near the village of Bobongara are now interpreted as related to oscillations of sea level linked with Heinrich Events that have been identified in the north Atlantic (Yokoyama and Esat, 2011).

The palaeo-sea level Psl_t, inferred from an emergent marine terrace is derived based on the following equation:

$$Psl_t = tu - (h + d) \qquad (5.1)$$

where t is the time (age) of the emergent coastal deposit of interest; u is the uplift rate independently determined from, for example, the MIS 5e shoreline; h is the elevation of the feature of interest APSL; and d is the water depth at the time of its formation. Uncertainties implicit in this calculation include inferred water depth of palaeo-sea-level indicator at time of its formation, longer-term changes in tidal amplitude associated with coastal evolution, the age-uncertainty term associated with a ^{14}C or U-series age, and the assumption of a constant rate of uplift.

The prominence of terrace VII (MIS 5e) on the Huon Peninsula has rendered its use as the benchmark for determining uplift rates for different sections along the coast (Figure 5.8a). Nevertheless, details of the morphostratigraphy of that terrace assemblage remain unresolved. Particularly anomalous are deposits from Aladdin's Cave, 90 m below the MIS 5e terrace, which appears to record a lower sea-level

excursion that occurred during the MIS 6 transgression associated with the penultimate deglaciation (Esat *et al.*, 1999). Heads of *Porites* and Faviid corals found
in growth position within the cave, indicate relatively shallow water at time of
formation (< 12 m). The corals yielded a mean age of 130 ± 2 ka (2σ-uncertainty).
At face value the ages are problematic because they imply sea level fell by over 60 m
after 134 ± 1 ka, a mean age based on dating of corals from the shallow fore-reef
position of reef terrace VIIb at the nearby Kwamba section. Reef Terrace VIIb at
Kwamba occurs at 220 m APSL and Aladdin's Cave at 100 m APSL. Oxygen-isotope
and Sr/Ca analyses on *Porites* corals from Aladdin's Cave suggest sea-surface
temperatures of approximately $22 \pm 2°C$ at 130 ± 2 ka compared with modern temperatures of $29 \pm 1°C$ (McCulloch *et al.*, 1999). These data imply that during Termination II sea level fell significantly and then rose rapidly between 130 and 128 ka.

    Issues critical for assessing the validity of the palaeo-sea-level record implied for
the penultimate deglaciation as deduced from the Aladdin's Cave coral succession
include preservation state of replicate coral specimens for U-series dating, concordance of replicate U-series ages, and the morphostratigraphical context of fossil
corals. In addition, the reliability of the U-series mean coral age of 134 ka from
Terrace VII representing an early onset for the last interglacial maximum is a further
consideration in the palaeo-sea-level history of this region. The corals within
Aladdin's Cave could potentially represent a *fenster* (window) within the lower
portion of the Terrace VII structure (lower fore-reef, keep-up reef facies). Accordingly, if the *in-situ* context of the Aladdin's Cave corals is beyond doubt, then an
alternative question may centre on the validity of the inferred spike in sea level based
on the 134 ka mean age from the reef front of Kwamba Terrace VIIb. It is noted that
Esat *et al.* (1999) also reported ages of 108–115 ka from Aladdin's Cave. Clearly,
additional research is needed from this locality as the EPICA Dome C ice-core record
casts doubt on the existence of a major sea-level spike preceding the last interglacial
maximum (Masson-Delmotte *et al.*, 2010).

    The interpretation of environments during the last interglacial highstand is also
unclear. The existence of an inner fringing reef and an outer barrier reef, separated by
a lagoon, has been subject to several interpretations. It has been used to invoke two
high points within the interglacial, although when compared with modern reefs at
the site, such an interpretation is not necessary (Stein *et al.*, 1993). In this respect, the
complex biodiverse reefs of the Indo-Pacific are more difficult to interpret than the
uplifted reefs of Barbados in the Caribbean, because zonation of corals can be more
clearly reconstructed in the latter (particularly the branching *Acropora palmata* which
is characteristic of the reef crest). Although general patterns in the growth forms
of corals and their relationship to environmental gradients have been proposed
(Chappell, 1980), there remains controversy about the detail of the MIS 5e record
at Huon (Blanchon, 2011a).

    One of the complexities in reconstructing a palaeo-sea-level record from the Huon
Peninsula, as with other rapidly emerging coastlines, is superimposition of younger

highstands on coastal features formed during sea-level lowstands. Accordingly, the Huon Peninsula does not represent solely a unilinear trend of offlapping reefs that decrease in age from the highest (oldest) reef progressively down to modern sea level. Submarine terrace features formed at times of sea level lower than present may be only poorly preserved once they have been uplifted significantly above present sea level. Subaerial depositional processes may in part cover the former terraces, rendering it extremely difficult to obtain reliable stratigraphical information for inferring former sea level, and for obtaining materials for dating. In addition, superimposition of a younger coralline facies on the older, now emergent former submarine terrace, if sampled for dating, will yield materials relating to the younger, more recent highstand, rather than the older reef that formed at a time of lower sea level. Thus, knowledge of some of the lower sea-level events may remain undefined.

As with all emergent coastal complexes, assumptions necessary for reconstructing a record of relative sea-level changes from the Huon Peninsula have involved an estimate of the elevation of the sea surface during the last interglacial maximum at 125 ka (MIS 5e) and the assumption of a constant rate of uplift of the peninsula. As global estimates of ice-equivalent sea level for the last interglacial have ranged between 3 and 9 m APSL (Veeh, 1966; McCulloch and Esat, 2000; Siddall *et al.*, 2006), the overall uncertainty introduced in the quantification of other sea-level events for the middle and late Pleistocene may amount to ± 4 m. While this is less significant for the determination of former sea level as derived from the Huon Peninsula in view of the high rates of tectonic uplift, it is significant when extrapolating the Huon Peninsula data to more stable tectonic settings.

5.6.2 Barbados

Since the early palaeo-sea-level investigations of Mesolella (1967, 1968) and Mesolella *et al.* (1969), which examined the emergent Pleistocene coral reefs of Barbados, the island has featured prominently in investigations of Quaternary sea-level changes (Broecker *et al.*, 1968; Steinen *et al.*, 1973; Hays *et al.*, 1976; Shackleton and Matthews, 1977; Bender *et al.*, 1979; Gallup *et al.*, 1994, 2002; Schellmann and Radtke, 2004a,b). Interest in the Barbados record centres on its correlation with the southwestern Pacific palaeo-sea-level record derived from the Huon Peninsula (Steinen *et al.*, 1973). Barbados is located in the Caribbean Sea approximately 150 km east of the Lesser Antilles. Covering an area of 430 km^2, the island is on a fore-arc ridge, representing the subaerial component of an accretionary prism resulting from the North American Plate being subducted beneath the Caribbean Plate (Schellmann and Radtke, 2004a,b; Figure 5.10). Mud diapirism since late Eocene time is responsible for ridge formation. Approximately 85% of the island is covered by Pleistocene coralline limestone that ranges between 15 and 130 m in thickness, extending up to 300 m APSL.

The coralline limestone formed from a series of fringing reefs that have developed as the island progressively emerged during the Pleistocene (0.23–0.38 m/ka).

The northern part of the island has a sea-level record extending beyond 600 ka, whereas the island's southern sector has an emergence record younger than 400 ka (Schellmann and Radtke, 2004a,b). Two prominent, high coastal cliffs extend around the island, with the First High Cliff trending east–west on the southern part of the island and the Second High Cliff trending north–south on the northwestern part of the island (Figure 5.10). Prominent cliffs represent boundaries between individual reef complexes and have been modified by subsequent sea-level changes and sub-aerial processes (see Figure 3.12a). In southern Barbados the individual terraces are generally less than 1 km in cross-section and preserve diverse coral faunal assemblages with the genera *Acropora*, *Montastrea*, *Diploria*, *Porites*, *Agaricia*, and *Siderastrea* commonly represented. The emergent reef terraces show consistent facies associations irrespective of age, with a zone of calcarenite in the fore-reef zone, and well-defined reef-crest and back-reef zones and a lagoonal facies in the back-reef zone (Mesolella, 1967; Schellmann and Radtke, 2004a,b).

A detailed record of sea-level highstands for the past four glacial cycles has been derived from the Barbados reef complexes based on U-series and ESR dating (Schellmann and Radtke, 2004a,b). The record indicates that sub-Milankovitch sea-level oscillations were responsible for the formation of three morphologically distinct reef terraces. They include the Maxwell Hill Terrace (22 m APSL), Rendez-vous Hill I (36 m APSL), and Rendezvous Hill II (39 m APSL). Uplift-corrected sea levels for the two younger MIS 5e terraces depend on the assumed palaeo-sea level at time of formation of Rendezvous Hill II (Schellmann and Radtke, 2004a,b). If Rendezvous Hill II formed at 6 m APSL then uplift-corrected sea levels for the Rendezvous Hill I and Maxwell Hill Terraces are 4 m APSL and 8 m BPSL, respectively.

In view of the slower rates of uplift of Barbados, coastal terraces occur at lower elevations than on the Huon Peninsula. The last interglacial Rendezvous Hill Terrace in southwestern Barbados, for example, occurs at 60 m APSL, implying an averaged long-term uplift rate of 0.43 mm/ka assuming a 6 m APSL sea surface for the last interglacial maximum at 125 ka. Holocene beach ridges and emergent reefs commonly reach up to 2 m APSL around extensive sectors of the island's coastline and in places are backed by shoreline notches and visors that occur up to 7 m APSL. At Inch Marlowe Point on the island's southern coastline in an area characterised by slower rates of uplift, middle Holocene shore platforms occur only up to approximately 0.5 m APSL (Schellmann and Radtke, 2004a,b).

In the South Point area in southern Barbados, three well-defined coral terraces indicate that sea level during MIS 5c ranged between 22 m BPSL (104 ka), 17 m BPSL (104 ka), and 10 m BPSL (105 ka) (Schellmann and Radtke, 2004a,b). These inferred palaeo-sea levels are based on an assumed 6 m APSL value for the last interglacial maximum, and accordingly, lower sea levels would be inferred if a lower value was assumed for the MIS 5e sea surface. Sea level for interstadial MIS 5a was identified at 17 m BPSL at 75 ka and 18 m BPSL at 85 ka.

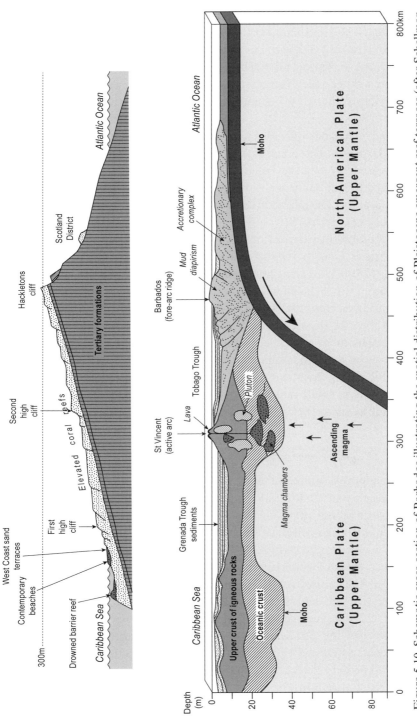

Figure 5.10 Schematic cross-section of Barbados illustrating the spatial distribution of Pleistocene emergent reef terraces (after Schellman and Radtke, 2004a).

Three well-defined coral terraces also formed during the penultimate interglacial (MIS 7). The Kendal Hill and Vauxhall Terraces revealed uplift-corrected palaeo-sea levels at 10 m BPSL at 222 ka and < 1 m BPSL at > 204 ka. An additional highstand at 4 m APSL, dated at 224 ka, was identified from a smaller, unnamed terrace to the west of Vauxhall Terrace. These estimates place MIS 7 sea level significantly higher than previously inferred from marine-based oxygen isotope records (Chappell and Shackleton, 1986).

Three sea-level highstands were also identified for the pre-penultimate interglacial (MIS 9) from Barbados. The Kingsland Terrace, with a surface elevation of 73 m APSL, and an age of 289 ka, places uplift-corrected sea level at 1 m APSL. An older (300 ka), unnamed terrace to the west of Kingsland Terrace with a surface elevation at 82 m APSL indicates an uplift-corrected palaeo-sea level of 7 m APSL, and the Balls Church Terrace (91 m APSL) with an age of 334 ka indicates a palaeo-sea level of 8 m APSL. Finally, two terraces of MIS 11 age with resolvable ages of 398 ka (St David's Terrace – 110 m APSL) and 410 ka (notionally correlated with the Kent Terrace – 120 m APSL) suggest uplift-corrected palaeo-sea levels at 11 and 18 m APSL, respectively (Schellmann and Radtke, 2004a,b).

An important lesson from the Barbados record is that uplift may not have been constant during the Pleistocene. Besides the obvious differential uplift of individual reef terraces such as those associated with MIS 5e which range from 20 to 60 m APSL, the terraces in the Christ Church area are represented by warped anticlines. The detailed ESR dating investigations of Schellmann and Radtke (2004a,b) have revealed that long-term uplift rate in the Clermont Nose emergent reef section on western Barbados has changed between 0.24 and 0.45 m/ka during the past 500 ka.

5.6.3 Convergent continental margins: Chile

Emergent marine terraces are particularly well developed along large sectors of the collision-style coastline of Chile and Peru due to regional aridity and limited fluvial erosion, aiding their preservation. The marine terraces of the Coquimbo–La Serena area on the central Chilean coastline were first described by Charles Darwin during his voyage around the world (Darwin, 1845, 1896). At Coquimbo Bay he noted six well-defined gently seaward sloping terraces (designated A–F) with heights estimated at approximately (A) 8 m, (B) 21 m, (C) 37 m, (D) a degraded terrace whose elevation was not reported, (E) 92 m, and (F) 111 m APSL. He also noted fossil molluscs within the terrace sediments and that the fossils had modern equivalents along the coastline. Darwin suggested that the terraces resulted from gradual uplift with minor episodes of more rapid uplift associated with earthquakes. Steep scarps backing each terrace were attributed to periods of cessation of uplift and marine erosion before continued uplift removed the terrace from the influence of marine agencies. He did not invoke eustatic sea-level changes to explain the origin of the features, as the influence of Quaternary glaciations was not then appreciated. Darwin also suggested that the rate of uplift was greater inland than closer to the coast based on the well-defined seaward slopes of the marine terraces. D'Orbigny (1842, p. 94) also examined the marine terraces of the Chilean coastline and suggested that their formation had been 'great and sudden'.

In the Hornitos area of northern Chile, four marine terraces have been identified within a narrow coastal plain on the footslopes of the Coastal Cordillera, a linear series of mountains that extend up to 2,000 m above sea level (see Figure 5.8c). The marine terraces have been identified based on morphostratigraphy and their ages assigned by U-series, AAR, and ESR dating of fossil marine molluscs (Ortlieb *et al.*, 1996). The terraces represent well-defined, seaward-dipping marine abrasion platforms developed on Jurassic volcanic rocks, and covered with thin successions of cross-bedded shelly gravels and alluvial sediments. The seaward edges of the terraces are represented by former marine cliffs, and the oldest and highest terrace is covered by distal alluvial fan sediments of the Coastal Cordillera. The youngest Pleistocene terrace surface ranges in elevation from 18 to 30 m APSL and is draped by a composite deposit of sand presumed to have been deposited during marine MIS 5e and 5c (Ortlieb *et al.*, 1996). U-series dating of fossil molluscs by TIMS, yielded ages consistent with MIS 5c (105.3 ± 1.0 ka to 109.2 ± 0.9 ka), while conventional alpha-counting on different shells yielded an age of 124 ± 7 ka, suggesting a correlation with the last interglacial maximum (MIS 5e) and implying a mixed assemblage of shells (Ortlieb *et al.*, 1996). Landward of the first Pleistocene terrace, a more degraded and narrower terrace occurs, the landward limit reaching up to 50–55 m APSL. The terrace also preserves a composite stratigraphical section and was attributed to two highstands during a single interglaciation (MIS 7). The third oldest terrace is the widest terrace and is up to 80 m APSL and is mantled by a cover sand typically less than 2 m thick with numerous relict beach ridges. A sample of the bivalve mollusc *Eurhomalea rufa* from the oldest and most landward terrace (Terrace IV at 90 m APSL) yielded an age of 217 ± 9 ka. Ortlieb *et al.* (1996) concluded that acceleration in subduction motion occurred in the later Quaternary based on a comparison of inferred uplift rate as determined on the youngest terrace with the highest terrace of 170 m APSL of presumed early Pleistocene age (age undefined).

Many of the marine terraces along the coastline of Chile illustrate the remarkable erosional power of the sea in forming abrasion platforms on resistant bedrock during sea-level highstands (Paskoff, 1970). Where abrasion platforms are particularly wide (> 1 km), the question arises of whether the features formed in more than one sea-level highstand in view of the resistance of the bedrock. Aminostratigraphy has identified mixed-age populations of fossil shells within the marine terrace cover sands, consistent with successive terrace occupation by sea-level highstands (Leonard and Wehmiller, 1992). An additional difficulty in studying the terraces is confidently distinguishing the products of marine or terrestrial processes. Some of the highest and most landward terraces may be bedrock-covered pediments flanking the Coastal Cordillera rather than marine terraces, highlighting the problem of convergence in landform development.

Rates of tectonic uplift associated with subduction vary spatially along the Chilean coastline. Time-averaged rates of coseismic uplift along the coastline are significantly less than the Huon Peninsula, Papua New Guinea, and are typically 100–150 mm/ka (Ortlieb *et al.*, 1996). The last interglacial shoreline shows variability in altitude resulting from contrasting rates of tectonic uplift associated with subduction of the Nazca Plate beneath the South American Plate. In the Hornitos area, on the northern coastline of Chile, the last interglacial shoreline occurs up to 36 m APSL. Elsewhere along the Hornitos Embayment, it occurs at elevations ranging between 18 and 30 m APSL (Ortlieb *et al.*, 1996). The oldest, highest, and most landward terrace of presumed latest Pliocene to early Pleistocene age also reveals significant variation in height. Between Iquique and Antofagasta, it occurs between 200 and 220 m APSL, and at Antofagasta, it (Antofagasta Terrace) occurs at 100 m APSL, implying a long history of differential uplift (Paskoff, 1970).

5.7 Vertical crustal movements associated with glacio-isostasy: Scandinavia

As the notion that regions of the Earth's surface had undergone uplift became more widely accepted in the nineteenth century, it was recognised how widespread such movements have been. It has long been clear that formerly glaciated regions underwent rebound following melt of continental icesheets. On return from the voyage aboard HMS *Beagle*, during which he had seen convincing evidence of uplift in Chile, Darwin visited Glen Roy in Scotland where he saw horizontal terraces at more than 300 m APSL on either side of the valley. He concluded that the entire valley had been inundated by the sea and that these terraces, popularly known as the 'parallel roads' of Glen Roy (see Figure 2.7a), represented former shorelines that had been episodically uplifted, an idea which he proposed in a paper accepted by the Royal Society shortly before he was elected a Fellow (Darwin, 1839). The site played an important part in the suite of evidence that Louis Agassiz drew upon to substantiate his Ice Age theory. Agassiz, however, realised from the lack of marine fossils that these were not shorelines formed at sea level, but developed instead around the margin of ice-dammed proglacial lakes, as later described by Thomas Jamieson (1863). These apparently horizontal lake shorelines display subtle gradients, broadly consistent with the trend of nearby marine shorelines (the Main Rock Platform) that attest to glacio-isostatic adjustments since they formed during the Younger Dryas (Dawson *et al.*, 2002). In passive margin regions that were formerly glaciated, glacio-isostasy represents the dominant mechanism for coastal uplift. Since the LGM, the sea-level records of these regions have been characterised by a relative fall in sea level. These regions occur within Zone I as defined by Clark *et al.* (1978) and include northeastern Canada, Greenland, Iceland, Svalbard, and the British Isles (see Figure 2.11).

Significant coastal landscape changes accompanied progressive deglaciation of the Fennoscandian Icesheet which commenced from about 13 ka ago (^{14}C years). The region is a classic example of Quaternary landscape change and many studies have examined relative sea-level histories of this region with valuable reviews presented by Donner (1995) and Lambeck *et al.* (1998). The Fennoscandian Shield was almost entirely covered by ice during the Weichselian glaciation with the southern limit of the icesheet extending through northern central Germany, Poland, and Belarus. The northwestern margin is poorly defined where the limit of ice is likely to have extended onto the continental shelf of Norway and in the northeast into what is now the southern Barents Sea. Fennoscandia is situated on the western portion of the Eurasian Plate. The bedrock geology of the formerly Weichselian ice-covered region comprises crystalline rocks of the Scandinavian–Baltic Shield, the Caledonides to the west, inliers of Proterozoic formations within the shield area, and Palaeozoic and Mesozoic formations to the southeast.

Accompanying deglaciation, a sequence of landscape changes involving formation of ice-marginal lakes and seaways developed, and at the smaller scale, numerous isolation basins. The latter have featured prominently in quantifying relative sea-level history of the region based on the record of varves and timing of basin isolation (Sandgren *et al.*, 2005). In the early deglacial phase, the *Baltic Ice Lake* formed on the southern and eastern margins of the Fennoscandian Icesheet approximately 12.3–10.3 ka ago (^{14}C years) and lasted as a distinct feature for about 1,500 years (Donner, 1995). Transition from the Baltic Ice Lake to the *Yoldia Sea* was established by 10 ka ago as a result of a marine connection through the Närke Strait and was manifested by an influx of brackish water in southern Finland and saline water in the present area of Stockholm. Continued emergence led to the formation of a sill in the Närke Strait area with subsequent development (9.6–9 ka ago) of a significantly larger lake, termed *Ancylus Lake*, named after the freshwater gastropod *Ancylus fluviatilis*. A decreasing rate of glacio-isostatic compensation in concert with a rapid rate of postglacial sea-level rise, resulted in incursion of marine waters and formation of the *Littorina Sea* after 7 ka as indicated by presence of the marine gastropod *Littorina littorea*, which requires a salinity of at least 8.1‰ (Donner, 1995).

A transect across Norway, Sweden, and Finland to Estonia (Figure 5.11) reveals that maximum uplift since melting of the Late Weichselian Icesheet occurred in the mid-northern Gulf of Bothnia and represents approximately 230 m of uplift in the past 9,000 years (see Figure 2.3). It is not possible, however, to quantify total recovery since onset of glacio-isostatic compensation in view of the absence of shoreline successions directly relating to this interval. Thus the maximum value for uplift within the region represents an underestimate of isostatic compensation. Coastal deposits associated with the *Tapes I* shoreline at 7 ka ago occur up to 140 m APSL. The Late Weichselian to Holocene relative sea-level curves from northern Ångermanland, Sweden, and the Rovaniemi-Pello area of Finland (Donner, 1995) reveal a dramatic decline in rates of uplift and hence relative sea-level fall between 9 and 7 ka (Figure 5.12).

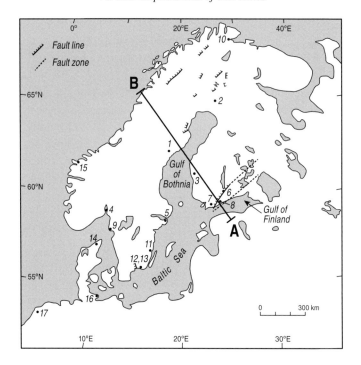

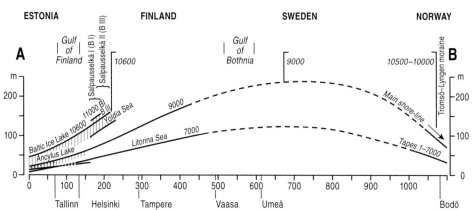

Figure 5.11 Transect across Norway, Sweden, Finland, and Estonia showing the relative trend in uplift since melting of the late Weichselian Scandinavian Icesheet (modified after Donner, 1995).

In the more marginal regions, removed from the former icesheet, relative sea-level curves are dominated by postglacial sea-level rise, such as in the Netherlands (Lambeck *et al.*, 1998).

The broad shape of uplift, as defined by *isobases* (lines of equal uplift) for elevation of coeval raised beach deposits, is that of an asymmetric but slightly

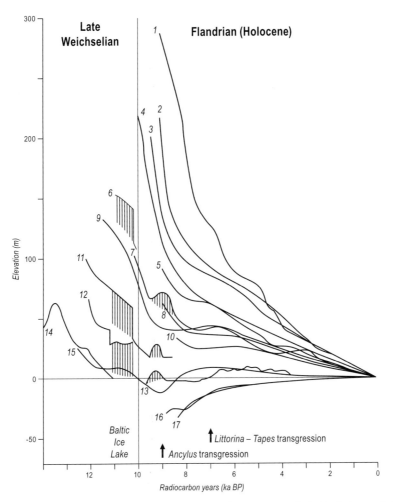

Figure 5.12 Relative sea-level curves representing late stages of deglaciation of the late Weichselian Scandinavian Icesheet and associated postglacial rebound of Scandinavia. Numbers corresponding with each relative sea-level curve correspond with field site numbers in Figure 5.11 (after Donner, 1995).

flattened dome structure with the greatest postglacial uplift registered in the eastern Scandinavian Mountains, representing the last region to experience deglaciation. The present-day spatial pattern is not representative of the entire period of uplift accompanying deglaciation, as localised short-term ice advances such as during the Younger Dryas may have changed icesheet geometry and hence crustal loading. It is thought that the locus of maximum uplift has shifted farther north in the Gulf of Bothnia (Flint, 1961; Donner, 1995). The region today is characterised by a significant negative gravity anomaly, indicating that the glacio-isostatic response to deglaciation will continue for a significant time into the future, as realised by

Niskanen (1939). The northern Gulf of Bothnia is likely to become a large isolation basin without a marine connection in ~8,000 years time. Thus, relaxation time for complete crust–mantle readjustment to deglaciation of the Weichselian Icesheet may be as much as 21 ka.

Development of faults and a higher incidence of seismicity for Fennoscandia have been inferred for the early phase of Late Weichselian deglaciation (Donner, 1995; Mörner, 2003). Although many regional faults within Scandinavia may be of Precambrian age, having developed within the Fennoscandian Shield, or associated with the Tertiary Alpine Orogeny, some faults were initiated during the early deglacial phase accompanying destruction of the Weichselian Fennoscandian Icesheet. Several fault scarps have been dated at ~9 ka.

Detailed glacio-hydro-isostatic modelling of crustal rebound in Scandinavia suggests that maximum ice thickness during the Weichselian glaciation may have been approximately 2,000 m (Lambeck *et al.*, 1998), in contrast to the value of 3,400 m proposed by Denton and Hughes (1981). The ice limits as defined by the mapping of moraines for the Saale (MIS 6) and Elster (MIS 8) glaciations are over 300 km farther south of the Weichselian ice limit, implying significantly larger icesheets for these earlier glaciations and, therefore, greater magnitudes of crustal compensation.

5.8 The Mediterranean Basin

The Mediterranean Sea is microtidal, with relatively restricted wave activity. Much of its 46,000 km of coastline is rocky, particularly comprising limestones, and it preserves a sequence of fossil shorelines (Stewart and Morhange, 2009). The Mediterranean lies largely outside the area directly affected by icesheets during successive glaciations, despite extensive ice over the Alps. Elevated shorelines have been recognised from this region, initially being correlated on the basis of height (Zeuner, 1945).

After 1948, with recognition of an older Calabrian shoreline, the sequence Calabrian, Sicilian, Milazzian, Tyrrhenian, and Versilian was considered a stepped sequence of terraces with the older, more landward. The last interglacial shoreline, termed the Tyrrhenian, has been widely described from around the Mediterranean on the basis of its morphology, lithostratigraphy, geochronology, or distinctive fauna (Stewart and Morhange, 2009). Tropical fauna, termed the Senegalese because of its present occurrence in West Africa, has provided a distinctive fossil assemblage. Most conspicuous is *Strombus bubonius*, but other elements include *Conus testudinarius*, *Natica lacetea*, *Natica turtoni*, and *Cantharus viverratus*. A last interglacial age for this assemblage has been indicated by U-series ages on an associated coral, *Cladocora caespitosa*, which has then been used to constrain Aminozone E, derived from AAR ages on the molluscs of the genera *Glycymeris* and *Arca* (Hearty, 1986; Hearty *et al.*, 1986).

The Mediterranean has been influenced by convergence of the African Plate to the south on the European Plate to the north. The African Plate is moving northward at a rate of 4 mm/year, and the present continental collision followed complete subduction of the Mesozoic Tethys oceanic lithosphere. During the Messinian Salinity Crisis about 5.96 Ma, convergence of the plates coupled with vertical tectonic movements resulted in closure of the gateway to the Atlantic, with a dramatic fall in water levels (exceeding 800 m) and a period of hypersalinity and deposition of evaporites. Since that time, interaction of the African and European Plates, together with the Arabian Plate to the east, and the smaller Anatolian and Apulia Plates, has resulted in a complex geotectonic setting. Plate movements are continuing at rates of up to 30–40 mm/year, with vertical displacements of up to 1–2 mm/year. Further variations in sea level caused by salinity or thermal adjustments may have been important because of the enclosed nature of the Mediterranean.

Much of the western Mediterranean appears to have been relatively stable since the middle Pleistocene. For example, Pleistocene shoreline deposits on Mallorca and other Balearic Islands reflect successive reoccupation during Quaternary sea-level highstands, but the last interglacial is identified a few metres above present (Butzer, 1962). Much of the eastern coast of Spain on the Iberian Peninsula appears to have been stable over the late Quaternary, except in the areas around Gibraltar which have been uplifted in response to the zone of convergence between the African and Eurasian Plates. Mapping of shoreline deposits of the last interglacial maximum (MIS 5e) and the late Pleistocene interstadial (MIS 5c) indicate fastest rates of uplift centred in the narrow Gibraltar Strait region, where MIS 5e shoreline deposits at Tarifa reach up to 20 m APSL (Zazo et al., 1999; Figure 5.13). In the area immediately surrounding Gibraltar, 30 km east-northeast of the Strait, MIS 5e deposits occur up to 5 m APSL and reach only up to 2 m APSL between Marbella and Torremolinos some 60–100 km east from Gibraltar. These latter regions appear to be in a zone of subsidence. Similarly, last interglacial shoreline deposits to the northwest of Tarifa progressively decline in elevation. In the Tarifa area, differential faulting has preferentially uplifted crustal blocks such that the MIS 5c shoreline deposits, which at the time of their deposition formed at lower sea levels to those of MIS 5e, have been uplifted to higher points in the landscape than their MIS 5e equivalents. This is attributed to the presence of NE–SW sinistral and NW–SE dextral (older faults) strike–slip fault complexes delimiting regions of uplift and subsidence.

Elsewhere in the Mediterranean, shorelines have been vertically displaced by tectonic movements, particularly associated with the Hellenic arc in the east and the Calabrian arc in the west. For example, Calabria and northeast Sicily have been uplifting at rates of up to 2.4 mm/year (Antonioli et al., 2006a). However, the highest rates of uplift appear to have been associated not with plate margins, but with an intra-plate rift in the Gulf of Corinth. These areas of neotectonism are considered in greater detail below.

5.8.1 Italy

Ongoing regional differential uplift of peninsular Italy and Sicily is in part a response to subduction of the African Plate beneath the Eurasian Plate which commenced

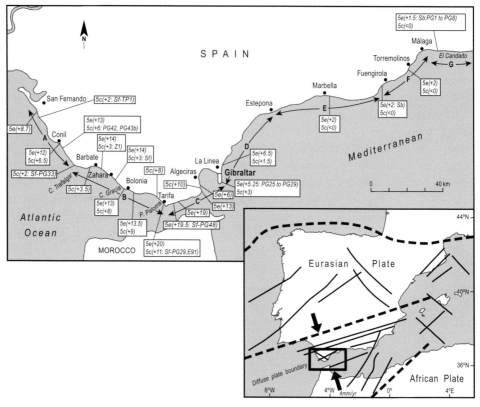

Figure 5.13 Map of the southern Iberian Peninsula showing differential elevations of MIS 5e and 5c coastal successions relating to subduction of the African Plate beneath the Eurasian Plate (modified from Zazo *et al.*, 1999, with permission from Elsevier).

between Cretaceous and Paleogene time, together with movements along the Calabrian arc. Subduction has given rise to a series of Vulcanian- and Strombolian-style volcanoes along the western margin of peninsular Italy and Sicily, and a back-arc basin represented by the Tyrrhenian Sea. A westward-dipping subduction zone of the Adriatic–Ionian lithospheric slab has resulted in crustal delamination and extension, as well as volcanism along the Tyrrhenian coastline. The Benioff Zone associated with the descending lithosphere is approximately 100 km beneath Calabria and extends down to 500 km beneath the central Tyrrhenian Sea. The island of Sardinia appears to be located on a relatively stable crustal block. A major thrust front trending NW–SE delimits the eastern margin of the southern Apennines with a series of horst and graben structures on the eastern margin of the Apennine Range in the Adriatic Foreland. To the southwest of this feature, along the Ionian Sea coastline of southern Italy, last interglacial shoreline successions occur between 75 and 142 m APSL and are consistently higher than along the entire Tyrrhenian and Adriatic coastlines (Hearty, 1986; Bordoni and Valensise, 1998; Ferranti *et al.*, 2006).

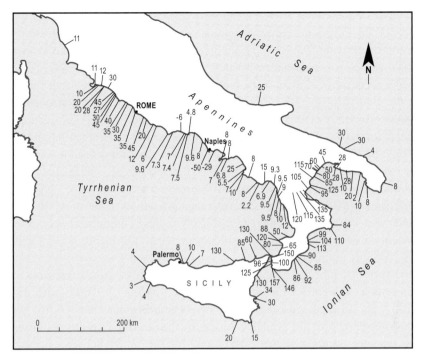

Figure 5.14 Map of peninsular Italy with Tyrrhenian (last interglacial; MIS 5e) shoreline elevations indicated for specific field sites (after Bordoni and Valensise, 1998).

Coastal successions of last interglacial age (MIS 5e, the classic Tyrrhenian stage) have been extensively studied in Italy and Sicily and differential uplift and subsidence of this shoreline recorded (Bordoni and Valensise, 1998; Ferranti et al., 2006). Along the coastline of the Tyrrhenian Sea, last interglacial successions show evidence for uplift with marine terraces and shoreline deposits ranging from 2.2 to 150 m APSL on the Italian Peninsula and up to 130 m APSL on northeastern Sicily (Figure 5.14). The highest record of MIS 5e shoreline successions is in southern Calabria where they reach up to 175 m APSL (Ferranti et al., 2006) and correspond with a zone of high seismicity as illustrated by the 28 December 1908 M = 7.1 Messina Earthquake in which 90,000 people died (Bonini et al., 2011).

Sicily shows evidence for differential uplift with coastal successions containing *Strombus bubonius* at contrasting elevations around the island's coastline. The highest rates of uplift, as deduced from elevation of last interglacial (MIS 5e) shoreline deposits, are on the northeastern sector of the island with an average rate of uplift of 0.92 mm/year (Antonioli et al., 2006b); it is geographically related to the eastern flank of Mount Etna. The northwestern sector of the coastline is characterised by slower rates of uplift with an average rate of uplift of 0.22 mm/year and localised subsidence in some areas (~40 mm/ka). The entire southern sector of Sicily does not appear to preserve MIS 5e shoreline features and may reflect subsidence associated with the Gela frontal thrust system related to regional subduction in the Calabrian arc.

An additional complication in reconciling rates of Holocene with late Pleistocene tectonism relates to glacio-hydro-isostatic adjustments. The late Holocene relative sea-level record from Calabria, Italy, for example, shows limited evidence for tectonic uplift. Glacio-hydro-isostatic adjustments at the scale of the Mediterranean basin, in response to decay of the Scandinavian Icesheet, and regional subsidence in the peripheral forebulge of the former icesheet, have largely overridden ongoing effects of tectonic uplift in Calabria during the Holocene, accounting for the limited uplift evident within the region during that time (Pirazzoli *et al.*, 1997). This contrasts with the longer-term record experienced during the late Pleistocene, which reveals last interglacial (MIS 5e) marine terraces ranging from 50 to 157 m APSL (Bordoni and Valensise, 1998). Recent emergence is likely to have commenced only in the past 3 ka when the rate of regional, isostatically induced subsidence was exceeded by ongoing tectonic uplift processes.

Local zones of subsidence also occur on the Italian Peninsula, restricted to the Roccamonfina, Phlegraean Fields and Vesuvius Volcanic provinces with displacement of –6 to –50 m recorded along the Tyrrhenian Sea coastline. On the Adriatic Sea coast, fewer records are available, but along the Strait of Otranto last interglacial marine terraces occur at 4–8 m APSL and further north between Bríndisi and Térmoli reach up to 25–30 m APSL. A complex history of tectonic vertical movements is evident at Pozzuoli, in the Phlegraean Fields volcanic complex in Italy. Here, the Roman columns at what has been called the Temple of Serapis, have long been known for the borings by marine molluscs that occur at elevations up to 7 m APSL (see Sections 1.4.4 and 7.8 and Figure 3.13a).

5.8.2 Greece

Greece and the Greek islands, within the broad region from the Ionian Sea across to the Aegean Sea in the eastern Mediterranean, are also affected by subduction processes of the African Plate beneath the Eurasian Plate. A series of volcanoes defines the Inner Hellenic Arc, and the Aegean Sea region represents part of a back-arc basin. To the south of mainland Greece, the Outer Hellenic Arc subduction complex is situated approximately 100 km south of Crete and is defined from east to west by the Strabo, Pliny, and Matapau Trenches. The region is characterised by a complex array of normal, reverse, and thrust faults.

The last interglacial (MIS 5e) shoreline in Greece is defined by emergent marine terraces and shoreline notches at varying elevations from 10 m APSL on peninsular Greece to 14 m APSL on the island of Crete (Ferranti *et al.*, 2006). The most tectonically active region in the Mediterranean is associated with the Gulf of Corinth. The gulf setting is anomalous in a wider regional context as it appears to be the only area within Greece undergoing crustal extension in a north–south direction.

The Gulf of Corinth is an asymmetric half-graben approximately 120 km long and up to 30 km wide with a maximum water depth of approximately 920 m. The southern portion of the gulf seafloor is significantly steeper than the northern margin, reflecting differential fault movements along normal faults bordering the gulf which have been active during at least the past 5 Ma (Stiros and Pirazzoli, 1998). The southern margin of the Gulf of Corinth on the Peloponnese coastline has undergone rapid uplift, and emergent marine terraces have been the subject of study since the early work of Depéret (1913). Controversy concerning the nature of the terraces is related to whether some of the features represent a formerly continuous surface disrupted by normal faults, or whether they are discrete marine terraces related to sea-level highstands of widely differing ages. Twenty terraces can be discriminated between Corinth and Xylokastro, rising to more than 400 m APSL, as a result of uplift over the past 500 ka (Keraudren and Sorel, 1987). Initially correlated on the basis of elevation, the availability of numerical ages for several terraces has enabled a revision (Amijo *et al.*, 1996; Westaway, 2002). Near Corinth, the Old Corinth marine terrace of MIS 7 age and the New Corinth marine terrace of Tyrrhenian age (last interglacial, MIS 5e) can be discriminated. The Kalamaki terrace contains the last interglacial index fossil gastropod *Strombus bubonius* occurring at approximately 80 m APSL. The terrace rises to the northwest, reaching up to 160 m APSL at Melissi, approximately 25 km from Corinth, and is around 180 m elevation near Xylokastro. Uplift rates of up to 1.5 mm/year have been experienced since the middle Pleistocene.

At the eastern extremity of the Gulf, the Perachora Peninsula has undergone a distinct tectonic history. It consists of Mesozoic limestones, flysch deposits and basic volcanics, overlain by Plio-Pleistocene sand and limestones. There are several distinct terraces; MIS 11 at 280–360 m, MIS 9 at 120–245 m, MIS 7 at 80–120 m, and MIS 5e at 35–80 m, each tilted to the west (Maroukian *et al.*, 2008). On the basis of four raised shorelines of Holocene age, uplift movements of 0.8 ± 0.3 m with a return period of 1,600 years have been inferred from this peninsula, apparently a faster rate than the average since the last interglacial (Pirazzoli *et al.*, 1994a). However, recognition of distinct notch levels is compounded by lithological and weathering influences and their significance for determining coseismic history has been disputed by Kershaw and Luo (2001).

The pattern of long-term tectonic behaviour around the Gulf of Corinth is also shown in the relative elevations of shoreline features of Holocene age. Along the southern shore of the Gulf, there is evidence from Mavra Litharia of Holocene emergence of more than 9 m, whereas an outcrop on the shore of the Gulf at Platanos, rising more than 12 m APSL, contains borings by Holocene gastropods indicating uplift of more than 2 mm/year (Pirazzoli *et al.*, 1994c). On the northern shore of the Gulf, a submerged notch at about 50 cm BPSL appears to provide evidence of subsidence along this coast (Evelipidou *et al.*, 2011).

5.9 The Caribbean region

The West Atlantic reef province has been a significant area for the reconstruction of sea level. The Great Bahama Bank is the most extensive modern carbonate bank and represents an important analogue for ancient carbonate banks. The bank and other Bahamian islands have been considered stable, and Pleistocene environments within them have been used as evidence for inferring past eustatic stands of the sea. The fossil reefs of southern Florida and those on the island of Bermuda contain a similar sequence of carbonate environments which have been interpreted in terms of their sea-level record. Elsewhere in the Caribbean, similar environments are observed on other islands, many associated with the West Indian island arc, and which appear to have undergone varying degrees of vertical displacement associated with tectonic activity. Distinctly offset from the Leeward Islands, Barbados has undergone a history of gradual uplift (as described above), comparable to, although less uplifted than, the sequence of reef terraces on the Huon Peninsula.

5.9.1 Southern Florida and the Bahamas

The Bahamas comprise more than 700 islands on a series of carbonate platforms, rising from deep water in the northern West Indies. The banks are underlain by accumulations of carbonate sediments several kilometres thick, dating back as far as the Jurassic. The centre of the platforms comprises sand, much pelletal or grape-stone in origin, and abundant mud, some of which appears to have been chemically precipitated (Milliman *et al.*, 1993). On the eastern, windward margins of these platforms there are high-energy environments comprising coral reefs, oolitic shoals, and islands of skeletal sands. The distribution of sedimentary environments appears to reflect these physical controls and the ecological communities that result. Lower-energy, muddy facies occur in the shallow water of the interior of the Great Bahama Bank, distant from the exposed bank perimeter, or in the shelter behind islands (Ball, 1967). The islands which have formed in the Bahamas occur on these windward, eastern margins (Hine and Neumann, 1977). They are composed of Pleistocene reef facies, with extensive lithified aeolianite ridges, formed as fossil dunes, reaching elevations of as much as 63 m on Cat Island (Beach and Ginsburg, 1980; Dominguez and Mullins, 1988).

Southern Florida also contains a suite of Pleistocene reefal carbonates that lie parallel to the modern and Holocene reefs. The reef unit, known as the Key Largo Limestone, is best exposed along the Florida Keys and was comprehensively described by Hoffmeister and Multer (1968). The outer margin of the Keys contains remnants of reef crest units and patch reefs are preserved, for example at Windley Key. The reef facies is overlain by oolitic sand, termed the Miami Oölite or Miami Limestone, which reaches elevations of up to 6 m, resembling the relative distribution of reef and sand shoals seen in the Bahamas (Hoffmeister *et al.*, 1967). Several

U-series ages, at least one at 128 ka which can be considered reliable, indicate formation of the reef unit during the last interglacial (Multer *et al.*, 2002).

Reefs of last interglacial age are prominent on many of the Bahamian islands (Newell and Ribgy, 1957). The significance of these was outlined by Neumann and Moore (1975), who described an accretionary reef unit to an elevation of around 3 m APSL, and a prominent notch that was eroded into older limestones and which occurred at around 6 m elevation. Detailed high-precision U-series ages of corals within the reef unit have provided an important data set for determination of timing and duration of the MIS 5e highstand (Chen *et al.*, 1991), particularly on San Salvador (Colby and Boardman, 1989; Hearty and Kindler, 1993), and Eleuthera (Hearty, 1998).

5.9.2 Other Caribbean sites and more tectonically active islands

Bermuda has played an important role in discriminating late Quaternary carbonate sequences. The islands are the northernmost extension to which coral reefs reach in the Atlantic Ocean, bathed by warm waters of the Gulf Stream. Most of the land exposed above sea level on the Bermuda platform is composed of aeolianite, over a core of pre-Quaternary volcanic rocks. There are impressive fossil carbonate dunes separated by *terra rosa* palaeosols. The initial interpretation by Sayles (1931) was that these dunes formed during glacial periods when lower sea level exposed the platform providing a source of sediment that was blown into the dunes. This view was challenged by Bretz (1960), who considered it more likely that the dunes formed during periods of high sea level. This latter interpretation has been confirmed and detailed litho- and chronostratigraphical studies have shown that dune units grade into interbedded shallow-marine, beach and reef units (Land *et al.*, 1967; Vacher and Rowe, 1997). Corals of last interglacial age are found in the Devonshire member of the Rocky Bay Formation which occurs up to around 2 m APSL. An older penultimate interglacial shoreline is also present, assigned to the Belmont Formation. Each of these marine shorelines is overlain by younger dunes (Harmon *et al.*, 1981; Hearty *et al.*, 1992).

On the Cayman Islands, late Pleistocene deposits are represented by the Ironshore formation. The reef facies within this is found round the margin of each of the three islands, reaching 2–3 m elevation (Woodroffe *et al.*, 1983; Jones and Hunter, 1990). Reef crest *Acropora palmata* can be seen in growth position in a few locations, but over much of its extent the reef facies comprises a broader range of corals that are less diagnostic of the reef crest setting. A prominent notch has been eroded into older Tertiary limestones on Cayman Brac (see Figure 3.12b), with a mean elevation of 6.4 m APSL. A similar notch at this elevation appeared only weakly developed on Grand Cayman, although more recent studies have indicated that it occurs at several sites (Vézina *et al.*, 1999; Coyne *et al.*, 2007). The Cayman Islands appear to have formed as three separate fault blocks that have undergone independent tectonic histories in the past; however, the similarity of the elevation of late Pleistocene depositional and erosional evidence on each of the islands implies that differential movement has not occurred in the past 120 ka. A late Pleistocene reef facies equivalent to the Ironshore Formation occurs at a similar elevation on the Swan Islands on the opposite side of the Cayman Trench (Ivey *et al.*, 1980).

A similar record of sea level is observed on the Yucatan peninsula on the east coast of Mexico. Corals, U-series dated to last interglacial, occur in the reef unit that is up to 3 m APSL, and this unit is overlain by bedded sand (Szabo *et al.*, 1978). Blanchon *et al.* (2009) have re-examined sites and proposed that there is evidence for a rapid rise in sea level at the end of the last interglacial with corals also found at this higher elevation of around 6 m for a short period at the culmination of the interglacial.

An equivalent last interglacial reef terrace is found along the north coast of Jamaica, but in this case the terrace is faulted and sections have been displaced up to 10 m elevation (Hendry, 1987). Quaternary vertical movements have been much more marked in the active plate-margin setting of the Greater Antilles, with as many as 28 reef terraces, reaching up to heights of 400 m on either side of the Windward Passage between Cuba and Haiti (Horsfield, 1975). The late Quaternary history of those on Haiti has received further study (Dodge *et al.*, 1983).

Pleistocene reef limestones occur on the islands of the Netherlands Antilles. Regional uplift has resulted in five emergent reefs on Curaçao, the largest of the islands, the oldest of Plio-Pleistocene age. The Hato unit of the lower Terrace corresponds with MIS 5e, and rises to 6–12 m APSL. A high-energy reef crest occurs on the eastern windward side of the island, composed of the coral *Acropora palmata* and the coralline algae *Porolithon*, although this shoreline is presently without a modern reef (Herweijer and Focke, 1978; Pandolfi *et al.*, 1999; Schellmann *et al.*, 2004).

5.10 Divergent spreading-related coastlines: Red Sea

The coastlines of the Red Sea, Gulf of Suez, and Gulf of Aden, trend parallel with a divergent spreading centre that extends along the axis of the central basin for some 3,500 km, demarcating the constructive plate boundary between the African and Arabian Plates. The rift system was initiated in late Oligocene time and has undergone a transformation from shallow continental depressions to a moderately deep-water oceanic seaway. At its widest the Red Sea is approximately 300 km and the seafloor attains depths exceeding 2,000 m. Approximately 40% of the Red Sea, however, is shallower than 100 m. Emergent coral reef terraces and shell beds extend along much of the Red Sea and Gulf of Suez coastlines, punctuated in places by *wadis* (erosional depressions) that have truncated the reefs by episodic fluvial processes during the late Quaternary (Plaziat *et al.*, 1998). The emergent fossil reefs are very well preserved due to regional aridity and absence of rivers providing terrigenous sediment to the region. The Red Sea, Gulf of Aden, and Gulf of Suez reveal

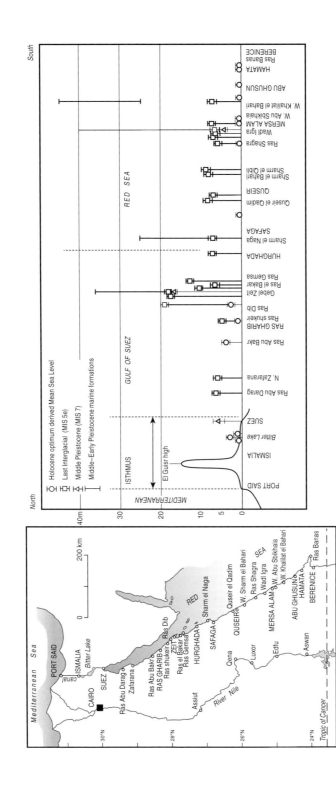

Figure 5.15 Map of the Red Sea region showing elevations of last interglacial (MIS 5e) shorelines at selected locations based particularly on reefal facies (after Plaziat *et al.*, 1998 and Lambeck *et al.*, 2011, with kind permission from Springer Science + Business Media B.V.).

contrasting stages of rift-related development. The Gulf of Aden has evolved from a rift to drift phase in its development, while the Red Sea is in a nascent phase of post-rift development with the onset of ocean basin formation. The Gulf of Suez remains in the syn-rift phase of its development. The last interglacial reefal facies, with overlying shoreline facies, reveal a surprising consistency in elevation of between 5 and 8 m APSL along much of the Red Sea coastline (Lambeck *et al.*, 2011). In the southwestern region of the Gulf of Suez between Ras Dib and Ras Gemsa, however, localised neotectonism has uplifted MIS 5e reefal deposits to between 11 and 19 m APSL (Figure 5.15). Uplift in this region may be related to strike–slip movement on the Aqaba–Dead Sea transform fault. Shells of the gastropod *Melanoides turberculata* from a marine terrace 26 m APSL at Wadi Nahari, with a U-series age of 212 ka, indicate a longer history of uplift along the Red Sea coastline (Plaziat *et al.*, 1998). The last interglacial comprises the Abdur Limestone in Eritrea, where there are two reef units assigned to MIS 5e, at an important archaeological site (Bruggemann *et al.*, 2004).

5.11 Pacific Plate

The Pacific Ocean covers almost an entire hemisphere of the Earth and it is dominated by the Pacific Plate. Although most of this is ocean lithosphere, scattered islands on the plate display several types of vertical displacement, and where the Pacific Plate converges on other plates there are active island arcs which are characterised by even more rapid displacement. In this section, several island types will be examined and the nature of vertical movement as deduced from elevation of last interglacial (MIS 5e) deposits will be considered. In addition, Japan and New Zealand, whose active tectonic behaviour results from their location on the convergent boundaries of the plate, will be described.

5.11.1 Pacific islands

The islands that are scattered across the Pacific Ocean have been subject to varying degrees of neotectonism. Some, such as the Galapagos in the eastern Pacific, represent young or still active volcanoes. Others, such as the Fijian Islands, are a complex group which have undergone a range of vertical movements associated with diverse micro-plate dynamics, largely independent of the Pacific Plate (Hamburger and Everingham, 1988; Nunn, 1998).

There are several distinct linear chains of islands apparent in the Pacific, many of which also contain numerous seamounts that do not reach the sea surface. The Hawaiian Islands represent this intra-plate setting and have developed as a series of volcanoes related to a hotspot, as described earlier in this chapter. Plate motion results in gradual movement of such volcanic edifices into deeper water, providing a mechanism that partially explains the subsidence deduced by Darwin. As a consequence the island types proposed by Scott and Rotondo (1983) can be seen in or along these chains. Darwin realised that drilling through an atoll would represent an effective way to test his coral-reef theory, and such drilling was undertaken at Funafuti in the Ellice Islands (now Tuvalu) in the 1890s. The coring recovered more than 300 m of shallow-water carbonate sediments. Pleistocene limestones are found below 26 m and above them are carbonates of Holocene age (Ohde *et al.*, 2002). Drilling after the Second World War on the Micronesian atolls of Bikini and Eniwetok revealed thicknesses of more than a kilometre of carbonates over volcanic basement, providing convincing support for Darwin's subsidence theory of atoll development (Guilcher, 1988). U-series dating of Pleistocene limestones from Eniwetok has revealed an unconformity between late Pleistocene limestone and the overlying Holocene reefs, termed the Thurber discontinuity after the early attempts to date it using U-series (Thurber *et al.*, 1965). Further solutional unconformities are encountered in deeper reef units. Where these have been dated, such as on Eniwetok and Mururoa, the last interglacial limestone has been shown to be underlain by older reef units deposited during preceding highstands (Szabo *et al.*, 1985; Camoin *et al.*, 2001).

Subsidence explains why the last interglacial is only exposed on a few islands in the Pacific and why it is encountered at depth in boreholes into atolls and other modern reefs. The Thurber discontinuity occurs at depths of 10–20 m below the modern rim of many atolls, such as Tarawa Atoll in Kiribati (Marshall and Jacobson, 1985) and several atolls in the northern Cook Islands (Gray *et al.*, 1992). The last interglacial reef facies has been found at similar depths beneath atolls in the Indian Ocean, such as the Cocos (Keeling) Islands and the Maldives (Woodroffe *et al.*, 1991b; Woodroffe, 2005). Although subsidence appears to be the primary reason why these last interglacial limestones are presently below sea level, some lowering and reshaping of the surface may result from solution and karst weathering which, it has been argued, has accentuated lagoon morphology on many atolls (Purdy and Winterer, 2001).

On several islands in the Pacific, last interglacial reef limestone is exposed above sea level. It occurs at 2–3 m on Christmas Island in eastern Kiribati (Woodroffe and McLean, 1998). Similar limestone is described on Anaa in French Polynesia, which is an island that is considered to have migrated onto the uplifted arch, a consequence of lithospheric flexure that surrounds Tahiti (Pirazzoli *et al.*, 1988). Lithospheric flexure has also resulted in uplift of other islands. Makatea is an uplifted limestone island reaching a maximum elevation of 113 m, with the last interglacial well above modern sea level (Montaggioni *et al.*, 1985). The term *makatea* has been extended to a class of islands where an elevated limestone rim surrounds a central volcanic core; ironically the island of Makatea does not have the volcanic core exposed and so is not included in this classification. The islands of Mangaia, Atiu, Mitiaro, and Mauke in the southern Cook Islands are makatea islands and these appear to have been uplifted as a result of loading by the relatively young (Pleistocene) volcanic island of Rarotonga (Spencer *et al.*, 1987). U-series dating has confirmed a last interglacial age for a prominent reef limestone on each of these, overlying a lower reef unit of penultimate interglacial age on the latter three of these islands. The last interglacial limestone reaches up to 20 m on Mangaia, 12 m on Atiu and Mauke, and 10 m on Mitiaro, implying that differential flexure of islands has persisted over the most recent glacial cycle (Woodroffe *et al.*, 1991a).

The stratigraphy and morphology of reefs at different sites around New Caledonia varies according to uplift or subsidence history (Cabioch *et al.*, 1999). A terrace of last interglacial age occurs at around 10 m elevation on the southeastern corner of Grande Terre, at Yate, and at this site modern reefs are relatively restricted as a consequence of the lack of substrate at suitable depths for reef growth (Andréfouët *et al.*, 2009). On the other hand, coring of the barrier reef at Amédée, on the western margin of the south of Grande Terre, indicates subsidence which has enabled the broad lagoon and associated reefs to form. The last interglacial and several older reefs are recorded at increasing depths, suggesting that the rate of subsidence, which has been 0.16 mm/year since MIS 5e, has not been constant over longer timescales (Cabioch *et al.*, 2008b).

5.11.2 Hawaii

The Hawaiian Islands preserve a rich record of Quaternary relative sea-level changes in the form of emergent coral reef, beach, marine benches, and aeolianite successions as well as submerged shoreline notches and platforms (Hearty, 2002; Fletcher *et al.*, 2008). The relative sea-level histories of the islands have been governed in part by the motion of the Pacific Plate over a stationary hotspot and associated flexural isostasy, caused by loading by the Big Island (Grigg and Jones, 1997; Fletcher *et al.*, 2008). Palaeo-sea-level interpretations deduced from the Hawaiian Islands have, accordingly, been complicated by these tectonic processes as well as the presence of emergent gravel deposits, some of which have been interpreted as the product of tsunamis (Moore and Moore, 1984) rather than raised beach deposits. Contrasting interpretations concerning rates of uplift associated with flexural isostasy have led to divergent opinions about ages and palaeo-sea-level significance of some coastal landforms. The 24–30 m emergent reef succession at Kaena Point on westernmost Oahu, for example, dated at 532 ka (Szabo *et al.*, 1994), was interpreted as reflecting a long history of moderate uplift. In contrast, Hearty (2002) suggested that the deposit reflected deposition during MIS 11 with a sea level some 20 m APSL and that the rate of uplift of Oahu occurred at a significantly slower rate.

Reefal and coastal successions of penultimate (MIS 7) and last interglacial age (MIS 5e) on the shelf and coastal plains of Oahu have been referred to as the Waianae and Waimanalo Formations, respectively (Stearns, 1974; Sherman *et al.*, 1999; Fletcher *et al.*, 2008). The successions are dominated by framestone and rudstone units. Corals from the Waianae Reef have been dated to between 206 and 247 ka based on U-series dating, and in terms of the island's uplift history of

0.03–0.06 mm/year suggest a sea level of between 9 and 20 m BPSL during formation of the Waianae Reef (Fletcher *et al.*, 2008). Correlative sediments are widely developed on the shelf surrounding Oahu. The last interglacial maximum on Oahu is represented by the Waimanalo Reef, which is widely developed around the island. The reef succession comprises *in-situ* coral–algal framestone, bindstone, and grainstone facies which extend from 8.5 to 12.5 m APSL. Corals from the formation have been dated to 131–114 ka (Szabo *et al.*, 1994) and 134–113 ka (Muhs *et al.*, 2002).

On Oahu, Molokai, Maui, and Kauai, aeolianites of the Leahi Eolianite rest unconformably on the Waimanalo Formation and have been correlated with interstadials (MIS 5a–d) based on aminostratigraphy (Fletcher *et al.*, 2005) and to MIS 5e by Hearty *et al.* (2000). The correlative Leahi Reef on the outer shelf of Oahu, dated at 110–82 ka (Szabo *et al.*, 1994), was sampled from between 25 and 30 m BPSL and does not show effects of subaerial exposure, indicating that during this time sea level remained above 25 m BPSL and that the interstadial reef continued its seaward accretion.

Both the MIS 7 and MIS 5c/a reefs surrounding Oahu display prominent shoreline notches and visors between 18 and 25 m BPSL, termed the Kaneohe Shoreline by Stearns (1974). The age of this erosional feature remains undefined. Similar prominent shoreline notches have been identified onshore in lithified aeolianite and a spectacular example of two superposed notches at 6.7 and 8.2 m APSL at Waimanalo, Oahu were described by Stearns (1985; see his figure 2.12), but have more recently been destroyed by quarrying.

5.11.3 Japan

The main islands of Japan, Hokkaido, Honshū, Shikoku, and Kyūshū, are located on the Eurasian Plate. Their eastern margins are adjacent to a subduction zone that involves the descending northwestern portion of the Pacific Plate and the Philippine Sea Plate. Absolute rates of plate movement are approximately 80 and 40 mm/year for Pacific and Philippine Sea Plates, respectively (Ota and Kaizuka, 1991). To the southwest of the main islands, the Ryukyu Islands are associated with the subduction zone involving the northwest trending Philippine Sea Plate. The main Japanese and Ryukyu islands have experienced long histories of tectonic uplift as attested by a series of well-developed emergent marine terraces (Ota and Kaizuka, 1991; Ota and Omura, 1992; Ota, 1996; Radtke *et al.*, 1996).

Considerable variation is evident in elevation of MIS 5e shoreline features on the main islands of Japan (Figure 5.16). The highest recorded elevations for individual sites tend to occur on the southern islands of Kyūshū (90 m APSL), Shikoku (181 m APSL), and southern Honshū (160 m APSL) (Ota and Kaizuka, 1991; Ota and Omura, 1992) in geographical settings close to the Nankai Trough. The emergent MIS 5e shoreline features at 160 m on southern Honshū occur on the northern Irō Peninsula and near the zone of convergence of the Japan Trench and Nankai Trough.

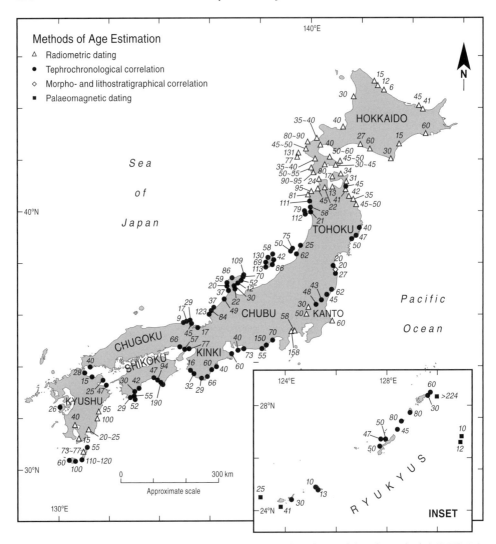

Figure 5.16 Map of Japan showing the differential elevations of last interglacial (MIS 5e) shoreline successions (after Ota and Omura, 1991).

Terraces of last interglacial age (MIS 5e) are well represented on the Ryukyu Islands and occur at contrasting elevations that relate to proximity to the subduction zone and the azimuth of plate movement of the descending slab. The highest reported MIS 5e reef terrace is on the island of Kikai at 224 m APSL (Ota, 1996). Kikai is approximately 80 km northwest of the Ryukyu Trench and the direction of the descending Philippine Sea Plate is normal to the trench axis. Elsewhere in the Ryukyu Islands the MIS 5e reef terraces occur at lower elevations and in part relate to their location relative to the descending Philippine Sea Plate. In the case of Kita-Daito and Minami-Daito Islands, both located on the Philippine Sea Plate, fragmentary MIS 5e reefal deposits occur up to 10–12 m APSL. Inferred rates of uplift since MIS 5e for the Ryukyu region range between 1.8 m/ka for Kikai to only 0.03–0.05 mm/ka for Kita-Daito and Minami-Daito Islands.

5.11.4 New Zealand

The North and South Islands of New Zealand are situated within a geotectonically complex region, having undergone tectonic activity associated with the Kaikoura Orogeny and Quaternary volcanism. The North Island is located on the eastern boundary of the Indo-Australian Plate, to the west of Hikurangi Trench, which represents the southern continuation of the Tonga–Kermadec subduction complex (Figure 5.17). In this region, the northwesterly dipping subducted slab of the Pacific Plate descends beneath the Indo-Australian Plate at about 50 mm/year (Berryman *et al.*, 1992). South of the North Island, the Hikurangi Trench joins with an offset of the Alpine Fault, a regional range-bounding transform fault that traverses much of the western South Island in a NE–SW direction. To the southwest of the South Island, the Alpine Fault intersects with the Puysegur Trough in which regional subduction of the Indo-Australian Plate is occurring beneath the southernmost part of the South Island at approximately 40 mm/year (Figure 5.17). Thus, the majority of the South Island of New Zealand is situated on the Pacific Plate. As a result of these tectonic processes, extensive sectors of the New Zealand coastline display evidence for coastal emergence in the form of well-developed marine terraces. Elsewhere, localised subsidence and tectonic down-warping are evident, as in northern Hawke's Bay and the Whakatane coastline in the Bay of Plenty on the North Island (Ota *et al.*, 1989; Williams, 1991). Much uplift appears to be coseismic as attested to by the general geomorphological character of Holocene marine terraces.

There has been a long history of study of vertical displacements around the New Zealand coast (Cotton, 1916). However, much of the early description and correlation of Quaternary terrace elevations relied on comparison with chronology and elevations from the Mediterranean. This changed following a review by Gage (1953), who recognised the significance of neotectonism. A major study of the Quaternary barrier sequence at Kaipara in Northland by Brothers (1954), although also correlating terrace levels for this relatively stable site with those in the Mediterranean, identified that the Pliocene–Pleistocene sequence had developed through the successive re-occupation of shorelines over that time period. In contrast to the relatively stable Northland region, Chappell (1975) showed that there was evidence of warping and variable rates of uplift along the Bay of Plenty, constraining the age of terraces on the basis of tephra layers (tephrochronology). This study also compared flights of terraces in New Zealand with those on the rapidly uplifted Huon Peninsula in Papua New Guinea. Although the Huon terraces are constructional and the New Zealand terraces are almost all erosional, comparison between uplifted

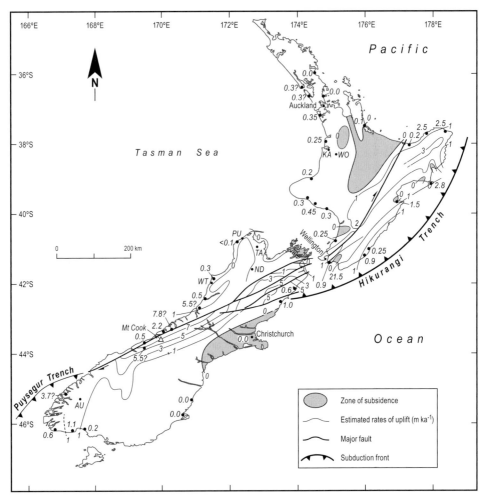

Figure 5.17 Regional plate-tectonic setting of the North and South Islands of New Zealand showing rates of vertical crustal movements (after Williams, 1991, with permission from Elsevier, with additional data modified from Pillans, 1990).

areas in New Zealand and Papua New Guinea has provided the basis for several correlations (e.g. Bull and Cooper, 1986). Elevation of terraces around New Zealand and implications for local uplift or subsidence have been reviewed by Pillans (1990) and Berryman and Hull (2003).

The Otago region in the southeast of the South Island appears relatively stable, so much so that Blueskin Bay, near Dunedin, was chosen as the reference site by Gibb (1986) against which other sites were compared on the basis of their Holocene relative sea-level history in order to compile a postglacial sea-level curve for New Zealand. Much of the rest of the South Island has been tectonically active; there are prominent sequences of terraces on the west coast; those in Fiordland reaching up to several hundred metres above sea level. It has not been easy to date the terraces, so correlations have often been tentative (Williams, 1982). Recent earthquake activity that has devastated Christchurch has shown that even those areas that were not considered tectonically active can be subject to vertical movement along previously undetected faults.

The tectonic behaviour of the North Island is complex. The accumulation of thicknesses of up to 4,000 m of Pliocene–Quaternary sediment in the Wanganui Basin has resulted in continued subsidence, but may also have contributed to lithospheric flexure and uplift rates of up to 0.6 m/ka in adjacent areas (Pillans, 1991; see also Section 6.14.2). Rapid rates of uplift are well known from the North Island. An earthquake in 1931 in Hawkes Bay devastated Napier and uplift of 2.7 m was observed. There are many Quaternary terraces along this east coast, with the last interglacial terrace occurring at up to 300 m above sea level, indicating average rates of uplift of 3–4 mm/year at some sites.

Situated on the southeastern sector of the North Island of New Zealand, Cape Turakirae has been of interest in studies of neotectonism since the early observations of Charles Lyell in 1868. In 1855 a M = 8.2 earthquake resulted in more than 2.5 m of local uplift of the Cape Turakirae area. The uplift is associated with movements along the Wairarapa Fault, a NE–SW range-bounding fault. The Cape Turakirae area is represented by a narrow, seaward-sloping coastal plain backed by a high cliff that delimits the seaward side of the last interglacial (MIS 5e) emergent marine terrace (Figure 5.17). Emergence implied by the elevation of the MIS 5e terrace is further demonstrated by up to six shingle gravel (sandstone and greywacke clasts) storm beach ridges of Holocene age preserved across the plain. The oldest and most inland ridge with a radiocarbon age of 6,500 years BP based on the dating of shell contained within the gravels has been uplifted to between 5 and 29 m APSL, reflecting differential uplift of this shoreline complex (Darby and Beanland, 1992). The uplifted (relict) beach ridges are significant for evaluating recurrence of major and potentially devastating earthquakes within the region. Growth of each ridge is terminated by sudden coseismic uplift accompanying high-magnitude earthquakes, explaining the absence of intermediate ridges. Interpretation of the palaeoseismic record is based on the assumption that relative sea level within the region has remained stable for the past 6,000 years (i.e. within ± 0.5 m of its present position). A re-examination of the beach ridges by Moore (1987) indicated that the ridge at 2.5 m APSL represents the modern storm ridge which is continuing to develop and that the next landward ridge represents the storm beach that was uplifted in the 1855 earthquake. The gravel beach extends up to 6 m APSL, and radiocarbon dating of 13 shell samples indicates that the structure formed from 1450 to 1855 AD. Although longer-term maximum rates of uplift of 1.2–1.4 mm/year, inferred from the last interglacial terrace, appear to be slower than values derived from Holocene beach ridges (0.8–4.5 mm/year), the geologically short duration of the

Holocene record means that uplift from a single disproportionately high magnitude earthquake event, may give an erroneous impression of the longer-term rate of uplift.

In the Pakarae River area in the northeastern sector of the North Island, seven marine terraces have formed since culmination of the postglacial marine transgression approximately 6,700 (^{14}C) years ago (Ota *et al.*, 1991). Onset of terrace formation appears to have been initiated by coseismic uplift. The terraces represent marine abrasion platforms and are covered by a thin veneer of shelly sands. Radiocarbon dating of marine and estuarine shell, suggest that episodic uplift to isolate each of the terraces from marine influences occurred at 6,700 (Terrace 1), 5,500 (T2), 3,900 (T3), 2,500 (T4), 1,600 (T5), 1,000 (T6), and shortly after 600 years ago for Terrace 7. The surface of the oldest terrace occurs some 27 m APSL and a maximum rate of uplift within the region is 8 mm/year. Successive movements on a fault of possible reverse character 5 km offshore have been attributed as the cause of coseismic uplift. The uplift appears to be associated with a 20-km-long dome structure that is truncated by the Pakarae Fault on its southern termination.

5.12 Synthesis and conclusions

Vertical movements have been experienced at a range of spatial and temporal scales throughout the Quaternary and they have complicated interpretation of relative sea-level histories on many of the world's coastlines. Depending on rates of crustal processes and reliability of palaeoenvironmental inferences, neotectonic contributions to relative sea-level changes may be comparatively easy to quantify where it comprises rapid uplift, as on many convergent plate boundaries at continental margins or on island arcs. In some cases, confident identification and dating of coastal successions of last interglacial age (MIS 5e; 130–118 ka) provide an important benchmark for quantifying crustal movements during the late Pleistocene, and a basis for extrapolating Earth surface movements into the late–middle Pleistocene. Contrasting rates and patterns of localised uplift, associated with subduction, contrast with more subtle isostatic compensation in response to deglaciation or local or regional loading of the lithosphere; longer-term processes related to localised mantle plumes or epeirogeny may also be evident. Long-term rates of coseismic uplift are commonly relatively constant, for many subduction settings, and within the uncertainties of geochronologically derived data sets. The fastest rates of uplift in subduction settings have been documented in Taiwan, reaching up to 5.5 m/ka. In contrast, regions that have experienced deglacial isostatic compensation appear to show exponentially declining rates of postglacial recovery.

Tectonically uplifted regions commonly preserve evidence for multiple sea-level highstands (interglacials and interstadials) and some sites extend back to at least MIS 11 (403 ka). In some cases, comparisons of the respective elevations of radiocarbon-dated Holocene and U-series dated late Pleistocene deposits provide greater confidence to interpretation of vertical displacement of shorelines. Elsewhere,

Holocene and Pleistocene rates of vertical displacement are not so easily reconciled, and, as recent earthquake events such as those in Aceh, Haiti, Christchurch, and Sendai so tragically indicate, our understanding of these processes is still incomplete. Recognition of palaeo-shorelines will continue to provide information on long-term neotectonism, past sea-level changes, and a framework for consideration of future land–sea trends.

6

Pleistocene sea-level changes

... you only had to fix an altimeter on the handlebar of your bicycle to become a specialist in one of the most difficult branches of geology!
(Maurice Gignoux, *pers. comm., in Beckinsale and Chorley, 1991, p. 85*)

6.1 Introduction

The introductory quotation by Maurice Gignoux provides an insight into the level of knowledge attained about Pleistocene sea levels by the middle of the twentieth century. It also illustrates, albeit facetiously, the presumed ease with which evidence for Pleistocene sea levels could be derived. Extensive tracts of the Mediterranean coastline had been surveyed and a rigid framework of Quaternary sea-level changes established within the fourfold model of alpine glaciation defined by Penck and Brückner (1909). Accordingly, before the development of geochronological methods, morphologically well-defined marine terraces at different elevations APSL were assigned ages based on correlations with interglaciations identified by Penck and Brückner. The presence of index fossils, such as the gastropod *Strombus bubonius*, was also used to distinguish the relative age of terraces (Gignoux, 1913). The Mediterranean sea-level record was progressively adopted as a global framework for assigning ages to sea-level events from other continents (e.g. southern England, northern France, Portugal, and Jersey, summarised in Zeuner, 1952; southern Australia: Ward and Jessup, 1965; Ward, 1966; Chile: Paskoff, 1970). With developments in geochronology, a huge literature has subsequently shown that the nature and timing of Quaternary relative sea-level changes is more complex than originally inferred from emergent marine terraces of the Mediterranean.

This chapter examines the nature, timing, and history of Pleistocene sea-level changes. It begins by examining sequential changes that heralded the end of the late Cainozoic ice-free world and led to progressive build-up of northern hemisphere and Antarctic icesheets. The contrasting nature of early, compared with middle and late Pleistocene records, is then examined. The principal lines of evidence for the present understanding of Pleistocene sea-level changes are outlined, and the broad position of sea level for different time intervals is discussed. The chapter also examines global variability in palaeo-sea-level records and their relative reliability. Finally, high-resolution and long records of Pleistocene sea-level changes are considered, as well as the environmental responses to sea-level changes.

In this chapter, Pleistocene sea levels are discussed in the context of the isotopic nomenclature introduced by Shackleton and Opdyke (1973) and individual events are termed Marine Isotope Stages (MIS) as advocated by Shackleton (2006; see discussion in Chapter 4). In this context, oxygen-isotope evidence for continental ice volumes as deduced from marine and ice cores have enabled reconstruction of long records of Pleistocene relative sea-level changes. Oxygen-isotope stratigraphy provides a framework for inferring relative sea-level changes; however, such records are not directly linked to coastal landforms. It is likely that future researchers will resolve coastal records in greater detail, and consider inferences about Quaternary relative sea-level changes based on correlations with isotopic records as potentially limited, in a similar manner to the current view of correlating models of alpine glaciation *sensu* Penck and Brückner (1909) to assign ages to marine terraces.

The Pleistocene epoch experienced the later stages of transition from an ice-free world of earlier Cainozoic time to the current icehouse regime of land-based ice-sheets. Major and irreversible changes in plate-tectonic configuration of continents during the Cainozoic played a significant role in development of the Antarctic Icesheet, leading to the current icehouse world and, ultimately, substantial vertical fluctuations of the ocean surface. Although during the Quaternary Period mean annual temperatures between glaciations and interglaciations differed by only 5–10°C for near-coastal and inner continental environments, respectively, substantial variations in sea level accompanied these temperature changes. During the Pleistocene, the geographical extent of continental icesheets varied significantly, resulting in profound sea-level changes. For the past 800 ka, global sea level oscillated over a vertical range of approximately 130 m every 100 ka (Bintanja *et al.*, 2005). Feedback effects resulting particularly from glacio-hydro-isostatic adjustment processes, as well as local tectonism, mean that major ice-volume changes on the surface of the Earth (glacio-eustatic, ice-equivalent sea-level changes) have been expressed as relative sea-level changes, so that different geographical regions have experienced unique sea-level histories.

6.2 Prelude to the Pleistocene

The transition from the Pliocene to the early Pleistocene heralded a major change in Earth history with continued progressive development of continental-scale icesheets in the northern hemisphere and Antarctica. The increase in continental ice volume during the Pliocene represents the latter stages of a major cooling trend that was initiated at the end of the Early Eocene Climatic Optimum (Zachos *et al.*, 2001; Figure 6.1). The composite $\delta^{18}O$ record from 40 Deep Sea Drilling Project (DSDP)

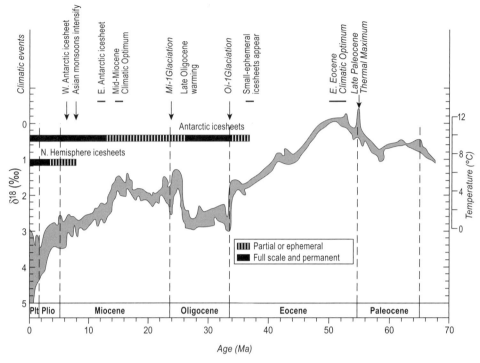

Figure 6.1 Global cooling trends as inferred from oxygen-isotope records on benthic foraminifera from more than 40 Deep Sea Drilling Project (DSDP) and Ocean Drilling Project (ODP) cores (adapted from Zachos *et al.*, 2001). The graph illustrates the progressive fall in global temperatures that heralded development of land-based icesheets and transition of an ice-free world of the Early Eocene Climatic Optimum to the Quaternary Ice Age.

core sites is predominantly based on the benthic foraminifera *Cibicidoides* and *Nuttallides*. The composite $\delta^{18}O$ stack of core records reported by Zachos *et al.* (2001) reveals that the range in $\delta^{18}O$ from the Cainozoic was 5.4‰, of which 3.1‰ of the isotopic fractionation was attributed to cooling of deep-ocean waters. Approximately 1.2‰ of the fractionation equated with formation of the Antarctic Icesheet and the remaining 1.1‰ to growth of the northern hemisphere icesheets.

Oxygen-isotope analyses of benthic foraminifera *Cibicidoides* and *Uvigerina* from core DSDP-607 from the western flank of the Mid-Atlantic Ridge has yielded a 3.2 Ma record of climate change (Raymo, 1992; Figure 6.2). The oxygen-isotope data provide compelling evidence for progressive drawdown of global sea level by possibly as much as 66 m to present levels, representing the later stages of transition from an ice-free world (hothouse) to the 'Quaternary Ice Age'. Estimates suggest that some 55–60 m of ice-equivalent sea level is presently stored in the Antarctic Icesheet (Denton *et al.*, 1971; Hsü and Winterer, 1980) and that a further 5 m (Chappell, 2002) to 7.3 m (Cronin, 2010) of ice-equivalent sea level is represented by the Greenland Icesheet.

From the end of the Jurassic, plate-tectonic processes appear to have played a significant role, through complicated but interrelated events that resulted in development of the Antarctic Icesheet. This involved both opening and closure of marine gateways critical for transfer of sensible heat around the surface of the planet by ocean currents (Zachos *et al.*, 2001). Global sea levels have been lower than present for much of the past 2.7 Ma (Figure 6.2) except during several interglaciations warmer than the present Holocene interglaciation (Raymo, 1992; Burckle, 1993). A time-averaged position of sea level for the Quaternary of approximately 50–60 m BPSL also represents the mid-point between MIS 5c and the MIS 2/1 (Holocene) boundary, based on the $\delta^{18}O$ spectra reported by Raymo (1992; Figure 6.2).

High sea levels in the Cretaceous flooded extensive inland areas of continents, such as the Eromanga and Carpentaria Basins in central Australia, attesting to a world free of permanent continental icesheets but with seasonal ice (Frakes *et al.*, 1995) and a likely global sea level of 250–300 m APSL (Haq *et al.*, 1987; Hallam, 1992). Fast rates of seafloor spreading were a major factor in the high sea levels of the Cretaceous (tectono-eustasy) (Flemming and Roberts, 1973). Global sea levels subsequently fell as rates of seafloor spreading slowed, changing the volume of the ocean basins to accommodate water.

Fragmentation of the supercontinent Pangaea commenced at the end of the Jurassic and resulted in a reconfiguration of continents which ultimately affected ocean circulation and transmission of sensible heat on the surface of the planet (both in terms of shallow-water ocean surface currents and deeper water, thermohaline circulation). For development of the Antarctic Icesheet, this sequence of events began with onset of passive rifting of the Gondwanan continents, Australia and Antarctica at about 160 Ma ago (Veevers and Eittreim, 1988; Veevers *et al.*, 1991). The present position of Australia resulted from its northward movement (~6–7 cm/year), while Antarctica remained largely in its present position. The rifting regime changed to a drifting phase approximately 45 Ma ago, resulting in formation of the Southern Ocean. The separation of Australia from Antarctica represented a necessary precursor for development of the Antarctic Circumpolar Current, an important surface current encircling Antarctica and preventing warmer surface waters from lower latitudes reaching the southern continent, thus permitting formation of an icesheet at sea level. The opening of the Drake Passage between the southern tip of South America and the Antarctic Peninsula at 23 Ma ago (Beu *et al.*, 1997) finally established the Antarctic Circumpolar Current.

Several other significant geotectonic events contributed to the transition from the hothouse of the early Cainozoic to the 'icehouse' world of the Quaternary. Collision of India with Eurasia (about 35 Ma ago) and reorientation of Africa with Europe (about 22 Ma ago) led to cessation of throughflow of warm equatorial waters of the former Tethys Sea to the region now represented by the Mediterranean Sea. This, together with closure of the Isthmus of Panama between the Americas (~3 Ma ago),

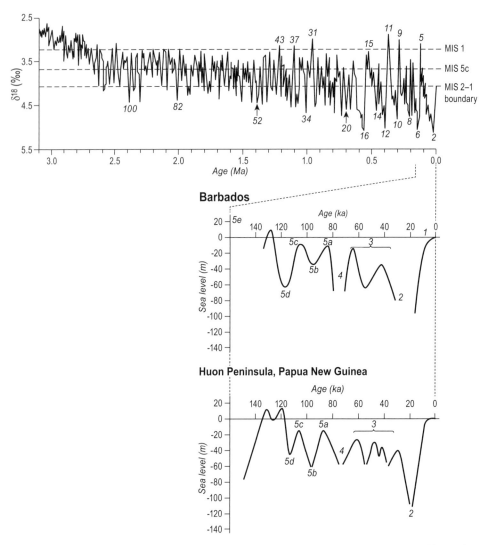

Figure 6.2 Core DSDP-607 spans the past 3.2 Ma. The dashed horizontal lines indicate the $\delta^{18}O$ values for the Holocene (MIS 1; top line), late Pleistocene MIS 5c (middle line, 105 ka ago) and the MIS 2/1, late Pleistocene–Holocene boundary (bottom line, 11.5 ka ago) as determined at this core site. MIS are indicated for some of the glacials (even numbers) and interglacials (odd numbers). The $\delta^{18}O$ values for MIS 5 (actually substage 5e), 9, 11, 31, 37, and 43, as determined at this core site, imply sea levels higher than present for these interglacials. Glacio-eustatic sea-level curves for the last glacial cycle (late Pleistocene and Holocene) from Barbados (Steinen *et al.*, 1973) and Huon Peninsula, Papua New Guinea (Aharon, 1984) illustrating the general pattern of sequential decline of sea level with time until attainment of lowest sea levels during the LGM (22–20 ka ago). The Marine Isotope Stages (MIS) of the last glacial cycle are also indicated on the Barbados and Huon Peninsula sea-level records. The records of the last glacial cycle are correlated with the benthic oxygen-isotope record from equatorial Atlantic core DSDP-607 (Raymo, 1992).

resulted in more intensified equator–pole temperature gradients. Closure of the Isthmus of Panama had the effect of intensifying the northward flow of the Gulf Stream which in turn bolstered formation of North Atlantic Deep Water (NADW) and enhanced evaporation of warmer, less saline water, favouring icesheet development.

A further major fall (regression) in global sea level by as much as 50–100 m is registered in the Upper Miocene (5.96–5.33 Ma) and coincided with an event termed the *Messinian Salinity Crisis* (Hsü *et al.*, 1973, 1977; Roveri and Manzi, 2006). At this time, the Mediterranean Sea was isolated from the world ocean rather like a barred basin with only restricted ingress of marine waters into the basin. A possible cause for isolation of the basin was a change in the nature of subduction-related intra-plate volcanism, with asthenospheric upwelling of magma causing some 1 km of uplift along the African and Iberian continental margins (Duggen *et al.*, 2003). Intense evaporation of seawater within the Mediterranean Basin resulted in widespread desiccation and the formation of large evaporite and sapropel deposits within sub-aerially exposed seafloor (e.g. up to 1 million km^3 of gypsum, halite, and other salts). The salinity of the world oceans was reduced by 6% resulting from slow ingression of marine waters into the Mediterranean Basin. This enabled seawater to freeze at a higher temperature and therefore favoured further development of icesheets at sea level. A major transgression re-flooded the Mediterranean Basin at the start of the Pliocene some 5 Ma ago. Upper Pliocene ice-rafted debris appears in North Atlantic deep-sea sediments indicating presence of northern hemisphere icesheets at sea level by 3 Ma ago (Cronin, 2010).

The oxygen-isotope evidence from Core DSDP-607 reveals a progressive increase in global ice volume and a concomitant fall in ice-equivalent sea level for the past 3.2 Ma (Raymo, 1992; Figure 6.2). The core record shows an overall increase in ice volume between 3.2 and 2.6 Ma that heralded onset of the Quaternary Period. Between 3.1 and 2.95 Ma, cold intervals have more negative δ^{18}O values implying either warmer bottom waters (~3.5°C) at the core site or higher sea levels (lower ice volumes). Evidence from sequence stratigraphy suggests a highstand some 60 m APSL at 3 Ma (Haq *et al.*, 1987).

At about 1.2 Ma ago a notable transition (the middle Pleistocene Transition, MPT) is evident in marine oxygen-isotope records from the previous 41-ka-dominated cycles of the early Pleistocene to the 100-ka-dominated cycles that have characterised middle and late Pleistocene time (Shackleton *et al.*, 1990; Head and Gibbard, 2005; Clark *et al.*, 2006; Rial *et al.*, 2013). This trend is also evident in Core

Caption for Figure 6.2 (*cont.*) The 100-ka cycles appear to commence at about 1 Ma ago (the Mid-Pleistocene Transition) and the previous 2 Ma are dominated by the 41-ka cycles characterised by inferred lower-amplitude sea-level oscillations. The progressive drawdown of sea level in response to formation of the Antarctic Icesheet is evident in the basal 600 ka record in this core (modified from Murray-Wallace, 2007, with permission from Elsevier).

DSDP-607, where amplitude in the isotopic signal ($\delta^{18}O$) in the higher frequency 41-ka cycles is less pronounced than in the 100-ka-dominated cycles, implying that the latter were characterised by lower sea levels at times of glacial maxima and higher sea levels during interglaciations. Core DSDP-607 indicates that MIS 5e, 9, and 11 were warmer than the current Holocene interglacial and accordingly experienced higher sea levels (Raymo, 1992). Based on a comparison of several deep-sea cores, Shackleton (1987) concluded that continental glaciation is likely to have been greater in MIS 6 than in MIS 2, and that MIS 12 and 16 experienced more extensive ice cover and lower sea levels. It is also evident that inferred sea levels for many of the early Pleistocene interglaciations that post-date 2.5 Ma were closer to that registered in interstadial MIS 5c (about 105 ka) of the late Pleistocene (Figure 6.2). The EPICA Dome C ice-core record indicates that MIS 2, 12, and 16 were the most intense glacial maxima and that MIS 5e and 11 are the warmest interglacial maxima (Masson-Delmotte *et al.*, 2010).

Post-depositional taphonomic and diagenetic processes potentially impact on the resolution of marine sediment-based oxygen-isotope records for making inferences about palaeo-sea level. Thus, the greater degree of complexity evident in isotopic records from deep-sea cores for the past four glacial cycles compared with older records of the past 1 Ma, may relate to combined effects of sediment compaction, bioturbation, differential preservation states of microfossils, and other diagenetic processes within the marine sediment. This is one of the classic uncertainties in spectral analysis of palaeoenvironmental data sets (i.e. are geochemical variations, observed with time, real or portraying a filtered and simplified account of 'reality'?).

6.3 Pleistocene icesheets

Many studies have examined the spatial extent of Quaternary icesheets and valley glaciers and detailed summaries are given elsewhere (Lowe and Walker, 1997; Dyke *et al.*, 2002; Ehlers and Gibbard, 2004a,b,c). Based on mapping of terminal moraines and glacier trimlines, as initially advocated by Penck and Brückner (1909), a detailed understanding has progressively emerged of the geographical extent and thickness of former icesheets and valley glaciers. Collectively this has provided insight into the magnitude of sea-level fall required to account for water sequestered in ice during glaciations. Geomorphological and stratigraphical evidence from many regions reveals that the geographical extent of icesheets and valley glaciers varied between different glacial maxima, as independently shown from oxygen-isotope records from deep-sea cores (Shackleton, 1987) and ice cores (Masson-Delmotte *et al.*, 2010). For example, the most recent glacial maximum approximately 21 ka ago was not necessarily characterised by the most extensive or thickest ice coverage. The southern limit of ice during the Last Glacial Maximum (LGM, MIS 2, also called the Devensian) was over 200 km farther north in many areas of southeastern Britain, than during MIS 12 (Anglian) (Bowen *et al.*, 1986). Similar contrasting patterns in extent of glaciation are evident from southern hemisphere settings such as Tasmania where the maximum extent of ice was experienced approximately 1 Ma ago (Colhoun *et al.*, 2010).

In view of erosional effects of multiple glaciations, the most detailed knowledge of the former extent of continental ice relates to the most recent glaciation. The LGM represents the interval of maximum icesheet development during the last glacial cycle and the most recent maximum development of continental icesheets in the Quaternary. Based on average sedimentation rates in deep-sea core V28–238 from the equatorial Pacific Ocean, MIS 2 extended from approximately 27 to 11 ka ago (Shackleton and Opdyke, 1973). The maximum phase of icesheet development, however, is considered to have occurred between 23 and 19 ka ago (Mix *et al.*, 2001). This event coincides with the insolation minimum at 21 ka. The Antarctic EPICA Dome C ice core brackets the last glacial between 35.6 and 11.6 ka and indicates that MIS 2 represents the coldest glacial maximum since MIS 12 (Masson-Delmotte *et al.*, 2010).

At their maximum extent during the LGM, continental icesheets extended over vast areas. The Laurentide and Cordilleran Icesheets extended across much of northern North America (Figure 6.3). The southern limit of the Laurentide Icesheet extended south of the present day Great Lakes region and over 50% of the icesheet exceeded 3,000 m in thickness (Dyke *et al.*, 2002). The northernmost sector of the Laurentide Icesheet may have been contiguous with the Greenland Icesheet in the form of the Innuitian Icesheet, extending across the present day islands of Axel Heiberg and Ellesmere in the Arctic Circle. Collectively, the Laurentide, Cordilleran, and Innuitian Icesheets at their maximum development accounted for approximately 74 m of ice-equivalent sea level (Peltier, 2004). Maximum development of the Laurentide Icesheet was attained at approximately 22–19 ka, although local timing of maximum ice accumulation is *diachronous* (time-transgressive; Dyke *et al.*, 2002).

In Europe, the Eurasian Icesheet during MIS 2 extended continuously from Ireland, northern England, Wales, and Scotland in the west, towards the Taimyr Peninsula in northern Russia (Figure 6.3). Its northern limit was near the Arctic Ocean close to the southern margin of the Nansen Basin and extended to Franz Josef Land and Svalbard (Siegert and Dowdeswell, 2004). The icesheet was likely to have been continuous across the North, Baltic, and Barents–Kara Seas, extending across all of Scandinavia, with its southern margin extending across Germany, Poland, and Belarus (Svendsen *et al.*, 2004). In Scandinavia, the icesheet attained a maximum thickness exceeding 2,000 m within the northern Gulf of Bothnia (Boulton *et al.*, 1985).

The Antarctic Icesheet was significantly larger than today during the LGM and was broadly coincident with the edge of the continental shelf (Anderson *et al.*, 2002; Davies *et al.*, 2012). The West Antarctic Icesheet adjacent to the Antarctic Peninsula is most likely to have accommodated the additional ice during MIS 2, with a wider extent of ice across the Ross and Weddell Seas. Huybrechts (2002) has suggested that the volume of the Antarctic Icesheet during MIS 2 may have been up to 29% greater

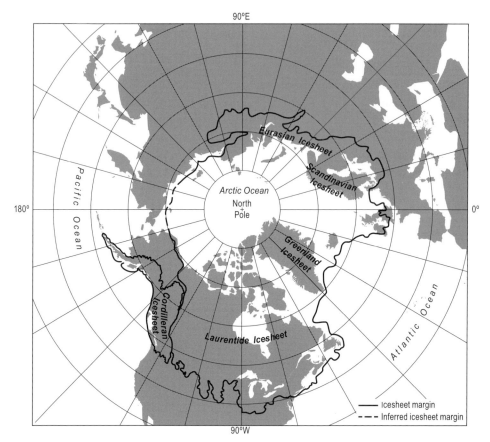

Figure 6.3 Extent of northern hemisphere continental icesheets during the LGM (MIS 2) at approximately 21 ka ago (adapted from Stokes and Clark, 2001).

than the present icesheet. Based on icesheet modelling, the EPILOG programme (Environmental Processes of the Ice-Age: Land, Oceans, Glaciers) concluded that between 14 and 21 m of ice-equivalent sea level was represented in the Antarctic Icesheet during the LGM, and that, globally, the total icesheets during MIS 2 equated with between 118.2 m (minimum) to 130.4 m (maximum) ice-equivalent sea level (Clark and Mix, 2002). During MIS 2, the Antarctic Peninsula Icesheet represented approximately 1.7 m of ice-equivalent sea level (Davies *et al*., 2012).

6.4 Early Pleistocene sea levels

Despite representing approximately 70% of Quaternary time, few coastal landforms and sedimentary successions of early Pleistocene age are preserved at a global scale. Accordingly, only a limited knowledge of early Pleistocene sea-level changes has been derived directly from coastal deposits. Although early Pleistocene coastal successions are uncommon, factors that have aided their preservation include cal-crete development in arid and Mediterranean landscapes, producing protective mantles to surface erosion, as well as areas that have remained unglaciated or subject to minimal fluvial erosion.

Deposits of presumed early Pleistocene age can rarely be dated with confidence, as outlined in Chapter 4. The time interval from 2.6 Ma to 780 ka is beyond the range of most geochronological methods, and finding materials suitable for dating by methods that are applicable to this window of Earth history represents a further difficulty. Commonly, physicochemical requirements or assumptions of relevant geochronological methods are not upheld in their field contexts, rendering it extremely difficult, if not impossible, to derive numerical ages for deposits. As a result, many attempts to assign ages to sequences of presumed early Pleistocene age have relied on palaeontological evidence (e.g. First Appearance Datums (FADs) of palaeontologically significant fossil taxa and their biostratigraphical correlation with the marine record). Other attempts have been based on Lyellian percentages of fossils (i.e. percentage of molluscan fossil assemblages represented by extant forms). Such approaches, however, are of low temporal resolution and may span several marine isotope stages. More recently, $^{87}Sr/^{86}Sr$ analyses have been applied to date early Pleistocene successions (James *et al.*, 2006; Wehmiller *et al.*, 2012).

Indirect evidence for early Pleistocene sea levels has been derived from oxygen-isotope analyses on benthic foraminifera (Shackleton *et al.*, 1990; Raymo, 1992; Figure 6.2). Equatorial Atlantic core DSDP-607 reveals that for the interval 3.1–2.9 Ma, sea levels remained significantly above the level attained during the Holocene highstand (Raymo, 1992). During the interval 3.1–2.75 Ma, sea levels progressively fell in response to build-up of the Northern Hemisphere and Antarctic Icesheets. The $\delta^{18}O$ spectral signature for inferred sea-level changes for the interval from approximately 2.75 to 1.35 Ma reveals a continued fall in sea level. During this period, the averaged position of sea level (i.e. between interglacials and glacials) is below that attained in the late Pleistocene (marine oxygen-isotope-defined) intersta-dial MIS 5c at 105 ka ago. A weak trend of sea-level rise from about 2.45 to 2 Ma is evident, followed by a renewed period of fall to about 1.35 Ma. For the interval from 1.35 Ma to the end of early Pleistocene time at 780 ka, the averaged position of sea level remained below that of MIS 5c; however, the amplitude of change between glacials and interglacials increased from that characterising the earlier record. The $\delta^{18}O$ spectral signature for inferred sea-level changes in Core DSDP-607 also reveals that the magnitude of sea-level change from glacials to interglacials was approxi-mately half that characterising the past 780 ka.

A further consideration relates to duration of interglaciations during the 41-ka-dominated cycles of the early Pleistocene. As these highstand events are likely to have only been of a few thousand years' duration, it is possible that they did not leave as prominent an imprint on coastal landscapes as the past two interglaciations (last interglacial and the present Holocene interglacial; both exceeding 10 ka in duration).

Many of the examples of sedimentary successions assigned to early Pleistocene time occur within regions that are tectonically relatively quiescent or have experienced moderate uplift. In this context, records from passive margin coastlines such as southern Australia and the United States Atlantic coastal plain are particularly important. Southeastern England has also featured prominently in discussions of early Pleistocene sea-level changes in the context of the Crag Group and its relation to long-term regional landscape evolution. In a global context, in instances where early Pleistocene successions have been preserved, their geographical extent and detail of fossil preservation are quite remarkable. The reliable dating of these successions, and particularly their correlation with MIS, remains a major challenge, as well as the unequivocal derivation of ice-equivalent sea level at time of their formation. In the examples reviewed, crustal uplift processes have in part contributed to their present elevation above sea level, but their high elevation in the present landscape also reflects a eustatic component in the early development of the Antarctic Icesheet.

6.4.1 Roe Calcarenite, Roe Plains, southern Australia

The Roe Calcarenite, which covers most of the Roe Plains (~5,800 km^2) in Western Australia, refers to a sheet drape of richly fossiliferous, semi-indurated, medium- to coarse-grained calcarenites and coquinas of early Pleistocene age (Ludbrook, 1958, 1978; Lowry, 1970; Figure 6.4). This early Pleistocene coastal complex is particularly extensive and is generally 1.5–2 m thick (Lowry, 1970). The landward transgressive feather-edge of the Roe Calcarenite occurs up to 33 m APSL at Madura, 40 km inland from the present coastline, and dips gently seaward. The landward limit occurs at the foot slope of a prominent escarpment termed the Hampton Range, a subaerially modified relict seacliff (Lowry, 1970).

The Roe Calcarenite contains a diverse molluscan faunal assemblage with over 250 species including the shallow-water cockle *Katelysia* sp. and the gastropod *Batillaria* (*Zeacumantus*) *diemenensis*, which frequent intertidal to shallow subtidal environments. *Katelysia* is a shallow infaunal mollusc that occurs in sandy shores in the lower littoral zone (Beesley *et al.*, 1998) at a depth of <1 m in relation to mean sea level and is therefore a useful palaeo-sea-level indicator. Several species of molluscs and the foraminifer *Marginopora vertebralis* indicate warmer than present water temperatures at time of deposition of the Roe Calcarenite.

Ludbrook (1978) assigned an early Pleistocene age (undifferentiated) to the Roe Calcarenite based on percentage of molluscan faunal assemblage represented by extant forms (Lyellian statistics), compared with the Red Crag of East Anglia. AAR analyses on *Katelysia* from the Roe Calcarenite suggest that the succession is of a similar age to the Jandakot Member of the Yoganup Formation of the Perth Basin (Murray-Wallace and Kimber, 1989). The Jandakot Member was assigned informally a Plio-Pleistocene age by Mallett (1982) based on the presence

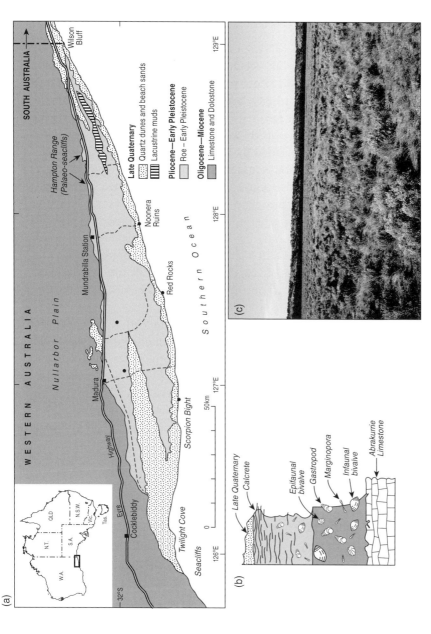

Figure 6.4 The Roe Plain, Western Australia showing extent of the Late Pliocene–early Pleistocene Roe Calcarenite (adapted from James *et al.*, 2006); (**a**) map of the Roe Plain and the head of the Great Australian Bight, (**b**) representative section of the Roe Calcarenite, and (**c**) view looking northeast to the Hampton Scarp, a subaerially modified, relict sea cliff of early Pleistocene age (photograph: Colin Murray-Wallace).

of phylogenetically primitive forms of the foraminifer *Globorotalia truncatulinoides*, which in deep-sea cores has a First Appearance Datum just below the base of the Olduvai Chron at approximately 1.87 Ma (Haq *et al.*, 1977; Berggren *et al.*, 1980). $^{87}Sr/^{86}Sr$ analyses suggest an age between 3.4 and 2 Ma for the Roe Calcarenite (James *et al.*, 2006).

The high landward elevation (33 m APSL) of the Roe Calcarenite in Australia has represented an enigma as the succession occurs between the Gawler and Yilgarn Cratons in a tectonically stable region (Lowry, 1970). Given that interglacial sea levels of the past 2.6 Ma are unlikely to have exceeded that obtained in the present Holocene interglacial by more than a few metres (Shackleton, 1987), the emergent nature of the Roe Calcarenite in part is likely to relate to a higher eustatic sea level between about 3.4 and 2.7 Ma, as indicated by the Sr-isotope inferred age of shells from the formation (James *et al.*, 2006). In addition, some minor tectonic uplift and regional tilting of the Roe Plains and Eucla Basin may have occurred (Sandiford, 2007). The undeformed nature of the Paleogene–Neogene successions underlying the Roe Calcarenite implies gentle, epeirogenic uplift of the Eucla Basin. A possible mechanism for the basin inversion may have involved lithospheric flexure in response to loading induced by Cainozoic post-rift marine sediment accumulation on the southern continental margin, as well as effects of continental-scale tilting associated with subduction-related subsidence of the northern margin of the Australian continent (Sandiford, 2007).

Geographically extensive coastal successions of Pliocene and early Pleistocene age are preserved in the Murray Basin and Coorong Coastal Plain of southern Australia and are discussed in Section 6.14.1 in the context of long-records of sea-level changes. Other examples of coastal and marine successions of early Pleistocene age in Australia occur in isolated settings reflecting chance preservation. The Point Ellen Formation on southern Fleurieu Peninsula and Kangaroo Island (Ludbrook, 1983; Milnes *et al.*, 1983) and the early Pleistocene Memana Formation on Flinders Island (Sutherland and Kershaw, 1971; Murray-Wallace and Goede, 1995) represent isolated examples of early Pleistocene coastal deposits from which it is difficult to infer palaeo-sea level in view of neotectonism.

6.4.2 The Crag Group, southeastern England

The richly fossiliferous Crag Group and, in particular, the Red Crag Formation, have featured prominently in discussions of early Pleistocene regional landscape development of southeastern England (Jones 1981, 1999; McMillan *et al.*, 2011). The Crag Group comprises four Crag formations which developed on the margins of a marine incursion in East Anglia, representing an extension of the North Sea that commenced in the early Neogene. The shell-rich successions represent portions of deltas formed on the margin of the basin. The shallow-marine or coastal Red Crag Formation has been genetically related to widespread planation surfaces on bedrock up to 190 m APSL and some 100 km west-southwest of its confirmed outcrop limit in southeastern England, to support a marine rather than subaerial origin for regional denudation and development of discordant drainage. Numerous flat-topped hills on a variety of lithologies were inferred to be the result of 'marine trimming' and linked to the early Pleistocene 'Calabrian transgression' responsible for deposition of the Red Crag Formation (Wooldridge and Linton, 1955). Identification of the Lenham Beds in Kent (Figure 6.5) with fossiliferous sands infilling solution pipes within chalk up to 190 m APSL provided additional evidence for a regional marine transgression.

The Red Crag Formation comprises two distinct sedimentary facies (Dixon, 1979) that collectively represent a shoaling-upward succession. The lower unit is represented by megaripple sands formed during a period of high current activity in water depths of approximately 15–25 m. The upper unit comprises shallow subtidal to intertidal sediments with well-defined medium-scaled ripples, tidal channels, and sand-flat facies. Calcareous worms, bryozoans, sponges, barnacles, and molluscs are common, many abraded and fragmental. The lower portion of the formation contains a cooler water fauna of North Pacific origin such as the gastropod *Neptunea angulata* and the benthic foraminifer *Elphidiella hannai* (Funnell, 1995). The top of

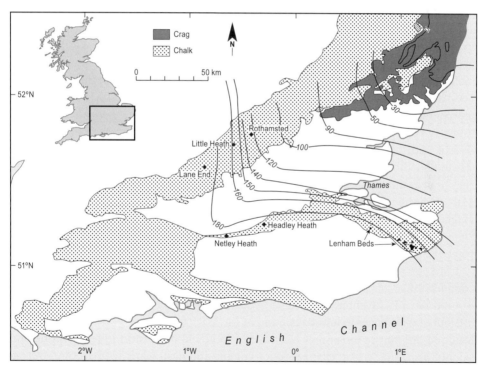

Figure 6.5 Distribution of the early Pleistocene Crag deposits of southeastern England including the Red Crag. The contours represent the base of the Red Crag succession in metres APSL (after Jones, 1999).

the formation is represented by bioturbated *Arenicola* flats. The base of the succession occurs up to 90 m APSL near Stevenage and has been correlated westwards up to approximately 190 m APSL at Lane End and Netley Heath (Jones, 1999). Given that early Pleistocene sea level (formerly assigned to late Pliocene time) was likely to have been at 20 m APSL at time of deposition of the Red Crag (Haq *et al.*, 1987), crustal warping and uplift within southeastern England is implied since its deposition, involving uplift within the London Basin and down-warping in East Anglia (Figure 6.5). The Red Crag Formation has been correlated with MIS 104 to MIS 86 (Funnell, 1995) and questionably to MIS 103–82 by McMillan *et al.* (2011). The base is broadly coincident with the Gauss to Matuyama geomagnetic polarity reversal at 2.6 Ma and the redefined base of the Quaternary. Thus the re-defined Quaternary places the succession firmly in the early Pleistocene.

6.5 The middle Pleistocene Transition

The middle Pleistocene Transition (MPT) represents a significant period of change in the spectral character of relative sea-level fluctuations during the Quaternary from lower-amplitude, higher-frequency 41-ka glacial cycles to higher-amplitude, lower-frequency 100-ka glacial cycles (Shackleton *et al.*, 1990; Head and Gibbard, 2005; Clark *et al.*, 2006; Cronin, 2010; Figure 6.6). The transition is from essentially a two-state system of glacials and interglacials that occurred in the 41-ka-dominated cycles of the early Pleistocene, to the more complicated spectral signature that includes stadials and interstadials, as well as glacials and interglacials, of the 100-ka-dominated cycles of middle and late Pleistocene. The post-MPT glaciations are characterised by an additional, approximately 50 m of ice-equivalent sea-level lowering (Clark *et al.*, 2006).

It is generally regarded that the MPT occurred between 1,500 and 600 ka ago (Head and Gibbard, 2005; Clark *et al.*, 2006; Cronin, 2010), although some variation is evident in timing of the MPT based on interpretations of different isotopic records. MIS 22 (about 880–870 ka) is considered the first major cold event similar in magnitude to glaciations that characterised the later Pleistocene (Head and Gibbard, 2005), and is likely to represent the first major episode of glacio-eustatic sea-level fall comparable in magnitude to the LGM.

Clark *et al.* (2006) suggested that the increasing amplitude of glacial cycles commenced at about 1,250 ka ago, based on a study of stacked marine oxygen-isotope records (average of 57 $\delta^{18}O$ records of Lisiecki and Raymo, 2005). They attributed differences in inferred timing of the MPT by various workers as a result of localised differences in isotopic records (e.g. regional differences in temperature and salinity obscuring global changes in ice volume and deep-water temperatures during the interval of the MPT).

Several hypotheses have been advanced for the origin of the MPT, but in many respects it remains an enigma. No changes are evident in the Milankovitch orbital

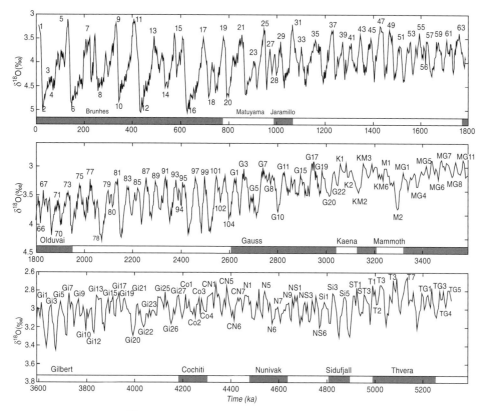

Figure 6.6 The LR04 $\delta^{18}O$ (benthic) stack for the past 3 Ma based on a compilation of 57 marine oxygen-isotope records (Lisiecki and Raymo, 2005). The palaeomagnetic timescale is also indicated. The MPT began at about 1,250 ka ago corresponding with MIS 37 (Clark *et al.*, 2006).

forcing variables over the course of the MPT and the 100-ka (eccentricity) cycle is actually no stronger in the past 700 ka than before the MPT. This has been variously described as the *100-ka problem* (Williams *et al.*, 1998) or the *eccentricity myth* (Maslin and Ridgwell, 2005), as the timing between successive deglaciations range between 119 and 87 ka over the past 700 ka and changes in eccentricity of the Earth's orbit do not appear to have triggered the transition. In particular, a mechanism is needed to explain persistence of icesheets spanning much of the 100-ka cycles identified by spectral analyses which post-date 1.2 Ma. Explanations have tended to centre on feedback effects within the Earth's climate system in response to ongoing cooling that heralded the start of the Quaternary, and the long-term secular trend in decreasing atmospheric pCO_2. The general ice-volume increase identified in spectral analyses reflects prolonging of glacial periods and hence an increase in time available for icesheet accumulation.

The size of the northern hemisphere icesheets has been regarded as a critical element in establishing the change from the 41-ka- to 100-ka-dominated cycles identified in spectral analyses of oxygen-isotope time-series data. Clark *et al.* (2006) have suggested that an increase in thickness of the northern hemisphere icesheets occurred following glacial erosion of deeply weathered bedrock before the MPT. They note that icesheets do not appear to have been any less extensive before the MPT, yet an increase in ice volume is implied following the MPT. Removal of regolith may have provided a more resistant substrate on which icesheets could be firmly anchored, enabling them to persist for longer intervals (i.e. > 41 ka) such that obliquity (41 ka) and precession (27 ka) cycles appeared less significant in spectral analysis of isotopic data.

The MPT has also been regarded as a manifestation of the long-term cooling trend initiated in the Neogene, as extensive formation of sea ice enhances planetary albedo. Other mechanisms suggested for the MPT have involved the global carbon cycle, the formation of the Greenland–Scotland submarine ridge and its influence on thermohaline circulation, and distribution of gas hydrates in continental shelf and slope environments, resulting in release of methane to enhance global warming (Maslin and Ridgwell, 2005). Perhaps each of these processes contributed towards the MPT. The abruptness of the MPT raises concerns as to the integrity of palaeoclimatic and sea-level inferences derived from spectral analyses of time-series isotopic data. In particular, the 100-ka cyclicity evident in spectral records post-dating the MPT may potentially give the erroneous impression that eccentricity forcing (100 ka) cycles are the driver of climate and sea-level changes since the start of middle Pleistocene time, when in fact these are a response to other processes.

6.6 Middle Pleistocene sea-level changes

The middle Pleistocene sea-level record is characterised by seven glacial cycles of broadly 100 ka duration. Since the early work of Emiliani (1955), numerous oxygen-isotope records have been used for subdividing middle Pleistocene time (e.g. Shackleton and Opdyke, 1973; Shackleton *et al.*, 1990; Bassinot *et al.*, 1994; Jouzel *et al.*, 2007; Masson-Delmotte *et al.*, 2010), and countless analyses based on marine cores from all of the ocean basins, under the auspices of various Ocean Drilling Programs, have replicated these benchmark studies (e.g. von Grafenstein *et al.*, 1999 and Pierre *et al.*, 1999 in the Mediterranean Sea). While oxygen-isotope results from different core sites may reveal contrasting magnitudes of isotopic excursions with time, reflecting local variations in temperature and salinity, the broad timing of glacial to interglacial transitions is remarkably consistent. The record from the Mediterranean basin, for example, shows amplitudes in $\delta^{18}O$ of 3–3.5‰, which are greater than the average for the world ocean and reflect local temperature and salinity variations (von Grafenstein *et al.*, 1999); however, the timing of the $\delta^{18}O$ excursions is in accord with results from other ocean basins.

Three approaches have been adopted for subdividing the middle Pleistocene and characterising its sea-level history; the first involves oxygen-isotope records from deep-sea cores, the second is based on EPICA Dome C ice-core records, and the third derives a sea-level record based on salinity changes using cores from the Red Sea. The marine-based oxygen-isotope chronology has been the longest-established, initially defined from deep-sea core V28–238 (Shackleton and Opdyke, 1973). It has been outlined above and assigns stage boundaries related to the assumed rate of sedimentation in deep-sea cores, which in the case of core V28–238 involved a linear interpolation of a mean rate of 170 mm/ka (Shackleton and Opdyke, 1973). Although the general sequential changes in successive glacial cycles and their amplitudes appear similar in various cores, and can be compared with the record derived from ice cores, each does suggest slightly different ages for the boundaries.

Ice-core records have featured prominently in the climatically defined stratigraphy of the middle Pleistocene. The EPICA Dome C ice core from Antarctica, for example, reveals that in ascending order, the first glacial cycle of the middle Pleistocene commenced with MIS 19 at approximately 790 ka (Masson-Delmotte *et al.*, 2010; Schilt *et al.*, 2010), which preceded the actual lower boundary of the middle Pleistocene which is the Brunhes–Matuyama geomagnetic polarity reversal at 780 ka (Shackleton *et al.*, 1990). The second glacial cycle extended from MIS 17 (720 ka) to MIS 16 (628 ka) and lasted 92 ka and ended at Termination VI as defined from the marine-based oxygen-isotope record (Broecker and van Donk, 1970). The third glacial cycle commencing with interglacial MIS 15 is characterised in the EPICA Dome C record by two distinctively broad peaks in both the δD and $\delta^{18}O$, MIS 15.5 (628–603 ka) and MIS 15.1 (580–560 ka), respectively, and concludes with MIS 14 at approximately 530 ka ago. This was followed by the fourth glacial cycle that commenced with a more subdued interglaciation (MIS 13) which began at MIS 13.3 (530 ka) and concluded with glacial MIS 12 at approximately 425 ka ago (Termination V of Broecker and van Donk, 1970). The fifth glacial cycle of the middle Pleistocene, in the older terminology preceding oxygen-isotope stratigraphy was referred to as the *anti-penultimate* glacial cycle and commenced with MIS 11. In EPICA Dome C core this cycle commenced at 424.6 ka and ended at Termination IV at 335.6 ka with a duration of approximately 89 ka. The sixth cycle, termed the *pre-penultimate* glacial cycle in the earlier literature, extended from 335.6 ka and concluded at Termination III at 245.6 ka and lasted approximately 90 ka. The *penultimate glacial cycle*, the seventh and last cycle of middle Pleistocene time, commenced at the beginning of MIS 7 (245.6 ka) and ended at the onset of deglaciation (Termination II) following MIS 6 (132 ka) and accordingly had an approximate duration of 113.6 ka (Masson-Delmotte *et al.*, 2010).

A continuous record of sea-level over the past five glacial cycles has been compiled from analysis of cores from the Red Sea (Siddall *et al.*, 2003; Rohling *et al.*, 2009, 2010). The Red Sea method is based on changes in residence time of water in the highly evaporative Red Sea basin that occur as sea level varies. The Red Sea exchanges seawater with the open ocean through the Strait of Bab-el-Mandab which contains a sill, called the Hanish Sill, that is only 137 m deep. Cross-sectional area was only about 2% of that existing today, during lowstands of sea level, with widths of only 1–3 km across the broadest of the narrow water bodies (Lambeck *et al.*, 2011). The Red Sea is bounded by deserts, and there is almost no dilution from rivers. Evaporative loss amounts to around 2 m/year, resulting in high salinities and a heavy δ^{18}O signature as a result of concentration of ^{18}O with respect to the lighter isotopes ^{17}O and ^{16}O. At lower sea levels the smaller dimensions of the entrance at Bab-el-Mandab impose further hydraulic constraints on exchange. Response times are short, less than a century. The compilation combines analyses on foraminiferal carbonate and bulk sediment data from three central Red Sea cores, core KL11 which spans the interval 0–360 ka, MD92–1017 which spans 0–470 ka, and KL09 which covers 0–520 ka.

Evaporation results in particularly saline water in the extreme northern end of the basin, which flows southwards as a dense, cool, salty layer, overlying a relatively stagnant layer of still denser water, which is formed during the winter months in the Gulf of Suez (Siddall *et al.*, 2004). The sediment record in cores from the Red Sea contains aplanktonic intervals that represent fully glacial periods when Red Sea salinities were in excess of the lethal limit for planktonic foraminifera of around 49 psu (practical salinity units, which approximates 49‰). Initial interpretations of middle Pleistocene changes in the Red Sea were based on abundance of different foraminifera and presence/absence ratios (Rohling, 1994). In the aplanktonic sections of the cores, representing sea-level lowstands, recon-structions use bulk sediment-based values. Modelling using an oceanic general circulation model is broadly consistent with core-based reconstructions, with isostatic modelling implying that, at the peak of the last glacial, sea level was 105 m lower than it is now, with around 30 m of water over the Hanish Sill (Biton *et al.*, 2008). The record, shown in Figure 6.7 (after Rohling *et al.*, 2012), is considered to have a 1σ uncertainty of ±6.5 m on the sea-surface elevation, and on the basis of a revised U-series chronology (Rohling *et al.*, 2009, 2010). Initial calibration against EPICA Dome C ice-core records was considered lagged because of ice-volume and sea-level responses to temperature change, so a shift of a few thousand years was used to correct for this (Rohling *et al.*, 2010). It is constrained to the orbital time scale with a ±1.5% age uncertainty (~250 years) for age of interglacials independently determined from U-series chronologies on corals and speleothems (Siddall *et al.*, 2006).

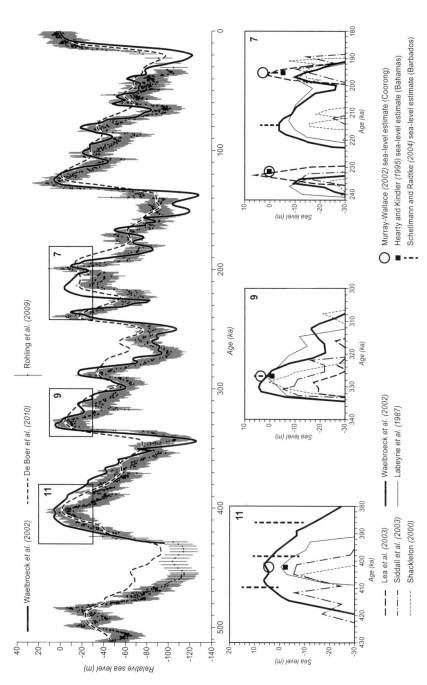

Figure 6.7 Middle and late Pleistocene sea-level changes as reconstructed from a series of sources. The upper diagram compares several different approaches for reconstructing global sea-level trends for the past 400–500 ka after Rohling *et al.* (2012). The dots, with 2σ uncertainties, represent reconstructions from cores in the Red Sea, described in Rohling *et al.* (2009). The black line with grey uncertainty buffer is based on the coral-calibrated deep-sea benthic δ¹⁸O records by Waelbroeck *et al.* (2002), and the dashed line is the model-based convolution of benthic δ¹⁸O records reported by De Boer *et al.* (2010). The lower panels show sea-level estimates for each of the interglacials MIS 11, 9, and 7, after Siddall *et al.* (2006). Note that the estimates by Siddall *et al.* (2003) use the Red Sea data, but that the results in the upper panel, following Rohling *et al.* (2009), adopt a revised chronology.

6.7 Sea-level highstands of the middle Pleistocene

The most continuous record of middle Pleistocene ocean volumes, resolvable at millennial timescales, comes from oxygen-isotope ratios from deep-sea and ice cores. This has formed the basis for recognition of MIS, and interglacials, during which the sea has been high, have been given odd numbers (7, 9, 11, etc.). The record can be divided into substages that can be identified during some interglacials, which are either decimalised (5.1, 5.2, etc.) or lettered (5a, 5b, etc.). These highstands are also observable in the EPICA Dome C ice-core record which reveals that MIS 11 and 9 are likely to have experienced the highest sea levels of the middle Pleistocene, and that MIS 13, 15, and 17 appear to be characterised by intervals of lower sea level, possibly of comparable magnitude to MIS 7 (Masson-Delmotte *et al.*, 2010). These highstands are also detectable in coastal sedimentary and landform evidence, such as fossil reefs, wave-cut features, and cave deposits. In combination, the continuous records of sea-surface variation and chronostratigraphy of coastal deposits provide insights into past sea level, indicating that some interglacials appear to have had multiple highstands of similar magnitude (Siddall *et al.*, 2006). The following discussion examines palaeo-sea-level records of successive interglacial highstands commencing with MIS 11. Discussion of earlier middle Pleistocene sea levels is presented in the context of long Pleistocene records in Section 6.14.

6.7.1 Marine Isotope Stage 11

Numerous marine oxygen-isotope records indicate that MIS 11 (420–360 ka) represented the warmest of recent interglaciations, with smaller ice volumes and higher sea levels. Core DSDP-607 from the Mid-Atlantic Ocean, for example, reveals that over the past 500 ka, stage 11 $\delta^{18}O$ values were more negative than for MIS 9, 7, 5e, and the current Holocene interglacial (Raymo, 1992). Interest in MIS 11 has stemmed from recognition that it experienced apparently similar orbital parameters to the current Holocene interglacial and, accordingly, is regarded as a potentially important analogue for future climate change (Cronin, 2010; Rohling *et al.*, 2010). MIS 11 is also characterised by the highest amplitude deglacial warming from a preceding glaciation in the past 500 ka (Droxler and Farrell, 2000; Cronin, 2010) and spanned two precession cycles (Raymo and Mitrovica, 2012). Besides higher isotopically inferred sea levels, MIS 11 was also characterised by warmer sea-surface temperatures in high latitudes, stronger thermohaline circulation, and extension in the geographical range of coral reefs. Terrestrial records reveal generally warmer conditions for MIS 11 with extension in geographical range of molluscs, various tree species, and greater intensity of weathering and pedogenesis (Burckle, 1993). MIS 11 has also been regarded as the longest interglaciation since the Brunhes–Matuyama boundary (780 ka) with a duration of approximately 32.6 ka (424.6–392 ka; Masson-Delmotte *et al.*, 2010).

Similarly, based on study of deep-sea cores (ODP cores 980 and 983), McManus *et al.* (2003) concluded that MIS 11 lasted for approximately 30–40 ka depending on the age interpolation models applied to these cores.

Recognition that MIS 11 was longer than the last interglacial maximum (MIS 5e) and potentially characterised by significantly higher sea levels led to numerous attempts to define ice-equivalent sea levels for this interval based on empirical evidence from relict shoreline features. Based on the analysis of shelly deposits from southeastern England, Bowen (2003) estimated MIS 11 sea level at 10.8 ± 0.7 m APSL. Laterally persistent, low-angle, planar cross-laminated skeletal carbonate sands, interpreted as the product of shallow subaqueous deposition, occur on the top of limestone headlands (20 m APSL) of North Eleuthera, Bahamas (Hearty *et al.*, 1999). A succession of cave infill sediments comprising rounded clasts of reworked limestone (presumed water worn) and superposed calcarenite with entire fossil marine shells and coral fragments were also surveyed at 21.6 m APSL at Dead End Cave, Bermuda (Hearty *et al.*, 1999). The conglomerate and coarse calcarenite are bounded by cave flowstone units dated by U-series at $680 \pm _{130}^{\infty}$ ka (basal flowstone) and an upper flowstone with ages derived from the top and bottom of the flowstone as 127.7 ± 1.6 ka and 420 ± 30 ka, respectively. As Bermuda and the Bahamas have long been regarded as tectonically stable (Land *et al.*, 1967), Hearty *et al.* (1999) concluded that these coastal successions were deposited in MIS 11 when sea level was higher by $\sim 20 \pm 3$ m APSL at, or before, 420 ± 30 ka. It was hypothesised that a partial collapse of the East Antarctic Icesheet was necessary, presuming that the Greenland and West Antarctic Icesheets (~ 12 m of ice-equivalent sea level) also contributed to the inferred sea-level rise.

A re-examination of palaeo-sea-level records from Bermuda and the Bahamas has revealed that the elevation of the MIS 11 deposits is due in part to glacio-isostatic adjustment (Raymo and Mitrovica, 2012). Up to 7.4 m (Bahamas) and 11.9 m (Bermuda) of the shoreline elevation APSL may be related to glacio-isostatic compensation following collapse of the glacial forebulge in the northwestern Atlantic Ocean, associated with decay of the North American icesheets. Modelling, incorporating a range of values for mantle viscosity, thickness of lithosphere, and duration of low ice volumes within interglacial MIS 11 low ice volumes, suggests that during MIS 11 eustatic (ice-equivalent) sea level reached between 6 and 13 m APSL.

Despite claims for a sea level up to 20 m APSL for MIS 11 (Hearty *et al.*, 1999; Olson and Hearty, 2009), globally there is a paucity of geomorphological and stratigraphical evidence to confirm unequivocally higher sea levels of this magnitude for this time interval. Similarly, there is contradictory evidence from oxygen-isotope records for the magnitude of warming during MIS 11. Deep-sea core records reveal that sea-surface temperatures in the high-northern and southern latitudes of the Atlantic Ocean were not significantly different from other interglacials of the past 500 ka (Hodell *et al.*, 2000). Oxygen-isotope evidence from Core PS1243 in the southwestern Norwegian Sea also reveals that MIS 11 was characterised by lower ice volumes than the current Holocene interglacial (Bauch *et al.*, 2000).

The global paucity of stratigraphical evidence for higher sea levels in MIS 11 may relate, in part, to the generally low preservation potential of such deposits, although absence of evidence is not evidence of absence. Erosional processes during the four glacial cycles following MIS 11 may have largely obliterated evidence for former higher sea levels. However, evidence from other regions characterised by long records is not consistent with higher sea levels in MIS 11 (e.g. with tectonic uplift-corrected, ice-equivalent sea levels of the Coorong Coastal Plain in southern Australia –3 m; Barbados –1, 5, or 18 m; Sumba Island, Indonesia $\geq$ 1 m; Pirazzoli *et al.*, 1993b; Murray-Wallace *et al.*, 2001; Schellmann and Radtke, 2004a,b). In the case of Barbados, the range of values depends on assumed elevation of the sea surface during the last interglacial maximum in calculation of uplift rates (0, 2, and 6 m APSL, respectively; Schellmann and Radtke, 2004a,b; see Chapter 5).

The suggestion of higher sea levels during MIS 11 also presents a difficulty in terms of the Croll–Milankovitch Hypothesis and has been termed the *Stage 11 Paradox*. The amplitude of climate variation between glacial MIS 12 and interglacial MIS 11 (~420 ka) is at a maximum at the very time when astronomical forcing is at a minimum (Maslin and Ridgwell, 2005).

6.7.2 Marine Isotope Stage 9 – the pre-penultimate interglacial

The EPICA Dome C ice core indicates that MIS 9 had a duration of 18.8 ka, commencing at 335.6 ka and ending at 316.8 ka (Masson-Delmotte *et al.*, 2010), whereas oxygen-isotope records from deep-sea cores show a single peak at 331 ka (Siddall *et al.*, 2006). Few coastal landforms of this age have been documented, most likely relating to their lower preservation potential. In the Coorong Coastal Plain of southern Australia, emergent coastal barrier shoreline complexes of this age, termed the West Avenue Range have been dated to 315 ± 25 ka (Huntley and Prescott, 2001) and 342 ± 42 ka (Huntley *et al.*, 1993) using thermoluminescence on quartz from calcarenite, and at 382 ± 73 ka by 'whole-rock' AAR (Murray-Wallace *et al.*, 2001). An uplift-corrected sea level of 1 m BPSL for MIS 9 is evident based on elevation of back-barrier lagoon facies (Murray-Wallace, 2002). Corals of this age (334 ± 4 to 306 ± 4 ka) have been dated from Henderson Island (Stirling *et al.*, 2001), which has been subject to an unknown degree of uplift precluding an estimate of sea level at that time. Sea level within a few metres of present is inferred from the Bahamas (Hearty and Kindler, 1995), Grand Cayman (Vézina *et al.*, 1999), and Barbados (Schellmann and Radtke, 2004a,b).

6.7.3 Marine Isotope Stage 7 – the penultimate interglacial

Although coastal successions relating to the penultimate interglacial (MIS 7; 245.6–233.6 ka as defined in the EPICA Dome C ice core; Masson-Demotte *et al.*, 2010) are more commonly preserved than successions relating to earlier middle Pleistocene interglacials, they are significantly less abundant, globally, than the last interglacial maximum (MIS 5e). Shoreline successions of MIS 7 are more commonly found on tectonically active, emergent coastlines. Observations from marine-based and ice-core oxygen-isotope records suggest a relative sea level of –20 m for MIS 7 (Shackleton, 1987). The EPICA Dome C record, for example, shows MIS 7 to be the coolest interglacial during the past 400 ka with an average temperature approximately 2°C cooler than present (Masson-Delmotte *et al.*, 2010). U-series and ESR dating of corals from Barbados suggest that MIS 7 sea levels, corrected for tectonic uplift, ranged between –20 and –6 m, assuming that during the last interglacial sea level approximated present levels. If an MIS 5e sea level of 6 m APSL is assumed in estimation of uplift rate, then MIS 7 sea levels ranged between –10 and 4 m APSL (Schellmann and Radtke, 2004a,b).

There is evidence for up to three sea-level peaks separated by episodes of substantially lower sea level (Siddall *et al.*, 2006); the older between 240 and 230 ka, followed by a prolonged peak from around 220 ka, with the most recent 201–193 ka (Figure 6.7), inferred from speleothems in Argenterola Cave in Italy (Antonioli *et al.*, 2004). An upper limit for sea level of –10 m is inferred from a speleothem that grew throughout the period 209–139 ka in a cave in the Bahamas, this being the roof of the cave in which it was found (Toscano and Lundberg, 1999). Reefs of MIS 7 age have been dated from several Pacific Islands, including Tonga (Taylor and Bloom, 1977), the Cook Islands (Woodroffe *et al.*, 1991a), and the Hawaiian Islands (McMurty *et al.*, 2010; see also Chapter 5). In southern Australia, in the far-field of former icesheets, MIS 7 sea levels have been identified at –6 m (MIS 7a at 200 ka) for the Dairy Range, and 0 m (MIS 7e at 250 ka) for the Reedy Creek Range of the Coorong Coastal Plain, 2 m APSL for the 'Older Pleistocene' marine beds in northern Spencer Gulf and 1 m APSL for estuarine shelly facies within an aeolianite complex at Peppermint Grove in the Swan Coastal Plain near Perth, Western Australia (Murray-Wallace, 2002), although Hearty and O'Leary (2008) note that this deposit may correlate with either MIS 7 or 9.

6.8 Middle Pleistocene sea-level lowstands

Defining the position of glacio-eustatic (ice-equivalent) sea level during glacial maxima of the middle Pleistocene is particularly challenging in view of the paucity of palaeo-sea-level indicators exposed in terrestrial settings or identified in cores from continental shelf environments. Much of the likely stratigraphical evidence for middle Pleistocene glacial age sea-level lowstands is confined to continental shelf settings. In the high wave-energy continental shelf setting of New South Wales in southeastern Australia, for example, storm reconcentrated shell lags indicate sea level between 120 and 130 m BPSL for the LGM, and the presence of shells of MIS 6 and MIS 8 age at lower stratigraphical levels within many of the vibrocores points to broadly similar sea levels for the last two glacial maxima of the middle Pleistocene (Murray-Wallace *et al.*, 1996b, 2005).

The general inability to define middle Pleistocene glacial lowstands from reliable proxy sea-level indicators has placed a greater reliance on oxygen-isotope stratigraphy from long time-series marine and ice-core records. Accordingly, many deep-sea and ice-core records have provided opportunities to derive first-order estimates of the magnitude of middle Pleistocene sea-level lowstands during glacial maxima, for example, core V28–238 from the southwestern equatorial Pacific (Shackleton and Opdyke, 1973), as well as composite stacks of oxygen-isotope records from different ocean basins (e.g. Prell *et al.*, 1986; Raymo, 1992; Bassinot *et al.*, 1994), and oxygen-isotope records from ice cores (e.g. Jouzel *et al.*, 2007; Masson-Delmotte *et al.*, 2010). Variability in amplitude of oxygen-isotope excursions registered from different core sites also provides insights into localised temperature and salinity variations within the oceans, providing a valuable basis for evaluating intensities of middle Pleistocene glacials and their ice-equivalent sea levels.

The magnitude of relative sea-level fall during glacial cycles has been quantified using palaeosalinity to determine sea-level lowstands in the Red Sea for the past 500 ka by Rohling *et al.* (1998), based on foraminiferal faunal changes with time in marine core MD921017. They estimated lowstands for the last five glacial maxima at 139 ± 11 m BPSL (MIS 12, ~440 ka); 134 m to 122 ± 9 m BPSL (MIS 10); 120 ± 8 m BPSL (MIS 8); 125 ± 6 m BPSL (MIS 6) and 120 ± 5 m BPSL (MIS 2). These data are consistent with modelling based on a compilation of 57 oxygen-isotope records which suggest global sea level during glacial maxima for the past 700 ka was 120 ± 10 m (Bintanja *et al.*, 2005). Rohling *et al.* (1998) noted that MIS 12 in core MD921017 is represented by an 'aplanktonic zone' and that the magnitude of inferred sea-level fall exceeded that of the LGM by some 15% suggesting, as previously concluded by Shackleton (1987), that MIS 12 was a more intense event characterised by larger ice volumes. A more detailed and better calibrated chronology using oxygen-isotope analysis of foraminifera from cores in the Red Sea is presented by Rohling *et al.* (2009) over the past 500 ka, with a high signal to noise ratio, as described above (see Figure 6.7). However, the absence of foraminifera during the highly saline glacial means that the record is reliant on bulk sediment analyses during these times, and values are based on assumptions about hydraulic modelling of exchange through the Strait of Bab-el-Mandab.

6.9 Late Pleistocene sea-level changes

The geological record reveals substantial differences in the rate of change in ice volume with time, and as a consequence glacio-eustatic (ice-equivalent) sea-level change during glacial cycles. The most recent glacial cycle (commonly termed the last glacial cycle), for example, was characterised by an overall gradual increase in ice volume commencing at the end of the last interglacial maximum (Figure 6.8). It extended over a period of some 96 ka (i.e. from the end of the last interglacial maximum at approximately 116 ka through a period with a sawtooth pattern of

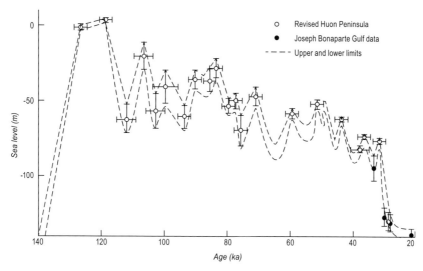

Figure 6.8 Age–depth plot of late Pleistocene sea-level indicators and inferred sea-level changes showing data from the Huon Peninsula, Papua New Guinea and the Joseph Bonaparte Gulf, northwestern Australia. The envelope delineates the range of palaeo-sea-level observations (from Lambeck and Chappell, 2001, reprinted with permission from AAAS).

sequential decline to maximum development of continental icesheets at 22–21 ka; the LGM: Figure 6.8). The record reveals a progressive increase in ice volume punctuated by periodic rises in sea level during *interstadials* (brief intervals of warmer climate during a general trend of glacial cooling). The deglaciation following the LGM, however, was characterised by rapid sea-level rise (described in Chapter 7), for brief periods attaining rates approximately 20 times faster than the rate of sea-level lowering during icesheet development (Fairbanks, 1989; Peltier and Fairbanks, 2006). Marine sediment and ice-core-based oxygen-isotope records reveal a broadly similar pattern of progressive sea-level fall accompanying growth of icesheets and rapid sea-level rise at times of deglaciation for earlier glacial cycles of the middle Pleistocene (780–128 ka). Maximum sea-level lowering during the LGM at Barbados was approximately 130 ± 5 m at approximately 26 ka (Peltier and Fairbanks, 2006).

Late Pleistocene sea-level curves have commonly been derived from restricted geographical regions. Although such records are site-specific and may have involved an assumption of a sea level 6 m APSL for the last interglacial maximum (MIS 5e at 125 ka) and a constant rate of tectonic uplift since that time, they show broad similarities globally (Steinen *et al.*, 1973; Aharon, 1984; Lambeck and Chappell, 2001). Sea-level curves derived from coral terraces of Barbados (Atlantic Ocean) and Huon Peninsula, Papua New Guinea (southwest Pacific Ocean) show a striking similarity in the long-term general trend of sea level, predominantly reflecting glacio-eustasy (see Figure 6.2). These broad patterns of sequential change have also been corroborated by oxygen-isotope records of ice-volume change from marine and ice cores (Shackleton and Opdyke, 1973; Hays *et al.*, 1976; Martinson *et al.*, 1987; Masson-Delmotte *et al.*, 2010). Local variations in palaeo-sea level reflect glacio-hydro-isostatic feedback effects on continental shelves of contrasting widths, and in the near-field of former icesheets, progressive adjustments in the crust–mantle system in response to fluctuating ice volumes (e.g. collapse of glacial forebulges and glacio-isostatic uplift in areas formerly covered by ice). These trends are further complicated by *steric effects* on sea level (i.e. changes in water volume accompanying changes in surface water temperatures and salinity).

As a 0.1‰ change in $\delta^{18}O$ in deep-sea core records equates to a 10 m change in ice-equivalent sea level, the magnitude of eustatic sea-level fluctuations may be indirectly inferred from the isotopic record of deep-sea cores, with due allowance for local variations in isotopic signals in response to differential water temperatures and salinity. The timing of these changes may be inferred from average rates of sedimentation revealed in deep-sea cores and ice accumulation in ice-core records. Radiocarbon dating towards the top of cores for the past 50 ka, U-series dating for the past 400 ka, and recognition of the Brunhes–Matuyama geomagnetic polarity reversal at 780 ka for longer records (i.e. the change from normal to reversed magnetic polarity which generally represents the first magnetic reversal to be observed down from the top of cores) have been instrumental in dating these records.

The last glacial cycle commenced with an abrupt fall in sea level at the end of the last interglacial maximum (MIS 5e) at about 116 ka and the level fell by approximately 20–30 m by 111 ka (Tzedakis *et al.*, 2012). As classically defined from marine core records, MIS 5 has five subdivisions (see Figure 6.2). Following the last interglacial maximum, eustatic sea levels fell during the stadial of substage 5d and rose during interstadial 5c to –10 m BPSL at ~105 ka. This was followed by a subsequent fall in sea level during the stadial of 5b and another rise during the subsequent interstadial 5a at 82 ka (–15 to –20 m BPSL). Interstadial 5a was followed by the stadial of Stage 4, which in turn was followed by the interstadial of Stage 3. Interstadial MIS 3 sea levels for the interval 70–30 ka range between 90 and 45 m BPSL, with a modal sea level of about 65 m BPSL. An overview of some of the early controversies relating to MIS 3 sea levels is discussed later in this chapter.

The LGM (MIS 2) followed MIS 3 and represents the period of lowest sea level during the last glacial cycle. Although isotopic records bracket the LGM between about 11 and 29 ka, sea level maintained its lowest level between 30 and 19 ka (Lambeck *et al.*, 2002a; Peltier and Fairbanks, 2006). At this time, continental areas were significantly larger resulting from the lower sea levels. In Australia, land bridges extended between the mainland and Tasmania as well as Papua New Guinea. The Sunda Shelf provided a land connection between Borneo, Java, Sumatra, Malaya, and Vietnam, and Beringia represented a major landbridge between the North American and Asian continents (see Figure 6.13).

6.9.1 *The last interglacial maximum (MIS 5e)*

The sea-level highstand of the last interglacial maximum (MIS 5e; 128–116 ka) is one of the most enduring and significant coastal geomorphological expressions of eustatic sea-level changes in the more recent geological record (Murray-Wallace and Belperio, 1991; Ota, 1994; Stirling *et al.*, 1995, 1998; Zazo *et al.*, 1999; Muhs, 2002; Ferranti *et al.*, 2006; Hearty *et al.*, 2007). U-series dating has been instrumental in defining the duration of the highstand associated with this interglacial (Edwards *et al.*, 1987a,b; Chen *et al.*, 1991; Stein *et al.*, 1991; Stirling *et al.*, 1995; Eisenhauer *et al.*, 1996; Muhs, 2002; Thompson *et al.*, 2011). For much of the world's coastlines, sediments and landforms of this age occur above present sea level, landward of their more recent Holocene equivalents, and in many regions have been used as a regional datum to quantify long-term vertical crustal movements that have occurred over the past 125 ka, as outlined in Chapter 5 (Figure 6.9). Many coastal landforms of the last interglacial are somewhat larger in extent than their Holocene equivalents (e.g. cross-sectional width of shore platforms and volumes of sand within coastal barrier successions), relating to apparently greater duration of this highstand compared with the current Holocene interglacial for coastlines situated in far-field locations. Along tectonically active coastlines in the tropics, extensive reef complexes have formed with broader cross-sections than Holocene examples. Similarly, along high wave-energy coastlines, such as southeastern Australia, last interglacial barriers are substantially larger coastal landforms (Thom *et al.*, 1992) and relict intertidal mudflats of the eastern margins of Gulf St Vincent and Spencer Gulf in southern Australia are larger features than their modern, Holocene equivalents. As with the current Holocene interglacial, the spatial scale of coastal landforms that developed during the last interglacial maximum, will relate to proximity of coastlines to Pleistocene icesheets. Accordingly, coastlines in the far-field of Pleistocene icesheets will have experienced a longer period of last interglacial highstand, permitting a longer history of sediment deposition. In contrast, coastlines located closer to Pleistocene icesheets will have had significantly less time for large-scale coastal features to develop. The differential timing of the MIS 5e highstand, resulting from the spatially variable response to glacio-hydro-isostatic processes, also accounts for some of the uncertainty in defining its duration based on U-series dating of coral-reef successions.

In the coastal realm, the last interglacial maximum has also attracted considerable attention because many sedimentary successions of this age are host to fossil faunas indicating warmer mean annual temperatures than present. The Senegalese warm-water fauna in the Mediterranean Basin including *Strombus bubonius* is now confined to the central West African coast from Western Sahara to Angola and Cape Verde Islands. Its disappearance from the Mediterranean is attributed to global cooling which culminated in the LGM (Hillaire-Marcel *et al.*, 1986). Similarly, the arcoid bivalve *Anadara trapezia* (Deshayes, 1840), popularly known as the 'Sydney Cockle', had a considerably wider geographical range in southern Australia

Figure 6.9 Shoreline features associated with the last interglacial maximum (MIS 5e – 125 ka); (**a**) relict intertidal flats of last interglacial age adjacent a backing cliff in northern Gulf St Vincent, southern Australia. The relict calcreted surface was reoccupied as supratidal flats during the past 7,000 years; (**b**) a partially eroded aeolianite coastal barrier capped by a pervasive calcrete, Still Bay, South Africa; (**c**) shallow subtidal shelly facies of the last interglacial Glanville Formation capped by a calcrete, Kangaroo Island, southern Australia; (**d**) an emergent marine terrace at Cobija Village, 130 km north of Antofagasta on the central coast of Chile. The seaward edge of the terrace is approximately 15 m APSL (photographs: Colin Murray-Wallace).

during the last interglacial maximum (Murray-Wallace *et al.*, 2000) as well as some reef-building corals (McCulloch and Esat, 2000).

6.9.2 Timing and duration of the last interglacial maximum

Marine oxygen-isotope records reveal that the last interglacial maximum equates with the period of lowest $\delta^{18}O$ values in the late Pleistocene, generally interpreted to mean smaller ice volumes and higher sea levels than present. This interval, termed MIS 5e, with a duration originally considered to be about 11 ka, is broadly equivalent to the Eemian Stage of Europe (Shackleton, 1969). Subsequent studies, however, based on $\delta^{18}O$ analyses of benthic foraminifers *Cibicidoides wuellerstorfi*, *Uvigerina peregrine* and *Globobulimina affinis* from deep-sea core MD95–2042 near Portugal, revealed that the base of the Eemian as defined by palynology is significantly younger than identified by marine oxygen-isotope evidence. The base of the Eemian has been defined by the increasing presence of Mediterranean and Eurosiberian trees and the decline in steppe vegetation which corresponds with an age of 126 ka (the plateau of MIS 5e). In contrast, the MIS 6–5e transition is at about 132 ka in core MD95–2042 (Shackleton *et al.*, 2003).

U-series dating has played a central role in delineating timing and duration of the last interglacial maximum based on dating of fossil coral-reef successions (Veeh, 1966; Edwards *et al.*, 1987a,b; Chen *et al.*, 1991; Stein *et al.*, 1991; Muhs, 1992; Zhu *et al.*, 1993; Stirling *et al.*, 1995; Eisenhauer *et al.*, 1996; Muhs, 2002; Thompson *et al.*, 2011). Technological improvements in measurement of ^{234}U and ^{230}Th by inductively coupled plasma mass spectrometry (ICP-MS) has improved precision of measurements so that greater emphasis can be placed on accuracy of the ^{234}U and ^{230}Th ages (i.e. deviation from 'true' age). Earlier measurements on corals based on conventional α- or γ-counting yielded larger uncertainties and required larger samples (see Chapter 4). Laser ablation techniques have also significantly reduced the requirements for large samples (Eggins *et al.*, 2005).

High-precision U-series dating from Bermuda, the Bahamas, Hawaii, and Australia suggest that sea level during the last interglacial maximum (MIS 5e) was at least as high as present between 128 and 116 ka (Muhs, 2002). In reviewing the literature, Muhs (2002) assessed the reliability of U-series ages from Bermuda, the Bahamas, Hawaii, and Australia by screening ages based on the criterion of back-calculated initial ^{234}U/^{238}U values to assess whether dated corals have exhibited closed-system behaviour. On this basis, ^{234}U/^{238}U values should be in accord with the range for seawater and living corals. Based on this criterion, the U-series ages deemed reliable fell between 128 and 116 ka for the last interglacial maximum.

U-series dating of fossil corals (*Montastraea annularis*) from Great Inagua and San Salvador, in the Bahamas, yielded numerical ages consistent with morphostratigraphy of the reef complex and suggesting that during the last interglacial maximum (MIS 5e), sea level attained 4 m APSL at 123 ka, 6 m APSL at 119.2 ka, and was at present sea level between these periods of higher sea level (Thompson *et al.*, 2011). The intervening 0 m value is associated with a well-defined marine abrasion surface developed on coralline sediments that is widespread within the Bahamas.

6.9.3 Global estimates of last interglacial sea levels – the sanctity of the 6 m APSL datum?

Controversial issues in the study of MIS 5e sea levels have included: (1) the maximum elevation of eustatic (ice-equivalent) sea level attained during the interglacial, (2) validity of the 6 m APSL datum as the widely cited highest point of MIS 5e sea levels, (3) the notion of bipartite sea-level stands during MIS 5e, (4) the pattern of sea-level variation during this interval, and (5) duration of the interglacial maximum.

Early studies of last interglacial coastal successions tended to report a single value for palaeo-shoreline deposits rather than attempting to quantify relative sea-level variations during the entire highstand (Mesollela *et al.*, 1969; Stearns, 1984; Murray-Wallace and Belperio, 1991). Historically, this partly related to the age-resolving power of U-series dating, the requirement for a large sample mass for analysis, and a tendency to restrict analyses using this technique to fossil corals. Many of the sites examined did not contain morphostratigraphical evidence spanning the interglacial maximum and, accordingly, changes in relative sea level could not be resolved.

To some extent, a single highstand and reference datum for MIS 5e emerged from the early work of Veeh (1966), who compiled palaeo-sea-level observations from fossil corals from sites considered stable in the Pacific and Indian Oceans remote from plate boundaries. The notion of a 6 m APSL reference level for the MIS 5e highstand appears to have been derived from a *de facto* average based on observations ranging between 2 and 9 m APSL. Subsequent work by Neumann and Moore (1975) further consolidated the 6 m APSL datum for MIS 5e based on work in the northern Bahamas (Grand Bahamas and Abaco) which identified prominent shoreline notches at 5.9 m APSL and U-series ages with a mean age of ~125 ka on corals. Similarly, in Hawaii, Ku *et al.* (1974) found a 5 m APSL reef and notches at 6.7 and 8.3 m APSL which they suggested accorded with the 6 m APSL sea level. Many workers have subsequently assigned a 6 m APSL reference level for last interglacial ice-equivalent sea level in neotectonic studies, to quantify the magnitude of crustal uplift since deposition of these shoreline successions as well as inferring the position of interstadial sea levels based on the determined rates of uplift since MIS 5e (as described in Chapter 5).

The notion of two sea-level highstands in MIS 5e was originally derived from tectonically, rapidly uplifting coastlines such as Papua New Guinea and eastern Indonesia, where well-dated coral-reef deposits are physically separated (Chappell and Thom, 1977; Chappell and Veeh, 1978). On Atauro Island, for example, Reef II identified by Chappell and Veeh (1978) has a maximum crest 65 m above low-water level. The reef wedge comprises an inter-fingering sequence of large corals in growth position (framework) and cobble pavement of regressive gravels that truncates broken framework corals. A disconformity near the crest of Reef II separates the first of two transgressive events that terminated approximately 130 ka from a minor sea-level oscillation before the 120 ka peak. On Huon Peninsula, Papua New Guinea, geomorphological and stratigraphical evidence was used to distinguish an early phase of reef development (Reef VIIa; 138 ± 5 ka) from a late phase (Reef VIIb, 118 ± 2 ka), corresponding to uplift-corrected high sea-level culminations of 5 ± 5 m and 6.5 ± 4 m APSL, respectively (Aharon and Chappell, 1986).

A compound sea-level peak is not evident in oxygen-isotope records from deep-sea cores. Detailed analysis of cores where sedimentation rates were high, and mixing and bioturbation low (and hence potential for high temporal resolution), indicate consistently that only a single peak is present (Shackleton, 1987). Similarly, the EPICA Dome C ice-core record from Antarctica reveals a single peak for MIS 5e (Masson-Delmotte *et al.*, 2010).

In many respects, the regional pattern of relative sea-level changes during MIS 5e is likely to have been similar to the present Holocene interglacial, as concluded on the basis of geophysical modelling (Nakada and Lambeck, 1989; Lambeck and Nakada, 1992). Intuitively, the overall structure of relative sea-level records for both inter-glaciations for far-field locations such as southern Australia should show develop-ment of an early highstand followed by a progressively falling sea surface in response to hydro-isostasy (Lambeck *et al.*, 2012). Empirical evidence from Holocene coastal successions in southern Australia supports this scenario for the past 7 ka since culmination of the post-MIS 2 transgression (Belperio *et al.*, 2002). In a compilation of global sea-level data from regions showing a high degree of tectonic stability, however, Hearty *et al.* (2007) found the converse situation; namely that relative sea-level histories revealed the highest stands of MIS 5e sea levels during the latest portion of the highstand (Figure 6.10). However, the compilation by Hearty *et al.* (2007) is based on numerous sites with potentially contrasting relative sea-level histories.

The significance of glacio-hydro-isostatic processes has commonly been under-appreciated in interpretation of palaeo-sea-level observations from last interglacial shoreline successions. As with earlier highstands, and the Holocene highstand, elevation of MIS 5e successions represents a single observation in time, determined

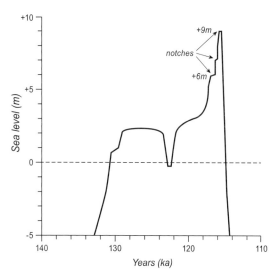

Figure 6.10 A composite reconstruction of the inferred broad pattern of relative sea-level changes during the last interglacial maximum (MIS 5e) based on empirical evidence (facies architecture analysis) from numerous coastal sedimentary successions (adapted from Hearty *et al.*, 2007).

by several variables that include the eustatic (ice-equivalent) sea level at time of formation, local tectonism, and the combined effects of glacio-hydro-isostasy. In the case of the latter, antecedent conditions are as significant as changes since deposition of the shoreline succession. This is particularly evident in near-field regions, where incomplete relaxation of the Earth's crust–mantle system to the penultimate glaciation (MIS 6) impinges on the relative sea-level record of MIS 5e. Thus global variability observed in the elevation of the last interglacial shoreline will also be a function of glacio-hydro-isostatic processes having occurred before, during, and following the interglacial (Lambeck *et al.*, 2012). Accordingly, duration of the highstand will vary spatially, particularly in relation to proximity to Pleistocene icesheets. In far-field sites, such as the coral reefs of Western Australia, a longer highstand is registered in contrast to the near-field Caribbean record (Lambeck *et al.*, 2012).

Statistical modelling to determine the probability of last interglacial global sea level suggests that at the 95% confidence level, global sea level (as distinct from local sea levels) peaked at 6.6 m APSL during MIS 5e, and that there is a 67% probability that sea level exceeded 8 m APSL but was unlikely to have exceeded 9.4 m APSL (Kopp *et al.*, 2009). The data set analysed by Kopp *et al.* (2009) included 42 localities with age estimates ranging between 140 and 90 ka. Onset of deglaciation at 135 ka and attainment of the highstand at 127 ka were implicit in their analysis. An extensive review of last interglacial shoreline successions ($n = 890$ records) was undertaken by Pedoja *et al.* (2011). Based on considerable variation in elevation of MIS 5e shorelines, they concluded that no continental platforms can be considered stable, and that since MIS 5e most regions showed evidence for uplift (0.22 mm/year global average) that may relate to plate-tectonic processes at a global scale. Dutton and Lambeck (2012) have re-asserted that sea level was between 5.5 and 9 m at its peak in MIS 5e (as discussed in Chapter 5).

6.10 Interstadial sea levels of the last glacial cycle (MIS 5c and 5a)

During the latter part of the last interglacial *sensu lato* (MIS 5), sea level fell dramatically from its highest stand (128–116 ka). Between about 116 and 80 ka ago, global sea level oscillated with two prominent highstands leaving their imprints during the warm interstadials MIS 5c (~105 ka) and MIS 5a (~82 ka). Marine oxygen-isotope records providing indirect evidence for former ice volumes at these times, consistently show that both highstand events were significantly below present sea level, although the relative sea-level records of some regions suggest that sea level at these times may have been close to that obtained during the last interglacial maximum. This is more evident in near-field locations where glacio-isostatic adjustment processes have significantly influenced the relative sea-level records such as those obtained from the Caribbean region (Coyne *et al.*, 2007), the United States Atlantic Coastal Plain (Cronin *et al.*, 1981), and Bermuda (Vacher and Hearty, 1989). In contrast, palaeo-sea-level records of tectonically active settings such as the Huon

Peninsula (Chappell *et al.*, 1996a), Barbados (Schellmann and Radtke, 2004a,b), and the Ryukyu Islands, Japan (Konishi *et al.*, 1974; Sasaki *et al.*, 2004) all provide a consistent picture of lower sea levels during interstadials MIS 5c and MIS 5a. Based on an assumed value of 6 m APSL for an ice-equivalent sea level for MIS 5e at San Nicholas Island in coastal California, sea levels of 2–6 m APSL for MIS 5c (~100 ka) and 11–12 m BPSL for MIS 5a (~80 ka) were determined (Muhs *et al.*, 2012). Similarly, evidence inferred from tectonically highly stable settings such as Spencer Gulf in southern Australia, point to MIS 5c and MIS 5a palaeo-sea levels of 8 and 14 m BPSL, respectively (Hails *et al.*, 1984a,b).

The United States Atlantic Coastal Plain has long been the subject of coastal stratigraphical and geochronological investigations (Cronin *et al.*, 1981; McCartan *et al.*, 1982; Wehmiller and Belknap, 1982; Szabo, 1985; Wehmiller *et al.*, 2004). The coastal plain is situated on the trailing edge, passive continental margin of the North American continent and is bounded by a wide continental shelf which exceeds 100 km in cross-section. The emergent Quaternary succession represents the latter part of a subsiding wedge of Mesozoic and Cainozoic sediments typical of passive continental margins. Post-rifting subsidence (~20–40 mm/ka) characterises the continental slope sediment wedge and uplift appears to be largely confined to the area represented by the on-land Quaternary succession. The coastal plain, which extends from southern Florida to New Hampshire, preserves a series of emergent terraces backed by erosional scarps. The terrace successions include transgressive–regressive facies associations with relict brackish-water and marine facies as well as barrier island and beach deposits (Cronin *et al.*, 1981). Individual terraces dip seaward (1–2°) and represent broad plains. Unconformities of regional extent within sedimentary successions relate to periods of relative sea-level fall and coastal emergence.

Several difficulties are inherent in inferring palaeo-sea level from the US Atlantic Coastal Plain. Although corals suitable for U-series dating are common, their solitary nature hinders inferences about former sea levels. In addition, it is difficult to quantify with precision indicative meaning of the proxy sea-level indicators as discussed in Chapter 3, and accordingly a relatively large uncertainty must be included in estimates of palaeo-sea level (Cronin *et al.*, 1981). In addition, a complex regional stratigraphical terminology has frustrated some attempts at geological correlation.

The palaeo-sea-level record of the US Atlantic Coastal Plain has represented an anomaly in global reconstructions of late Quaternary sea levels, being inconsistent with the palaeo-sea-level records derived from emergent reef settings, such as Barbados and the Huon Peninsula, Papua New Guinea. For much of the coastal plain there appears to be a paucity of highstand deposits relating to the last interglacial maximum (MIS 5e; 125 ka), yet MIS 5 coastal deposits, particularly those relating to MIS 5a (~85–80 ka), are common (Wehmiller *et al.*, 2004). Cronin *et al.* (1981) noted that during the interstadials of MIS 5, relative sea level on the US Atlantic Coastal Plain was at 6.5 ± 3.5 m at 94 ka and 7 ± 3 m at 72 ka (correlated with MIS 5c and 5a, respectively). In contrast, sea-level records from Barbados (Schellmann and Radtke, 2004a,b) and the Huon Peninsula (Bloom *et al.*, 1974) suggest that sea levels were between 21 and 15 m BPSL for MIS 5c and between 25 and 13 m BPSL for MIS 5a. Marine oxygen-isotope inferred sea levels are consistent with these empirical observations (Chappell and Shackleton, 1986; Shackleton, 1987). U-series ages determined on solitary corals from Atlantic Coastal Plain successions appear younger for the interstadials MIS 5c and 5a, being around 94 ± 5 and 72 ± 5 ka, respectively, compared with 105 ± 5 and 82 ± 5 ka from Barbados (Cronin *et al.*, 1981). Cronin *et al.* (1981) speculated that if the seemingly anomalous U-series ages are not the result of diagenetic processes (e.g. uranium mobilisation), then age differences may relate to differential glacio-isostatic effects on the relative sea-level record of this region.

In re-examining the ages and palaeo-sea-level record of the US Atlantic Coastal Plain, Wehmiller *et al.* (2004) noted that U-series ages on fossil corals overwhelmingly correlated with MIS 5a, with ages between 85 and 80 ka. All of the sedimentary units in Virginia, South Carolina, and Georgia from which these samples were collected occur from 0 to ~6 m APSL and are only between 3 and 5 m below early Stage 5 deposits, particularly those of the last interglacial maximum (125 ka), although the latter are not common. It is difficult to explain the paucity of coastal units relating to the last interglacial maximum compared with many other emergent coastal records around the world. It may have resulted from erosional processes occurring during the interstadial highstands (MIS 5c and 5a); however, a geophysical mechanism is required to explain the empirical field observations.

The palaeo-sea-level record for the US Atlantic Coastal Plain has commonly been evaluated in relation to sea-level records from Bermuda, which has long been regarded as tectonically stable in view of its intra-plate setting. Coralline deposits at Fort St Catherine on Bermuda contain *in-situ* corals at elevations between 1 and 2 m APSL dated at 78–83 ka using $^{234}U/^{230}Th$ (Ludwig *et al.*, 1996). At face value, these data imply a significantly higher ice-equivalent sea level for MIS 5a than otherwise indicated by the Huon Peninsula or Barbados successions or from oxygen-isotope inferred palaeo-sea levels from deep-sea cores (Chappell and Shackleton, 1986; Shackleton, 1987). The last interglacial (MIS 5e) record for Bermuda shows that samples cluster between 1 and 4 m APSL (Muhs *et al.*, 2002; see Section 5.9.2) and in this context, it would appear that differences in the elevation of MIS 5e and 5c deposits on Bermuda are similar to the US Atlantic Coastal Plain. As Bermuda is an oceanic island its relative sea-level history will not be subject to quite the same influence of hydro-isostatic processes as are likely for the US Atlantic Coastal Plain with its wide continental shelf. In view of this, some other regional set of processes must explain the similarity of the Stage 5 highstand records of these two regions (Wehmiller *et al.*, 2004).

The contrasting relative sea-level record from the US Atlantic Coastal Plain compared with the Barbados, Huon Peninsula, and oxygen-isotope inferred sea-level histories, most likely relates to its proximity to the former Laurentide Icesheet, coupled with the wide continental shelf. This physical setting would enhance glacio-isostatic and hydro-isostatic processes, influencing the relative sea-level records of the region. The influence of glacio-hydro-isostatic processes will vary in time and space as a function of whether ice or water represent the dominant load on the crust–mantle system (Horton *et al.*, 2009; Peltier, 2009; Raymo and Mitrovica, 2012). The forebulge was located to the northeast of the Florida Peninsula and its collapse was manifested by a lateral redistribution of viscous mantle material within the asthenosphere, and the space created by decay of the forebulge progressively increased the capacity of the North Atlantic Ocean basin to accommodate water introduced from melting of the Laurentide, Cordilleran, and other icesheets. This may have resulted in less time to leave a morphostratigraphical signature on the US Atlantic Coastal Plain.

In a compilation of interstadial highstand records across the western North Atlantic Ocean, Potter and Lambeck (2003) noted that, relative to present sea level, MIS 5e shoreline deposits showed little spatial variation in elevation within tectonic- ally stable locations. In contrast, deposits of MIS 5a age ranged between 3 m APSL (Virginia and South Carolina) to an uplift-corrected value of –19 m for Barbados. They concluded that these contrasting relative sea-level observations can be recon- ciled in terms of the timing of collapse of the peripheral bulge associated with the North American Icesheet *sensu lato*. Geophysical modelling revealed that the observed gradient in MIS 5a shoreline deposits required a comparatively small, northerly located icesheet during early development of the North American Icesheet in MIS 5, a large ice volume during the LGM and a high viscosity (about 2×10^{20} Pa s) for the lower mantle. Significantly, this implies that the observed gradient in MIS 5a successions in the northwestern Atlantic Ocean is directly a function of continuing collapse of the glacial forebulge associated with the former Laurentide Icesheet.

Hydro-isostatic processes would also have an expression on the Atlantic Coastal Plain but at a lower magnitude than the glacio-isostatic signature (Walcott, 1972). Thus, with the redistribution of water from melting icesheets to the ocean basins, the geologically rapid return of water to the continental shelves and ocean basins would, in turn, have a loading effect, counteracting the processes of glacial unloading and forebulge collapse. This effect may ultimately be manifested in a relative sea-level fall late in the highstand record of the coastal plain. Oxygen-isotope records suggest a rapid fall in ice-equivalent sea level following the last interglacial maximum possibly by as much as 50 m within a period of less than 10 ka to the stadial substage 5d low sea level at ~110 ka (Chappell and Shackleton, 1986) and, more conservatively, 20–30 m by 111 ka (Tzedakis *et al.*, 2012). Hydro-isostatically induced flexural uplift of the coastal plain following the post-substage 5e regression may have enhanced the opportunity for marine erosion during the interstadial highstands of substages 5c and 5a, leading to widespread destruction of previously deposited substage 5e sediments on the US Atlantic Coastal Plain.

The Ironshore Formation on Grand Cayman, British West Indies, is a succession of six unconformity-bounded bioclastic limestones with, in places, interbedded *terra rossa* and calcrete palaeosols. The formation provides evidence for deposition during six sea-level highstands during the later Quaternary, correlated with MIS 11, 9, 7, 5e, 5c, and 5a (Coyne *et al.*, 2007). Evidence for deposition during MIS 5c and 5a is based on $^{234}U/^{230}Th$ dating of the corals *Montastrea anularis*, *Acropora cervicornis*, and the gastropod *Strombus gigas* from shallow subtidal facies.

Despite the relatively close proximity to the Oriente Transform Fault, a spreading centre (Mid-Cayman Rise), and the Cayman Trench, Grand Cayman has long been presumed to be tectonically stable based on elevation of a shoreline notch at 6 m APSL developed in Miocene dolostones of the Cayman Formation (Woodroffe *et al.*, 1983b; Coyne *et al.*, 2007). The age of the shoreline notch, critical to interpretation of palaeo-sea level for other Pleistocene highstands is based on correlation with sediments of last interglacial age (2–3 m APSL) abutting the basal portion of the coastal cliff in which the notch is developed, and comparable elevation of the notch with similar features on the islands Cayman Brac (see Figure 3.12b), Bermuda, and the Yucatan Peninsula (see Section 5.9). However, the age of the shoreline notch in the strictest sense remains undated. The presumed age of the shoreline notch carries the implicit assumption that the last interglacial (MIS 5e) sea surface attained an elevation of 6 m APSL at Grand Cayman.

The stratigraphical evidence presented by Coyne *et al.* (2007) suggests that interstadials MIS 5c and 5a attained elevations of 2–5 m APSL (110–95 ka) and 3–6 m APSL (87–74 ka), respectively. Curiously, notwithstanding the inherent uncertainties of quantifying palaeo-sea levels, these data at face value suggest that relative sea level was higher during MIS 5a than MIS 5c at a time known to be characterised by greater global ice volumes based on glaciological and marine oxygen-isotope records (Chappell and Shackleton, 1986).

The sea-level highstand for Unit E (MIS 5c) of the Ironshore Formation is based on isolated occurrences of the gastropod *Strombus gigas* which in itself is not an accurate palaeo-sea-level indicator, given its present-day occurrence at a range of water depths. Similarly, the paucity of *in-situ* fossils in Unit F (MIS 5a) also rendered assessments of palaeo-sea level difficult. Based on the presence within the skeletal carbonate sands of the trace fossils *Conichnus* (bioturbation by sea anemones) and *Ophiomorpha* (burrowing shrimps) up to 3 m APSL, Coyne *et al.* (2007) concluded that sea level at time of deposition of Unit F may have ranged between 3 and 6 m APSL. In this context, the highstands for MIS 5c and 5a, although relatively securely dated, are less well constrained in terms of accurately quantifying eustatic (ice-equivalent) sea level.

6.11 Interstadial sea levels during MIS 3

Knowledge of sea levels during MIS 3 (64–32 ka) is based on uplift-corrected estimates of palaeo-sea level derived from emergent coastal complexes, stratigraphical evidence from shallow-marine, continental shelf sediment cores, and indirect evidence inferred from oxygen-isotope records (marine sediment and ice cores). As sea level during this interval was at least 50 m BPSL, and for much of the interval ranged between 60 and 80 m BPSL (Chappell *et al.*, 1996a; Lambeck and Chappell, 2001), evidence for interstadial highstands of MIS 3 are more easily identified in tectonically active settings that experience moderate to high rates of crustal uplift (> 0.5 m/ka), exposing marine sediments above present sea level.

Several studies undertaken in the 1960s and 1970s suggested that interstadial sea levels attained heights close to, or above, present sea level based on radiocarbon dating of fossil shells and corals (Curray, 1961; Shepard, 1963; Milliman and Emery, 1968; Emery *et al.*, 1971), and controversies continue (Doğan *et al.*, 2012a,b; Bekaroğlu, 2012). Palynological and glaciological evidence suggested that global climate was cooler than present at 30 ka with larger ice volumes, implying lower sea levels than suggested based on radiocarbon dating of emergent marine shell beds. Thom (1973) summarised 188 radiocarbon ages on coastal deposits from which sea levels close to present at ~30 ka had been inferred. Chappell (1982) subsequently found that about 70 of these ages, which could not be invalidated on other grounds (e.g. fossils from the same deposit yielding U-series ages of 100 ka, or poor replication in radiocarbon measurements on different fossils from the same deposit), plotted over a wide depth range between 48 m BPSL and 8 m APSL (Figure 6.11). Commonly, the radiocarbon ages represented under-estimates of the 'true' age. As contamination by 1% ^{14}C with a modern activity on a fossil of last interglacial age (125 ka) will yield a finite, apparent radiocarbon age of 37 ka, many of the reported radiocarbon ages suggestive of an interstadial MIS 3 age (Giresse and Davies, 1980) are very likely to reflect differential levels (0.5–2%) of contamination by modern ^{14}C which could not be removed during sample pre-treatment. It does not mean, however, that all radiocarbon ages that fall within the interval 30–50 ka are unreliable. Clearly, rigorous stratigraphical and geomorphological evidence, where the internal facies architecture, lateral continuity of sediments and *in-situ* context of fossils can be confidently established in outcrop, and preferably additional independent methods of dating, such as U-series ages, are needed to assess the validity of such ages.

In many respects, the most reliable evidence for palaeo-sea level for MIS 3 is derived from tectonically uplifted coastlines. Early estimates for MIS 3 from the Huon Peninsula, Papua New Guinea implied that sea level oscillated between 24 and 62 m BPSL with tectonically corrected sea-level highstands at 39 ± 6 m BPSL at 40 ± 3 ka and 41 ± 1 m BPSL at 31 ± 2.5 ka, respectively (Bloom *et al.*, 1974; Chappell and Veeh, 1978). These suggested a disparity between isotopically inferred palaeo-sea levels and values derived from observational evidence (Chappell and Shackleton, 1986; Chappell *et al.*, 1996a). The revised sea-level curve based on subsequent field surveys and high-precision U-series measurements has revealed that the previously derived ages for MIS 3 were too young. In particular, the revised ages from reef II are some 3–12 ka older; from reef IIIb, 2–4 ka older; and from reef

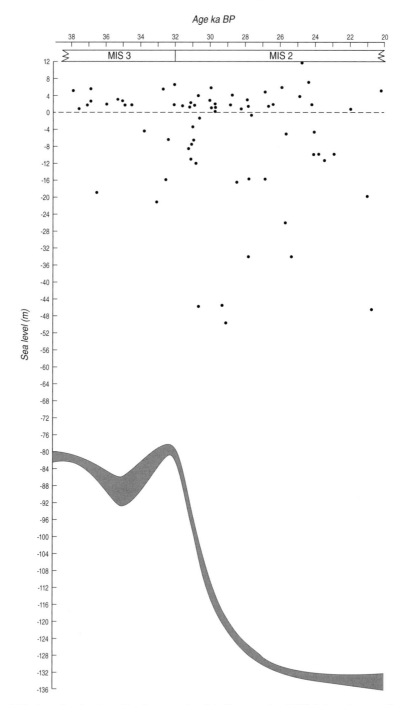

Figure 6.11 Age–depth plot of palaeo-sea-level indicators for MIS 3 based on radiocarbon ages compiled by Thom (1973) and Chappell (1982), compared with a portion of the late Pleistocene sea-level curve of Lambeck and Chappell (2001).

IIIa ~7–17 ka older. In view of the uplift history of the Huon Peninsula, this means that sea levels at time of reef-terrace formation were lower than previously determined on the basis of the younger U-series ages (Chappell *et al.*, 1996a). The revised data set has revealed that between 64 and 32 ka, sea levels ranged between 65 and 45 m BPSL, and at 30 ka may have attained levels between 84 and 73 m BPSL (Chappell *et al.*, 1996a). This illustrates the significance of inferred uplift rate for palaeo-sea-level estimation and is more consistent with values previously independently inferred from the marine oxygen-isotope record (Chappell and Shackleton, 1986).

Konishi *et al.* (1970) reported a tectonically uplift-corrected palaeo-sea level of 40 m BPSL at 42 ka based on an average of five U-series ages on coralline limestone from the central Ryukyu Islands, Japan. Subsequent work by Sasaki *et al.* (2004) on the coral-reef complex on the northern part of Kikai Island, central Ryukyus, identified three sea-level hemicycles culminating at ~52, 62, and 66 ka based on high-precision U-series dating. The palaeo-sea-level record inferred from the off-lapping, drowned coral-reef succession suggests a minimum relative sea-level rise of 6.8 and 6.4 m at 52 and 62 ka, respectively. Although Sasaki *et al.* (2004) did not attempt to quantify ice-equivalent sea level, taking into account inferred palaeo-water depths of coral-reef successions, the U-series ages of *in-situ* corals and long-term uplift rates for Kikai Island of 1.7 m/ka (Konishi *et al.*, 1970) suggest minimum palaeo-sea levels of ~46 m BPSL at 52 ka and 55 m BPSL at 62 ka.

As the island of Barbados is characterised by only a moderate rate of tectonic uplift (0.27 m/ka; Schellmann and Radtke, 2004a,b), reef terraces relating to MIS 3 do not occur above present sea level in southern Barbados. Based on modelling of uplift rates, Mesolella (1968) suggested that a terrace relating to the latter part of MIS 3 at 30 ka would have formed in a palaeo-sea level of 26 m BPSL.

Based on studies of benthic foraminifera from marine vibrocores from Gulf St Vincent, South Australia, highstands of 22.5 m BPSL at 40 ka and 22 m BPSL at 31 ka were suggested by Cann *et al.* (1988), with an intervening regression to 28 m BPSL. The palaeo-sea-level estimates for MIS 3 were based on linear regression of population ratios of the foraminifers *Elphidium macelliforme* and *E. crispum*. In the modern environments of Gulf St Vincent, *Elphidium macelliforme* tends to frequent deeper water, whereas *E. crispum* is more common in shallower water and, accordingly, ratios of these species were used to infer palaeo-sea level. The age of late Pleistocene sediments was determined by radiocarbon and AAR dating. Based on degree of isoleucine epimerisation and aspartic acid racemisation, and a model of apparent parabolic kinetics, a late Pleistocene numeric age of 46.5 ± 9.1 ka was determined for fossil specimens of the shallow-water mollusc *Katelysia rhytiphora*, indicating that the sediments are older than Holocene but younger than the last interglacial. These results contrast with radiocarbon and AAR-dated deposits from the Lacepede Shelf of southern Australia, which indicate that sea level was between 58 and 62 m BPSL for much of the period from 74 to 40 ka (Hill *et al.*, 2009).

The inferred palaeo-sea levels for MIS 3 from Gulf St Vincent are considerably higher than the revised sea-level record from the Huon Peninsula (where sea level for the interval 70–30 ka was 50–60 m BPSL: Chappell *et al.*, 1996a; or possibly up to 80 m BPSL: Yokoyama *et al.*, 2001), implying that much of the present Spencer Gulf and Gulf St Vincent was subaerially exposed at this time. The North Greenland Ice Core Project (NGRIP) oxygen-isotope record (GICC05 timescale; Austin and Hibbert, 2012), however, reveals the presence of 13 short-term (~1–2 ka) Greenland Interstadial (GI) warming events in the traditionally marine-defined MIS 3 (64–32 ka; Shackleton and Opdyke, 1973). These events are not identified in marine sediment oxygen-isotope records due to the variable effects of bioturbation. The amplitude of interstadial warming is up to half that of a full glacial cycle, implying rapid sea-level changes within a short period (see Section 6.12). The magnitude of sea level rise of up to 10–15 m during these events (Chappell, 2002) may have been sufficient to flood the southern portion of Gulf St Vincent and hence provide an explanation for the seemingly ambiguous record. The succession may in part correlate with GI 8 at 38 ka ago, based on the ages reported by Cann *et al.* (1988).

In the near-field setting of the United States Atlantic Coastal Plain and Texas Gulf Coast, MIS 3 coastal successions indicate a sea level at approximately 20–15 m BPSL (Wellner *et al.*, 1993; Rodriguez *et al.*, 2000). The relative sea-level records of this region are significantly influenced by glacio-hydro-isostatic adjustments related to fluctuations of the Laurentide Icesheet, as previously illustrated in the discussion of MIS 11 sea levels (see Section 6.7.1).

Knowledge of long-term (at the scale of ka) relative sea-level changes during MIS 3 has been significantly enhanced by oxygen-isotope analyses from ice cores which reveal a significantly higher temporal resolution than deep-sea cores. Greenland and Antarctic ice-core records show significant fluctuations in ice volume and atmospheric temperatures, particularly during the last glacial cycle. The Greenland record, in particular, reveals abrupt temperature increases of approximately 8–16°C (Svensson *et al.*, 2008). Although representing a proxy indicator of climate and sea-level changes, the data generated reveal changes in relative sea level of as much as 25 m within a few thousand years, associated with Dansgaard–Oeschger Events (Dansgaard and Oeschger, 1989). Dansgaard–Oeschger events involve significant isotopic shifts in $\delta^{18}O$ and are explored in greater detail in Section 6.13; events 8, 12, 14, 16/17 are particularly well-defined in the last glacial cycle, but remain undetected in most coastal sedimentary successions.

During the last glacial cycle, the Red Sea increasingly adopted the characteristics of a barred-basin with the bedrock sill in the Straits of Bab-el-Mandab reducing to a width of no more than 6 km during the LGM at 21 ka, and with shorter widths at some places (Lambeck *et al.*, 2011). Oxygen-isotope measurements on the planktonic foraminifer *Globigerinoides ruber* and the benthic foraminifera *Cibicides mabahethi* and *Bulimina marginata* suggest that several interstadial sea-level highstands occurred within the Red Sea during MIS 3 in response to meltwater derived primarily from northern hemisphere icesheets with only a minor contribution from Antarctica. A core GeoB5844–2 from a water depth of 963 m in the northern Red Sea reveals evidence for relative sea-level changes during MIS 2 and 3 between 10 and 80 ka (Arz *et al.*, 2007). In a similar study from the Red Sea, Siddall *et al.* (2003) quantified the magnitude of ice-equivalent sea-level changes for the interval 70–25 ka, based on oxygen-isotope analyses on the fossil foraminifer *Globigerinoides ruber*. They noted that sea level repeatedly oscillated by up to twice the present volume of the Greenland and West Antarctic Icesheets. The data also reveal sea-level rises associated with Heinrich Events, described in the next section.

6.12 Late Pleistocene interstadial sea levels: Dansgaard–Oeschger and Heinrich Events

Studies of ice cores have provided a greater opportunity to resolve short-term climate changes than previously possible with deep-sea sediments or from coastal sediment-ary successions (Blockley *et al.*, 2012). The combined effects of slow rates of sedimentation in some ocean basins, bioturbation and the possible incompleteness of some marine sedimentary records due to turbidites and erosional scouring, have reduced the age-resolving capacity of many marine core records.

Greenland ice cores reveal numerous abrupt changes in $\delta^{18}O_{ice}$ during the last glacial cycle suggestive of rapid changes in atmospheric temperature of between 8 and 16°C. The magnitude of these changes is approximately 50–75% of that registered in the transition from fully glacial to interglacial conditions (Cronin, 2010). These periodic episodes of rapid change in atmospheric temperature have been termed Dansgaard–Oeschger Events after Willi Dansgaard and Hans Oeschger who first identified them in the ice core record (Figure 6.12), and may have been accompanied by sea-level changes of several tens of metres. Dansgaard–Oeschger Events typically extend over approximately 1,470 years, but have ranged between 500 and 2,500 years (Dansgaard *et al.*, 1969, 1982, 1993; Dansgaard and Oeschger, 1989; Oeschger *et al.*, 1984). Each event is initiated by a rapid decrease in $\delta^{18}O_{ice}$ from approximately –44‰ to –38‰ in the span of several tens of years. As a 1‰ shift in $\delta^{18}O_{ice}$ is equivalent to an atmospheric temperature change of approximately 1.5°C, the abrupt changes in $\delta^{18}O_{ice}$ signify rapid climate change. This is followed by a gradual and fluctuating increase (to more negative values) in $\delta^{18}O_{ice}$ in a sawtooth pattern, reflecting a progressive cooling over a period of several hundred years (Dansgaard and Oeschger, 1989). Twenty-one Dansgaard–Oeschger Events have been identified in the ice-core records of the NGRIP (Svensson *et al.*, 2008). These events have also been referred to in the nomenclature of stadials and interstadials [Greenland Stadial (GS) and Greenland Interstadial (GI)] in reference to intervals of rapid warming followed by the onset of gradual cooling and return to colder, stadial conditions during the last glacial cycle. Quasi-periodic events of ~1,500 years duration in the Holocene have been termed Bond cycles (Bond *et al.*, 1992).

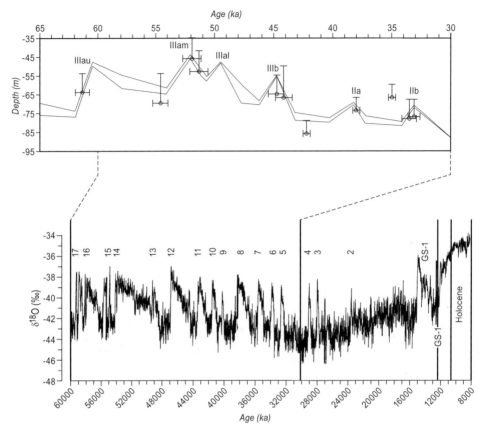

Figure 6.12 The North Greenland Ice Core Project (NGRIP) oxygen-isotope record for a portion of the late Pleistocene (60–8 ka ago back in time from the year 2000 AD) showing timing of Greenland Interstadials (numbered peaks after Davies *et al.*, 2012). The Greenland Interstadials are compared with the envelope of the Huon Peninsula sea-level curve based on dating of fossil corals for the late Pleistocene cold interstadials of MIS 3 (Chappell, 2002).

North Atlantic deep-sea sediments reveal the presence of discrete layers of lithic fragments (e.g. angular quartz, limestone, and dolomite) termed Ice Rafted Debris (IRD), which appears to precede the warming phases of Dansgaard–Oeschger Events. This terrigenous–clastic sediment was shed by many icebergs as they melted, and these episodes have been termed Heinrich Events by Bond *et al.* (1992) after their discoverer, Heinrich (Heinrich, 1988). The North American (Laurentide) and European Icesheets (northwestern Europe, Iceland, and Britain) represent the source areas of the IRD. In deep-sea cores, individual Heinrich layers are several centimetres thick and their thickness is in part a function of their proximity to source area.

The emergent coral reefs of Huon Peninsula, Papua New Guinea, reveal evidence for short-term sea-level changes associated with Heinrich and Dansgaard–Oeschger

Events. The relatively rapid emergence of the Kanzarua (2.7–2.9 m/ka) and Bobongara (3.2–3.4 m/ka) reef complexes assisted in the preservation of these GI-related sea-level events. Palaeo-sea level was defined from elevation of the catch-up reef terraces and their ages based on the oldest dated corals associated with each reef. Six discrete phases of sea-level rise and subsequent fall between 9 and 20 m lasting some 1–2 ka for the period 65–30 ka (MIS 3) were identified (Chappell, 2002; Figure 6.12). In particular, sea-level maxima at 33, 38, 44.5, 52, and 58–60 ka were identified in the Huon Peninsula record. The sea-level highstands at 38, 44.5, and 52 ka correspond with GIs 8, 12, and 14, respectively. A sea-level highstand at 60 ka may correlate with GI 17 (Svensson *et al.*, 2008). As the Huon Peninsula is in the far-field of Pleistocene icesheets, the uplift-corrected palaeo-sea-level record will largely reflect ice-equivalent, glacio-eustatic sea-level changes. Importantly, the record reveals that millennial-scale sea-level changes in addition to orbitally driven (Croll–Milankovitch scale) changes are registered in the emergent reef record (Yokoyama and Esat, 2011). The metre-scale coseismic uplift events that characterise the Huon Peninsula preclude identification of smaller scale (< 3 m) GI sea-level changes. Another significant conclusion from the Huon Peninsula record is that, provided a broad understanding of reef-terrace age is derived from U-series dating, a more precise age can be assigned from the Greenland ice-core record. Thus, Huon Peninsula reef terrace IIIau dated at 61.4 ± 0.6 ka (2σ-uncertainty) correlates with GI 17 at 58.5 ka (Chappell, 2002).

The Huon Peninsula palaeo-sea-level record for the last glacial cycle implies potential multiple sources for meltwater returned to the world ocean associated with these oscillations, as complete melting of the Greenland Icesheet would only raise sea level by about 7.3 m (Cronin, 2010). Stratigraphical analyses of the geometry of the reef terraces at Huon Peninsula and their dating by U-series reveals that rates of sea-level rise associated with these GIs did not exceed the more rapid phases of deglacial sea-level rise following the LGM.

6.13 Eustatic sea levels during the Last Glacial Maximum (MIS 2)

The lowest sea levels of the LGM (MIS 2) were attained between 26 and 20 ka, slightly after maximum icesheet development (Lambeck *et al.*, 2002a; Peltier and Fairbanks, 2006). The CLIMAP (Climate: Long-Range Investigation, Mapping and Prediction) minimum and maximum estimates of last glacial sea-level lowstand are 127.5 m BPSL and 163 m BPSL, respectively (Clark and Mix, 2002). Icesheet modelling, taking into account probable maximum thickness and lateral extent, placed MIS 2 sea level between 118.2 and 130.4 m BPSL (Denton and Hughes, 2002; Huybrechts, 2002).

During the LGM, many continental shelves were subaerially exposed (Figure 6.13). Australia had a land area approximately 25% larger than present and land bridges joined the mainland to Tasmania and Papua New Guinea.

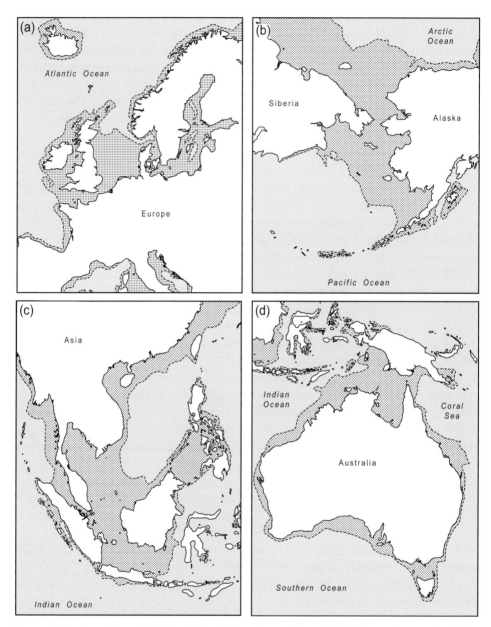

Figure 6.13 An outline of the Last Glacial shoreline at about 22 ka for various parts of the
world. At this time coastlines were situated close to the shelf-break of the continental shelves
at approximately 120 m BPSL. (a) Northwestern Europe, showing land bridges connecting
Britain through the North Sea; (b) the Bering Strait, connecting Asia and North America via
Beringia; (c) the Sunda Shelf; and (d) Australia (including Tasmania) and New Guinea.

Similarly, the Sunda Shelf (Sundaland) provided a land connection between Borneo, Java, Sumatra, Peninsular Malaysia, and Vietnam, subaerially exposed the Gulf of Thailand, and closed the gateway (Sunda Strait) between the South China Sea, Java Sea, and the Indian Ocean, affecting thermohaline circulation (Broecker, 2010). Beringia also represented a major land bridge during the LGM and provided an important land corridor for migration of biota between North American and Asian continents (Elias and Brigham-Grette, 2001).

The position of sea level at its lowest point during the LGM at 22–20 ka is relatively well defined from several independent lines of evidence that includes direct stratigraphical evidence from sediment cores from continental shelves, inferences drawn from modelling the areal extent and thickness of icesheets at time of their maximum development, and inferred ice volumes derived from oxygen-isotope measurements from deep-sea and ice cores. The most important evidence for defining the maximum extent of sea-level lowering during the LGM is most likely to come from proxy sea-level indicators such as shallow-water coastal facies on continental shelves (particularly fixed sea-level indicators from far-field sites as discussed in Chapter 3). However, there has been a tendency in the literature to assign a global significance to such estimates, when, as with all estimates of former sea level, they are site-specific and subject to glacio-hydro-isostatic influences.

Areas regarded as having the greatest tectonic stability have been favoured in studies attempting to define position of sea level during glacial maxima, particularly regions in the far-field of former icesheets such as Australia and oceanic islands. In these regions, effects of glacio-isostatic adjustment processes are significantly reduced as they experienced minimal, if any, direct glaciation and are not located in zones of former glacial forebulges, and the relative sea-level signal is predominantly eustatic in nature (ice-equivalent sea level) as evident in the record from Barbados (Peltier and Fairbanks, 2006).

Early indications of position of the LGM shoreline were derived from dredged and cored samples from continental shelf settings, and in some instances, significant uncertainties are related to precise stratigraphical context of materials, as well as limited data replication. Commonly, interpretations of palaeo-sea level were derived from single samples and therefore provided only a generalised understanding of maximum sea-level lowering (Van Andel and Veevers, 1967; Veeh and Veevers, 1970). Subsequent studies provided time-series data for position of sea level based on radiocarbon and U-series dated palaeo-sea-level indicators (Fairbanks, 1989), as well as critical assessments of validity of inferred sea levels for the LGM within the context of geophysical modelling of glacio-hydro-isostatic adjustment processes (Yokoyama *et al.*, 2000, 2001; Lambeck *et al.*, 2002b; Peltier and Fairbanks, 2006). The following discussion briefly reviews some of the early evidence for sea level during the LGM. This is followed by an overview of the current understanding based on the more recent dating of palaeo-sea-level indicators and geophysical modelling, and more detail is presented in Chapter 7.

Early estimates placed full glacial sea level at about 90 m BPSL using a range of geomorphological evidence (Daly, 1934). From mapping relict lowstand fluvial channels on the Sunda Shelf (South China Sea, Java Sea, and Straits of Malacca), Umbgrove (1929) concluded that Pleistocene glacial low sea levels and, by implication, the Last Glacial shoreline, were close to 100 m BPSL.

Developments in radiocarbon and U-series dating were instrumental in quantifying timing of glacial lowstands. Curray (1961) reported an uncorrected radiocarbon age of $19,300 \pm 300$ years BP on a specimen of *Strombus grannulatus*, a shallow-water marine mollusc from the outer continental shelf of Mexico in a present water depth of 114 m. Van Andel and Veevers (1967) reported an uncorrected radiocarbon age on the shallow-water mollusc *Chlamys senatorius* of $16,910 \pm 500$ years BP at a depth of 130 m BPSL from the Timor Sea. Fossil marine shells between 130 and 120 m BPSL between 15 and 13 ka (^{14}C years) were reported from the Atlantic continental shelf of the USA by Milliman and Emery (1968). They also reported older ages on marine shells closer to timing of the LGM, but their shallow elevation in relation to present sea level implied that they are *remanié* (reworked) fossils transported landward across the continental shelf since the LGM.

A specimen of the coral *Porites bernardi* encased within red algae and sediment was dredged from a depth of between 111 and 103 m BPSL on the outer continental shelf of Guinea, West Africa, and yielded an uncorrected radiocarbon age of $18,750 \pm 350$ years BP (McMaster *et al.*, 1970). Other indicators of LGM low sea levels in this region included two prominent lowstand delta complexes (Orango and Nunez Deltas) with their seaward edges close to the 90 m isobath and numerous relict stream channels traversing the shelf.

Relict shells of the shallow-water species *Mactra chinensis* and *Pecten albicans* dredged from skeletal carbonate sands at a depth of 140 m BPSL in the East China Sea have yielded an uncorrected radiocarbon age of $23,260 \pm 600$ years BP, and fossil specimens of the brackish-water shell *Ostrea gigas* from a depth of 112 m BPSL yielded an uncorrected radiocarbon age of $15,200 \pm 850$ years BP (Emery *et al.*, 1971).

Observations have been made at several locations around Australia. The Arafura Sea, north of Darwin, comprises a shelf that was subaerially exposed, with fluvially incised channels and drowned reefs towards the shelf edge, indicating a sea-level lowering > 130 m between 21 and 14 ka (Jongsma, 1970). Several terraces have been identified in echosounder and seismic profiles at depths of 64, 90, 130, 170–175, and ~200 m BPSL. Radiocarbon dating of 'shallow water fossiliferous algal rock' (Jongsma, 1970, p. 150) yielded an uncorrected age of $18,700 \pm 350$ years BP obtained from a drowned reef between 130 and 175 m BPSL. On this basis, Jongsma (1970) concluded that MIS 2 attained a lowstand of as much as 175 m BPSL.

Veeh and Veevers (1970) reported an uncorrected radiocarbon age of $13,660 \pm 220$ years BP and a U-series age of 17 ± 1 ka on an aragonitic specimen of the coral *Galaxea clavus* from 175 m BPSL on the forereef slope of the Great Barrier Reef, but uncertainties remain concerning the palaeo-sea-level significance of this sample, as the species has been recorded in exceptional circumstances living in water depths of between 25 and 75 m BPSL. A calcarenite sample, interpreted as beachrock, was obtained at 150 m BPSL. The sediment comprising bioclastic skeletal carbonate grains including *Lithothamnium* yielded a radiocarbon age of $13,860 \pm 220$ years BP (^{14}C years) and was suggested to represent 'a precise indicator of former water level' (Veeh and Veevers, 1970, p. 536). They concluded that sea level was some 175 m below present between 17 and 13.6 ka, some 45 m lower than previously considered, and rose to 150 m BPSL soon after 13.6 ka.

A palimpsest succession of richly fossiliferous, mixed quartz–skeletal carbonate sands occurs on the outer continental shelf of New South Wales, and preserves fossil marine molluscs of shallow-water affinity including *Glycymeris radians* and *Pecten fumatus* (Ferland and Roy, 1997). The shell-rich sedimentary successions occur in present water depths between 120 and 150 m BPSL and have been assigned to the LGM based on radiocarbon and AAR dating (Phipps, 1970; Ferland *et al.*, 1995; Ferland and Roy, 1997; Murray-Wallace *et al.*, 1996, 2005). Radiocarbon ages on fossil *Glycymeris* and *Pecten* range between 26 and 17 ka (Phipps, 1970; Ferland and Roy, 1997). Sediments retrieved from the more landward portion of the outer shelf in present water depths of 123–130 m reveal significant reworking of shell assemblages. This is reflected in random and non-systematic changes in amino acid D/L values (a measure of fossil age; see Chapter 4) with increasing depth down core. This is in contrast to cores obtained from deeper water, which revealed systematic step profiles in increasing extent of racemisation with depth in core for the different lowstand sediment packages. Reworking of shells in the shallower-water cores is attributed to storm-related reconcentration at times of glacial lowstand. A conservative estimate would place LGM sea level between 120 and 130 m BPSL based on dating of the shoreface species *Glycymeris*, yielding ages between 20 and 26 ka. The data also imply that the penultimate glacial maximum (MIS 6) attained a similar sea level to the LGM (MIS 2).

Results from the Joseph Bonaparte Gulf in north-western Australia are particularly significant in constraining sea levels during the LGM in view of the long-term tectonic stability of this region (Yokoyama *et al.*, 2000, 2001). Yokoyama *et al.* (2000, 2001) suggested a eustatic sea-level lowstand of 125 ± 4 m BPSL based on radiocarbon dating of shallow-water and littoral-dwelling marine molluscs from a series of marine cores. Maximum sea-level lowering occurred before 22 ka and ended abruptly at about 19 ka (calendar years) followed by an inferred rapid sea-level rise of ~10–15 m within a few hundred years (equivalent to a reduction in global ice volume of about 10%). This interpretation is based on elevation of the transition between marginal marine and brackish-water facies assuming absence of diastems or erosional discontinuities within the sedimentary successions. Based on lower and upper limits of brackish-water sediments in a reference core GC5, Yokoyama *et al.* (2000) concluded that deposition of this facies commenced at 21,280 cal years BP and lasted for about 3 ka. The lowstand brackish-water facies is characterised by dwarf specimens of the benthic foraminifera *Ammonia beccarii*, *Elphidium* spp. and the euryhaline ostracod *Cyprideis australiensis*. Taking into account corrections for glacio-hydro-isostasy, Yokoyama *et al.* (2000) concluded that sea level (ice-equivalent) during the LGM was between 135 and 130 m BPSL, corresponding to a grounded ice volume in excess of current-day glacial ice volumes by some $52 \pm 2 \times 10^6 \, \text{km}^3$. These inferred LGM sea levels from Joseph Bonaparte Gulf were evaluated critically by Shennan and Milne (2003). Notwithstanding the inherent difficulty of accurately determining elevations of the top of marine cores, difficulty arises in exact correlation of different sedimentary facies between cores and the magnitude of uncertainty associated with palaeo-sea-level estimations based on different sedimentary facies. In their critique, Shennan and Milne (2003) noted some discrepancies between cores in estimates of relative sea level during the LGM that ranged between 97.5 ± 2 and $115.3 \pm 2 \, \text{m}$ BPSL and concluded that the most likely explanation is the presence of a hiatus in some cores. Despite these difficulties, the Joseph Bonaparte data remain important for defining sea level during the LGM. Assigning larger uncertainties to the different palaeo-sea-level indicators is more likely to resolve apparent inconsistencies in estimation of glacial lowstands.

The U-series dated coral record from Barbados reported by Peltier and Fairbanks (2006) tracks sea level from approximately 26 ka, almost continuously through to 11 ka ago, and the complete record also describes postglacial sea-level rise through to approximately 6 ka (including previously reported data of Fairbanks, 1989; see Figures 2.5 and 7.1). The data set includes U-series ages on the corals *Acropora palmata*, *Montastrea annularis*, *Porites asteroides*, and *Diploria* sp. The coral palaeo-sea-level indicators plotted by Peltier and Fairbanks (2006) were in a form that accounted for the 0.34 mm/year uplift that characterises the forearc ridge (accretion-ary prism) setting of Barbados. The corals are relational sea-level indicators (see Chapter 3) and, accordingly, may live in a range of water depths (e.g. *Acropora palmata* up to 5 m water depth and *Montastrea annularis* in up to 20 m water depth, see Section 3.3.2). The Barbados coral record reveals that sea level at the time traditionally regarded as the LGM (~23–19 ka) exceeded 120 m BPSL and rose above 120 m BPSL by 20 ka. A short hiatus between approximately 22 and 20.5 ka renders it difficult to precisely state when palaeo-sea level rose above 120 m BPSL. The U-series ages also reveal that the point of maximum lowering of sea level of $130 \pm 5 \, \text{m}$ BPSL occurred at 26 ka, some 5 ka earlier than previously understood.

Thus, mutually consistent evidence suggests that during MIS 2 at the time of maximum glaciation, sea level was at least 120 m BPSL, and estimates that are considered more reliable, acknowledging associated uncertainties in measurement, have commonly ranged between 120 and 135 m BPSL (Yokoyama *et al.*, 2000; Lambeck and Chappell, 2001; Peltier and Fairbanks, 2006). Estimates of palaeo-sea level during the LGM have been fundamental in calibrating oxygen-isotope records and assessing the relationship between oxygen-isotope fractionation, ice volume, and temperature variations in the transition from glacial to interglacial conditions (Shackleton and Opdyke, 1973; Chappell and Shackleton, 1986). Estimation of palaeo-sea level during MIS 2 is complicated by uncertainties associated with icesheet thickness and geographical extent, vertical movements in shoreline successions associated with tectonism and glacio-hydro-isostatic adjustments, post-mortem transport of fossils, and inherent variability in some palaeo-sea-level indicators. Notwithstanding these issues, estimates of eustatic sea level during the LGM centred on 21,000 years ago suggest a level some 125 ± 5 m BPSL corresponding with a land-based ice volume of between 4.6 and 4.9×10^7 km^3 (Fleming *et al.*, 1998). The period of postglacial sea-level rise from the end of the LGM at 19 ka (sidereal years) to the beginning of Holocene time following the end of the Younger Dryas at 11.5 ka is examined in Chapter 7.

6.14 Long records of Pleistocene sea-level highstands

Few coastlines around the world preserve multiple sea-level highstand successions in view of the variable effects of weathering, erosion, and regional denudation. Extensively glaciated areas such as northern Europe, Canada, and the northern USA preserve only fragmentary records of interglacial coastal successions. Humid regions characterised by vigorous fluvial processes have experienced significant erosion and destruction of highstand records at times of glacial lowstands. In tropical regions with extensive reef-building corals, tectonic uplift has favoured preservation of evidence for former sea level, the rate of uplift having a direct influence on the resolution of former sea levels, particularly short-term events associated with GIs (Chappell, 2002). In arid regions regional calcrete development on sedimentary carbonates has preserved highstand packages of sediment from erosion.

In the following discussion, long records of Pleistocene highstands, defined here as spanning at least the past four glacial cycles, are briefly reviewed from regions where there is a clear morphostratigraphical expression of relative sea-level changes. In this context, the Coorong Coastal Plain and Murray Basin of southern Australia, and the Wanganui Basin, North Island, New Zealand are particularly important. Early investigations by Sprigg (1948, 1952) and Hossfeld (1950) on the Coorong Coastal Plain, and Fleming (1953) in the Wanganui Basin established that these regions represent archives of multiple Pleistocene highstands, which challenged the traditional notion of fourfold glaciation in the Pleistocene (Penck and Brückner, 1909). Collectively, their research revived an interest in the Croll–Milankovitch Hypothesis as an explanation of long-term climate and sea-level changes.

6.14.1 Coorong Coastal Plain and Murray Basin, southern Australia

In a global context, the Coorong Coastal Plain and Murray Basin, in southern Australia, represent one of the world's longest Pleistocene interglacial highstand records in the form of a succession of emergent coastal barriers (Sprigg, 1948, 1952;

Hossfeld, 1950; Murray-Wallace *et al.*, 1998, 2001; Bowler *et al.*, 2006; McLaren *et al.*, 2011; Figures 6.14 and 6.15). The coastal plain is situated on a passive, trailing-edge continental margin and many of the barriers extend along the length of the coastal plain, from the Murray River mouth region in the north to southwestern Victoria, a distance exceeding 300 km. Landward from the modern coastline, the broad, flat coastal plain, with an average seaward gradient of 0.5%, extends for approximately 100 km to the early Pleistocene East Naracoorte Range (MIS 21 or 25), a prominent coastal barrier landform. The emergent shoreline features were first described by Woods (1862), who attributed them to a progressive fall in sea level (see Figure 1.5). Detailed mapping by Hossfeld (1950) and Sprigg (1952), who termed the features 'Dune Ranges', led to the appreciation that at least 13 interglacials had occurred during the Quaternary, and that individual barrier landforms reflected a succession of sea-level oscillations associated with Pleistocene glacial cycles.

The mixed calcareous–siliceous barrier shoreline complexes of the coastal plain formed in response to high wave and wind energy that characterises this coastline. The barriers comprise aeolianite (coastal transverse and parabolic dunes) with inter-bedded beach and back-barrier coastal lagoon facies, as well as numerous palaeosols, as illustrated in the facies architecture of the last interglacial Woakwine Range (Murray-Wallace *et al.*, 1999). The relict barriers attain maximum surface elevations of 30–40 m above their adjacent lagoonal flats, and are 1–2 km wide. The barriers may occur up to 10 km apart and are subparallel to the modern coastline. Several are composite features, having formed in more than one interglacial, in view of the slow uplift rate and the return of interglacial sea levels to a broadly common elevation. The barriers increase in age landwards (Huntley *et al.*, 1993, 1994; Murray-Wallace *et al.*, 2001).

An older succession of early Pleistocene to Early Pliocene barriers extends over 400 km farther inland from the Naracoorte Range towards the margin of the Murray Basin (Figure 6.14). The succession termed the Loxton Sand (Brown and Stephenson, 1991; Alley and Lindsay, 1995; Kotsonis, 1999) refers to fine-grained glauconitic, micaceous and shelly sand that covers a substantial part of the Murray Basin (Figure 6.14). The upper part comprises cross-bedded calcareous sands with shelly debris deposited as arcuate coastal barriers in a regressive strandplain. Although their ages have yet to be reliably determined, preliminary ^{87}Sr/^{86}Sr ages reveal that they extend back to at least 6.86 ± 0.36 Ma (McLaren *et al.*, 2011).

Although the Loxton Sand contains the Early Pliocene (Kalimnan) foraminifer *Elphidium pseudonodosum*, it is likely that the most seaward barriers of the succession are of early Pleistocene age. The series of coastal barriers have been associated with draw-down of sea level accompanying onset of Quaternary glaciation, and progressive growth of the Antarctic Icesheet (Kotsonis, 1999; Bowler *et al.*, 2006). Approximately 170 coastal barriers have been identified with crest to crest spacings of 2–3 km and surface relief of up to 20 m (Bowler *et al.*, 2006). The features occur up to 500 km inland from the modern coastline and up to 60 m APSL at their landward limit in the

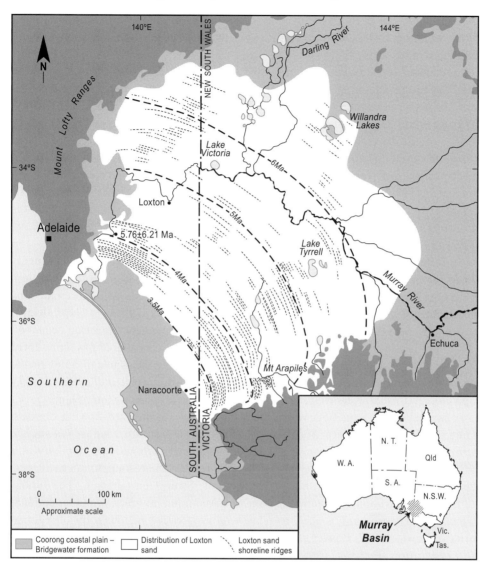

Figure 6.14 Distribution of the Early Pliocene to early Pleistocene Loxton Sand, Murray Basin, southern Australia, illustrating the general geometry of the barrier shorelines and their inferred ages (linear accumulation model; Kotsonis, 1999). The Coorong Coastal Plain, an extensive region of temperate carbonate sedimentation (Pleistocene Bridgewater Formation) with well-developed coastal barriers, is situated seaward of the town of Naracoorte (after Murray-Wallace, 2002, with permission from Wiley). The succession between Naracoorte and the modern coastline is younger than 1 Ma. Strontium isotope age of Loxton Sand is after McLaren *et al.* (2011).

Murray Basin. In parts of the Murray Basin the succession occurs at higher elevations resulting from localised uplift (e.g. 150 m APSL on the Padthaway High;
McLaren *et al.*, 2011; Figure 6.14).

A younger series of coastal barriers of the Bridgewater Formation of middle to
late Pleistocene age occur seaward of the Loxton Sand, which relate to the eccentricity (100-ka cycle) component of orbital forcing (Huntley *et al.*, 1993, 1994; Murray-
Wallace *et al.*, 2001; Figure 6.15). Accordingly, the more closely spaced and subdued
ridges of the Loxton Sand are correlated with the earlier, obliquity-dominated
insolation cycle (41 ka). Isotopic records derived from deep-sea cores suggest that
the transition between the earlier obliquity-dominated cycle to the eccentricity component occurred at approximately 1.2 Ma (Shackleton *et al.*, 1990; Raymo, 1992;
Clark *et al.*, 2006) and this broadly accords with the morphostratigraphy of the
Coorong Coastal Plain and Murray Basin, as well as luminescence (Huntley *et al.*,
1993, 1994) and AAR dating (Murray-Wallace *et al.*, 2001).

A lithological change from siliciclastic sediments of the Loxton Sand, Murray
Basin, to the temperate carbonate sediments of the younger Bridgewater Formation,
Coorong Coastal Plain, is evident. The Bridgewater Formation is represented by
aeolianites of middle to late Pleistocene age, as originally defined at Cape Bridgewater in western Victoria (Boutakoff, 1963). A similar sediment transition occurs on
the Swan Coastal Plain, near the city of Perth, Western Australia, and is represented
by a change in a seaward direction from the siliciclastic Bassendean Sand to the
coastal aeolianite sequence of the Tamala Limestone (Kendrick *et al.*, 1991; Hearty
and O'Leary, 2008). Kendrick *et al.* (1991) discuss several possible interrelated causes
for the lithological transition, which include a stronger southward-flowing Leeuwin
Current along the Western Australian coastline, with associated higher sea-surface
temperatures. This, they suggest, resulted in enhanced carbonate productivity at a
time of reduced terrigenous sediment supply to the inner continental shelf, and
coastal deposition mainly within estuaries of the Swan Coastal Plain. The similar
timing of the carbonate transition in southern Australia, however, implies a regional
climatic explanation and is likely to relate to a progressive enhancement of regional
aridity (Bowler, 1982; Bowler *et al.*, 2006).

Preservation of these large-scale coastal landforms is the result of epeirogenic
uplift and regional calcrete development. Emergence of the coastal plain has resulted
from the combined effects of regional uplift of the Murray Basin in response to plate-
tectonic interactions between the Indo-Australian and Pacific Plates as part of
regional uplift of the southern Australian margin, and at a more local scale, crustal
doming in the late Quaternary around Mount Burr and Mount Gambier–Mount
Schank volcanic complexes (0.13 m/ka) (Murray-Wallace *et al.*, 1998; Sandiford,
2003; Figure 6.15).

The relict coastal barriers have been dated by luminescence and AAR (whole-
rock) and a consistent picture of barrier formation during interglacial sea-level
highstands has been derived (Huntley *et al.*, 1993, 1994; Murray-Wallace *et al.*,

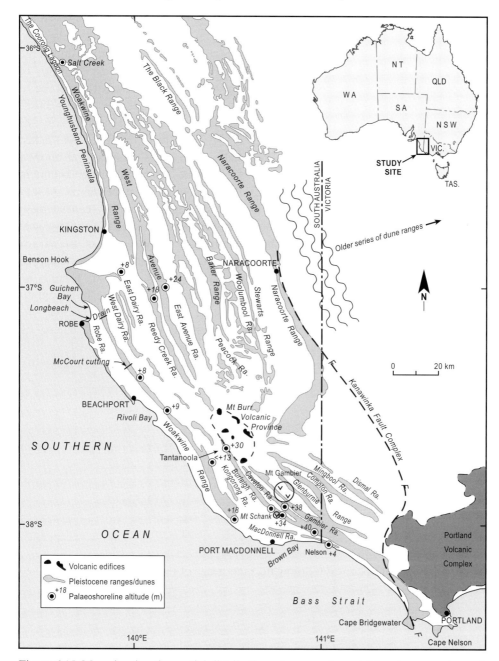

Figure 6.15 Map showing the spatial distribution of emergent coastal barrier successions of the Pleistocene Bridgewater Formation on the Coorong Coastal Plain of southern Australia. The names of each of the coastal barriers (dune ranges represented by shading) are indicated (after Murray-Wallace, 2002, with permission from Wiley).

2001). The Brunhes–Matuyama geomagnetic polarity reversal at 780 ka between the East and West Naracoorte Ranges has also constrained the rate of coastal development (Idnurm and Cook, 1980). An average rate of uplift of the central portion of the coastal plain based on the altitude of back-barrier lagoon facies, an indicator of former sea level, of 12 successive relict coastal barriers in a transect from Robe to Naracoorte is slow in global terms being only 0.07 mm/year (Figure 6.16). Accordingly, intertidal back-barrier lagoon facies of the last interglacial maximum in this region only occurs up to 9 m APSL, an order of magnitude lower than for the last interglacial successions on the Huon Peninsula, Papua New Guinea, and imply a eustatic (ice-equivalent) sea level close to present for the last interglacial maximum in this region. The slow rate of tectonic uplift of the Coorong Coastal Plain means that mainly evidence for former interglacial highstands is preserved. The coastal plain succession indicates a general comparability in the height of successive interglacial highstands. Interglacial sea levels did not deviate by more than ± 6 m of present sea level for middle and late Pleistocene time, suggesting that the sea surface returned repeatedly across the continental shelf to a similar regional elevation. There is no evidence for 'warmer interglacials' (Burkle, 1993) characterised by sea levels significantly higher than obtained in the current Holocene interglacial in this record.

In a transect from Robe to Naracoorte, the ages and uplift-corrected palaeo-sea levels for the middle to late Pleistocene dune ranges are; Robe Range II (MIS 5a; 17 m BPSL), Robe Range III (MIS 5c; 9 m BPSL), Woakwine Range I (MIS 5e; 2 m APSL),

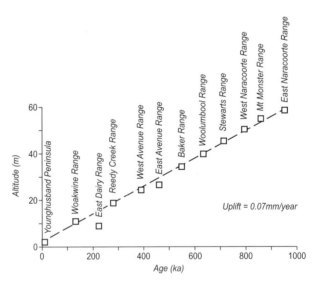

Figure 6.16 Plot of shoreline elevation against their age of the Pleistocene coastal barriers of the Coorong Coastal Plain in southern Australia, based on elevation of back-barrier lagoon facies, a proxy for mean sea level, in a transect from Robe to Naracoorte (after Murray-Wallace, 2002, with permission from Wiley).

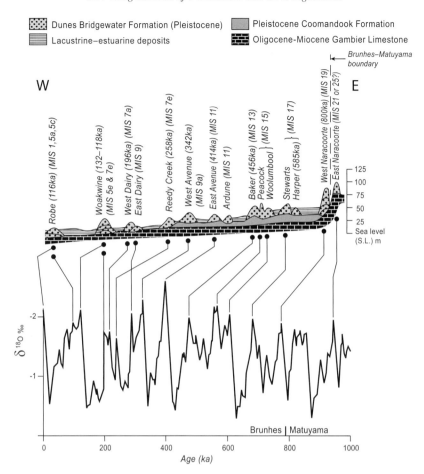

Figure 6.17 Age of the coastal barriers of the Coorong Coastal Plain, southern Australia (after Murray-Wallace *et al.*, 1998), in relation to the oxygen-isotope record of Imbrie *et al.* (1984).

Dairy Range (MIS 7a; 6 m BPSL), Reedy Creek Range (MIS 7e; 0 m APSL), West Avenue Range (MIS 9; 1 m BPSL), Ardune–East Avenue Range (MIS 11; 3 m BPSL), Baker Range (MIS 13; 1 m BPSL), Peacock–Woolumbool Ranges (MIS 15; 0 m APSL), Harper–Stewarts Ranges (MIS 17; 3 m BPSL), West Naracoorte and Black Ranges (MIS 19; 0 m APSL), and the East Naracoorte and Coonalpyn Ranges (MIS 21 or 25; < 3 m APSL) (Murray-Wallace, 2002; Figure 6.17).

6.14.2 *Wanganui Basin, New Zealand*

The broad geotectonic context of New Zealand and the Wanganui Basin (Figure 6.18) were outlined in Chapter 5. The ovoid-shaped (200 × 200 km) Wanganui Basin, a back-arc basin, preserves up to 4,000 m of Pliocene–Quaternary sediments. The basin fill provides a detailed record of relative sea-level changes in the form of 47 cyclothems that extend back to MIS 100 at approximately 2.5 Ma ago (Carter and Naish, 1998). Coastal cliffs provide excellent exposures of the principal stratotypes of the New Zealand Pleistocene stages (Nukumaruan 2.6–1.6 Ma; Castlecliffian 1.2–0.4 Ma, and Haweran 0.4–0 Ma; Bowen *et al.*, 1998). The sedimentary successions are essentially undeformed with regional dips of 2–6° and appear today above present sea level due to basin inversion resulting from uplift of the adjacent Greywacke Ranges which traverse the southeastern margin of the North Island of New Zealand (Pillans, 1994). Uplift rates have been determined at 0.2–0.7 m/ka and appear to increase inland (Pillans, 1983, 1990). The longer-term history of the basin involved accumulation of sediments during basin subsidence. Several techniques have been used for establishing ages of depositional sequences, including biostratigraphy, tephrochronology, fission-track dating, palaeomagnetism, radiocarbon, and AAR dating (Pillans *et al.*, 1994; Bowen *et al.*, 1998; Carter and Naish, 1998).

The basin fill is dominated by interglacial highstand successions and their bounding surfaces represent erosional discontinuities formed during glacial age lowstands (Carter *et al.*, 1999). A representative depositional sequence for an interglacial, which may attain a thickness of up to 12 m, includes a transgressive shallow-water, shoreface or tide-dominated succession with cross-bedded shelly conglomerates that overlie an erosion surface (Pillans *et al.*, 1994). These sediments are overlain by a mid-cycle condensed shell bed, which is in turn overlain by deeper water shelf siltstones with depauperate shelly assemblages. An erosional surface caps the shelf siltstones heralding onset of a younger cyclothem. For younger cyclothems preserved on the marine terrace sequence, interglacial highstand successions may be capped by a range of terrestrial sediments that include peats, fluvial sediments, dune sand, loess, and tephra. The total thickness of sediment packages covering abrasion platforms increases with age partly due to aeolian accession of loess and tephra. In view of the variable thickness of cover sediments, palaeo-sea level cannot be inferred from the modern land surface and requires excavations to identify potential palaeo-sea-level indicators, as well as elevation of the base of backing cliffs.

The Haweran marine beds, the youngest portion of the Wanganui Basin sedimentary record (approximately the past 400 ka), is represented by morphologically well-developed marine terraces (Figure 6.18). Each terrace is defined by a subhorizontal marine abrasion surface covered by a late-stage transgressive facies, an interglacial highstand succession, and to their landward limit, backed by a relict sea cliff, representing maximum inland location of the shoreline associated with the marine terrace. Highstand successions are commonly draped by cover sands and palaeosols. Between Hawera and Wanganui, up to 12 emergent marine terraces have been identified (Pillans, 1983, 1990; Figure 6.18). The terraces extend parallel with the modern coastline as much as 100 km, and up to 20 km inland, where the highest terrace surface reaches about 300 m APSL. In a transect near Waverley in the area of

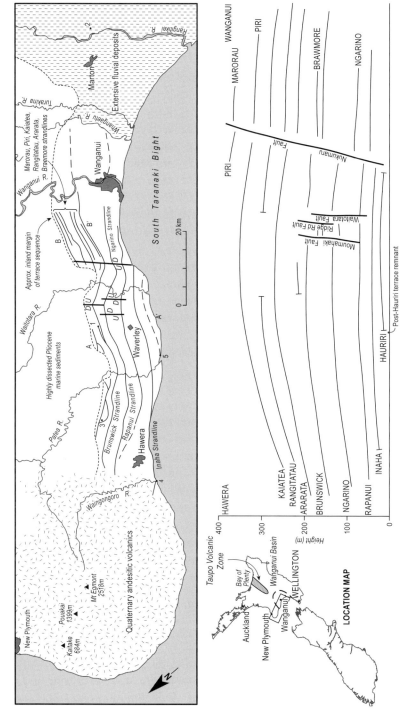

Figure 6.18 Pleistocene marine terraces of the Wanganui Basin, North Island, New Zealand. Regional deformation (warping) of the terraces is evident (after Pillans, 1983).

maximum uplift rate, individual terrace forms in cross-sectional profile range between 800 m (Braemore Terrace; early MIS 7?) to approximately 4.6 km (Ngarino Terrace; late MIS 7; Pillans, 1983). The last interglacial (MIS 5e) Rapanui Terrace is approximately 4 km in cross-sectional profile.

Individual terraces have been assigned names and, in ascending order of surface elevation and age, include the Rakaupiko Terrace (60 ka; MIS 3), Hauriri Terrace (80 ka; MIS 5a), Inaha Terrace (100 ka; MIS 5c), Rapanui Terrace (120 ka; MIS 5e), Ngarino Terrace (210 ka; MIS 7), Brunswick Terrace (310 ka; MIS 9), Ararata Terrace (400 ka; MIS 11), Rangitatau Terrace (450 ka; MIS 11?), Kaiatea Terrace (520 ka; MIS 13), Piri Terrace (600 ka; MIS 15), and Marorau Terrace (680 ka; MIS 17; Figure 6.18).

Linear regression of strandline height (as defined by elevation of relict sea cliffs above present sea level for the MIS 11 and MIS 5e shorelines) against uplift rate of individual terraces at South Taranaki, reveals that the Inaha (MIS 5c) and Hauriri (MIS 5a) Terraces formed when sea level was at 11 and 20 m BPSL, respectively (Pillans, 1983). The Rakaupiko Terrace of presumed MIS 3 age formed at approximately 24 m BPSL. A reassessment of sea level during MIS 3 from the Huon Peninsula record (Chappell *et al.*, 1996a), however, calls into question the age of this terrace feature. Apart from short-term sea-level rises associated with GIs, sea level remained at 60 m BPSL for much of MIS 3. Thus, insufficient time may have been available to erode a well-defined marine abrasion platform due to short-term (1–2 ka) rises in sea level associated with GIs.

In view of the regional doming evident in the Wanganui Basin, shoreline relation diagrams cannot precisely model palaeo-sea level for the terraces older than the last interglacial. However, in the most general sense, the data suggest that for the older interglacials (MIS 7, 9, 11, 13, 15, and 17), the sea surface returned to a broadly common datum, an observation also evident for the Coorong Coastal Plain of southern Australia, where coastal barrier features have been described that can be correlated to the Wanganui Terraces (Murray-Wallace *et al.*, 2001). The Wanganui Basin sequences also permit an assessment of the magnitude of earliest Pleistocene glacio-eustatic sea-level oscillations. For the interval 2.6–2.2 Ma, sea level fluctuated within a range from 110 ± 20 m (MIS 100–99) to 25 ± 10 m APSL (MIS 76–75) (Naish, 1997). During this interval, glacio-eustatic sea-level oscillations averaged 74 ± 24 m.

6.14.3 *Sumba Island, Indonesia*

Situated north of the Java Trench in a zone of subduction and regional andesitic volcanism, Sumba Island, part of the Indonesian archipelago, preserves a sequence of at least 11 raised coral reef terraces (Figure 6.19). At Cape Laundi on the northern coast of Sumba Island, the terraces range between 100 and 500 m wide and extend up to 475 m APSL (Pirazzoli *et al.*, 1993b). Individual terraces represent geomorphologically distinct features which are backed by prominent changes in slope,

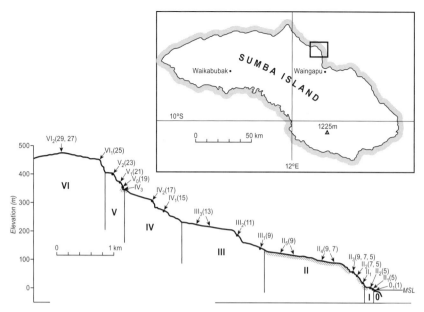

Figure 6.19 Topographical transect of reef terraces near Cape Laundi, Sumba Island, Indonesia, showing initial terrace stages identified by Jounannic *et al.* (1988) in Roman numerals, and inferred marine oxygen-isotope stages (in parentheses), based on U-series and ESR ages (adapted from Pirazzoli *et al.*, 1993b).

erosional scarps and, in places, shoreline notches. U-series (α-counting) and ESR dating of corals derived numerical ages permitting direct correlations with MIS 5e (Reef Complex I$_2$ at 19 ± 1 m APSL), MIS 5 and 7 (Reef Complex II$_2$ at 50 ± 5 m APSL), MIS 5, 7, and 9 (Reef Complex II$_3$ at 62 ± 5 m), MIS 9 (Reef Complex III$_1$ at 145 ± 10 m APSL), and MIS 15 (Reef Complex IV$_1$ at 275 ± 10 m APSL) (Pirazzoli *et al.*, 1993b). Based on an inferred long-term rate of tectonic uplift for Cape Laundi of 0.49 ± 0.01 mm/year, Pirazzoli *et al.* (1993b) suggested that the age of the highest terrace was likely to correlate with MIS 27 at about 0.99 Ma. Subsequent U-series dating by TIMS yielded broadly confirmatory ages (Bard *et al.*, 1996). The ages reported by Bard *et al.* (1996) also indicated that the last interglacial maximum (MIS 5e) started earlier, at 135–130 ka, which significantly precedes the insolation maximum in orbital forcing at 128 ka.

Some of the older terrace surfaces appear to have been reoccupied during successively younger interglacials because of the moderate rate of uplift (e.g. terrace II$_2$ being occupied by a sea-level highstand in MIS 7 and 5e). The polycyclic origin of many of the terraces may in part explain the narrower cross-sectional width of the modern coral reef platform (about 75–100 m) compared with the older raised terraces. If the age of the oldest and highest terrace surface at Cape Laundi corresponds with MIS 29 and 27, inferred from the long-term uplift rate, then the reef complexes provide some information on the likely preservation potential of coral reef terraces in the longer geological record. Despite the passing of some 29 isotope stages with 13 interglacial sea-level highstands in the past 1 Ma, only 6 geomorphologically distinct reef terraces are preserved, notwithstanding that some of the reef terraces represent palimpsest carbonate surfaces.

6.15 Synthesis and conclusions

A coherent and integrated understanding of the nature of Pleistocene sea-level changes requires multiple lines of evidence (e.g. relict shoreline deposits, and oxygen-isotope data from marine and ice cores within a rigorous geochronological framework, supplemented where possible by results from magnetostratigraphy and biostratigraphy) from a globally diverse range of settings (i.e. near-, intermediate- and far-field from the continental icesheets). Preferably, multiple geochronological methods should be applied to the sedimentary successions to critically evaluate the integrity of analytical results for each method. In addition, knowledge of the relative contribution of tectonically induced vertical crustal movements and the feedback effects of glacio-hydro-isostasy is essential in deriving a meaningful under-standing of palaeo-sea level (Lambeck *et al.*, 2010), and in particular, glacio-eustatic (ice-equivalent) sea level.

One of the paradoxes of palaeo-sea-level investigations is that although the most rigorous studies of former relative sea level are derived from single regions, where the data can be normalised, the notion of eustatic (ice-equivalent) sea level requires a consideration of global data sets in order to derive a more complete understanding of change. A further consideration from geodesy is that current relative sea-level observations represent mere instants in time, in view of the continuing complex response of the crust–mantle system to fluctuating ice and water volumes. An obvious illustration is the continuing uplift of Scandinavia following melt of ice-sheets; an uplift regime that will continue for several thousand years into the future.

In keeping with the longer-term geological record, not all the pages of Earth history are preserved in one locality in the context of relative sea-level changes. Additional complexities in trying to reconstruct Pleistocene relative sea-level changes relate to the contrasting sensitivities of different geomorphological environments in preserving evidence of former sea level. Emergent reef sequences are more likely to provide a record revealing the nuances of Pleistocene sea-level changes compared with a more filtered perspective from coastal barriers or other forms of depositional systems. Continental denudation, particularly by glacial, fluvial, and coastal pro-cesses, has removed many traces of former coastal successions through repeated cycles of sea-level changes, accounting for the 'imperfection of the geological record' as noted by Darwin (1859). Accordingly, few coastlines preserve direct evidence for Pleistocene sea-level changes, in the form of highstand successions older than 200 ka. The Coorong Coastal Plain of southern Australia, and the Wanganui Basin,

North Island, New Zealand represent two examples of remarkable preservation with highstand records spanning much of the Quaternary. Cape Laundi, Sumba Island, Indonesia, and the Huon Peninsula, Papua New Guinea also preserve long records of Quaternary sea-level changes. The Wanganui Basin record is based on marine abrasion platforms with cover sequences of shallow-water shelly facies, and the Coorong Coastal Plain succession is represented by coastal barrier and associated back-barrier lagoonal facies. The Cape Laundi reef record shows a greater sensitivity to preserving interstadial successions, partly due to the inherent sensitivity of coral growth to relative sea-level changes and the rate of tectonic uplift in this setting (0.49 mm/year). In contrast, the Wanganui Basin and Coorong Coastal Plain, both characterised by slower rates of tectonic uplift, provide less information about interstadial sea levels.

Particularly detailed records of Pleistocene sea-level changes have been inferred from ice-volume records of deep-sea and ice cores. Oxygen-isotope records of Pleistocene ice-volume changes have now been constructed from deep-sea sediments from all of the world's ocean basins (Imbrie *et al.*, 1984; Prell *et al.*, 1986; Lisiecki and Raymo, 2005) as well as from ice cores from Antarctica and Greenland (Lemieux-Dudon *et al.*, 2010; Masson-Delmotte *et al.*, 2010). As previously stated in this chapter, the results of this vast body of research have yielded a consistent geohistorical framework for defining the timing and relative magnitude of ice-volume changes, and by inference, glacio-eustatic sea-level changes during successive glacial cycles. While coastal records of relative sea-level changes provide a direct proxy, sedimentary successions generally represent discrete events such as the culmination of sea-level cycles in the form of sea-level highstands or lowstands, and restricted portions of transgressive and regressive events, as noted in Chapter 2. Accordingly, such successions only chronicle narrow vertical bands of relative sea level. In view of this, oxygen-isotope records have greatly assisted the delineation of the broad pattern of sea-level changes between highstands and lowstands. Investigation of oxygen-isotope variations during the last glacial cycle by Shackleton and Opdyke (1973) provided early insights into the pattern of global sea-level changes between discrete highstands (last interglacial maximum, MIS 5e) and the late Pleistocene warm interstadials (MIS 5a, 5c, and 3) and the lowstand of the LGM. Their early observations have been replicated in many subsequent investigations, involving time-series analysis of $\delta^{18}O$ data.

Time-series analysis of oxygen-isotope data can provide insights into the nature of ice volume fluctuations, and their potential impact on glacio-eustatic sea-level changes, not immediately apparent from coastal sedimentary records. For example, the highstands of MIS 5e, MIS 7e, MIS 9e, and MIS 9c appear to have followed periods of rapid deglaciation, while MIS 11c and MIS 17 appear to have followed a more prolonged episode of deglaciation (Tzedakis *et al.*, 2012).

Development of oxygen-isotope stratigraphy has led to a consistent picture of Pleistocene sea-level changes, and a framework for validating the Croll–Milankovitch

Hypothesis of insolation-driven long-term climate change (Hays *et al.*, 1976; Barbante *et al.*, 2010). The longer, marine-derived records, although lacking the resolution of ice cores from Greenland or Antarctica, reveal the progressive change from an ice-free world with high sea levels before the Quaternary, to a world characterised by an increasing magnitude in the fluctuating volume of continental icesheets, and as a corollary, significant change in sea level. The substantial advances in geochronological methods have provided a fundamentally important basis to quantify rates of sea-level changes. These advances have been bolstered by an increasing appreciation of the complex response of the Earth's crust–mantle system to fluctuating masses of continental icesheets and volume of ocean basins.

Ice cores afford a level of temporal resolution not available in marine sediment cores which, owing to the effects of bioturbation, provide a filtered temporal perspective of long-term ice-volume change. The Greenland Ice Core record, for example, reveals the 17 GI events during the past 60 ka (Davies *et al.*, 2012). These changes correspond with rapid atmospheric temperature shifts of between 8 and 16°C and represent approximately 50–75% of the temperature changes registered in the transition from fully glacial to interglacial conditions (Cronin, 2010). Accordingly, Dansgaard–Oeschger events may potentially have been accompanied by sea-level changes of several tens of metres. A future challenge in coastal stratigraphy will be identification of evidence for abrupt relative sea-level changes during the last glacial cycle that relate to Dansgaard–Oeschger events.

Collectively, palaeo-sea-level records for the Pleistocene reveal that early Pleistocene time (2.59 Ma to 780 ka) was characterised by interglacial–glacial sea-level oscillations of approximately 50 m amplitude, recurring every 41 ka. The oscillations were of lower amplitude than the glacial cycles of the middle and late Pleistocene, which resulted in up to 120–130 m of sea-level fall during maximum glaciation (Peltier and Fairbanks, 2006).

The marine oxygen-isotope record shows that in the most general sense, interglacial sea levels were broadly comparable within a band of ± 10 m. MIS stage 11 (420–360 ka) is likely to have been the warmest interglacial in the past 500 ka, and as a corollary may have experienced significantly higher sea levels (Burkle, 1993). The suggestion that MIS stage 11 was characterised by a sea level 20 m APSL (Hearty *et al.*, 1999) remains controversial, as few coastlines have been found to preserve anomalously high coastal deposits of this age, and not all oxygen-isotope records show evidence for an inferred sea level higher than the last interglacial maximum for MIS stage 11.

One of the long-standing challenges has been to resolve relative sea-level changes within interglacials in a manner that has been possible for the Holocene. Until the more recent advances in U-series dating that led to significantly smaller analytical uncertainties (e.g. ± 1 ka on an age of 125 ka), this quest was not possible. However, as with all geochronological methods, the highly specific requirements for the appropriate application of U-series dating means that many late Pleistocene coastal successions remain poorly constrained geochronologically. A related challenge has been to find appropriate field localities which reveal sufficient detail in the facies architecture to confidently resolve relative sea-level changes within interglacial highstand successions. Improvements in ESR dating have also enabled identification of sub-Milankovitch scale sea-level oscillations from the coral reef record from Barbados (Schellmann and Radtke, 2004a,b), and evidence for an abrupt rise in sea level during MIS 5e has been described from within reefs on the Yucatan Peninsula (Blanchon *et al.*, 2009).

Another approach to defining sea-level changes during interglacials has been based on oxygen-isotope analyses on foraminifera from marine cores. High-resolution sampling of planktonic foraminifera from central Red Sea marine cores KL09 and KL11 reveal that during MIS 5e ice-equivalent sea level rose by up to 1.6 m per century from about 123.5 ka ago (Rohling *et al.*, 2008). The Greenland Icesheet would completely melt if this inferred rate of sea-level rise extended over approximately four centuries.

The record of Pleistocene sea-level changes provides a particularly detailed and informative template to understand present environmental changes. They provide the context within which to examine postglacial sea-level changes, described in Chapter 7, and background against which to consider potential future change. The long time-series data available from marine and ice-core records, supplemented by the much shorter instrumental records, form the basis for considering the likely magnitude of future sea-level rise in an enhanced Greenhouse Earth, which is discussed in Chapter 8.

7

Sea-level changes since the Last Glacial Maximum

Has the change of the relative level of land and sea been accomplished by an upward movement of the land, or by the recession of the sea? Has the shift been slow and equable with regard to time, or by fits and starts with long pauses between, or by a slow movement interrupted by pauses? Has the time embraced by the whole series of phenomena been long or short, geologically speaking? What have been the general and particular circumstances and results of the whole movement?

(Chambers, *1848, pp. 3–4*)

7.1 Introduction

At the peak of the Last Glacial Maximum (LGM; ~22–20 ka) sea level was ~130–120 m lower than present as a result of the volume of water that was sequestered in icesheets. As discussed in Chapter 6, the lowest sea levels of the LGM at maximum development of continental icesheets exposed many continental shelves (see Figure 6.13). That lowstand and the present Holocene highstand represent end-points of a glacio-eustatic sea-level hemicycle. During the first half of this time interval, sea level rose rapidly through increase in ocean volume, until it was close to present levels; thereafter, during the Holocene epoch (the past 11,500 years since the end of the Younger Dryas), the rate and general pattern of relative sea-level change was spatially variable and differed according to proximity to former icesheets, isostatic adjustment processes, tectonism, and localised climatic changes (e.g. steric changes related to localised sea-surface temperatures and salinity).

Reefs have proved to be one of the best indicators of postglacial sea-level rise, particularly through drilling into relict reef features that occurred along what were to become the foreslopes of modern reefs. Radiocarbon and U-series derived reef chronologies have provided a very detailed picture of relative sea-level changes associated with deglaciation at the end of the LGM (Fairbanks, 1989; Yokoyama *et al.*, 2001; Peltier and Fairbanks, 2006). The pattern of sea-level rise accompanying ice melt has been reconstructed primarily from dating of reefs that tracked closely behind the rising sea (see Chapter 3). The initial reconstruction (see Figure 2.5) was based on drilling of offshore reefs around Barbados (Fairbanks, 1989). Analysis of the deeper corals from this site, combined with a model of continental deglaciation, indicated that the LGM may have begun at this site as early as 26,000 years ago (Peltier and Fairbanks, 2006; Gehrels, 2010a). The sea-level reconstruction for the later part of this record was determined simultaneously by drilling through the Holocene reef on the Huon Peninsula in Papua New Guinea on the other side of the world (Chappell and Polach, 1991). Subsequently, comparable long records of postglacial sea-level rise have been derived from Mayotte in the Comoros Islands, and Tahiti in the Society Islands (Bard *et al.*, 2010; Camoin *et al.*, 2012), with only a slightly shorter record from the Abrolhos Islands off the coast of Western Australia (Eisenhauer *et al.*, 1993). Independent support for the reconstruction of this pattern of sea-level rise comes from dating of muddy shoreline deposits such as mangrove wood and peat on the continental shelf off northwestern Australia (Yokoyama *et al.*, 2000), and on the Sunda Shelf (Hanebuth *et al.*, 2000).

Knowledge of the general pattern and rates of relative sea-level changes during Holocene time has been greatly aided by radiocarbon and, more recently, U-series dating of proxy sea-level indicators (Pirazzoli, 1991, 1996). Numerous studies of coastal sedimentary successions have been undertaken around the world using radio-carbon dating to derive chronologies of relative sea-level changes (Smith *et al.*, 2011). The earlier literature, particularly pre-1980s papers, noted discrepant relative sea-level histories in different geographical regions but was unable to offer a geohistorically coherent solution for contrasting datasets (Kidson, 1982; Hopley, 1987). This led to considerable debate on the nature of postglacial relative sea-level changes, particularly whether sea levels higher than present had been attained during the Holocene.

Geophysical modelling has significantly enhanced understanding of the nature of postglacial (post-MIS 2) relative sea-level changes and reaffirmed that numerous factors have contributed to these changes over contrasting temporal and spatial scales (Nakada and Lambeck, 1989; Lambeck and Johnston, 1995; Milne and Mitrovica, 1998; Mitrovica and Milne, 2002; Peltier, 1998, 2002a,b; Lambeck *et al.*, 2010). For the transition from the LGM to the Holocene, the dominant signal is eustatic (global, ice-equivalent) sea-level change, principally from decay of continental-scale icesheets. Hence the term glacio-eustasy has been used to describe these processes. Superimposed on the glacio-eustatic signal is the spatially variable effect of glacio-isostasy and hydro-isostasy. The magnitude and style of glacio-isostatic and hydro-isostatic response to differential water and ice loads has been shown to vary spatially and temporally around the globe and is primarily a function of icesheet dynamics and rheological properties of the crust–mantle system. The basis for the model-predicted regional variability of postglacial sea-level changes was initially described by Walcott (1972) and Clark *et al.* (1978), as outlined in Chapter 2 (see Figure 2.11).

One of the recurring themes in the literature on relative sea-level changes since the LGM has been whether the pattern of sea-level rise was smooth, punctuated by a sequence of steps, or characterised by well-defined oscillations, as examined in

Chapter 2. Model-generated sea-level curves appear unrealistically smooth, but it is not clear whether aberrant points in data sets represent genuine oscillations in sea level or inherent noise. Historically, attempting to decipher these issues has represented a major challenge, constrained by the age-resolving power of radiocarbon dating, vertical confidence limits of proxy sea-level indicators, and the statistical validity of curve-fitting to empirical observations of relative sea-level data. Despite recent advances in geochronology and refinements in sampling techniques, these complexities continue to confound interpretations of relative sea-level changes, particularly when caution is not exercised in interpreting indicative meaning of proxy sea-level indicators or inappropriate curve-fitting routines are adopted to derive continuous lines through discontinuous evidence.

7.2 Deglacial sea-level records of marine transgression

The record of postglacial sea-level rise has been derived, in many instances, from several sites using only a few types of evidence. Reconstructions of Holocene relative sea-level changes are likely to be more reliable when derived from a range of types of evidence from one site with a near-continuous record. Fossil coral reefs have been the primary source of information, and postglacial curves derived from different locations produce a relatively similar picture of deglaciation, enabling robust interpretations of eustatic (ice-equivalent) sea-level changes that appear broadly correlated with icesheet melting histories from the LGM (22–20 ka) to the early–mid Holocene when most of the northern hemisphere icesheets had melted.

The first comprehensive record was derived from submerged reefs around Barbados, an island located on the forearc ridge approximately 150 km east of the Lesser Antilles as described in Chapter 5. Submerged reefs had been detected around many Caribbean islands, but investigations focused on Barbados (Macintyre, 1967, 1972; Macintyre et al., 1991). Based on radiocarbon dating of *Acropora palmata*, the coral typical of reef crests, and considered likely to have grown in the upper 5 m of the water column (see Chapter 3), Fairbanks (1989) obtained a lengthy time-series delineating postglacial sea-level rise from a series of 16 cores drilled offshore through successive submerged reefs extending the record back to the LGM (see Figure 2.5). This early work was further explored with U-series dating, enabling the comparison of the two geochronological techniques (Bard et al., 1990a, 1990b) and geophysical modelling (Peltier and Fairbanks, 2006). The Barbados record reveals that during deglaciation following the LGM, sea-level rose rapidly at an average rate of ~10 mm/year. The *A. palmata* record at Barbados indicated several periods during which the sea rose particularly rapidly, termed meltwater pulses, in particular an event termed Meltwater Pulse 1A (MWP-1A) around 14 ka, and a second one, Meltwater Pulse 1B (MWP-1B), around 11.2 ka (Fairbanks, 1989). During MWP-1A, approximately 14,000 km^3 of continental ice melted per year and returned to the oceans; from Barbados data it was inferred that this occurred between 14.2 and 13.8 ka (Figure 7.1).

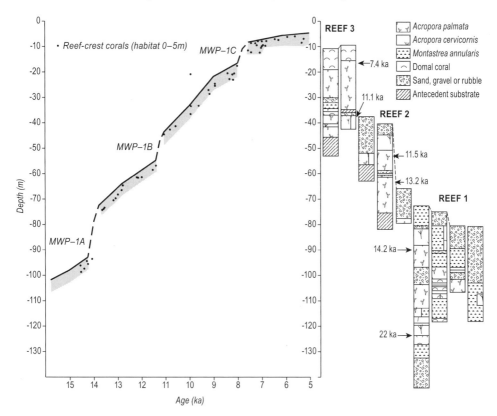

Figure 7.1 Postglacial sea-level rise as determined from drilling of submerged reefs around Barbados undertaken by Fairbanks (1989; see Figure 2.5). This summary of the three-step model (adapted from Blanchon, 2011, with simplified stratigraphy and age of reefs after Montaggioni and Braithwaite, 2009) uses reconstructions based on *Acropora palmata* reef-crest sequences (assumed to have formed within 5 m of the sea surface) corrected for an uplift rate of 0.34 mm/year.

Almost simultaneously, a record of reef growth was derived from radiocarbon dating of corals from a 52 m deep core drilled through the crest of a Holocene fringing reef complex near Sialum on the Huon Peninsula in Papua New Guinea (Chappell and Polach, 1991). This yielded a series of ages spanning the interval 11,000–7,000 ^{14}C years BP (13,300–8,000 cal yr BP), representing the later phase of postglacial marine transgression (Figure 7.2). Both Barbados and Huon Peninsula have experienced uplift which at the latter location is primarily co-seismic in nature with uplift rates typically up to 1.9 m per thousand years (see Chapter 5). By correcting for inferred uplift rates, the two records show a similar pattern, a linear trend in rapid sea-level rise over this time interval with relative sea level (uplift-corrected) rising from approximately 68 m BPSL to present sea level. A further record over this period was determined from a barrier reef on the northern side of

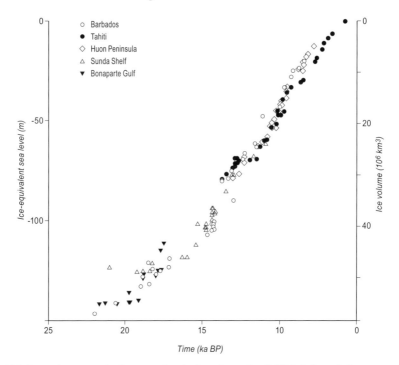

Figure 7.2 Ice-volume equivalent sea-level rise since the LGM inferred from coral, sedimentary, and palaeontological data. The records from individual sites have been corrected for glacio-hydro-isostatic effects. Equivalent grounded ice volumes (including ice grounded on the shelves) are shown by the scale on the right-hand side (after Lambeck *et al.*, 2010, with permission from Wiley).

Tahiti (Figure 7.2), which is a site that is experiencing subsidence (Montaggioni *et al.*, 1997). Other reef sequences also show a consistent record of growth over this period, for example, from the Abrolhos Islands (Eisenhauer *et al.*, 1993), Mayotte (Colonna *et al.*, 1996), and Vanuatu (Cabioch *et al.*, 2003).

Corals grow in a range of water depths and ages on fossil coral do not directly record the level of the sea surface, as discussed in Chapter 3, so greater confidence that the records derived from these fossil reefs really do indicate sea level was only possible when independent evidence could be found. This was provided by a postglacial sea-level record derived from siliciclastic sedimentary successions on the Sunda Shelf for the period 21–14 ka ago (Hanebuth *et al.*, 2000). The Sunda Shelf is located in a relatively stable geotectonic setting between the Indonesian archipelago and Vietnam. A composite record based on radiocarbon dating of organic remains from mangrove, tidal flat, and shallow-marine facies obtained from more than 50 sediment cores documents onset of deglacial sea-level rise, but does not record the final phases of late Pleistocene or early Holocene sea level. The radiocarbon chronology, expressed in calendar years, revealed four discrete phases in relative sea-level rise between 21 and 14 ka. Sea level rose from 116 to 114 m BPSL during the interval 21–19 ka. From 19 to 14.6 ka, sea level rose from 114 to 96 m BPSL. An accelerated rate of sea-level rise is evident for the interval 14.6–14.3 ka, when sea level rose from 96 to 80 m BPSL. The Sunda Shelf record indicates that after 14.3 ka, sea level rose more slowly and reached 64 m BPSL by 13.1 ka (Hanebuth *et al.*, 2000). The Sunda Shelf record is important because it corroborates evidence for the rapid rates of postglacial sea-level rise previously derived from coral reef settings, particularly the rapid rise around 14.6 ka which corresponds with Meltwater Pulse 1A reported in the Barbados record (Fairbanks, 1989).

Although individual colonies of coral can grow rapidly, massive corals averaging extension of around 10 mm per year and branching corals up to 100 mm/year, the reef itself is a porous aggregation of coral containing calcareous algae and loose calcareous sediment, incompletely lithified into a structure, and partially eroded by physical and biological processes. There is a limit to the rate at which reefs can accrete vertically, with studies of Holocene reefs indicating that they rarely accrete at more than 10 mm/year (Hopley *et al.*, 2007), and in most cases at rates considerably less than this. For this reason, the existence of meltwater pulses has been the subject of considerable controversy. There is now general acceptance of a rapid rise in sea level associated with MWP-1A around 14 ka (despite the slight discrepancy in timing between 14.2–13.8 and 14.6–14.3 ka), but there remains debate about the source of the meltwater, whether it was primarily from Antarctica (Clark *et al.*, 2002), or the northern hemisphere (Peltier, 2005). The period of rapid rise appears to have been between 14.6 and 14.3 ka, corresponding with the Bølling–Allerød warming, based on drilling from IODP (Integrated Ocean Drilling Program) cruise 310 which recovered cores off the northern coast of Tahiti. Sea level rose between 12 and 22 m (or at least 14–18 m) in this period, averaging a rate of 40 mm/year (Deschamps *et al.*, 2012).

Such periods of rapid sea-level rise were considered catastrophic for reefs by Blanchon and Shaw (1995), who interpreted three such phases, MWP-1A, MWP-1B, and MWP-2, the last of which they called catastrophic rise event-3, about 7.6–7.5 ka, prior to the cessation of the main phase of postglacial ice melt around 6,000 years BP. They considered that such periods of drowning resulted in coral reefs backstepping (see Figure 3.4), and a new phase of reef growth being initiated upslope of the older, drowned reef, except on those coasts that are uplifting (Montaggioni, 2005). The evidence from Tahiti indicates that reef growth continued through MWP-1A, despite the very rapid rise over an amplitude of 20 m, without backstepping (Camoin *et al.*, 2012). The existence of MWP-1B has been questioned based on dating of *Acropora* in the coral reef record from Tahiti (Bard *et al.*, 1996; Montaggioni and Braithwaite, 2009). It is possible that a shorter period of rapid rise is less likely to be recorded in the more diverse coral assemblages of Indo-Pacific reefs, but may be sufficient to cause backstepping in the *Acropora palmata*-dominated reef crests of the Caribbean (Blanchon, 2011b).

The discussion concerning a meltwater pulse in the early Holocene has become focused around what is known as the 8200 cooling event (Smith *et al.*, 2011). An abrupt rise in sea level in the early Holocene, attributed to the final drainage of proglacial Lake Agassiz–Ojibway, may have drowned several reefs that are presently submerged, with little or no modern coral growth on them. Submerged reefs are found throughout the Caribbean (Macintyre, 1988; Hubbard *et al.*, 2008) and southeastern Florida (Banks *et al.*, 2008). Independent evidence for an abrupt increase in sea level at 9,000–8,500 years BP has been interpreted from Asian deltas where sediments in incised valleys suggest decreases in sediment accumulation rates and abrupt deepening of the depositional environment, just before initiation of delta progradation in mid-Holocene (Hori and Saito, 2007). A sea-level curve reconstructed for Singapore indicated sea-level rise until around 8,000 years BP, followed by a 300–500-year still-stand centred around 7,700 years BP, with a subsequent rapid sea-level rise of 3–5 m at around 7,600–7,000 years BP (Bird *et al.*, 2007). High-resolution analyses of cores from Cambodia at the apex of the Mekong river delta were interpreted to indicate abrupt sea-level rise of more than 4–5 m which occurred in a short period around 8,500–8,400 years BP (Tamura *et al.*, 2009). Further evidence for rapid rise of sea level has been observed from the Swedish Baltic Sea (Yu *et al.*, 2007), at the base of Holocene sediments in Chesapeake Bay in the eastern USA (Cronin *et al.*, 2007), and in deltaic sediments in both Europe and North America (Törnqvist and Hijma, 2012; Li *et al.*, 2012). It appears that much of the rise can be attributed to rapid melt of the Laurentide Icesheet (Carlson *et al.*, 2008), although an alternative interpretation based on ice-melt modelling suggests that it might have come from the saddle between the Labrador and Baffin ice domes (Gregoire *et al.*, 2012).

In view of the significance of reconstructions of sea level from sites that are considered stable, it is perhaps surprising that there has not been a more concerted effort to derive a deglacial sea-level curve from reef environments around Australia. There were early estimates for the sea-level lowstand in the Australian region as outlined in Chapter 6; for example, shell material from around 130 m water depth from the Sahul Shelf (Van Andel and Veevers, 1967), a dredged *Galaxea clavus* coral colony and 'beachrock' from 175 to 150 m depth and dated at ~17,000 years BP for the southern Great Barrier Reef (Veeh and Veevers, 1970), and similar evidence at around 130 m depth from New South Wales (Phipps, 1970). More recently, evidence for successive lowstands has been reported from NSW based on shallow-water molluscs *Pecten fumatus* and *Placamen placidium* from vibrocores from the outer continental shelf in depths between 123 and 152 m (Ferland *et al.*, 1995; Murray-Wallace *et al.*, 1996b, 2005). A more detailed core record has been obtained from Joseph Bonaparte Gulf by Yokoyama *et al.* (2000, 2001a) who indicated that sea level was around 124 ± 4 m below present at ~ 20,000 years BP. A rapid rise around 19 ka is inferred from this research on the Sahul Shelf (Yokoyama *et al.*, 2000); it can also be detected on the Sunda Shelf (Hanebuth *et al.*, 2009), but is not detectable in the Barbados record which has much greater vertical errors associated with samples. Disappearance of reefs from the Marquesas Islands at about this time may also be related to this event (Cabioch *et al.*, 2008a).

The deglacial record from the Great Barrier Reef is fragmentary at this stage. Some researchers have argued for a highly episodic postglacial sea-level rise with up to nine periods during which sea level changed little, based on submerged terraces and drowned reefs, correlated with other evidence from New Zealand (Carter and Johnson, 1986; Carter *et al.*, 1986). A compilation of sea-level data from the Great Barrier Reef led Larcombe *et al.* (1995) to infer a 6 m regression (centred ~8,200 years BP). The evidence on which this was based was reviewed and rejected by Harris (1999), who considered a reversal of sea level at that time unlikely in that it would require a mechanism to extract water from the oceans. He also demonstrated that there were insufficient grounds for such an interpretation in view of vertical uncertainties associated with the depth at which material was recovered from cores. However, it is pertinent to note that there is now a body of evidence that indicates drowning of several submerged reefs at around this time; for example, radiocarbon ages on fossil *Acropora* of $8,460 \pm 100$ ^{14}C years BP from 28 m and $7,630 \pm 100$ ^{14}C years BP were obtained on submerged reefs at Cootamundra Shoal offshore from the Northern Territory (Flemming, 1986), and ages of around 7,000 years BP have been obtained on fossil reefs in 25–30 m water depth in the southern Gulf of Carpentaria (Harris *et al.*, 2007, 2008) and on the shelf around Lord Howe Island (Woodroffe *et al.*, 2010). Loss of such reefs may have been exacerbated by a decrease in sea-surface temperatures (SST). Evidence from Huon Peninsula in Papua New Guinea indicates that SST was 2–3°C cooler than modern SST in the early Holocene (McCulloch *et al.*, 1996), which when taken in combination with a reconstructed SST record from Vanuatu, implies a reduction of SST around 8,000–7,500 years BP, followed by an increase to present SST (Beck *et al.*, 1997). Such changes in temperature are likely to have had a pronounced effect on marginal reefs such as those in the turbid Gulf of Carpentaria, or at their latitudinal limit, as at Lord Howe Island. Submerged reefs are known to occur along the outer margin of the Great Barrier Reef (Hopley, 2006; Beaman *et al.*, 2008), and recent studies of some of these, including during IODP leg 325 indicates the potential to extract a much more detailed sea-level history from them (Abbey *et al.*, 2011).

7.3 Holocene relative sea-level changes in the far-field: Australia

In an effort to produce a global compilation of sea-level evidence, Rhodes Fairbridge wrote a far-sighted paper in 1961 that reviewed the data and mechanisms for sea-level change and endeavoured to compile evidence into a single Holocene sea-level curve (Figure 7.3). The curve he derived is shown in Figure 3.2, where it is emphasised how this contrasted with other compilations at that time. Fairbridge, with his colleague Curt Teichert, had described a series of sites in Western Australia where there were well-preserved sea-level indicators that supported a higher sea level than present

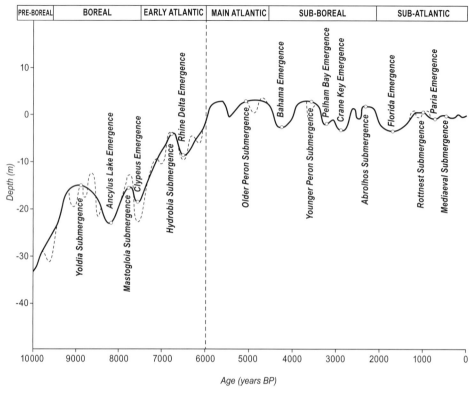

Figure 7.3 A simplified version of the sea-level compilation by Fairbridge (1961), showing a series of oscillations as a consequence of compiling evidence from around the world. Note the Older Peron, Younger Peron, Abrolhos, and Rottnest Submergences, based on highstand evidence from Western Australia, and the Bahamas, Crane Key, Pelham Bay, and Florida Emergences, based on sites in Florida and the Caribbean (adapted from Fairbridge, 1961).

during the mid and late Holocene. The existence of terraces that appeared to record postglacial sea level above present was known to Daly (1920) from both the west and east coasts of Australia. However, Fairbridge and Teichert had undertaken more detailed studies, particularly at Point Peron in Western Australia (Fairbridge, 1950), with similar evidence on the Abrolhos Islands and Rottnest Island (Teichert, 1950). Fairbridge's sea-level curve of global eustasy (Fairbridge, 1961) included evidence from emergent shorelines in Western Australia as key type localities for several highstands (submergences). These highstands contrasted with the Bahamas and Florida, where postglacial sea-level evidence was found below sea level. In the proposed eustatic curve that Fairbridge produced, he named a series of highstands after Australian sites (the Older Peron 3–4 m high at 5 ka, Younger Peron 3 m high at 3.9–3.4 ka, Abrolhos <2 m at 2.3 ka, and Rottnest <1 m at about 1 ka, submergences), which he implied were interspersed with lowstands based on evidence from sites in the Bahamas and Florida (Bahama, at –3 m around 4.3 ka, Crane Key at –2 m at 3.3 ka, and Pelham Bay at –3 m at 2.4–2.8 ka, and Florida at –3 m at 2 ka, emergences). This oscillating pattern of sea level contrasted with an alternative view that Godwin *et al.* (1958) had proposed, which maintained that the sea had risen up to its present level a few thousand years ago and been relatively stable since, and a widely held view that the sea had been rising to present at a gradually decreasing rate, which was adopted in the Netherlands (Jelgersma, 1961) and championed by Shepard from work in the Gulf of Mexico and elsewhere (Shepard, 1963; see Figure 3.2). It soon became evident that these Atlantic and Pacific data sets could not be reconciled into a single eustatic sea-level record, because the isostatic response of the Earth to changed ice and ocean volumes produced differing relative sea-level histories around the world. Thereafter, an initiative commenced, particularly through support of the International Geological Correlation Programme (IGCP, involving UNESCO and the IUGS), to address the geographical variation in sea-level records, and to attempt to produce an atlas of these variations (Bloom, 1977).

Recognition of hydro-isostatic adjustments, in addition to the glacio-isostasy that was apparent from areas that had been glaciated and were rebounding from the ice load, produced a broad pattern of variability around the globe that built on the geophysical projections of Walcott (1972), Clark *et al.* (1978), and others (see Figure 2.11). With an understanding that these processes underpinned the relative sea-level history at any site (see Chapter 2), it became clear that a different pattern of change could be anticipated at sites remote from former icesheets (far-field), as opposed to those that were close to icesheets, or had been covered by ice (near-field). In the following discussion, broad patterns of relative sea-level changes since the LGM are critically reviewed. The discussion focuses on regions that are representative of the general pattern of relative sea-level behaviour in terms of their distance from, or proximity to, the former Pleistocene icesheets. It reveals that there are significant contrasts in timing and patterns of relative sea-level changes in near- and far-field sites and that, accordingly, a single sea-level curve for the Holocene is unattainable.

By 9,000 years BP, about 80% of the volume of continental ice that occurred at the LGM had melted, and the sea was 20–25 m BPSL. The ICE-5G eustatic model indicates a sea level rising from 20–21 m BPSL at 9,000 years ago to 8 m BPSL 7,600 years BP (Peltier, 2004), although emerging evidence discussed above now supports a significant ice-melt event ~8.2 ka. The pattern of sea-level change during the Holocene no longer shows consistency from location to location. Although isostatic adjustments to altered ice and water loads would have been occurring during the deglacial phase, they were masked by the rapid melt, such that the overwhelming signal was glacio-eustatic. However, during the Holocene, these glacio-isostatic and hydro-isostatic components become prominent and contribute to the geographical variability in relative sea-level history.

Discussion of the pattern of Holocene relative sea-level curves that follows starts by looking at the far-field, because it is here that it might be anticipated that there would have been the least isostatic adjustment, and the clearest record preserved of the ice-equivalent eustatic trend. The attainment of modern sea level would be anticipated to be most clearly registered in far-field sites that have not been tectonic-ally active. Australia is one such landmass, and the evidence for an early Holocene highstand around Australia is examined (Sloss *et al.*, 2007; Lewis *et al.*, 2013).

Many far-field sites also record a fall in relative sea level following attainment of the early Holocene highstand due to hydro-isostatic adjustments. Hydro-isostasy involves subsidence of the ocean floor and subtle levering of continental shelves due to the geologically 'instantaneous' loading effects of water that has returned to the oceans following melt of icesheets. This is accompanied by landward migration of viscous mantle material away from the region of seafloor loading and results in formation of emergent shoreline deposits but without a reduction in volume of ocean basins. The highstand is explained by equatorial ocean siphoning at ocean islands (discussed in Chapter 2), and the magnitude of the highstand on larger landmasses is in part a function of the width of continental shelves (see Figure 2.12).

The detail of sea-level change around Australia during the Holocene has been the subject of ongoing debate. Located in an intra-plate setting and in the far-field of former icesheets, Australia experienced minimal direct glaciations, with ice cover restricted to the central highlands of Tasmania and the Snowy Mountains on the Australian mainland (Bowler, 1986; Colhoun *et al.*, 2010). Accordingly, the effects of glacio-isostasy have been very limited at a continental scale, although there is evidence for some local ongoing neotectonic processes (Murray-Wallace, 2002; Quigley *et al.*, 2010). Evidence described by Fairbridge from Western Australia (Fairbridge, 1950), and central to his attempted global compilation (Fairbridge, 1961), indicated that sea level had been higher than present in the mid-Holocene. This hypothesis was radical for its time, and was roundly rejected by researchers in the northern hemisphere, who had been deriving sea-level curves that rose at a decelerating rate up to present (see Figure 3.2). However, further dating of sea-level indicators from eastern Australia provided incontrovertible evidence that the sea had attained a level close to present by around 6,000 years ago (Thom and Chappell, 1975). The stratigraphical record from prograded coastal plains indicated that they had accumulated over the past 6,000 years with little if any change in sea level, in contrast to the decelerating pattern of sea-level rise experienced in the north Atlantic. An envelope of sea-level change was proposed by Thom and Roy (1983). However, whether the sea had been higher than present since reaching this level was a contentious issue along the east Australian coast, with evidence proposed in support of sea level having been above its present level in Queensland and Victoria (Gill and Hopley, 1972) refuted along other sections of the coast (Thom *et al.*, 1969, 1972).

7.3.1 Queensland

A significant contribution to the vigorous debate about whether a Holocene high-stand could be identified, its amplitude, timing, and geographical extent, was made by the Great Barrier Reef Expedition, undertaken by the Royal Society of London and the Universities of Queensland, in 1973. This expedition, led by David Stoddart, undertook particularly extensive mapping and survey of numerous reefs in the northern Great Barrier Reef. Results provided unequivocal evidence for higher sea level (Stoddart *et al.*, 1978; McLean *et al.*, 1978). A particularly important outcome of this expedition was recognition of the significance of large, flat-topped corals, called microatolls, whose upper surface was constrained by exposure at low tide (see Chapter 3), meaning that those that were not moated by local landforms on the reef top provided an indication of past sea level (Scoffin and Stoddart, 1978; Stoddart and Scoffin, 1979).

The Great Barrier Reef Expedition and the two decades following it was a time of active coring and dating on a series of reefs which led to clearer understanding of how different reefs have developed during the Holocene. Sea-level variations can be seen to have exerted a control on upward growth of reefs (Hopley, 1982) and on development of landforms on their surface (Hopley *et al.*, 2007). Microatolls have been key indicators which have enabled geographical patterns in relative sea level to be detected. An extensive sequence of microatolls from the continental margin of northeast Queensland was described by Chappell who suggested that sea level had fallen gradually over the past 6,000 years without evidence for oscillations exceeding the observational uncertainty of ± 0.25 m (Chappell, 1983d; Figure 7.4). The geographical variation in elevation of this evidence, together with other lines of sea-level evidence, such as the base of chenier ridges at Karumba in the Gulf of Carpentaria (Rhodes *et al.*, 1980), enabled the first evaluations of potential hydro-isostatic contributions to relative sea-level records across north Queensland (Chappell *et al.*, 1982, 1983).

Emergence on the inner shelf and stability or slight submergence on the outer shelf were inferred by Chappell *et al.* (1982), using the 5.5 ka isochron for radiocarbon ages,

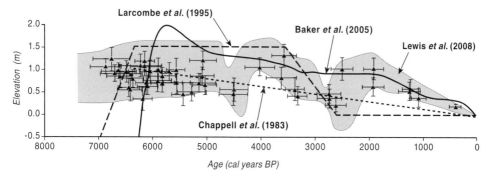

Figure 7.4 Radiocarbon ages on microatolls from northern Queensland. The data, collected primarily by Chappell and co-workers (Chappell *et al.*, 1983), were initially interpreted to indicate gradual sea-level fall to present. Shown are a number of subsequent interpretations, largely of these data, although supplemented by additional ages on other fixed biological indicators (not shown, but see Lewis *et al.*, 2008). The most recent interpretation is by Lewis *et al.* (2013).

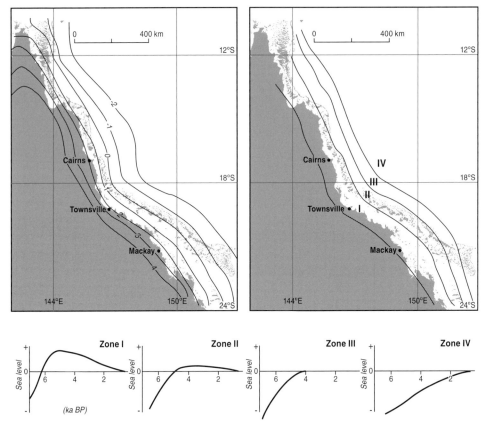

Figure 7.5 Variation in sea level due to hydro-isostatic modelling along the northern Great Barrier Reef, showing predicted Holocene sea-level highstand (in metres) at 6,000 years BP, and zones defined by characteristic sea-level curve and time at which sea level first attained its present level for ARC3 + ANT3B ice-melt model (adapted from Nakada and Lambeck, 1989, and Hopley *et al.*, 2007).

which was considered representative of when maximum sea level was attained. This was broadly consistent with a transect from the Gulf of Carpentaria to Britomart Reef, and accorded well with evidence summarised by Hopley (1982). Variation in width of the shelf enabled Nakada and Lambeck (1989) to model four zones that would lie progressively offshore and differ in terms of the time at which modern sea level was achieved and the presence and magnitude of emergence (Figure 7.5). These comprise zone I, which around Townsville extends to 65 km offshore, in which sea level reached present before 6 ka and experienced pronounced emergence; zone II, 65–140 km offshore, in which sea level reached present around 5.5 ka with minor emergence; zone III 140–200 km offshore in which present sea level was attained around 4 ka; and zone IV further seaward with sea level only reaching its present level recently (since 2 ka). This pattern remains broadly consistent with radiocarbon ages that have become available in the past 25 years, summarised by Hopley *et al.* (2007). However, these authors emphasise the many compounding factors which preclude precise sea-level definition, such as poor age and height resolution of sea-level indicators, incomplete spatial coverage, and the need to revisit modelling outputs (Hopley *et al.*, 2007).

Although there are several hundred radiocarbon ages from reefs on the Great Barrier Reef, compiled, for example, by Larcombe *et al.* (1995) and Hopley *et al.* (2007), these and other researchers indicate that they do not unequivocally define the precise nature of postglacial sea-level rise, largely because corals which have been dated were growing in water depths that cannot be determined with sufficient accuracy. Geographical variation across the reef indicates that outer reefs lagged behind those on the inner shelf and mainland, where there is independent evidence of marine transgression from ages on mangrove sediments beneath Hinchinbrook Island (Grindrod and Rhodes, 1984). To some extent, this may reflect hydro-isostatic flexure across the continental shelf, or there may be other reasons why various reefs lagged behind the rising sea surface; for example, reefs may keep up, or catch up with sea level (see Chapter 3), because of reef accretion rates, or lag behind deposition of mud in the nearshore (Larcombe and Woolfe, 1999).

The extent to which reefs across the shelf have reached sea level may result in an internal morphodynamic adjustment which can have an effect on the nature of evidence preserved along the mainland shore of north Queensland. Hopley (1984) proposed what he called a 'high-energy window'. He suggested that during the mid-Holocene, outer reefs had not caught up with sea level which had stabilised close to present level. As a consequence, higher-energy events such as storm waves could translate across the shelf producing storm ridges on the mainland that would be anomalously higher than could be formed by such events in the late Holocene when reef development was much more widespread and outer reefs had reached sea level. This is an example of an internal morphodynamic adjustment that might result in an apparent change in sea level, when the level of the sea had not changed in the open ocean. In this case, geomorphological evolution of the reef would have filtered wave energy and hence reduced the magnitude of run-up along the mainland coast. Such internal morphodynamic changes have been discussed by Chappell and Thom (1986), who caution against inferring a change of boundary condition (sea level) based on local evidence, when other changes within the system could cause similar phenomena. Dating of sequences of storm ridges at sites on Lady Elliot Island at the southern end of the reef (Chivas *et al.*, 1986) and Curacao Island north of Townsville (Hayne and Chappell, 2001) indicate little change in frequency or magnitude of tropical cyclones across the region over the past 5,000 years, although variability of storms is inferred from recent studies on the mainland (Nott, 2012). The high-energy window appears unlikely to have contributed to anomalous sea-level evidence; however, it provides an illustration of the complex task of differentiating changes in sea level within a region, from local effects which may distort interpretations of local levels.

Although north Queensland provides a rich coral record of past sea-level change, there remains vigorous discussion and disagreement on details of the Holocene highstand and the pattern of change over the past 6–7 ka. Different types of evidence favour different interpretations. At the time when support for a Holocene highstand was emerging from dating corals, other evidence, such as from mangrove deposits, appeared contrary to such an interpretation (Belperio, 1979). Detailed stratigraphical studies of Princess Charlotte Bay revealed a discrepancy between dated palynological reconstructions from mangrove muds beneath a prograded chenier plain that showed no evidence of a higher sea level (Chappell and Grindrod, 1984; Grindrod, 1985), and emerged fossil microatolls on adjacent offshore islands (Chappell *et al.*, 1983). Sea level above its present level from as early as before 7 ka is indicated by relict barnacle deposits (~ +1 m) at $7,380 \pm 240$ years BP (Lewis *et al.*, 2008), and U-series dating of fossil coral microatolls (+0.7 m) at $7,012 \pm 22$ years BP (Yu and Zhao, 2010) from Magnetic Island adjacent to Townsville. There are similar disparities further south along the reef; for example, emergence by as much as several metres was inferred from mangrove and chenier deposits in the macrotidal Broad Sound (Cook and Polach, 1973), which it is difficult to reconcile with radiocarbon dating evidence from shallow reef cores on reefs such as Redbill Reef, or coral and shell material from the southern Great Barrier Reef (Flood, 1983; Hopley *et al.*, 2007). In this case, mid–late Holocene amplification of tidal range in Broad Sound will need to be considered, in addition to possible tectonic displacement or subtle hydro-isostatic flexure.

Such diversity of interpretations have increased, rather than been resolved, in the subsequent years. For example, Chappell (1983d) considered that the simplest inference that could be drawn from microatoll data along the mainland coast was a smoothly falling sea to present levels (Figure 7.4). Recent compilations that build on these data, but include additional ages on encrusting oysters, barnacles, and tubeworms (Lewis *et al.*, 2008), and sea-level tendency based on a transfer function from foraminifera from Cleveland Bay (Woodroffe, 2009), have suggested a sustained highstand followed by a pronounced fall after 2,000 years BP. Lewis *et al.* (2008) inferred two oscillations of about ~1 m amplitude at approximately 4800 and 3,000 years BP, which they suggested might be related to Bond cycles (see Figure 7.4); however, these have been largely discounted as an artefact of data interpolation by Perry and Smithers (2011). A highstand of 1.0–1.5 m above present is currently the most accepted value for the region, but the disparity between different types of evidence and the geographical variation, reviewed by Lewis *et al.* (2013), imply that data quality remains insufficient to resolve between hypotheses or to validate detailed site-specific hydro-isostatic modelling.

7.3.2 Southeastern Australia

The coastline of southeastern Australia is part of a passive continental margin. Drilling and radiocarbon dating of many shell deposits from beneath sand barriers of southeastern Australia produced a wide suite of data that could provide a general interpretation of the final stages of marine transgression. A major limitation in the eastern Australian data set is the paucity of unambiguous proxy sea-level indicators from the intertidal zone for the period of postglacial sea-level rise. In addition, shell ages have large vertical uncertainties spanning several metres, as well as the need to recognise that radiocarbon ages record cessation of ^{14}C uptake in molluscs and not time of deposition. Many such radiocarbon ages formed the basis for the sea-level envelope that was described by Thom and Roy (1983, 1985), who recognised these various sources of uncertainty and considered an envelope more appropriate than a sea-level curve. They argued that a curve implied that the actual level of the sea was known with a precision which they considered unattainable. Although this postglacial sea-level record was compiled from a geographically wide region from southern Queensland to Victoria, the record reveals that sea level rose from 60 m BPSL at 13 ka to present sea level ~7 ka ago (Thom and Roy, 1985; Sloss *et al.*, 2007; Lewis *et al.*, 2013).

The most recent compilation of sea-level data from southeastern Australia was undertaken by Sloss *et al.* (2007), and is reviewed by Lewis *et al.* (2013). It contains many newly derived ages, particularly AAR ages, but continues to rely on many radiocarbon ages (now reported as calibrated ages) obtained during the intensive studies undertaken in the 1970s and 1980s. There has continued to be debate as to when the sea reached its present level and whether a Holocene highstand occurred. Recognition of fixed biological indicators (initially fossil tubeworms, but subsequently oysters and barnacles) appeared to provide evidence for sea level considerably above its present level from central NSW (Flood and Frankel, 1989). Radiocarbon ages on mangrove stumps exposed by coastal erosion near Bulli, initially reported by Jones *et al.* (1979), continue to be pivotal in arguments for higher sea level along this coast (Young *et al.*, 1993b). The compilation by Sloss *et al.* (2007) incorporates these and other evidence and suggests that sea level reached a highstand of 1.0–1.5 m APSL between 7,700 and 7,400 years BP, and remained at this elevation until 2,000 years BP when sea level gradually fell to its present position (Figure 7.6).

Based on radiocarbon ages on fixed biological indicators, such as worm tubes and oysters, from along the coast of New South Wales, a series of sea-level oscillations of ~1 m amplitude have been inferred during the Holocene (Baker and Haworth, 1997, 2000a, 2000b). The shape of the oscillating sea-level curve proposed by these authors results from statistical curve-fitting using a polynomial regression (Baker *et al.*, 2001a). A similar approach was adopted at other sites, both within Australia (e.g. Rottnest Island; Baker *et al.*, 2005), and elsewhere (Baker *et al.*, 2001b). Ages on such fossil fixed biological indicators appear to provide support for a mid-Holocene highstand both in New South Wales and from Queensland sites, but the compilation and plotting of ages has been criticised. Perry and Smithers (2011) have questioned the confidence that can be placed in interpreting tubeworm data, partly because statistical curve-fitting produces apparent trends through time periods for which there are no data. Disparate sites have been amalgamated despite local variation in

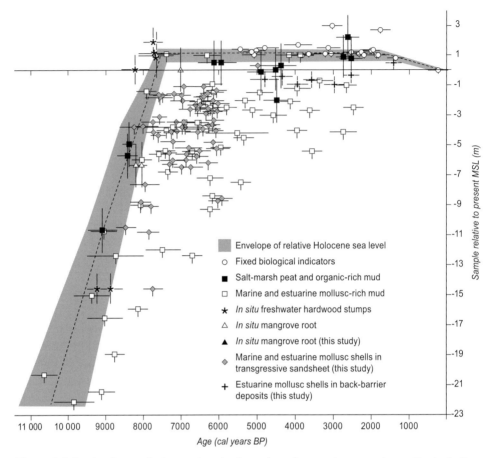

Figure 7.6 Sea-level compilation and revised envelope for southeastern Australia, including aspartic-acid racemisation-derived ages, with earlier radiocarbon ages on fossil evidence calibrated to sidereal years (from Sloss *et al.*, 2007. Reprinted by permission of SAGE).

the degree of exposure and without consideration of differential isostatic responses to water loading which could amount to as much as 1 m depending on exact location (Nakada and Lambeck, 1989). Further study will be needed to resolve details across the region.

7.3.3 *Other parts of the Australian coast*

Evidence from vibrocores from Spencer Gulf and Gulf St Vincent in South Australia reveals that sea level rose from 41 m BPSL at 10 ka (Cann *et al.*, 2006) and that transgression culminated around present level approximately 7,000 years ago (Belperio *et al.*, 1984a, 2002; Belperio, 1995) consistent with the record from southeastern Australia. Hydro-isostatic response, manifested by a seaward reduction in

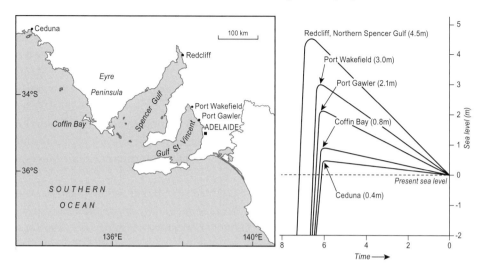

Figure 7.7 Postglacial sea-level curves from South Australia illustrating the effect of hydro-isostasy. Each curve represents a best fit of composite data from radiocarbon results on mangrove, samphire, intertidal and coquina facies with an average palaeo-sea-level uncertainty of ±40 cm and an age uncertainty of ±110 years (after Belperio, 1995 and Belperio *et al.*, 2002).

amplitude of the Holocene highstand and systematic offset in its timing is detectable in these gulfs, similar to that across the Great Barrier Reef shelf. The highstand ranges between 0.4 m APSL at Ceduna around 6 ka, to > 4 m at Redcliff in northern Spencer Gulf around 7 ka (Belperio, 1995; Chappell, 1987; Belperio *et al.*, 2002; Figure 7.7).

Variation in Holocene sea level in South Australia is revealed in sediment cores and backhoe excavations along Spencer Gulf and Gulf St Vincent (Belperio *et al.*, 2002). Transitions in sedimentary facies (seagrass, sandflat, mangrove, samphire, and chenier ridge) coinciding with various tidal levels were recognised, and dated, and compiled in terms of indicative meaning for sea-level position (see Figure 3.8) based on surveys related to modern equivalents (Burne, 1982; Belperio, 1993; Belperio *et al.*, 1983, 1984a; Harvey *et al.*, 1999). Vertical disparity along the gulfs with higher sea levels (> 4 m) recorded at the innermost sites (Redcliff) closely matches modelled outputs (Nakada and Lambeck, 1989; Lambeck and Nakada, 1990).

Compelling evidence for higher sea level in Western Australia during the Holocene stimulated Fairbridge's (1961) compilation of sea-level change around the world. Timing and magnitude of emergence remain subjects of debate, as elsewhere around Australia (Lewis *et al.*, 2013). For example, Semeniuk and Searle (1986) reported completely different sea-level reconstructions (i.e. no highstand, smooth-fall, highstand then fall) for three sites separated within a distance of 170 km near Perth, attributing discrepancies to local tectonism, although much of the Western Australian coastline is tectonically stable having developed on or adjacent to

Precambrian cratons (Yilgarn and Pilbara Shields). Since the pioneering work of Fairbridge (1961), Rottnest Island has continued to feature prominently in discussions of Holocene sea-level changes. Fairbridge interpreted the sedimentary record to indicate several sea-level stands higher than present in the Holocene, initially identifying evidence for four sea-level highstands with peaks at 3–5 m APSL at 6–4.6 ka (Older Peron Terrace), a peak 3 m APSL between 4 and 3.4 ka (Younger Peron Terrace), a peak 1.5 m APSL 2.6–2.1 ka (Abrolhos Terrace), and a peak 0.6–1 m APSL some 1.6–1 ka (Rottnest Terrace). Some evidence for these sea-level highstands was based on elevation of marine shell beds containing intertidal cockles *Katelysia scalarina* and *K. rhytiphora*.

The Rottnest Island record was re-examined by Baker *et al.* (2005), who radiocarbon-dated serpulid tubeworms on visors immediately above intertidal shoreline notches cut in the late Pleistocene Tamala Limestone (Figure 7.8). The fossil tubeworms at an elevation of 2.1 m APSL yielded ages of 5,960 and 5,750 cal years BP, whereas those above a lower intertidal notch (0.5–0.8 m APSL) yielded ages between 1,510 and 1,180 cal years BP. Baker *et al.* (2005) concluded that sea levels responsible for these accumulations could not be explained by hydro-isostatic or tectonic processes and attributed them to sea-level oscillations which they considered

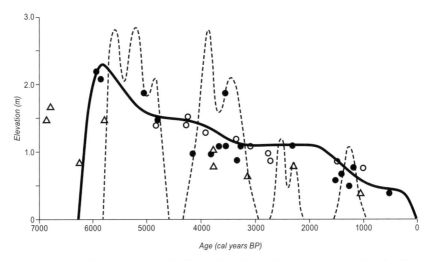

Figure 7.8 Plot of radiocarbon ages (calibrated to sidereal years) against altitude of serpulid worm cases (fixed intertidal biological palaeo-sea-level indicators) from Rottnest Island (solid dots), and adjacent sites in Western Australia (open dots). The oscillating Holocene sea-level curve of Fairbridge (1961), as deduced for sites in Western Australia (see Figure 7.3), is shown (dashed) as well as the octic polynomial curve of Baker *et al.* (solid). Triangles are ages on emergent coral pavement from the Abrolhos Islands, > 350 km to the north, from which Collins *et al.* (2006) inferred gradual sea-level fall. Uncertainties in inferred palaeo-sea levels cannot be quantified as the worm species does not currently live in the Rottnest Island area, preventing accurate delineation of its modern vertical range (after Baker *et al.*, 2005).

might be global in nature, noting similar Holocene highstands in southern Brazil and southeast Asia. Such an interpretation relies on statistical analyses of disjunct data. It could be argued that the data might equally be explained by a steadily falling sea surface (linear trend fit) from about 7,000 years BP despite significant gaps in data between 4,800 and 3,500 years BP. The sensitivity of different palaeo-sea-level indicators also needs to be considered: some will be more sensitive to short-term excursions of sea level, such as steric changes associated with El Niño events, than others. *Serpulid* worms may respond quickly to sea-surface fluctuations whereas other indicators might have a delayed response, or not even register short-term changes. By contrast with the oscillating Holocene sea level inferred by Baker *et al.* (2005), Collins *et al.* (2006) reported a high continuity, precisely dated and accurately surveyed record from emergent coral pavements in the Abrolhos Islands, which implies that sea level declined linearly from a highstand about 2 m above present around 7 ka, to the present (Figure 7.8).

There remain significant parts of the Australian mainland for which the details of Holocene sea level are inadequately known. Mangrove muds in cores from the Alligator Rivers region in the Northern Territory provide a record of the final stages of marine transgression and suggest that sea level attained its present position around 7,400 ± 200 cal years BP (Woodroffe *et al.*, 1987; Lewis *et al.*, 2013). The plains along the southern margin of van Diemen Gulf have formed since sea level has stabilised, and extensive mangrove forests occupied much of these estuarine environments from 7,000 to 5,500 years BP, in what has been called the 'big swamp' phase (Woodroffe *et al.*, 1985; see Figure 3.9). Similar 'big swamp' mangrove muds are known from tidal flats flanking King Sound (Jennings, 1975; Semeniuk, 1980, 1981, 1982) and the Ord River and Cambridge Gulf (Wright *et al.*, 1972; Thom *et al.*, 1975). However, tidal range exceeds 6 m in the Alligator Rivers region and at the Ord River, and reaches as much as 12 m in King Sound, providing too wide an elevational range for these fossil mangrove remains to provide more than a general indication of sea levels over the past few thousand years along this section of northern and northwestern Australia.

Although evidence in support of a higher sea level has been described from Victoria (Gill, 1983; Gill *et al.*, 1982), there is relatively little known about variation in elevation of the highstand on this part of the southern Australian coast, and also few data from Tasmania. It is unfortunate that this part of the record is so fragmentary. Although Australia lies far from the northern hemisphere icesheets, the southern parts occur within the expected intermediate field of Antarctic rebound, such that sites in southern Australia would be anticipated to be more strongly influenced by a glacio-isostatic sea-level signal, and hence with less pronounced highstands than would be expected for northern Australia (Nakada and Lambeck, 1988; Lambeck and Nakada, 1990). Absolute values for isostatic corrections are dependent on model parameters, with glacio-isostatic corrections varying only slowly with latitude from southern to northern Australia, but differential values for nearby sites may vary because of cross-shelf hydro-isostatic variations (Lambeck, 2002). Generally available evidence from the southern latitudes is less satisfactory, with variation in Holocene highstands along that part of the coast attributable to local or hydro-isostatic effects (Belperio *et al.*, 2002), and a more concerted effort, with more rigorous treatment of data recognising the need to discriminate on the basis of location rather than aggregating data across broad regions, will be required to quantify the north–south gradient of past sea level along the Australian margin (Lambeck *et al.*, 2010).

7.4 Holocene sea level across the Pacific Ocean

Continental margins may be subject to complex tectonic and glacio- or hydro-isostatic adjustments. With this in mind, Bloom proposed that oceanic islands might act as dip-sticks in the ocean (Bloom, 1967), providing sea-level records that are easier to interpret. Although return of large volumes of meltwater to the oceans has loaded the ocean floor, it has been suggested that small islands may be immune to local isostatic effects (Nakada, 1986). Many islands may record a Holocene highstand as a consequence of ocean siphoning, as water was redistributed from equatorial seas into collapsing forebulge regions peripheral to former ice margins (Mitrovica and Milne, 2002; see Figure 2.12).

Since the pioneering work of Reginald Daly, it has been clear that there are former shorelines preserved above present sea level on many islands across the Pacific Ocean (Daly, 1920, 1925; Wentworth, 1931). Before the advent of radiometric dating, their age could not be established, and it is now apparent that various shorelines identified at between 2 (the 'two-metre eustatic terrace') and 6 m APSL were in some cases Holocene (Newell, 1961; Newell and Bloom, 1970), but in others were Pleistocene, of last interglacial age or older. There has been an ongoing debate about whether a higher sea level can be detected during the Holocene across the Pacific (Figure 7.9), or whether evidence from corals of Holocene age that are incorporated in boulder conglomerate outcrops are the result of deposition from storms (see discussion in Chapter 3). An expedition to the Caroline and Marshall Islands (CARMARSEL expedition), specifically to resolve whether these were storm deposits or an indicator of higher sea level, reached no consensus (Shepard *et al.*, 1967; Newell and Bloom, 1970).

7.4.1 *High islands*

There seems little doubt that some interpretations of a Holocene highstand have been discounted because they did not fit the then-current paradigm, in the same way that researchers in North America or Europe discredited the purported emergent shorelines described by Fairbridge. For example, evidence for a sea-level highstand had been described from the Hawaiian Islands by Stearns (1935, 1978), but it was not widely accepted. The bench at Hanauma Bay on the island of Oahu was attributed to other geological processes (Bryan and Stephens, 1993), and when adjacent reef was

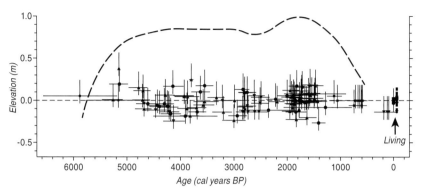

Figure 7.9 Evidence for stable sea level close to present from Christmas Island in the central Pacific (in the Line Islands, Kiribati). Radiocarbon ages are on fossil microatolls, which have been plotted relative to the elevation of modern counterparts (squares), or, for those in the island interior where there are no longer living microatolls, adjusted by the average offset of the entire field (after Woodroffe *et al.*, 2012b). Also shown is the generalised curve for sea level derived from French Polynesia by Pirazzoli and Montaggioni (1988): note this was based on conventional radiocarbon ages, and has not been adjusted for radiocarbon calibration.

cored and dated, radiocarbon ages were interpreted to indicate a reef that tracked gradually rising sea level up to present (Easton and Olson, 1976). This is a particularly instructive example of misinterpreting radiocarbon ages on coral. Radiocarbon ages from a sequence of six cores across the small fringing reef in Hanauma Bay record onset of reef growth around 7 ka with vertical accretion at an average rate of 2.9 mm/year of a *Porites*-dominated reef from 5.7 to 3.5 ka, with some gradual oceanward progradation. The record of reef growth derived from this pioneering reef coring exercise was inferred to track sea-level rise, but in fact the reef had adopted a catch-up reef growth mode (see Figure 3.4) and not a keep-up mode (Grigg, 1998). Evidence for a Holocene highstand of sea level is now widely recognised on Oahu, including the Hanauma Bay bench originally described by Stearns (Fletcher and Jones, 1996). Mid-Holocene emergence is apparent from ages on beach deposits on Kapapa Island (Figure 7.10a), as well as corroboratory evidence throughout the other principal Hawaiian Islands (Grossman and Fletcher, 1998). Although they are remote from plate margins and hence would not be expected to be tectonically active, the massive extrusion that is associated with these volcanic islands has imposed an enormous load on the Pacific Plate and the islands have experienced lithospheric flexure which has caused vertical displacement along the chain, as described in Chapter 5. The Big Island of Hawaii is undergoing rapid subsidence as can be seen from the depths at which recent reefs are found (Campbell, 1986). This has had flexural effects on other islands in the chain, as discussed in Chapter 5.

Holocene shorelines are known in considerable detail from several of the more major island groups in the Pacific. For example, a range of erosional and depositional types of evidence have been documented for several of the larger Fijian

(a)   (b)

(c)   (d)

Figure 7.10 (**a**) Evidence of a mid-Holocene beach on Kapapa Island, Oahu. Radiocarbon ages from shells in the lower left of the photograph have yielded ages of ~3,500 years BP (photograph: Colin Murray-Wallace); (**b**) conglomerate on the eastern rim of Tarawa Atoll, Kiribati. The lower unit comprises the octocoral *Heliopora* in growth position and reaching an elevation 20–40 cm above that to which corals presently grow. This is overlain by a storm unit of disoriented coral (photograph: Naomi Biribo); (**c**) eroded conglomerate on rim of Suwarrow Atoll in the northern Cook Islands. Corals and other biota (such as *Tridacna* clam immediately behind machete) can be seen in growth position, comprising a mid-Holocene reef at an elevation above that to which these organisms are presently able to grow (photograph: Colin Woodroffe); (**d**) a fossil microatoll of mid-Holocene age from Christmas Island in the central Pacific that is more than 9 m in diameter (photograph: Helen McGregor).

Islands, but these are probably related to differential vertical movements between islands (Nunn, 1990, 1998; Nunn and Peltier, 2001). Vertical movement is clearly a factor around the coast of New Zealand, where a pattern of Holocene oscillations of sea level had initially been interpreted from a chenier plain in the Firth of Thames by Schofield (1960). The stratigraphy of the plain was re-examined by Woodroffe *et al.* (1983a), who considered that there had been a gradual fall of relative sea level at this site, a view further substantiated by ground-penetrating radar (Dougherty and Dickson, 2012). A major compilation of sea-level data was undertaken by Gibb (1986), who considered Blueskin Bay near Dunedin to be the most tectonically stable site and compiled a master relative sea-level curve for that location and then corrected records for other sites around New Zealand in terms of vertical displacement by comparing them with that master record. There are many other island groups around the margin of the Pacific Ocean that can be anticipated to have experienced differential vertical movement of adjacent fault blocks which will have influenced the elevation at which evidence of emergence during the Holocene is found (e.g. on Guam and the Mariana Islands; Kayanne *et al.*, 1993; Dickinson, 2000).

It is also important to recognise differential flexural displacement in other settings. Thus, a sequence of emergent features have been described from Niue by Nunn and Britton (2004), but this island has been experiencing uplift as it migrates across the flexural bulge on approaching the Tonga Trench (Kennedy *et al.*, 2012). Evidence of emergence is recorded on many of the southern Cook Islands (Yonekura *et al.*, 1988; Woodroffe *et al.*, 1990a), but these are likely to have been subject to flexural uplift in response to Pleistocene loading by the island of Rarotonga. Rarotonga itself has been studied by several researchers (Schofield, 1970; Moriwaki *et al.*, 2006; Goodwin and Harvey, 2008) with little consensus on the distribution and elevation of sea-level evidence during the Holocene.

7.4.2 Atolls

Atolls would appear less vulnerable to flexural displacement and hence more suitable locations from which to derive mid-ocean sea-level records. Shallow drilling on several atolls has encountered Pleistocene reef limestone, often dated to the last interglacial, at depths of 10–20 m below the modern atoll rim: for example, Tarawa Atoll (Marshall and Jacobson, 1985), several atolls in the northern Cook Islands (Gray *et al.*, 1992), and Funafuti Atoll (Ohde *et al.*, 2002). This is consistent with gradual subsidence of the seafloor as an oceanic plate migrates away from a mid-ocean spreading centre; however, the extent to which lowering and reshaping of the surface results from subsidence or from solution which accentuated lagoon morphology on many atolls remains an issue of debate (Purdy and Winterer, 2001, 2006). Deeper drilling, such as on Eniwetak and Mururoa, indicates that the last interglacial limestone is underlain by older reef limestones deposited during preceding highstands, consistent with gradual subsidence (Szabo *et al.*, 1985; Camoin *et al.*, 2001). Prolific reef growth on atolls occurred as the Pleistocene margin was submerged by rising sea level around 8,000 years ago, after which the rim grew in an effort to catch up with sea level, as revealed in the case of Tarawa in the central Pacific Ocean (Marshall and Jacobson, 1985; Figure 7.10b). After reefs caught up with sea level, lateral progradation of the reef seems to have occurred, particularly in those situations such as Suwarrow, in the northern Cook Islands (Figure 7.10c), and beyond the flexural arch, where there are emergent reef flat and reef crest deposits with emergent fossil algal rims abandoned behind the modern reef crest (Scoffin *et al.*, 1985; Woodroffe *et al.*, 1990a).

Hydro-isostatic adjustments mean that details of relative sea-level history vary geographically across the Pacific, and that elevation of mid-Holocene reefs can also vary between atolls so timing of initiation of a reef flat in mid-Holocene varies too

(Lambeck, 2002). In particular, ocean siphoning is registered as a fall of sea level relative to many far-field remote islands (Mitrovica and Peltier, 1991; see Figure 2.12), although Clark *et al.* (1978) had foreshadowed that this would vary along the 180° meridian. Whether the sea-level curve at far-field sites peaked abruptly around 6,000 years ago and fell since then (through ocean siphoning) or whether there has been post-6,000 year melt, with a more gradual peak around 4,000 years is difficult to discriminate from atolls, because it was necessary for the reef rim to accrete to sea level before evidence would be preserved (Nunn and Peltier, 2001). However, atoll rims needed to 'catch up' with sea level.

Corals in growth position within the conglomerate platform on some atolls imply that it formed as part of a former reef flat (see discussion in Chapter 3). Fossil microatolls preserved in growth position within conglomerate are especially useful, as are other reef flat corals such as the octocoral, *Heliopora*. The value of such corals was recognised as early as the Royal Society expedition to Funafuti in the 1890s. David and Sweet (1904) described large *Porites* corals in growth position when mapping islands and reef flats around Funafuti Atoll, and considered that these indicated that the sea had been above its present level relative to the atoll in the past. Similarly, Cloud (1952) described outcrops of *Heliopora* in growth position, above the elevation that it presently reaches on the reef flat, as evidence of emergence on Onotoa in the Gilbert Islands (now Kiribati).

There have been several attempts to infer sea level either geographically (Grossman *et al.*, 1998) or at a site (eg. French Polynesia: Pirazzoli and Montaggioni, 1988; Funafuti: Dickinson, 1999), but these have rarely discriminated *in-situ* corals from more extensive conglomerate outcrops. Various sea-level indicators have been described for the Tuamotu atolls in French Polynesia, from which it has been inferred that sea level was above present for much of the period between 5 and 1.7 ka, after which it is considered to have fallen to its present level (Pirazzoli and Montaggioni, 1988; Figure 7.9). Large sea-level oscillations or abrupt changes appear unlikely, and studies that identify large anomalous fluctuations of sea level have generally been refuted. For example, evidence for an abrupt fall of sea level around 1300 AD inferred by Nunn (1998) has been criticised by Gehrels (2001), because ages were handled inconsistently and evidence from a wide geographical area was amalgamated without regard to spatial variability.

Radiocarbon ages on disoriented corals from outcrops of conglomerate record inputs of corals detached during storms. For example, the radiocarbon ages reported by Schofield (1977a) from Kiribati and Tuvalu appear to be from corals within conglomerate that were not in growth position. More recent studies indicate variation along the Kiribati–Tuvalu (Gilbert–Ellice) chain, from north (where conglomerate formation at sea level may have begun around 4 ka) to south (where it has been inferred that reefs reached sea level closer to 2 ka: McLean and Hosking, 1991). Ages from other atolls seem broadly comparable, with radiocarbon ages as old as 5,500 years BP reported from atoll surfaces in the Tuamotu Archipelago (Pirazzoli and Montaggioni, 1988). Several researchers suggested that island formation on atolls occurred as a result of this slight fall of sea level (Cloud, 1952; Schofield, 1977b; Dickinson, 2004).

It is important to differentiate units within the conglomerate platform (see Chapter 3), discriminating a lower unit that marks former reef flat surfaces and which may indicate higher than present sea level, from overlying storm deposits (Woodroffe, 2008). The conglomerate in the northern part of the Gilbert chain is composed of a lower unit that contains *Heliopora* in its growth orientation at a few localised sites (Falkland and Woodroffe, 1997; Woodroffe and Morrison, 2001; Figure 7.10b), overlain by an upper unit of disoriented cemented coral clasts.

There are several atolls on which last interglacial limestone is exposed at the surface (for example, Anaa in French Polynesia: Pirazzoli *et al.*, 1988; and Christmas Island in eastern Kiribati: Woodroffe and McLean, 1998; although neither of these have the extensive Pleistocene limestones found on Aldabra in the western Indian Ocean: Braithwaite *et al.*, 1973). These have undergone minimal, if any, subsidence over the late Quaternary, and appear particularly suitable sites to determine sea-level change. A near-continuous sequence of fossil microatolls has recently been described from Christmas (Kiritimati) Island in the central Pacific (Woodroffe *et al.*, 2012b). Christmas Island is an atoll with a large land area, comprising a ridge around its margin that encircles a low-lying former lagoon with numerous hypersaline lakes. This was an open lagoon with flourishing reefs throughout most of the past 5,000 years. Colonies of *Porites* thrived across the reef flats and throughout this lagoon, and were constrained by exposure at low tide, adopting microatoll morphologies similar to the many microatolls that occur on the modern reef flats. Radiocarbon and U-series dating of more than 100 fossil microatolls, many of which are several metres in diameter (the largest exceeding 9 m in diameter, Figure 7.10d) indicates a near-continuous record consisting of corals that have been constrained by sea level for the several decades to centuries of their life histories. The upper surface of fossil micro-atolls on the modern reef flat occurs within 0.1 m of the height of adjacent living counterparts (Figure 7.9). The more continuous sequence of fossil microatolls from the former lagoon lies within a slightly broader elevation range (0.4 m). Surveyed heights of different fields of microatolls indicate that they are at different elevations, which has been attributed to a substantial geoidal gradient over the 40 km from one end of the atoll to the other. However, when fossil microatolls are plotted relative to elevation of adjacent modern equivalents, a near-continuous sequence over the interval from 5.2 to 0.7 ka appears to preclude significant oscillations of sea level (Woodroffe *et al.*, 2012b).

7.5 Other far-field locations

Far-field sites appear most likely to provide insights into the pattern of ice-equivalent sea level, and there is a wider range of evidence from tropical than from more temperate locations, particularly because of the suitability of coral for dating

(Zong, 2007). Soon after the publication of Fairbridge's 1961 compilation it became evident that there was a Pacific pattern of sea-level change and an Atlantic pattern, which was particularly noticeable from the morphology and development of reefs (Adey, 1978). Before examining the Atlantic story, other far-field locations (the Indian Ocean and southern Asia) will be considered to briefly evaluate whether the Australian and Pacific examples are typical.

7.5.1 Holocene sea level in the Indian Ocean

Over one hundred years ago, Gardiner speculated that there had been a highstand of sea level in the Maldives archipelago based on what he termed 'reef rock', which he considered to be built as part of the growth of the reef, as opposed to boulder conglomerate which he inferred to be storm deposits (Gardiner, 1902, 1903). His most detailed descriptions were for Minicoy, the southernmost of the Laccadive (Lakshadweep) Islands, and he estimated that this phase of reef formation had occurred about 4,000 years ago, which appears a remarkably accurate estimate in light of subsequent radiocarbon dating (Siddique, 1980). Gardiner inferred from these conglomerates that sea level had been 7–8 m above present (Gardiner, 1931), but subsequent studies clearly discount this amount of emergence. Support for a highstand was perpetuated by Sewell (1935, 1936), but he recognised that the massive *Porites* corals had been deposited by storms, and that branching *Acropora* coral in vertical orientation may not actually have been in growth position. The most detailed reconnaissance was of Addu Atoll, the southernmost in the group, where it was considered that there was likely to have been emergence (Sewell, 1936; Stoddart *et al.*, 1966), but definitive evidence was lacking. Further investigation of Addu Atoll recorded an outcrop of *Heliopora* clearly in growth position, on the northern margin of one of the reef islands where local erosion had stripped the overlying beach sand from the former reef flat; a radiocarbon age of $2,710 \pm 85$ [14]C years BP (see Figure 3.5b) was determined, although it was not clear whether this was at the same level as modern reef flat corals, or slightly higher (Woodroffe, 1993a). The subtleties of inference necessary to resolve the actual height of the sea are such that it may be unrealistic to anticipate resolving the elevation of the sea surface to less than an accuracy of ± 0.5 m during periods of rising sea level, and ± 0.2–0.3 m for periods of highstand. For example, two recent studies have proposed detailed sea-level curves for the Maldives during the Holocene based on drilling and dating of corals (Figure 7.11), but they differ in their interpretation of whether sea level has been higher than present in the mid-Holocene (Gischler *et al.*, 2008; Kench *et al.*, 2009).

The Chagos archipelago in the central Indian Ocean comprises few well-developed reefs with only small islands on them. The Great Chagos Bank is largely submerged, reaching into the intertidal zone around only a small part of its margin and support-ing only eight small islands. The most extensive land is on the southernmost atoll, Diego Garcia, and on the atoll of Peros Banhos. Gardiner inferred that the sea had

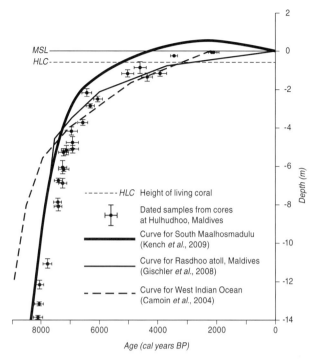

Figure 7.11 Sea-level data for the Maldives, showing different interpretations of data from the atolls within the Maldives and contrasted with a curve from the western Indian Ocean (adapted from Kench *et al.*, 2009).

been higher in Chagos (Gardiner, 1936), extending his interpretation of the significance of reef rock south from the Maldives. Conglomerates on the atoll of Diego Garcia have been described by Stoddart (1971), who tentatively inferred emergence, although no sites at which fossil coral was in its growth orientation were found. Subsequently, Holocene radiocarbon ages have been reported from Salomon largely on disoriented *Acropora* and *Porites*, none of which can be reliably related to a higher sea level (Eisenhauer *et al.*, 1999).

In contrast to the central Indian Ocean, there is convincing evidence for a Holocene highstand in the eastern Indian Ocean, and its absence in the western Indian Ocean (Woodroffe, 2005). Based on detailed mapping of conglomerate around the Cocos (Keeling) Islands (Woodroffe *et al.*, 1994), several *in-situ* microatolls have been radiocarbon dated and indicate that there has been a gradual fall of sea level from an elevation 0.5–0.8 m above present over the past 3,000 years (Woodroffe *et al.*, 1990b). Camoin *et al.* (2004) summarise reef studies from the western Indian Ocean where there appears to be no evidence for the sea having been higher than present during the Holocene, and on the basis of which they infer that sea level reached its present level 3–2.5 ka. The Seychelles are granitic, and appear likely to have been stable, with last interglacial reef found at elevations of up to 8.2 m above present (Israelson and Wohlfarth, 1999; Dutton and Lambeck, 2012). Corals from conglomerate on Mahé have U-series ages within the range 7.5–4.2 ka, but outcrops do not contain unequivocal evidence of emergence, in contrast to interpretation of marine cements in conglomerate on Farquhar and Saint Joseph, which suggested sea level was up to 1.2 m higher than present (Pirazzoli *et al.*, 1990).

The pattern of Holocene sea-level rise is modelled to show variation across the Indian Ocean, primarily because of dependence of hydro-isostatic response on distance from the African and Madagascar coasts and shoreline geometry. For example, predicted response would indicate that a sea-level highstand should be observed at Toliara in western Madagascar on the western margin of the ocean, but that sea level would have shown a decelerating rise to present at both Mauritius and Mayotte in the Comores, with modern level being attained a few hundred years apart at the latter sites (Lambeck *et al.*, 2010). There is relatively little sea-level evidence from Madagascar, although *in-situ* coral colonies have been dated between 3 and 2 ka at up to 2.5 m above modern (Camion *et al.*, 1997); a more complete record of Holocene emergence from as early as 6.7 ka has been reported from South Africa (Compton, 2001; Ramsay and Cooper, 2002), representing some of the only published data from the eastern margin of the African continent. Although modelling and observational data appear to show some accord, there are limitations in terms of the vertical accuracy with which dates from such reef environments can be interpreted.

7.5.2 Southeast Asia

Southeast Asia, and the adjacent Sunda Shelf, have experienced a sea-level history similar to that recorded for northern Australia. A drowned delta in the Straits of Malacca at a depth of 146 m below present may record the shoreline position at the LGM (Emmel and Curray, 1982). A large area of the Sunda Shelf was subaerially exposed at this time, and rivers flowed into a larger system, which has been given the name Molengraaff (or North Sunda) River in recognition of the pioneering research into this area undertaken by Molengraaff (1921). Sea level rose over this shelf, flooding much of the South China Sea, and a record of the transgressive shorelines has been reconstructed from coring (Hanebuth *et al.*, 2000, 2009). The final stages of the transgression have also been recorded from the Straits of Malacca, and it is clear that it reached a level above present, with emergence evident along much of the Thai-Malay Peninsula (Geyh *et al.*, 1979). A diverse range of observational data has been compiled from the Peninsula including raised intertidal notches or abrasion platforms, beachrock, and fossil oyster and coral (Tjia, 1977, 1996). These data have been interpreted to indicate a series of oscillations of sea level (Figure 7.12), similar to those previously proposed by Fontaine and Delibrias (1974) from scattered data along the coast of Vietnam.

A view persists that there have been several highstands during the mid and late Holocene, suggested to have reached a peak of 5 m above present at 5 ka, with a

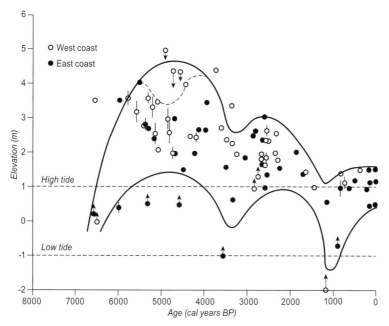

Figure 7.12 Envelope of sea-level curves for Peninsular Malaysia; index points with vertical error bars indicate a range within which sea level may have been, whereas arrows indicate directional evidence (based on Tjia, 1996; from Woodroffe and Horton, 2005, with permission from Elsevier).

second peak of 2.5 m at 4 ka, and 2 m at 2.5 ka (Fujimoto *et al.*, 1996; Woodroffe and Horton, 2005). However, there is poor constraint on many of the low points in these curves, such that the oscillations may be illusory (Pirazzoli, 1991). Dates from intertidal corals on the reef flat at Phuket indicate a sea level about 1 m higher than present at 6 ka, falling gradually to present (Scoffin and Le Tissier, 1998), and from Singapore, where corals have been dated that imply sea level was up to 3 m higher than present at 7–6.5 ka and fell to present after 3 ka (Hesp *et al.*, 1998). Emergence has been identified along the east coast of India (Banerjee, 2000) and in Sri Lanka (Katupotha, 1988; Katupotha and Fujiwara, 1988), and is inferred from evidence near Dhaka in Bangladesh, despite compaction in the broad Ganges–Brahmaputra delta (Rashid *et al.*, 2013). There are records of Holocene sea level from Indonesia (Thommeret and Thommeret, 1978), and from the Philippines (Maeda *et al.*, 2004; Omura *et al.*, 2004), but tectonic activity is likely in these and other island-arc or plate-margin settings.

7.6 Holocene relative sea-level changes: the intermediate-field

The contrast between an Indo-Pacific and an Atlantic sea-level curve, emanating from Fairbridge's (1961) compilation, was evident in the differences in the morphology of coral reefs, emphasised by Adey (1978), and reinforced by subsequent summaries of reef growth and morphology (Davies and Montaggioni, 1985). Fairbridge included sea-level evidence reported from Florida and the Bahamas, in particular. The Everglades of southern Florida span the ecotone between mangroves (dominated by *Rhizophora mangle* to seaward and *Avicennia germinans* to landward) and grass and sedge species of the freshwater wetlands. Pioneering studies by Scholl (1964) had described the stratigraphy and reported radiocarbon ages on mangrove peat. Although the zonation of modern communities, with propagules of *Rhizophora* establishing on muddy carbonate banks in Florida Bay, provided circumstantial evidence that mangroves might extend seawards, prograding across the Bay, this view was challenged by Egler (1952). He recognised that the several-metres thickness of mangrove peat underlying the wetlands exceeded the elevational range within which mangroves occurred, and must have been deposited as sea level had risen, implying that mangroves had expanded landwards into the sedgelands. This view was confirmed by more detailed stratigraphical and palynological studies when it was realised that calcareous muds beneath the peats in Florida were of freshwater (not marine) origin (Spackman *et al.*, 1966). This trangressive sedimentary sequence (freshwater marl–freshwater peat–mangrove peat) provided a constraint on sea level, indicating that it had risen gradually towards present (Scholl and Stuiver, 1967; Scholl *et al.*, 1969), a view perpetuated by further research in southern Florida (Enos and Perkins, 1979; Parkinson, 1989).

This was broadly consistent with a gradually rising sea level which had been compiled for the Gulf of Mexico (Shepard, 1960), and inferred from there to have occurred more widely (Shepard, 1961). Further studies of Holocene sea level based on mangrove deposits in the Caribbean included decelerating sea-level curves from the Cayman Islands (Woodroffe, 1981) and Jamaica (Digerfeldt and Hendry, 1987), which were similar to that derived by dating mangrove peat in cores from a wetland on Bermuda, indicating gradual rise to present sea level (Neumann, 1971). Much more detailed studies along the coast of Texas have further refined this pattern of rise (Rodriguez *et al.*, 2004; Milliken *et al.*, 2008; Wallace and Anderson, 2013), implying that average rates of rise decreased from 9 mm/year ~10 ka, to 5 mm/year ~8 ka, to 2 mm/year ~5 ka, and slowing to ~0.6 mm/year over the past 2 ka. Despite such apparent confirmation of a pattern of decelerating sea-level rise, there have been other studies that have reported evidence for a Holocene highstand along parts of the Texas and Alabama coasts (Morton *et al.*, 2000; Blum *et al.*, 2001). Evidence includes raised marsh deposits, wave-cut features, and beach ridges; the latter up to 2 m APSL, for example at Mullens Bayou in Texas. In contrast, there is abundant evidence that there has been subsidence of the Mississippi Delta, and the relative sea-level history beneath the delta implies the sea was 10 m lower around 8 ka. Subsidence of the order of 1.1 mm/year is implied (Törnqvist *et al.*, 2004), although the extent to which this reflects tectonic displacement, flexure of the shelf in response to sediment loading, or compaction remains a subject of debate (Blum *et al.*, 2008; Törnqvist *et al.*, 2008).

Radiocarbon-dating of coral from coring of west Atlantic reefs provides further independent evidence of past sea level. Radiocarbon ages on *Acropora palmata*, which occurs on reef crests and on shallow reef fronts, have been particularly important in the Caribbean, and enabled Lighty *et al.* (1982) to derive what they called a minimum sea-level curve. They compiled data on submerged reefs that occur in around 15 m water depth offshore of southern Florida and several Caribbean islands (Lighty *et al.*, 1978). This pattern of decelerating sea-level rise, observed widely throughout the Caribbean, has been supplemented by many more ages, from the Bahamas, south Florida, Belize, Panama, Jamaica, Puerto Rico, St Croix, Antigua, and Martinique, and has enabled what has been termed a 'corrected' curve by Toscano and Macintrye (2003; Figure 7.13). It represents the gradual response of this region to rebound following melt of the North American icesheets. There is an increasing body of evidence from the eastern Caribbean, particularly from St Croix, which indicates that the reef on the shelf edge was replaced by reef growth further landward around 7 ka (Hubbard *et al.*, 2005, 2008),

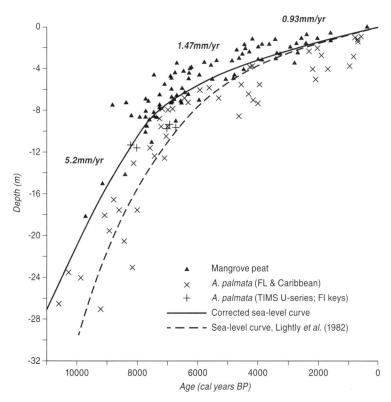

Figure 7.13 Sea-level curve for the West Indies based on a combination of mangrove peat and coral data (after Toscano and Macintyre, 2003, with kind permission from Springer Science + Business Media).

although it does not support the idea propounded by Lighty that the submerged reefs were restricted by inimical shelf water during the early Holocene.

There is no consensus, however, as to whether this sea-level curve also applies along the western margin of the Caribbean; where sea level appears to have been closer to its present level than the curve derived from islands further east implies. The Belize barrier reef has been studied in detail, and several preliminary studies included radiocarbon ages that were not consistent with the broader Caribbean trend. Coring and dating of mangrove sediments on Ambergris Cay and elsewhere implied sea level within 1 m of present by 4.5 ka (Woodroffe, 1988a). A slightly different sea-level history has continued to be inferred for Belize (Gischler and Hudson, 2004; Gischler, 2006), questioning whether the Toscano and Macintyre interpretation has misinterpreted data from Belize that appear to lie above their broader Caribbean curve. Blanchon (2005) argued that there has not been enough consideration given to whether dates from cores were on *in-situ* coral, pointing out that storms can deposit branches of coral in ridges above sea level, and favouring a stepped sea-level history based on further evidence from submerged reefs (Blanchon *et al.*, 2002). Regional discrepancies have been indicated between the curves for Bermuda, Barbados, Belize, and the broader western Caribbean, emphasising that these include a range of tectonic settings that may experience different isostatic adjustments. Further integration of modelling and observational data offers the prospect of resolving some of these issues (Toscano *et al.*, 2011). The broad trend for this intensively studied region remains one of decelerating sea-level rise until historic times, but the details will continue to be the subject of discussion.

7.7 Holocene relative sea-level changes and glacio-isostasy: the British Isles

A large proportion of the surface of the planet has undergone significant adjustment through glacio-isostasy, a concept that was introduced by Jamieson (1865) and has been widely discussed, for example by Daly (1934). A particularly well-studied example is the relative sea-level history of the British Isles, which is examined to demonstrate flexural processes. A broader range of responses is seen across other areas of Europe, with particularly rapid uplift in Scandinavia, and a complex pattern that has been variously interpreted around the Mediterranean Sea; these areas are addressed in Chapter 5 and discussed only briefly here. The North American continent has experienced a more complex history which is summarised briefly.

It has been clear for more than 150 years that there have been variable sea-level trends around Britain, many of which are ongoing (Figure 7.14). In Scotland, accounts of sea levels higher than present date back to at least the seventeenth century. Darwin speculated that the 'parallel roads' of Glen Roy, in the Grampian Highlands (see Figure 2.7a), marked former higher shorelines (although this interpretation was shown to be incorrect as discussed in Chapter 5). In England, Huxley (1878) remarked on submerged forests which contrasted with observations of sea

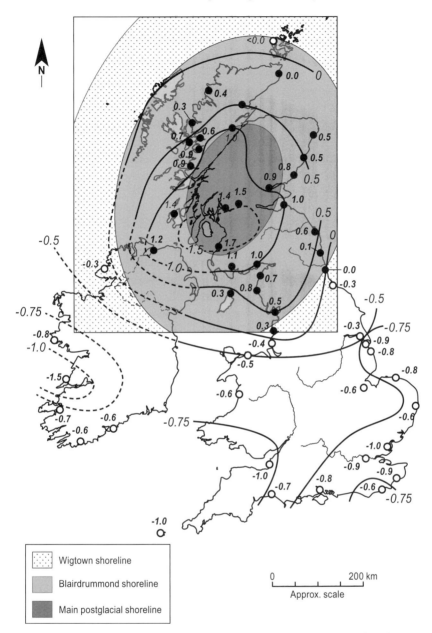

Figure 7.14 Contemporary rates of land motion in Britain (in mm/year), based on data reported by Shennan and Horton (2002b) but accounting for a regional sea-level rise of 0.14 m in the past century, ocean siphoning (0.3 mm/year), and possible ice-equivalent sea-level rise of 0.1 ± 0.1 mm/year (from Gehrels, 2010b, with permission from Elsevier). Postglacial uplift in Scotland results in declining elevations of Holocene shoreline features away from a centre of maximum uplift in the southeast Grampian Highlands, and regions are shown in which each of the Main Postglacial, Blairdrummond, and Wigtown shorelines are visible features (after Smith *et al.*, 2012).

levels higher than present further north. Jamieson (1865) had recognised that Scotland had been depressed beneath the weight of icesheets and realised that uplift would be proportionate to the amount of ice load (Jamieson, 1908), explaining decline in shoreline altitude away from the centre of former icesheets. However, Jamieson did not explain how shoreline terraces in such areas were formed, given that the rate of land uplift varied with distance away from the centre of the icesheet, and that rate of rise of sea level would vary over time. Wright (1914) inferred that shoreline terraces formed when rise of the land and rise of the sea took place at the same rate, and that contiguous terraces had been formed *diachronously* (time-transgressively). These important concepts, demonstrated in Scotland, were applied to other areas affected by glacio-isostasy, particularly Scandinavia (see Section 5.7).

7.7.1 Northern Britain

Evidence for sea-level change is particularly extensive around British coasts and the pioneering work of Jamieson and Wright provided opportunities to determine, and then model, patterns of glacio-isostatic uplift. Modelling approaches have been of two kinds reviewed by Smith *et al.* (2012): shoreline-based and glacial isostatic adjustment (GIA) based (see Figure 2.8). Shoreline-based models, as exemplified in Scotland, involved systematic surveys of elevation of former shorelines (for example, levelling landwardmost extent of estuarine mudflats, known as 'carse' in Scotland, and interpreting buried shorelines from pollen and microfossil analysis of cores), followed by statistical fitting of trends (through trend-surface analysis or kriging). GIA models assume earth-rheological and ice-load parameters and produce broad modelling outcomes that have been tested primarily through studies of isolation basins (Shennan *et al.*, 2006b). Shoreline reconstructions, such as that of Sissons (1963), are limited by temporal and spatial constraints, but GIA models are less constrained. Estimates of the ice load term vary, and have been based upon mapping of features such as glacial trimlines (Ballantyne *et al.*, 1998). However, details of the changing icesheet configuration have not been adequately incorporated into GIA modelling for the British Isles; temporal history of ice loading is only broadly understood both before and during the LGM. A recent integration of output from numerical ice-flow models, adopting minimal, median, and maximal extents for ice, produces a range of data-model fits, but also implies influence from Fennoscandian ice, and that trimlines may have formed beneath the ice rather than at its margin (Kuchar *et al.*, 2012).

The shorelines reached during and following decay of the British–Irish Icesheet are best known for northern Britain. Originally, shorelines formed at this time were known as the '25-feet', '50-feet', and '100-feet' shorelines, following mapping by the British Geological Survey in 1879. This terminology implied that these shorelines were horizontal, a concept incompatible with their formation in a glacio-isostatically affected area, as was evident from work in similarly glacio-isostatically affected areas, notably Scandinavia (e.g. de Geer, 1888). Accordingly, in the 1960s, Sissons and co-workers began the first systematic, detailed fieldwork on Scottish shorelines (Sissons *et al.*, 1966), employing mapping and stratigraphical investigations. Elevations for a prominent limit along the margin of the icesheet, the Perth Stage (reached ~14 ka), interpreted using trend-surface analysis, disclose the centre of glacio-isostatic uplift at the time to have occurred towards the southeast Grampian Highlands (Smith *et al.*, 1969).

Evidence for later shorelines is more complete (Cullingford and Smith, 1980), and a sequence of four episodes (HRSL 1 to HRSL 4) has been proposed by Smith *et al.* (2012). The first of these episodes commenced with a prominent shoreline reached during the Younger Dryas (Loch Lomond) Readvance. This shoreline comprises an erosional terrace, known as the Main Lateglacial Shoreline and is widespread around Scotland, having formed largely as a periglacial rock platform, the Main Rock Platform, with a cliff behind it. It rises to an elevation of around 12 m in the Firth of Lorn, and forms a bench, up to 40 m wide, cut into calcareous schists at 7–8 m above sea level around the Isle of Lismore. Cosmogenic dating supports the contention that it dates from ~12.9–11.6 ka, and it seems likely that it was rapidly abandoned as climate warmed after the Younger Dryas (Stone *et al.*, 1996). In central Scotland this erosional shoreline was widely overlain by 6 m or more of marine and estuarine deposits as relative sea level began to rise further. The culmination of this sea-level rise is marked by the High Buried Beach (estimated at ~11.8 ka) identified only in the Forth lowland, with a Main Buried Beach (~10.6–11.3 ka) and a Low Buried Beach (~9.3–10.2 ka) at lower altitudes. The lower 'beaches' are more widespread, each recording this fall in relative sea level.

After relative sea level had fallen from the Low Buried Beach, the second episode of Smith *et al.* (2012), HRSL 2, began with a marked rise, the Main Postglacial Transgression. This commenced around 9.6 ± 1.5 ka (Sissons, 1974), and culminated at the Main Postglacial Shoreline, formed at ~6.1–7.6 ka, reaching a height of 15 m (above Ordnance Datum).

A major tsunami, the Storegga Tsunami, occurred during this transgression (Dawson *et al.*, 1988; Smith *et al.*, 2004). Marked by a distinctive layer of sand in the sediments then accumulating, it has been dated at ~8.1 ka (Dawson *et al.*, 2011). HRSL 2 was followed by two further episodes, HRSL 3 and HRSL 4. In the first, around 5 ka, the Blairdrummond Shoreline was formed, and in the second the later Wigtown Shoreline was formed. The sequence of shorelines in episodes HRSL 2 to HRSL 4 can be followed into northern England and Ireland, where the more steeply sloping earlier shorelines are overlapped by more gently sloping later shorelines, reflecting declining glacio-isostatic uplift but continuing sea-level rise. Isobase models of shoreline altitudes in each episode indicate that the pattern and location of the centre of uplift did not change during the Holocene, with maximum uplift in the south-east Grampians (Smith *et al.*, 2012; Figure 7.14). While there is some evidence for dislocation of some shorelines along faults (e.g. Sissons, 1972; Smith *et al.*, 2010), a broad pattern of uplift is recognisable. Recognition of successive episodes outlined above indicates temporal patterns of change of metre-scale which appear to have been superimposed on the more general glacio-isostatic pattern that can be modelled by GIA.

7.7.2 Southern Britain

By contrast with Scotland, northern England, and Northern Ireland, there is ample evidence from southern England for a gradual rise to present, but with some evidence for a number of apparent fluctuations. The broad estuarine plains of southern England contain a stratigraphy that comprises peats interbedded with clays. The significance of interbedded mud and peat, in the Fens in eastern England, was recognised by Godwin and Clifford (1938). Later radiocarbon dating reinforced Godwin's view that pollen analysis indicated a sequence of sea-level oscillations (Godwin *et al.*, 1958). Reconstructions of Holocene sea-level rise were compiled for the Thames estuary by Devoy (1979), who recognised five transgressive sequences based on pollen and diatom analyses. Many focused dating studies have been undertaken; for example, in the Fenland region (Shennan, 1986a, 1986b, 1989), and in Morecambe Bay (Zong and Tooley, 1996). Following the meticulous work of Tooley (1978) in north-western England, considerably refined approaches were developed to delineate sea-level index points and discriminate sea-level tendency (Shennan *et al.*, 1983), which enabled a data set of more than 1,200 radiocarbon ages to be analysed systematically to provide validation of GIA modelling across Britain (Shennan and Horton, 2002a; Gehrels, 2010b).

A similar flexural history, although of lower amplitude, seems to have occurred in Ireland (Carter, 1982; Carter *et al.*, 1987). Recent development of a sea-level database for Ireland and new glaciological data on spatial extent, thickness, and deglacial chronology of the Irish Icesheet has enabled extension of GIA modelling to cover Ireland (Brooks *et al.*, 2008). Although the dominant signal is one of glacio-isostatic adjustment in response to glacial unloading in northern Britain, several processes acting in concert are responsible for contrasting relative sea-level histories including variable response to meltwater loading (hydro-isostasy) of the surrounding seas and Atlantic Ocean, and to a lesser extent, effects of unloading of the Fennoscandian and North American Icesheets (Lambeck, 1995).

Although empirical evidence and theoretical model predictions for relative sea-level change around the British Isles are generally in accord (Lambeck, 1995), there continue to be further refinements, both to modelling and to observational data sets (Bradley *et al.*, 2011). Sites in Scotland show a consistent, overall, relative fall in sea level following the LGM, in response to regional isostatic uplift, resulting from the progressive decay of the British–Irish Icesheet (Shennan, 1999; Shennan *et al.*, 2000b; Shennan and Horton, 2002b). According to Lambeck (1995), between 20 and 14 ka, modelling reveals that postglacial rebound largely kept pace with sea-level rise, resulting in a near-stationary shoreline for the North Sea coast in much of northern Britain. In contrast, progressive submergence of southern Britain was commencing at this time, with gradual submergence of the English Channel. Between 15 and 11 ka, an interval encompassing the Older and Younger Dryas, the shorelines of the North Sea coast remained relatively unchanged due to apparent dynamic equilibrium between postglacial sea-level rise and glacio-isostatic rebound. This interval witnessed progressive development of the North Sea and by 11 ka a large, shallow gulf had formed between Britain and France. Modelling of postglacial relative sea-level curves for southern Britain indicates that sea level has never been higher than present, although some researchers have inferred a higher sea level, for example from the dating of chenier ridges in Essex (Greensmith and Tucker, 1973). The most rapid subsidence is occurring in southwest England, where data have generally been scarce; coring at Slapton Sands and other south Devon sites indicated that sea level was around 21 m lower relative to present at these sites at 9 ka and around 8 m lower at 7 ka (Massey *et al.*, 2008), which is broadly in agreement with earlier studies by Heyworth and Kidson (1982).

Studies of relative sea-level history have a particular relevance in the context of contemporary and future sea-level rise. The pattern of flexure of Britain has been recognised since it was discerned from an analysis of tide gauge data by Valentin (1953). There is considerable accord between patterns of sea-level change shown by Holocene reconstructions and sea-level trends noted from tide gauges (Shennan and Woodworth, 1992). This continues to be researched (Woodworth *et al.*, 2009a), and the latest interpretations of crustal flexure across Britain (Shennan *et al.*, 2009, 2012; Gehrels, 2010b) can be compared with observations from GPS (Bradley *et al.*, 2009). The significance of considering present trends in the context of longer-term sea-level trajectories is considered in more detail in Chapter 8.

7.8 Europe

There has been a long tradition of sea-level study in the Netherlands (van Straaten, 1954; Jelgersma, 1961; van de Plassche, 1982). The Holocene trend has been one of gradual rise at a decelerating rate, although the details have been vigorously debated (van de Plassche and Roep, 1989). A similar pattern has occurred along the north coast of France, but spatial variability of up to 20 m is anticipated along the Atlantic coast of France because of the effects of glacio-hydro-isostasy (Lambeck, 1997). Radiocarbon ages on Holocene peats are broadly consistent with the modelling (Delibrias and Guillier, 1971).

Although the Mediterranean Sea played an important role in the initial identification of past highstands of the sea, its relative sea-level history remains a subject of debate and contention. Most of the Mediterranean is beyond the direct influence of former icesheets so the sea-level signal that would be expected for the past 7,000 years is one of gradual rise. It is characterised by low tidal amplitudes and restricted wave activity meaning that evidence of past sea levels has a higher probability of being preserved (Pirazzoli, 1996). Numerous studies using geological, geomorphological, biological, and archaeological evidence have provided insights into former sea levels. However, tectonic movements have occurred in many localities, and, being a nearly closed basin, variations in sea level from salinity or thermal adjustments, and wind stress anomalies, may be important on timescales of decades and centuries (Landerer and Volkov, 2013).

Areas that have been tectonically stable, such as Sardinia and northwest Sicily, can be identified from the position of the last interglacial shoreline (Tyrrhenian shoreline), which is clearly recognisable from morphological and lithostratigraphical evidence, and by the associated Senegalese fauna (Issel, 1914; see Chapter 5). At other sites there is evidence for long-term changes in uplift rates with the average rates for the past ~125 ka being up to twice those for the past 10 ka; for example, Calabria and northeast Sicily are uplifting at rates of up to 2.4 mm/year (Antonioli *et al.*, 2006). Sequences of erosional notches, such as those around Corinth, provide evidence of intermittent uplift, and isolated colonies of the coral *Cladocora* and the borings of *Lithophaga* provide further evidence for localised uplift in the Holocene (Pirazzoli *et al.*, 1994).

The complexity of land motion is illustrated by the well-known example of Pozzuoli, in the Phlaegrean Fields volcanic complex in Italy, where the Roman columns, illustrated by Lyell in his *Principles of Geology*, have marine borings that occur at elevations up to 7 m above sea level (Morhange *et al.*, 1999; see Figure 3.13a). Recent radiocarbon dating of organisms (*Lithophaga*, *Chama*, *Ostrea*, *Astroides*, and the vermetid, *Vermetus*) from the columns indicates that sea level reached 7 m above present at three periods: during the fourth century AD (mean age $2,230 \pm 25$ years BP, shortly following the most recent mention of the market in 384); the early Middle Ages (mean age $1,860 \pm 25$ years BP); and before the 1538 eruption of Monte Nuovo (mean age $1,205 \pm 40$ years BP: Morhange *et al.*, 2006). Archaeological evidence indicates that significant changes have occurred across the region – for example, the buried port of Troy (Kraft *et al.*, 2003), or at the ancient harbour of Marseilles (Morhange *et al.*, 2001). Subtle change, such as the vertical displacement of Roman fish tanks, whose relationship to the tidal range is well established, appears to be related to ongoing isostatic adjustment, which is spatially variable because of the variability in the coastline geometry and distance from former icesheets (Lambeck *et al.*, 2004). The evidence from Greece, Italy, Israel, or Mediterranean France has been interpreted to show an ongoing increase in ocean volume until about 3,000 years ago, since which it has been near-constant (Lambeck and Purcell, 2005).

Northern Europe was glaciated with the centre of ice accumulation over central Sweden. Rapid deglaciation resumed after the Younger Dryas, with total disappearance of ice by around 7.5 ka. Four main stages have been recognised in the Baltic region, related to retreat of the ice (Eronen, 1983): the Baltic Ice Lake about 10.5 ka, the Yoldia Sea about 10 ka, the Ancylus Lake stage (named after the mollusc *Ancylus fluviatilis*) in the early Holocene when a freshwater lake was dammed in the Baltic region as a result of uplift, and the final Littorina stage which reflects saltwater reconnection through the Kattegat, with the gastropod *Littorina* (broadly corresponding to the Tapes transgression in Norway). A number of transgressions have been described by Mörner (1976b). The pattern of uplift has been examined in detail using a variety of evidence (Mörner, 1979), but particularly through the dating of deposits in isolation basins (Eronen *et al.*, 2001), as discussed in Chapter 5.

7.9 The Americas

The west coast of the Americas, the Andes and Rocky Mountains, have been the focus of active tectonic movements associated with the plate margin that runs along much of those coasts. Vertical displacement associated with this seismic and volcanic activity has been described in Chapter 5. The east coasts of North America and South America are passive margins, also called trailing-edge coasts. Major rivers, draining from the mountains on the west of the continents, bring significant sediment loads which have accumulated along the eastern margin over millions of years. Substantial Pleistocene icesheets developed over Canada, particularly the Cordilleran and Laurentide Icesheets, and there has been emergence as the ice has melted (Newman *et al.*, 1980); a complex isostatic history is indicated in eastern Canada (Grant, 1970, 1980; Quinlan and Beaumont, 1982). The pattern of Holocene shorelines along the east coast of North America is intricately related to melt and glacial rebound history of these icesheets. Along this coast it has been presumed that the forebulge has migrated as icesheets melted (in contrast to Great Britain, where forebulge migration has not been detected), an idea proposed by Daly (1934).

A radiocarbon-based sea-level history was derived for the past 36,000 years from the Atlantic coast of the United States, initially using dredged and cored samples from the seabed over a widespread area (Milliman and Emery, 1968). The older ages in this compilation can be largely discounted because they are likely to be beyond range of radiocarbon dating and are not consistent with later results for MIS 2. The postglacial, however, was characterised by a smoothly rising trend (Bloom and Stuiver, 1963). Even when subsidence-corrected (Dillon and Oldale, 1978), these still produced a scattered pattern that relates poorly to later studies because of unselective sample choice.

A pattern of gradual submergence has been interpreted for much of the eastern US coast since the pioneering work on salt-marsh stratigraphy by Redfield (Redfield and Rubin, 1962; Redfield, 1967, 1972). Interpretations have varied from emergence in Maine (Bloom, 1963), little rise in the past 3 ka (Kaye and Baarghoorn, 1964), to gradual rise, for example in New Jersey (Rampino and Sanders, 1980; Miller *et al.*, 2009), Virginia, and other states (Belknap and Kraft, 1977; van de Plassche, 1990), with other interpretations inferring an oscillating sea level (Moslow and Colquhoun, 1981).

As in Britain, more pronounced emergence occurs to the north, with continuing uplift in Arctic Canada (Quinlan, 1985). Newfoundland records a pattern of uplift at its northern extremity, but crustal subsidence to its south (Daly *et al.*, 2007). Ice-load induced subsidence of 3–4 mm/year is anticipated in New Brunswick with slower rates for the Atlantic coast of Nova Scotia (Gehrels *et al.*, 2004). In Maine, highstand shorelines can be seen emergent at more than 60 m elevation, dating older than 14 ka. A lowstand is recorded in the Gulf of Maine at about 60 m below present ~12 ka (indicating uplift has exceeded sea-level rise by ~120 m over this time). This has been followed by sea-level rise, rapidly at first, then a 'slowstand' between ~11 and 8 ka, followed by decelerating rise to present (Kelley *et al.*, 2010, 2013).

The influence of glacio-isostasy along this coast appears amenable to geophysical modelling (Nakada and Lambeck, 1987; Peltier, 1996). A database of nearly 500 sea-level index points, identifying 16 separate zones along the coast for which there are data (a subset of which is shown in Figure 7.15), has been compiled for the region and used to examine the effectiveness of modelling (Engelhart *et al.*, 2011a; Engelhart and Horton, 2012). Despite improvement in the ability of GIA models to predict spatial and temporal patterns of relative sea-level changes, there remain significant discrepancies between model predictions and observational data. Recent analyses using ICE-5G and a viscosity model outlined by Peltier and Drummond (2008) appear to overestimate the amount of GIA-related subsidence. Although in good agreement with the gradually rising marsh-derived sea-level histories of the northeastern Atlantic coast, model output deviates and further south is as much as 10 m below the dated index points at 6 ka (Engelhart *et al.*, 2011a, 2011b).

The sea-level history of the Gulf Coast and southern Florida has been viewed as comprising a decelerating rise over past millennia, consistent with an intermediate-field setting as described in Section 7.6. Similarly in the West Indies, and in those sections of the Central American coast where sea-level studies have been undertaken, such as Belize, there does not appear to have been a Holocene highstand. This interpretation is challenged by recent reports of emergent shoreline evidence, for example from Texas (Blum *et al.*, 2001; see Section 7.4), and ongoing questions as to whether it is appropriate to extrapolate evidence from around the West Indian islands to the mainland.

In contrast, both observational and modelling studies indicate that a highstand was experienced along much of the South American coast (Martin *et al.*, 1986; Milne *et al.*, 2005). Along the east coast of South America there are sequences of sand barriers which indicate that sea level has been close to its present level for several thousand years (Bigarella, 1965). Present sea level was achieved much earlier than in the Caribbean (Cohen *et al.*, 2005). Many studies have inferred several oscillations of sea level (Martin and Suguio, 1992; Martin *et al.*, 1998), but some have been disputed (Lessa and Angulo, 1998). In reviewing reconstructions of Holocene sea level for Brazil, Angulo *et al.* (2006) recognise that the indicators used are problematic, and many have been misinterpreted. Modelled results by Milne *et al.* (2005) suggest that the highstand

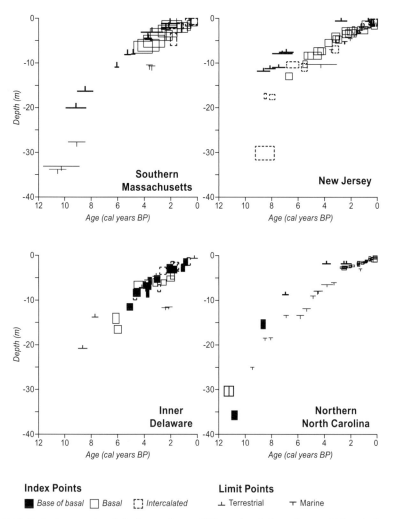

Figure 7.15 Relative sea-level index points, differentiated into base of basal, basal, and intercalated, and limiting ages for selected sites in the eastern USA: (**a**) southern Massachusetts, (**b**) New Jersey, (**c**) inner Delaware, and (**d**) northern North Carolina (after Engelhart *et al.*, 2011a).

occurred around 7 ka, reaching about 4 m at Rio de Janeiro in Brazil, but closer to 2.5 m in Santa Catarina, after which sea level declined. A Holocene highstand is inferred for most of the east coast, as it is for much of the southern hemisphere (Isla, 1989).

7.10 The past two millennia

Studies of sea level have taken on a new significance in the context of the widely held view that human activities have warmed the planet beyond the Croll–Milankovitch cycles that have dominated the Quaternary. There is convincing evidence, from tide gauges, supplemented in the past two decades by satellite altimetry, that the sea is rising, and much concern that such rise may be accelerating. These issues are examined in greater detail in Chapter 8. However, it is appropriate to conclude this summary of the past 20,000 years by asking to what extent the techniques that have been described in earlier chapters and applied by the studies described in this chapter can detect a rise over the past few centuries.

Most of the major northern hemisphere icesheets had disappeared by around 7 ka. It is less clear whether Antarctica was still a source of meltwater to the oceans. As background to its assessment of anthropogenic impacts on climate, and observed rise in sea level in historic times, the IPCC in its Fourth Assessment report, assumed that global sea-level change over the past 2,000 years had been negligible (Bindoff *et al.*, 2007). It therefore becomes pertinent to ask when the sea commenced rising and whether it has accelerated, particularly as some researchers consider that human activities might have been having a detectable effect on climate since widespread clearing of forests began in prehistoric times (Ruddiman, 2003). Relatively few lines of sea-level evidence, or broader palaeo-environmental reconstructions during the Holocene, have been adequately linked with historical and observational records to bridge the palaeo/modern gap. Acceleration in rate of sea-level rise has been proposed starting in the early 1800s on the basis of tidal observations from Amsterdam, one of the oldest tide records available. However, other lines of evidence imply that acceleration began a little later (as discussed in Chapter 8).

Four types of evidence that could be used to reconcile palaeo- and historical evidence are identified by Gehrels *et al.* (2011): salt-marsh sediments, coral micro-atolls, vermetid mollusc bioconstructions, and archaeological indicators. Of these, salt-marsh studies have so far been the most successful; they appear able to offer some indication of sea-level trends that can be linked, to some extent, with overlapping tide-gauge observations, and enable extrapolation beyond the observational record. Proxy records from salt marshes in eastern North America (Donnelly *et al.*, 2004; Gehrels *et al.*, 2004, 2005; Kemp *et al.*, 2009, 2011, 2012), Spain (Leorri *et al.*, 2008), and New Zealand and Tasmania (Gehrels *et al.*, 2012) suggest the timing of the change in trend occurred between 1880 and 1920. A particularly comprehensive review of sea-level reconstructions with correction for GIA has been undertaken by Kemp *et al.* (2011), who identify four phases of persistent sea-level trend from foraminiferal studies using a transfer function from sites in North Carolina. Here, sea level appears to have been stable until 950 AD, followed by a rise at 0.6 mm/year for 400 years, followed by another phase of relative stability, until the late nineteenth century, since when it has been rising at 2.1 mm/year (Figure 7.16). These authors use a semi-empirical modelling approach to indicate consistency with global temperature, and compare their reconstruction with other sea-level data for this period from around the world.

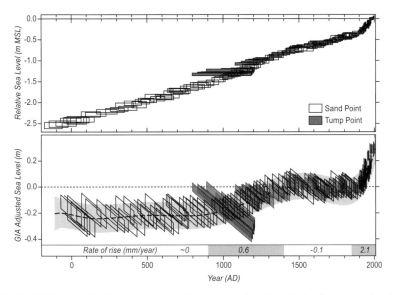

Figure 7.16 Relative sea-level reconstruction from microfossil analysis of two marshes in North Carolina with boxes representing sample-specific age and elevational or indicative meaning uncertainties (2σ). The same data adjusted for GIA rates inferred for these sites from modelling, from which apparent rates of sea-level rise are expressed relative to the average for the period 1400–1800 AD, which was stable or recorded a slight fall (after Kemp *et al.*, 2011).

Coral microatolls appear to offer opportunities to link fossil and historical trends. Although microatolls have been dated that lived in the past 2 ka (Goodwin and Harvey, 2008; Woodroffe *et al.*, 2012b), age and elevation are still too scattered to reliably conclude a trend. The upper surface of two living *Porites* microatolls was examined from the Cocos (Keeling) Islands, and a consistent pattern between the two sites was demonstrated with low-amplitude (~30 mm) fluctuations over the twentieth century (Smithers and Woodroffe, 2001). Palaeoenvironmental records from modern corals from the central Pacific have been linked, by wiggle-matching, with similar isotopic records from fossil corals, to develop a record of sea-surface temperatures going back several centuries (Cobb *et al.*, 2003), but such modern–fossil links have yet to be achieved for sea level using microatolls.

Archaeological evidence includes Roman fish tanks, precisely related to sea level because their inflow was tidally controlled by sluice gates (see Figure 3.13b), which indicate that sea level at 2 ka was similar to that at the start of the nineteenth century; a pattern that is also supported from constructions by *Dendropoma petraeum*, from sites on the coast of Israel and elsewhere (Lambeck *et al.*, 2004). Such data, coming from the Mediterranean Sea where tectonism has been significant, provide a rather incomplete picture of sea-level trends over the past two millennia, and it is clear that there is much scope to improve correlation between palaeo-sea-level and historical sea-level records (Gehrels and Woodworth, 2012), as examined in Chapter 8.

7.11 Unresolved issues in the postglacial record of sea level

Despite key questions that remain in linking palaeodata with historical data, our understanding of sea-level variations throughout the Holocene has increased greatly since the global compilation attempted by Fairbridge. It has ceased to be the goal of researchers to reconstruct a global 'eustatic' curve from local data. It is clear that glacio-isostatic adjustments have had wide impacts, and that hydro-isostasy has influenced many coasts, even in the far-field. Sites remote from former icesheets may preserve evidence from which a relative sea-level history can be deciphered that does not differ greatly from the 'ice-equivalent' pattern of change, but ocean siphoning implies that many have experienced a slight fall of sea level over the past few millennia, although ocean volume has not decreased. Below, some of the key unresolved issues that require further research are considered.

7.11.1 The elusive eustatic sea-level record

The quest for a global postglacial sea-level curve, as attempted by Fairbridge in his 1961 compilation, has been abandoned. Geographical variability in relative sea-level change during the Holocene has become clearly evident. Since Bloom initiated IGCP project 61, and began to compile sea-level evidence from around the world into an atlas (Bloom, 1977), and during the ensuing IGCP and INQUA projects, an enormous database of sea-level index points reinforces differences between locations. Research has focused on determining regional and local departures from general trends, rather than global patterns of change.

Recognition of 'zones' described by Clark *et al.* (1978), and Clark and Lingle (1979), as shown in Figure 2.11, appeared to offer the prospect of regional eustasy (e.g. Mörner, 1980), but deformation processes are such that almost all places have undergone some vertical displacement. Milne and Mitrovica (2008) indicated that many of the sites from which key sea-level histories have been derived, such as the Huon Peninsula and Barbados, have not been immune to isostatic adjustment which could be 15–20 m for LGM shorelines and 4–8 m for mid-Holocene shorelines. They used modelling to establish which sites might offer sea-level researchers the optimum chance to minimise such displacement, as did Lambeck *et al.* (2010) and Spada and Galassi (2012), indicating that sites such as the Seychelles may be suitable, although differential movement has been inferred, even for some of these islands (Pirazzoli *et al.*, 1990).

Paradoxically, as the palaeo-sea-level community has come to realise that 'eustasy is therefore merely a concept not a measureable quantity' (Gehrels, 2010a, p. 26), defining ice-equivalent ocean volume has become increasingly important for the geophysical modelling community, and an essential goal for climate science in the context of global warming. Details of late Holocene ice-melt history remain unresolved which is a complication in relation to GIA modelling. At present, GIA modelling is generally based on a small subset of sea-level observations; fits between model output and observations from one region can produce misfits elsewhere (Gehrels, 2010a). Solutions are not unique; either ice-equivalent sea-level input or viscosity parameters can be changed and model output will change accordingly. There remains contention as to the time at which ice melt ceased, and the pattern of Holocene relative sea level at any site is very sensitive to this input (Lambeck, 2002). Peltier (2000, 2002a) maintained that ongoing ice melt would eliminate observed mid-Holocene highstands on islands in the equatorial Pacific Ocean, and accordingly considered that there had not been any ice melt after 4 ka. An alternative view has been to allow for up to 2–3 m of sea-level contribution from Antarctica from 7 to 2 ka (Nakada and Lambeck, 1988; Fleming *et al.*, 1998). Contributions are likely to have come from West Antarctica, where the icesheet is still adjusting to Holocene temperature and sea level; melt from Marie Byrd Land alone could have contributed 0.2–0.3 m to global sea level since 7 ka (Lambeck *et al.*, 2010).

7.11.2 The question of sea-level oscillations

A feature of Fairbridge's 1961 compilation was that it generated an apparent series of oscillations. Recognition of geographical variability from site to site, particularly in response to glacio-isostasy, has meant that alternate emergences and submergences of the Fairbridge curve, with evidence drawn from the Caribbean and Australia, respectively, have clearly been refuted. However, there persists a view that there have been oscillations in sea level during the Holocene. For example, Schofield (1960) inferred sea-level oscillations from a chenier plain in New Zealand, and then attempted to correlate these oscillations with radiocarbon ages from atolls in the Pacific, extending inferred fluctuations across the Pacific Ocean (Schofield, 1977a). Further investigations at each of Schofield's sites have failed to support this inter- pretation, as discussed above.

Sequences of sandy ridges have also given rise to a number of sea-level curves with oscillations in South America (Martin *et al.*, 1998; Lessa and Angulo, 1998), although more recent studies have reduced the amplitude of inferred fluctuations (Angulo *et al.*, 2006). There has been a similar tendency to infer oscillations in southeast Asia, but Pirazzoli (1991) considered these largely illusory. The debate has been reinvigorated in Australia, with a number of publications using statistical curve-fitting through ages on fixed biological indicators to detect oscillations, which Baker *et al.* (2005) considered might be detected across continents, and could even be global.

Several studies found little support for oscillations of more than 0.5 m (e.g. Baete- man, 2008). Sea level is considered to have risen at varying, although rapid, rates during deglaciation, with MWP-1A widely recognised. There is growing support for a short period of 'instantaneous' sea-level rise of the order of 0.4 m in the far-field at 8.2 ka (Gehrels, 2010a), but by contrast there is less support for major variations in ocean volume in short periods during the mid and late Holocene (Kendall *et al.*, 2008). A reduction in ocean volume of ~0.5 m would require rapid growth of an icesheet of about $2 \times 10^5 \, km^3$ in volume (roughly equivalent to the British Icesheet at the LGM). Such rapid withdrawals of water from the ocean seem unlikely to have occurred.

Sea-level fluctuations during the Medieval Climatic Optimum and the Little Ice Age cannot be ruled out, but steric change is more likely to account for regional sea-level variability than a global phenomenon (Gehrels, 2010a). Although some regions may record effects of localised climatic changes as anomalies in their relative sea-level records (Goodwin, 2003), it remains challenging to distinguish such a signal from noise when looking in sedimentary records for sea-level response to short-term climatic changes. Local anomalies may be associated with steric effects, such as changes in sea-surface temperatures related to ENSO; detection and attribution of these is complex, frustrating their discrimination from otherwise monotonic glacio-hydro-isostatic adjustments.

It is also important to recognise constraints on interpretation of individual sea-level indicators. Although their indicative meaning may be understood, there remain vertical errors in terms of accuracy with which they can be related to any tidal level, as well as age errors associated with their dating. The examples reported in this chapter have been summarised and sea-level data and curves drawn as reported in the past, predominantly without sufficient attention to indicative meaning. Disparate field data rarely produce smooth trends resembling the discrete time steps of GIA model outputs. Where data are discontinuous (e.g. Figure 7.8), they can be used to support numerous different sea-level curves; indeed, even near-continuous sequences of sea-level indicators also permit conflicting interpretation (e.g. Figure 7.4). In few cases has the careful consideration of indicative meaning of each sample been explicit; such detail will be necessary in order to resolve sea-level changes at sub-metre, and ultimately centimetre scales.

Chappell and Thom (1986) challenge the view that it is necessary to invoke a change in sea level, as a boundary condition, to explain a difference in height between palaeo and modern sea-level indicators. Tidal planes are not horizontal across coastal landscapes; internal morphodynamics result in significant distortions. For example, the tidal amplitude can vary substantially from the open ocean to the tidal limit within an estuary, and tidal prism may be constrained within a lagoon or other embayment, particularly in macrotidal settings. In the macrotidal Bay of Fundy, apparent changes in sea level have been inferred that actually resulted from artificial reclamation of marshes (Graf and Chmura, 2010).

Detailed surveys of absolute elevation of upper surfaces of microatolls on Christmas Island in the central Pacific have indicated variation around the atoll because of a gradient in the geoid, together with attenuation of tidal amplitude in the former lagoon (Woodroffe *et al.*, 2012b). Knowledge of past environmental setting within which fossil sea-level indicators formed is therefore critical to interpretation of their indicative meaning for past sea level. Rarely does evidence enable unambiguous determinations of the open-ocean surface, and even then several explanations could account for discrepancies. Where oscillations of sea level appear to have occurred, it will be necessary to demonstrate that they are not attributable to observational uncertainties of the particular sea-level indicator. Until concurrent fluctuations can be demonstrated from geographically separated sites, it seems more likely that such adjustments are local phenomena rather than global trends. On this basis it will require much more meticulous reconstruction, preferably corroborated using several sea-level indicators and from adjacent sites, to constrain the magnitude of any fluctuations in sea level during the past few millennia.

7.12 Synthesis and conclusions

Quantifying the magnitude of eustatic (ice-equivalent) sea-level rise from the end of the LGM to early Holocene is complicated by variable glacio-hydro-isostatic responses of the crust–mantle system. Estimates from continental shelves in the far-field of Pleistocene icesheets suggest that sea level rose from approximately 130 to 120 m BPSL ~19 ka and attained present levels by ~7 ka. Far-field locations such as Australia record a stand of relative sea level higher than present as early as 7 ka ago with a progressively falling sea surface since that time due to hydro-isostasy. Elevation and timing of Holocene highstands varies spatially and temporally, not only around Australia but also across the Pacific Ocean. Variations in the Indian Ocean appear slightly more systematic, with a mid-Holocene highstand in the eastern Indian Ocean but a pattern of gradual rise up to present in the western Indian Ocean, where a level close to present was attained more recently. In the Caribbean, in closer proximity to former icesheets, a Holocene highstand is not seen. Instead, there has been gradual rise up to present, but mangrove and coral records may record subtle differences across the region from Bermuda to Barbados or from the eastern Caribbean to Belize. In those areas that were beneath major icesheets, rebound has resulted in shorelines that have been uplifted sometimes to several hundred metres, and isostatic history overwhelms detection of changes in eustatic sea level. There is a much improved understanding of the variability of Holocene shorelines and relative changes in sea level that can be inferred from them, particularly through application of geophysical modelling in conjunction with field data. The record of glacio-isostatic adjustment is particularly well documented from the British Isles, comprising domed uplift following disappearance of icesheets over Scotland where several emergent shorelines have been well mapped, resulting in an apparent relative fall in sea level. In southern England, present sea level was only attained within the past 2 ka. Despite many more ages and substantial data sets, there remain discrepancies between observations and model outcomes around the margin of North America. Details of the ice-equivalent 'eustatic' sea level, which is an important input into modelling, remain in contention, with differing views on timing, source, and amount of post-7 ka melting. Although isostatic adjustments explain formation of well-defined early Holocene highstands in far-field locations, there has continued to be debate about whether there have been discernible oscillations of sea level. Further resolving these issues is important because it underpins a clearer understanding of the likely regional and local variation in future changes of sea level, as discussed in the following chapter.

8

Current and future sea-level changes

... through geological eyes we might regard the present M.S.L.
[mean sea level] as ephemeral as a fleeting ray of sunshine on a
wintery afternoon.

(Fairbridge, *1961, p. 99*).

8.1 Introduction

The pattern of climate variation, and hence sea-level change, over the Quaternary has been characterised by large fluctuations, driven primarily by variations in the Earth's orbit, as emphasised in the preceding chapters. The past few millennia of mid and late Holocene, in contrast, have experienced relatively little sea-level change, a period of atypical stability. The trajectory of sea-level change has assumed a new relevance in recent decades with the perceived association between human-induced climate change and future sea-level rise. This chapter examines the issues associated with current sea-level trends and the likely pattern of future sea-level changes.

The cyclical changes in orbital parameters that are considered to have driven past climate, contributing to the Ice Ages and forcing significant fluctuations of sea level, appear to exert the principal constraint on future sea level over the long term. Relative sea-level stability during the present interglacial has enabled development of broad coastal plains (Hansen and Sato, 2012). The extensive delta plains of the Near East (Nile and Mesopotamia), for example, developed during this period, which provided conditions suitable for settlement of urban populations and the rise of civilisations (Stanley and Warne, 1993, 1994). In relatively stable coastal environments around Africa, Asia, and Australasia, wide coastal plains have developed, many of which have also become the focus of agriculture or industrialisation.

Past interglacials appear to have persisted for around 10,000–12,000 years. It was therefore accepted wisdom until recently that the world would shortly head towards a further glaciation (Kukla *et al.*, 1972). This no longer seems assured. The orbital characteristics which drive glacial–interglacial cycles are not especially favourable for imminent onset of glacial conditions. The eccentricity of the Earth's orbit is

369

particularly low, and will remain close to a minimum. In 25,000 years time the Earth will be circling the Sun in a nearly circular orbit. Low eccentricity serves to dampen the effect of precessional changes; conditions resemble those experienced in MIS 11 at a comparable part of the 412-ka orbital cycle. This implies delayed onset of cold conditions with significant icesheet expansion perhaps as much as 55 ka in the future, assuming that atmospheric carbon dioxide levels remain less than 250 ppm. However, with elevated CO_2 levels (CO_2 levels exceeding 400 ppm have now been observed, considered a consequence of human-induced greenhouse gas emissions, described below) it seems more likely that this period of ice-age conditions may be delayed (Berger and Loutre, 2002). Modelling the likely extent of ice under three sets of CO_2 levels (similar to those of the last interglacial, with elevated human-induced levels of 750 ppmv, and with 'natural' CO_2 levels of 210 ppmv), reveals a less-certain glacial onset (Figure 8.1). The contrast with the last interglacial is striking, with much lower amplitude variations in insolation at the upper atmosphere.

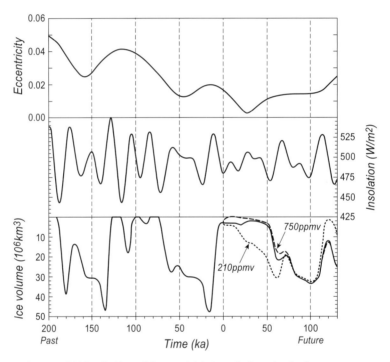

Figure 8.1 The past 200 ka (left) and future (right) variations in the key parameters that will determine sea level. The upper panel shows eccentricity of the Earth's orbit. The middle panel shows the incoming insolation at 65°N. The lower panel shows northern hemisphere ice volume (0 represents high sea level and ice volume increases downwards). The principal line continues CO_2 concentrations as were typical under the last glacial–interglacial into the future, but a constant concentration of 250 ppmv and a human-induced greenhouse scenario of 750 ppmv are also shown (after Berger and Loutre, 2002, reprinted with permission from AAAS).

It therefore seems likely that present interglacial conditions will persist for an uncharacteristically long period, perhaps resembling MIS 11 (Tzedakis *et al.*, 2012).

This argument was extended by Ruddiman (2003), who proposed that the Earth should already have begun to experience noticeable cooling into the next glaciation. Ruddiman inferred that this cooling had already been averted by human activities, as a result of early societies that had cleared forests to grow crops. The effect of industrial societies is now widely seen as contributing to global warming, as discussed below. Ruddiman, however, has proposed that even burning of forests by pre-industrial societies since the Neolithic, as far back as 8,000 years BP, has countered the anticipated cooling. This view is contentious; nevertheless, it highlights the increasingly widely held concerns that human actions are having a detectable effect on climate and are likely to alter climate in the future. Indeed, it has been suggested that humans have had a measurable impact on the environment since the Industrial Revolution, and that this should be recognised by assigning the period we live in to a new geological epoch called the Anthropocene (Zalasiewicz *et al.*, 2010).

Among the human-induced climate-change impacts in the Anthropocene it appears that sea level has also been affected. The extent to which the sea is already rising, whether it has been influenced by human activities, and how sea level will vary in the future, remains incompletely understood. There are major uncertainties about emission (and hence atmospheric greenhouse gas concentration) scenarios, but there is also a lack of detailed understanding of processes by which many other factors contributing to relative sea-level change will evolve in future.

These concerns with present sea-level trends and future sea-level trajectories involve several factors explored in preceding chapters in relation to Quaternary sea-level changes. However, there are also other components operating at shorter timescales or that result from human activity. In order to better understand current and future sea-level changes there are new, sophisticated techniques which provide insights into aspects of contemporary sea-level change, and offer considerable potential to understand changes at a scale that can better inform human endeavours in the coastal zone.

In the following sections, reconstruction of sea-level changes from tide gauges is reviewed. Extension of tide-gauge records back in time using geological proxies is examined; corroboration using sophisticated satellite and other monitoring systems is outlined; and modelling future scenarios is explored. Complications related to land movements are summarised, together with potential impacts of sea-level rise and the implications for adaptation measures that will be needed.

8.2 Historical sea-level change

It has been realised for more than half a century that the sea has been rising (Guttenberg, 1941). In looking for causes of the gradual rise indicated by tide gauges, it was considered likely that this was related to greenhouse gases, and what has been termed the greenhouse effect (Barnett, 1983). Concern about the potential impacts of sea-level rise was triggered by a US Environmental Protection Agency (EPA) study (Hoffman *et al.*, 1983; Barth and Titus, 1984), which has been a focus of concern ever since (Titus, 1986; Gornitz, 1991). The principal source of evidence is from ongoing observations at tide gauges around the world. Many authors have focused on trying to collate tide-gauge records to derive a global average of the rate of rise (Figure 8.2). Following a series of regional syntheses (Emery and Aubrey, 1985; Aubrey and Emery, 1986; Aubrey *et al.*, 1988), Emery and Aubrey (1991) produced an overview in *Sea levels, land levels and tide gauges*, in which they emphasised the tectonic variability of setting of the various tide-recording stations. Subsequently, other studies have produced summaries of the global mean sea-level signature. Several of these corrected for GIA using the ICE 3G model, deriving

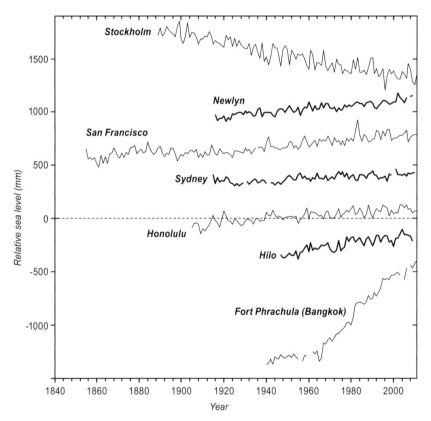

Figure 8.2 Records from selected tide gauges showing the variability of trend at different locations, as discussed in text (data downloaded from the Permanent Service for Mean Sea Level (PSMSL) 2013, 'Tide Gauge Data', retrieved 26 January 2013 from http://www.psmsl.org/data/obtaining/).

values for the global mean sea-level rise that range from 1.7 to 2.4 mm/year (Peltier and Tushingham, 1989; Douglas, 1991).

Extracting a global signature of what is considered to be 'eustatic' rise in sea level from individual tide-gauge records is complex. The tide gauges provide the most robust set of historical data about how sea level has changed, but continuous records of water level dating back more than a century are rare. A tide gauge has been operating at Brest in France since 1807 and at Swinoujscie in Poland since 1811. Further careful extraction of historical information, termed data archaeology, can recover even older records of high water level, for example back as far as 1679 for Brest, or back to 1774 for Stockholm in Sweden. Reliable tidal records have been collated by the Permanent Service for Mean Sea Level (PSMSL), which is one of the oldest scientific services, having been responsible for collection, publication, analysis, and interpretation of sea-level data from the global network of gauges since 1933 (Woodworth and Player, 2003).

However, tide gauges were not initially installed to detect long-term sea-level changes; they were generally intended as a guide for navigation, for example, to determine access to harbours dependent on stage of the tide. Methods of recording tidal data, continuity of record, and representativeness of water level actually recorded (whether an open ocean level, or that within a sheltered harbour) all compound analysis of tide records (Pugh, 2004). Records from each tide gauge need to be expressed in relation to a datum, generally a tidal zero. The lowest tide recorded (lowest astronomical tide, LAT) is commonly related to a nautical datum (chart datum) which forms the basis for navigational charts.

Observations of sea level usually use mean sea level (MSL), which is derived from analysis of tidal records themselves (preferably over at least one tidal nodal period of 18.6 years; Woodworth, 2012). There are many issues associated with changes to tide gauges and related datums, and these will become still more complex as the record from gauges needs to be assimilated into more global sets of observations acquired by satellite (or airborne) technologies and referenced to different datums. Tide-gauge records should be related to one or more permanent survey benchmarks in order that any gradual vertical movement can be detected (such as sinking of the wharf on which such gauges are commonly installed). GPS monitoring can provide important information on vertical movement of tide stations (Wöppelmann et al., 2009; Blewitt et al., 2010; Santamaría-Gómez et al., 2012). However, even continuous GPS will not capture all aspects of vertical land movement (VLM) because a component may be related to gravitational change. Another approach will make use of DORIS (Doppler Orbitography and Radio-positioning Integrated Satellite). Satellites orbit around the centre of mass of the Earth, meaning that observations are now acquired relative to a geoid (generally WGS84). It is complex to bring past and present monitoring into one frame of reference, ultimately all observations need to be referred to the International Terrestrial Reference Frame (ITRF).

Tide gauges do not provide ideal records of changes of the sea as a whole. Most of the sea surface is not monitored because tide gauges are located on the coast, most on continental shorelines, with relatively fewer on islands. The geographical bias apparent in the tide-gauge data set, with the majority of long records located in Europe or North America, has been repeatedly recognised as a constraint on how they can be used (Pirazzoli, 1993; Gröger and Plag, 1993). The southern hemisphere lacks long records; Fremantle commenced in 1897 and Sydney in 1886. There are only a few incidental additional sources of information: for example, a mark made at Port Arthur in Tasmania in 1841, which provides an indication that the trend at Sydney may also have been experienced in Tasmania (Hunter *et al.*, 2003).

Tide records show significant differences between sites (Figure 8.2). To some extent this can be understood in terms of broad geophysical factors. For example, Scandinavia is continuing to experience postglacial uplift, which explains why gauges in Stockholm have been recording a relative fall of sea level. Similarly, gradual subsidence of Hilo on the Big Island of Hawaii accounts for the relatively rapid rate of rise there, which appears faster than at Honolulu. The pattern of rise at Sydney may reflect a change in global MSL, but the record from San Francisco is likely to contain a tectonic component. On the other hand, the rapid rate of relative sea-level rise recorded around Bangkok (Fort Phrachula is on the Gulf of Thailand at the mouth of the Chao Phraya River) has been augmented by subsidence as a result of groundwater extraction. The issue of extracting the vertical component of land movement for locations is a major constraint; GPS measurements at or near tide stations are producing conflicting trends that indicate vertical movements other than GIA (King *et al.*, 2012a; Houston and Dean, 2012; Dean and Houston, 2013).

In deriving a global trend, the fundamental approach has involved selecting reliable tide gauges and determining trends from individual records, and then compiling these into a global data set. This was the approach adopted by Douglas (1991), and Peltier and Tushingham (1991). An alternative approach involves reconstruction techniques, such as those employed by Church *et al.* (2004), who determined the pattern of ocean variability derived from altimeter data available since 1993, and then used this spatial averaging to re-examine longer tide-gauge trends. There appears broad agreement that sea level has risen about 18 cm over the twentieth century. Church *et al.* (2004) indicated average sea-level rise of 1.8 mm/year for the past 50 years, but a similar rate was also suggested for an analysis over the full length of the twentieth century (Church and White, 2006). There has been relatively little variation in rates of sea-level rise determined from tide gauges although the methods adopted have changed (Spada and Galassi, 2012; Figure 8.3).

An overall pattern of rise is apparent in Figure 8.4, but it is surprising that the various analyses differ in detail although based on the same set of underlying data. Despite the apparent acceleration from tide gauge to altimeter (discussed in a later section), there has been little agreement as to whether acceleration can be detected.

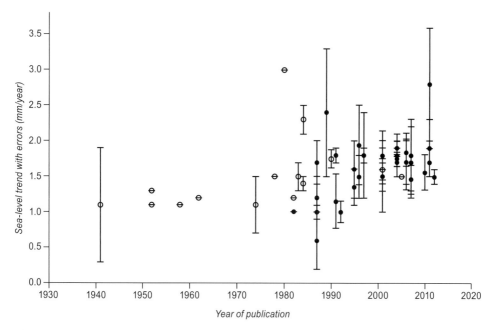

Figure 8.3 Estimates of rates of global MSL rise from tide gauge data over past decades, indicating those that included a correction for GIA (closed symbols) and those that did not (open symbols), plus uncertainties (from Spada and Galassi, 2012).

At first acceleration was not considered detectable (Douglas, 1992). Subsequent analyses observed acceleration in the early 1900s (Church and White, 2006), and also what has been interpreted as human-induced acceleration since 1850 (Jevrejeva *et al.*, 2008, 2009). In examining records from 57 US gauges over the past 82 years, a recent analysis found no acceleration (Houston and Dean, 2011). Similarly, the longer-record tide gauges from Australia and New Zealand do not show acceleration over the past century, although there may be a slight increase in the rate of rise in the past decade (Watson, 2011). However, detecting a trend is sensitive to the starting and finishing points selected within a time series (e.g. apparent deceleration of sea level for US gauges described above is attributed by Rahmstorf and Vermeer (2011) to choosing 1930 as a starting point).

There is a growing consensus about global warming. A consequence of rising temperature is global sea-level rise, resulting both from expansion of seawater in response to warming, and accelerated melting and disintegration of land-based ice. The threat of accelerated sea-level rise, and the likely future consequences for the many people who live, work, and relax in the coastal zone, provides particularly compelling arguments for urgent action to attempt to mitigate climate change. Ocean warming will continue for decades to centuries beyond the time during which efforts are made to reduce emissions, because of the slowness of ocean turnover, meaning

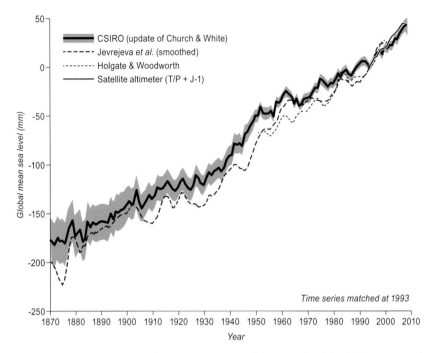

Figure 8.4 Global mean sea level from 1870 to 2008 with 1 standard deviation error estimates (data from Church and White, 2006, with updates from more recent papers by Church and co-workers), compared with a low-pass filtered record by Jevrejeva *et al.* (2006) and a shorter period of record of similar data by Holgate and Woodworth (2004). Data matched at 1993, after which satellite altimetry data has also been acquired (from Church *et al.*, 2010b, with permission from Wiley).

that even with the strongest mitigation measures, sea-level rise will continue beyond 2100 (sometimes called a commitment to sea-level rise), pointing to the unavoidable need for adaptation in coastal communities.

Globally averaged sea level is an integrated outcome of the Earth's heat budget and hence observations of contemporary sea-level change provide one of the more convincing lines of evidence for global warming, and can also serve as a constraint on climate models (Mitchum *et al.*, 2010). The threat posed by sea-level rise has been identified as one of the most inimical impacts of climate change by successive assessments of the IPCC. Low-lying coastal areas, particularly delta regions and small reef islands on coral atolls, were considered particularly vulnerable in the Third Assessment Report (TAR). The Fourth Assessment Report (AR4) broadened the range of climate-related impacts recognised as having the potential to affect coastal areas, acknowledging that less attention had been given to factors other than sea-level rise and that these will also cause significant impacts (Nicholls *et al.*, 2007).

8.3 Linking geological proxies with historical observations

The culmination of ice melt from the last glaciation is generally considered to have been around 6,000 years ago, after which major northern hemisphere icesheets had largely disappeared (only Greenland remaining). Debate continues about the extent to which there has continued to be ice melt from Antarctica (King *et al.*, 2012b). Ongoing adjustment in response to this ice melt has involved continual gradual isostatic subsidence of the ocean floor (at a rate of around 0.3 mm/year according to Peltier, 2004), as discussed in previous chapters. This, together with rebound in glaciated and forebulge regions, has resulted in an apparent relative sea-level fall over the mid to late Holocene, leaving evidence of emergent shorelines around many parts of the world that are far from former icesheets, a process that is termed ocean siphoning (Mitrovica and Milne, 2002; see Figure 2.12). There is much evidence from around the world, based on historical tide-gauge records and recent satellite altimetry, that the sea is presently rising (Gehrels, 2010a). Extending the historical tide-gauge record back in time and comparing it with the longer-term geologically derived record from proxy data provides insights into whether there has been acceleration of rates of change that could be attributed to human activities.

Recently, a number of researchers have used proxy indicators of sea level from salt marshes to extend sea-level reconstructions over past centuries and compare them with tide-gauge records. This approach has been most widely applied along the east coast of North America (Gehrels *et al.*, 2005; Donnelly *et al.*, 2004; Engelhart *et al.*, 2011a; Kemp *et al.*, 2011, 2012). Sediment core chronology is generally obtained by radiocarbon dating, extended into historical times with shorter-lived isotopes, such as ^{210}Pb, or marker layers, such as ^{137}Cs (see Figure 3.10), with further validation where the trend extrapolated from these salt-marsh proxies can be aligned with concurrent tide-gauge records during the twentieth century. Age uncertainty terms produce proxy records with a precision of ± 10 years in the nineteenth century and ± 5 years in the twentieth century, depending on sedimentation rate and sampling resolution. Vertical precision of individual palaeo-sea-level estimates of ± 5–20 cm are possible depending on site-specific factors such as tidal range. Limitations inherent in the dating methods used render it difficult to detect sub-centennial sea-level oscillations smaller than 10–20 cm in amplitude as cores are typically dated with only one age per decade on average (Gehrels *et al.*, 2005; Woodworth *et al.*, 2009b, 2011).

Figure 8.5 shows a compilation of several records and a comparison with the inferred longer-term late Holocene rate of change (Gehrels and Woodworth, 2013). The record from Nova Scotia (*a* in Figure 8.5) indicates considerable accord between salt-marsh and tide-gauge records during the twentieth century, although only one age and a late eighteenth-century pollen marker are available to constrain longer-term sedimentation (Gehrels *et al.*, 2005). The inferred acceleration is less firmly based than that proposed by Kemp *et al.* (2009, 2011) for North Carolina (*c* in

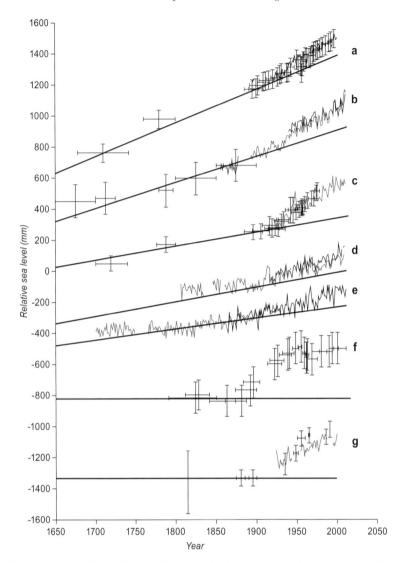

Figure 8.5 Recent sea-level changes from geological proxy data compared with the late Holocene background trend (solid linear trend line). Tidal records are shown by fine lines and salt-marsh data are shown as crosses that indicate age and altitude uncertainties. Sites are (a) Chezzetcook, Nova Scotia (tide-gauge record which underlies data is from Halifax); (b) Barn Island, Connecticut (tide gauge New York); (c) Sand Point (see Figure 7.16), North Carolina (tide gauge Charleston); (d) Brest (western France) and Newlyn (southern Britain) tide records; (e) Amsterdam and Den Helder (Netherlands) tide records; (f) Little Swanport, Tasmania; and (g) Pounawea, New Zealand (tide gauge Lyttelton) (from Gehrels and Woodworth, 2013, with permission from Elsevier).

Figure 8.5; see also Figure 7.16 and discussion in Chapter 7), where there is better age control on the pre-instrumental sedimentary record. Bayesian analysis of the North Carolina record appears to support a temperature-related acceleration of sea-level, although the interpretation relies on the accuracy of the GIA correction (re-considered below in Section 8.5.3) and an adjustment of pre-1000 AD global temperature. The approach has been extended into Iceland (Gehrels *et al.*, 2006), northern Spain (Leorri *et al.*, 2008), New Zealand and Tasmania (Gehrels *et al.*, 2008; Woodworth *et al.*, 2011).

Coral microatolls are another potential source of additional data on sea level, and may be used to track changes in sea level beyond the historical tidal records with a precision of a few centimetres (Gehrels *et al.*, 2011). In the central Pacific Ocean, undulations on the top of microatolls have been shown to be related to the ENSO phenomenon (Woodroffe and McLean, 1990; Spencer *et al.*, 1997). Smithers and Woodroffe (2001) inferred synchronous broad fluctuations of sea level in the eastern Indian Ocean, of about 20 years length of ~5 cm amplitude, using two microatolls from different parts of the reef flat around the Cocos (Keeling) Islands. These microatolls implied an average rate of sea-level rise of only 0.35 mm/year over the past century, considerably lower than the global mean rate for this period. The dead upper surface of microatolls can undergo erosion, in which case the record will deteriorate, but microatolls provide an indication of past water levels where large specimens survive (Lambeck *et al.*, 2010).

8.4 Satellite altimetry and sea level over recent decades

There is a range of technologies that enable a clearer perspective on sea-level changes over recent decades, which complement and extend the tidal records. Satellite altimetry offers a much higher-resolution approach to monitoring sea level than that available from tide gauges. Although early altimeters on satellites such as Seasat, Geosat, and ERS-1 provided some data, high-resolution altimetry commenced with the launch of TOPEX/Poseidon (T/P) with its radar altimeter in 1992 (Figure 8.6). For example, a comparison of Geosat- and Seasat-derived altimetry along broadly similar tracks yielded apparent changes in sea level that were clearly incompatible with what the tide-gauge record indicated, suggesting that calibration against tide gauges needed to be an important component of later altimetry methodologies (Wagner and Cheney, 1992). The T/P satellite had precision dual-frequency altimeters, a microwave radiometer and technologies enabling precise orbit determination. Subsequent altimeter satellites that have extended the sea-level observations include Jason-1 launched in 2001, and Jason-2 launched in 2008, with a further Jason 3 mission planned for launch in 2015. Satellite altimetry represents a significant advance over tide gauges because the altimeter 10-day repeat groundtrack samples the ocean surface between 66°N and 66°S (note some satellites such as ERS-1, ERS-2, and Envisat extend as far as 81°N, but their Sun-synchronous orbits are not as

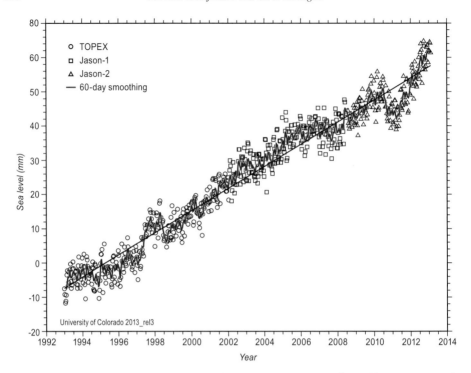

Figure 8.6 Global MSL change determined from satellite altimetry since 1992. Data comprises 2013 release 3, corrected for inverse barometer and GIA (see Nerem *et al.*, 2010), with seasonal signals removed. Linear trend is 3.2 ± 0.4 mm/year (*source*: University of Colorado 2013_rel3 http://sealevel.colorado.edu/).

efficient for precision). The signal that altimetry reveals is dominated by redistribution of ocean volume through dynamic oceanography (steric changes) rather than the net increase in ocean volume (mass).

There are important issues to consider when comparing altimetry with tidal records. For example, correlations must be examined between the different satellites during their overlap and sequencing (discontinuities occurred between major switches; e.g. between side A and side B in T/P, or between T/P and Jason records; Mitchum *et al.*, 2010). Overlap periods of 210 days between T/P and Jason-1, and 180 days between Jason-1 and Jason-2, enabled detection of significant problems, requiring calibration of altimeter data against tide-gauge records (Nerem *et al.*, 2010). The 10-day repeat orbits, once tidal corrections are applied, yield precisions of 4–5 mm. However, the signal also contains other information. There is a prominent seasonal component that appears to comprise a response to rainfall, with amplitude of about 15 mm. Variations in ocean volume, primarily from thermal expansion, also record local heating-related effects of the dynamic ocean surface. Early concerns revolved around whether warm north Atlantic waters were exerting an influence on the global signal (Cabanes *et al.*, 2001), and while this seems unlikely, it is clear that there are other local effects. For example, a major oscillation in 1997–1998 (Figure 8.6) relates to the pronounced El Niño then, probably because of enhancement of ocean volume as a result of increased tropical precipitation (Ngo-Duc *et al.*, 2005). A lower rate of rise in 2010 and 2011 may similarly be related to the prominent La Niña at that time.

The altimeter record since 1993 shows a greater rate of sea-level rise (3.2 ± 0.4 mm/ year in the record shown in Figure 8.6) than was apparent from the tide-gauge record for the decades preceding (Cazenave and Llovel, 2010; Meyssignac and Cazenave, 2012). An allowance is made for GIA subsidence of the ocean floor, but altimeter observations do show local rates of change that can be five times the global mean (Bindoff *et al.*, 2007). It remains questionable whether altimeter data cover a long enough period for decadal trends, such as those seen in tide-gauge data, to be detectable (it only covers one tidal nodal cycle of 18.6 years). It is notable that there has been a more rapid period of sea-level rise also seen in some tidal records for this period (Merrifield *et al.*, 2009), but it is too short to discriminate acceleration. Particularly rapid sea-level rise has been observed in the tropical western Pacific and the Southern Ocean, as well as around Indonesia (Han *et al.*, 2010). The Pacific Ocean shows considerable variation between different islands (Church *et al.*, 2006; Becker *et al.*, 2012).

However, there remain several cautions. First, two completely different techniques are being compared when tide records and altimetry trends are juxtaposed. Not only are the methodologies and their precision quite different, but they are also focused on geographically different parts of the ocean. The tidal records document coastal water levels, whereas the altimetry measures the water surface in the open ocean. Coastal regions are generally masked out of altimeter data because of compounding issues in coastal waters, although ironically it is these areas along the coast for which sea-level rise threatens the most serious consequences. The possibility that coastal waters and ocean waters might behave differently was explored by Holgate and Woodworth (2004) and it now appears unlikely (Prandi *et al.*, 2009); but nevertheless, there are insufficient measurements across coastal waters. Rapid rise in the western Pacific warm pool, in comparison with little change in the eastern Pacific appears likely to be an expression of interannual and decadal variability, such as ENSO or the Pacific Decadal Oscillation (Zhang and Church, 2012).

8.5 Sea-level enigma: volume, mass, and sea-level fingerprinting

Determining the cause of recent sea-level rise has, until recently, been controversial, because the various contributing factors did not appear to add up, giving rise to what Munk (2002) called the 'enigma of 20th century sea-level rise'. In contrast to the long-term record across the Quaternary during which the prime driver of sea level was exchange of water mass between ice and oceans, changes in mass are only one component of contemporary sea-level changes. In the current Holocene interglacial there is significantly less ice at the poles and hence a limited amount that can continue to melt (although the total volume of ice were it to melt would result in a rise of sea level of ~60 m, contributed only 1% from mountain glaciers (0.4 m), 11% Greenland (7 m), and the remainder from Antarctica). In this section, the concept of the enigma referred to by Munk is explored and it is shown that the 'missing' component has now been reduced.

There are two principal factors that contribute to sea-level rise; the first is an increase in mass through the addition of further water to the oceans, primarily from ice melt; the second is an increase in volume which results from what are called *steric* factors primarily reflecting expansion of water as it warms, and its density is reduced. Both of these are more complex than such a simple summary implies; both have been subject to significant revision over the past decade in an effort to resolve the enigma, and both are likely to become even better understood as a suite of new observing methodologies reduce uncertainties.

Complex regional and local effects complicate the determination of relative sea-level change at any one site. Foremost among these are subtle vertical movements. Glacial isostatic adjustment is the most conspicuous element and the component most frequently addressed. However, crustal deformation is not the only aspect of GIA and those regions for which only this component has been modelled cannot be regarded as fully GIA-corrected, because there is also a significant gravitational component (Tamisiea and Mitrovica, 2011). Ice exerts an attraction on the sea surface, and this gravitational component leads to further changes in the sea-surface configuration. Coastal regions in the far-field are subject to other subtle responses; the increased capacity of the ocean basins following collapse of forebulge regions and redistribution of the mantle back under polar regions, causes the slight fall in ocean level known as ocean siphoning and levering in coastal regions. GIA is reconsidered in Section 8.5.3, but other aspects of vertical land movement, such as subsidence are only recently getting the attention they deserve.

In recent decades the sea-level enigma has largely been resolved (Gregory *et al.*, 2013). Altimetry has enabled observation of seasonal variations, with the ocean surface higher in the northern hemisphere in summer than in winter as a consequence of the hydrological cycle, and an opposite situation in the southern hemisphere (i.e. surface higher in Austral summer). It is predominantly the steric changes which dominate this seasonal cycle and can be detected as regional variations in the sea surface (Landerer and Volkov, 2013); the addition of mass primarily from ice melt is spread more globally. The global sea-level rise budget since 1960 is considered well understood (Leuliette and Willis, 2011); the steric contribution has been approximately 0.7 ± 0.5 mm/year and the mass contribution was $\sim 0.8 \pm 0.5$ mm/year, roughly equivalent to the 1.6 ± 0.2 mm/year estimate for the total (Domingues *et al.*, 2008). Since satellite observations have been available it has been possible to refine the budget with greater precision. Tide-gauge calibration has corrected altimetry for any drift and indicates that sea-surface elevation uncertainties from Jason satellites are of the order of 4 mm. Monthly steric changes in the ocean surface can now be detected with verification from *Argo* floats with uncertainties of around 3 mm, although those parts of the ocean that are not monitored, such as the Southern Ocean and the abyssal ocean (below 4,000 m) may contribute as much as a further 1 mm/year. The twin GRACE satellites (Gravity Recovery and Climate Experiment) enable monitoring of the motion of mass, particularly terrestrial water storage and redistribution of water between land and ocean; uncertainties in these estimates are of the order of 2 mm, although it is important to note that coastal areas are largely masked out of calculations because of the complexity that lies in this transition zone between land–sea interface at the 500 km spatial resolution of measurement. Over the period 2005–2010, during which Jason, GRACE, and *Argos* have been concurrently operational, the budget appears to be resolvable at monthly time intervals (Leuliette and Willis, 2011).

Although it may no longer be regarded as an enigma, the fact that there continue to be errors in early estimates of the sea-level budget highlights the ongoing need for validation and cross-verification of different methodologies. Closing the budget has required, and will continue to require, a range of observations, including tide gauges, satellite altimetry, gravimetric monitoring, and ocean drifters, as well as GPS measurements and geophysical modelling. It is also important to realise that there are parts of the ocean that are not measured and therefore not included in the budget; these include high latitudes close to icesheets and the deep ocean that is not sampled by *Argo* floats. Whereas effort has been focused on the global budget, there is a pressing need for regional and local-scale projections that can underpin policy development and effective planning and management.

8.5.1 Thermal expansion – changing volume

More than 90% of the heat absorbed by the Earth over the past 50 years as a consequence of global warming is stored in the ocean. As the ocean heats it expands and this thermal expansion results in an increase in ocean volume. As outlined below, there is also an increase in ocean mass through the addition of further water (primarily from ice melt), but in this section thermosteric sea-level changes are discussed. Ocean volume is influenced by salinity (salty water expands more rapidly than fresh water) and pressure (with greater expansion at depth), but the overwhelming effect is related to temperature. A 1,000-m deep water column expands 1–2 cm for every 0.1°C rise in temperature.

Extensive compilations of ocean temperature data have been prepared by Levitus and colleagues (reviewed in Church *et al.*, 2010a; Johnson and Wijffels, 2011; Levitus *et al.*, 2012). Initial ocean temperature measurements were derived from ship-based observations, often vessels of opportunity. Since the 1960s more systematic observations were collected using expendable bathythermographs (XBTs). The World Ocean Circulation Experiment (WOCE) in the 1990s comprised a series of specific transects and, since 2004 considerably more extensive observations of temperature and salinity have been achieved through the use of *Argo* floats and other ocean drifters. These are deployed from oceanographic vessels across the oceans; the autonomous profiling drifters communicate with satellites when they are at the surface, but after deployment they sink to a designated depth within the upper 2,000 m of the ocean and drift passively for 9–10 days. They record data as they drift, then descend to 2,000 m prior to an ascent during which they profile temperature and salinity through the water column, transmitting their data on resurfacing. The *Argo* programme has provided almost ocean-wide data with a spatial coverage far more comprehensive than previous methods over the past decade (Lyman *et al.*, 2010). Since 2007 the complement of around 3,000 floats has been operational, providing hitherto unprecedented coverage of the global oceans.

Recent re-analyses of ocean temperature data have indicated that it contributes a greater proportion to sea-level rise than previously appreciated (~0.54 ± 0.05 mm/ year from 1955 to 2010: Levitus *et al.*, 2012). The old data have been corrected primarily through application of fall-rate corrections to XBT data. Domingues *et al.* (2008) attribute 0.52 ± 0.8 mm/year of sea-level rise over the period 1961–2003 to thermal expansion in the upper 700 m of water column. Acceleration is detected around 1970, and thermosteric contributions over the period of altimeter data since 1993 are considered to lie in the range of 0.8–1.5 mm/year (Church *et al.*, 2010a). A further complication is the interruption in the general warming during this period that has resulted from cooling associated with volcanic eruption of Mount Pinatubo in 1991, as a consequence of which increased aerosols reduced ocean temperatures (Church *et al.*, 2005), a cooling which may be equivalent to as much as 0.5 mm/year of sea-level rise (Church *et al.*, 2011a; Gregory *et al.*, 2012).

8.5.2 Ice melt and mass exchanges

The prime driver of steric changes are alterations to ocean volume through thermal expansion, which in the early IPCC assessments was seen as the principal component contributing to contemporary sea-level rise. However, it has become increasingly apparent that the ocean has also been experiencing increases in mass over these time intervals, and in recent decades there has been focused effort to quantify the contribution from ice melt (Pfeffer, 2011).

Perhaps the most convincing case for ice melt has been the well-documented decrease in sea-ice, which occurred at an average rate of 746 ± 127 km^3/year between 1994 and 2004 (Shepherd *et al.*, 2010). Most spectacular has been reduction of sea-ice cover in the Arctic Ocean (Figure 8.7). Various shipping routes, broadly termed the Northwest Passage, have become ice-free and navigable throughout the year, which required wintering in ice-bound straits in previous centuries. Satellite imagery provides a compelling record of the reduction of sea-ice, with the September minimum showing a marked decrease in extent, on the basis of which an ice-free Arctic Ocean has been forecast in decades to come (Stroeve *et al.*, 2007). Sea-ice is floating and

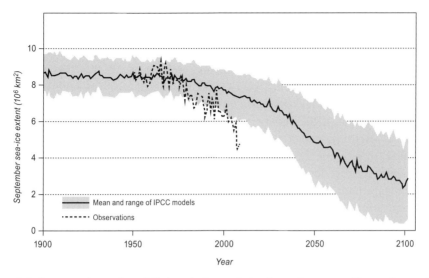

Figure 8.7 A comparison of modelled and observed Arctic sea-ice extent (September, end of summer minimum). The solid line is the ensemble mean of 13 IPCC AR4 models with the range of projections indicated by the shaded area. *Note*: this is floating sea-ice, so does not directly contribute to sea-level rise (adapted from Stroeve *et al.*, 2007, © American Geophysical Union, with permission from Wiley).

when it melts it makes negligible contribution to sea level, in contrast to grounded ice which adds to the mass in the ocean. The spectacular break-up of the Larsen B ice shelf in 2002 demonstrates a similar response in Antarctica; again this does not make a contribution to sea level (or only a negligible contribution directly and through freshening of the ocean which influences density), but the removal (and elsewhere thinning, perhaps from basal melting) of such a large ice shelf does have implications for ice streams previously contained behind the sea ice, with acceleration of several tributary glaciers observed after the Larsen B ice shelf broke up (Rignot *et al.*, 2004; Pritchard *et al.*, 2012).

More significant is the direct melting of grounded ice. In early assessments (including in the first IPCC reports), the rate of melt of the Antarctic Icesheet was considered very slow, and Antarctica was believed to contribute negatively to sea-level rise because of the accumulation of snow over the central areas of East Antarctica (Oerlemans, 1989). Similarly Greenland was considered to make only a small contribution to sea-level rise. However, this situation has been reconsidered in recent years, as there has been further evidence for rapid melt around the margins, and for surge of the lower parts of some Greenland glaciers. Not only have better observations of polar melt been possible, but better methodologies for assessing mass balance of icesheets have become available. Whereas balances were previously estimated from stakes sunk into the ice surface, broad synoptic assessments of icesheet morphology can now be made from satellites. Satellite radar altimetry provides some indication of their shape; however, the technique is not as effective as it is for the ocean surface, because of spatial variability of the radar footprint and dialectic properties of snow and ice. Airborne laser altimetry provides a finer scale with greater precision, but this is expensive and only available for limited parts of icesheets. Gravity measurements from GRACE are one of the most promising data sets. However, gravity fields comprise not only mass of ice, but also a component for isostatic adjustment of the underlying ground. This aspect is significant in both the Arctic and Antarctic, and it remains challenging to differentiate ongoing isostatic adjustment from ice melt (Steffen *et al.*, 2010).

Assessments of ice loss from different techniques, such as mass balance (Rignot and Kanagaratnam, 2006), satellite altimetry (Zwally *et al.*, 2005; Thomas *et al.*, 2006), and gravity from GRACE (Velicogna and Wahr, 2006a, 2006b) have been yielding conflicting estimates, largely because of uncertainties inherent in the different approaches. Rignot *et al.* (2011) compared over 8 years of estimates derived from differencing perimeter loss from net accumulation, and comparison of a dense time-series of time-variable gravity measurements to infer that Greenland and Antarctica contributed 900 Gt/year in total in 2009. Although this appears considerably greater than the contribution from small ice caps and mountain glaciers, there remain insufficient inventories of their melt to confirm this. For the period 2001–2005 over which estimates of the two sources are available, the small ice caps and mountain glaciers contributed 510 Gt/year whereas Greenland and Antarctica contributed 332 Gt/year (Pfeffer, 2011).

There is compelling evidence of recent acceleration of rapid melt particularly around the margin of Greenland. Rignot and Kanagaratnam (2006) indicated that Greenland was melting at a rate of 80 Gt/year in 1996, but 220 Gt/year in 2006. This acceleration is attributed to a series of dynamic changes, the physics of which are not yet fully understood, but the change in glacier behaviour appears to be moving northwards. The southern dome in Greenland is threatened both by increased melting and by accelerated discharge of glaciers. In addition to mass balance calculations and the conventional dynamics of icesheets which are incorporated into climate models, the behaviour of ice streams, iceberg-calving, basal melt rates and sliding, and penetration of heat beneath ice to increase rates of flow at marine outlets are all components of this more dynamic behaviour for which there are not yet adequate deterministic numerical models (Pfeffer, 2011). Rapid melt is not only occurring in Greenland, similar melt has been occurring along the Antarctic Peninsula, where changes in glacier dynamics are leading to more rapid calving of icebergs from their terminus (Lorbacher *et al.*, 2012; Balco *et al.*, 2013).

Ice melt seems to be accelerating, and is making a significant if still poorly quantifiable contribution of freshwater to the oceans, raising sea level. Mountain glaciers and ice caps are almost all melting rapidly and their contribution to ocean mass has been consistently recognised (Meier, 1984; Oerlemans and Fortuin, 1992). Total inventories are poorly known; the rate of mass loss has not only increased but it also appears to have accelerated from perhaps $0.46 \pm 0.52 \, \text{mm/year}$ sea-level equivalent 1900–1961 to perhaps 0.8–0.9 mm/year since then (Steffen *et al.*, 2010). The contribution will not be geographically constant across the global ocean, but because of the gravitational effects on the ocean surface, together with isostatic adjustments of the underlying solid earth, will result in a 'signature' in the geographical pattern of sea-level response (Spada *et al.*, 2012). Mitrovica *et al.* (2009) indicate the response that would be anticipated following rapid melt of the West Antarctic Icesheet, and suggest that on this basis it ought to be possible to fingerprint the source of contributing water. It is not always clear that such a fingerprint can be detected (Douglas, 2008); counter-intuitively the gravitational response involves a lowering of sea level adjacent to melting icesheets (Tamisiea and Mitrovica, 2011).

Melting ice is not the only contribution to altered ocean mass. There are indirect consequences of human actions that alter the various storages of freshwater on land. A seasonal variation in ocean altimeter levels has been mentioned above (Chen *et al.*, 2005). However, there are also consequences from sequestering of water in storages behind dams (Pokhrel *et al.*, 2012), and the reduction of groundwater through extraction. The likelihood that human contributions to sea-level rise were actually significant was raised by Sahagian *et al.* (1994), who suggested that effects such as groundwater extraction, irrigation, desertification, and deforestation might contribute as much as 0.5 mm/year to global sea-level rise, which was about 30% of estimates at that time.

Observations from GRACE have the potential to resolve spatial and temporal changes in some of these terrestrial water storages, with a broad 300–400 km footprint, with a vertical equivalent water level of the order of a few millimetres. The amount of water retained in reservoirs in the second half of the twentieth century may be equivalent to 0.4 mm/year sea-level change (Chao *et al.*, 2008). On the other hand, coastal waters register a 'self-gravitation-effect' with seasonal elevation of water level related to hydrological factors, particularly delivery of water by the big rivers, which can be recorded as a sea-level component of up to 10 mm in a year (Wouters *et al.*, 2011). In practice, although these changes in water storage may be significant, it seems likely that the water captured behind dams is of a similar magnitude to that mined from groundwater, such that the two contributions may cancel out when considered in terms of their effect on sea-level rise (Milly *et al.*, 2010). Of course, these effects may have their own geographical signature, and their impact on isostatic adjustments is largely unknown, but is considered to accentuate relative sea-level rise close to tide gauges, implying that global MSL estimates may be up to 10% higher than they should be (Fielder and Conrad, 2010). GRACE indicates that ice melt contributes $1.0 \pm 0.4 \, \text{mm/year}$ to global MSL rise, but hydrological signals account for regional variation in relative sea level (Riva *et al.*, 2010). Even the contribution of sediment eroded from the land and carried by rivers and glaciers to the sea may make a minor (~ 0.01 mm/year) contribution to sea-level rise (Syvistki and Kettner, 2011).

It now appears possible to infer ocean steric changes from the difference between ocean volume (measured by the ocean surface from altimetry) and ocean mass, measured through satellite-based gravity observations, as shown by the apparent closure of the budget for the period 2005–2010 (Leuliette and Willis, 2011). Suitable gravity measurements are already being acquired by the GRACE satellite that was launched in 2002 and enables mass measurements to a precision of about 1.8 mm of sea level, but still better results are anticipated from the Gravity Field and Steady-State Ocean Circulation Explorer satellite launched in 2009.

8.5.3 *Glacial isostatic adjustment and response of the solid Earth*

Steric effects that alter volume are well represented in climate models. The meltwater contribution from ice is incorporated, primarily as a mass balance, although as yet with the accelerated dynamics of ice streams poorly constrained (represented by a single term of up to 0.17 m additional sea-level rise by 2100 in AR4, but perhaps contributing ~0.1 mm/year with potential to accelerate through feedbacks). In contrast, GIA effects have been recognised for a considerable time (Mitrovica and Davis, 1995), but are derived from a different suite of modelling performed by geodynamicists and are rarely incorporated directly or transparently into climate model outputs. The surface of the ocean represents the upper boundary of the sea, but the solid Earth provides the lower boundary, and adjustments in the shape of the Earth are therefore also critical components of a full understanding of sea-level variations (Milne *et al.*, 2009; see Chapter 2). GIA models are generally used to capture the crustal deformation associated with visco-elastic response of the mantle to redistributed ice and water loads which occur over millennia. There is also an elastic response to changed loads which is modelled as an instantaneous response of the lithosphere (Tamisiea and Mitrovica, 2011). Significant effects on rotation of the Earth and polar wander are a further feature of these models.

The recognition of deformation of the Earth in response to varying loads has been modelled since the 1970s (Walcott, 1972). General numerical techniques using one-dimensional models were developed in the 1980s, and these have been progressively refined; for example, Peltier and Tushingham (1991) examined global sea level against a GIA background. There has been an expectation that global GIA models can correct for vertical movements at numerous sites around the world; however, it is important to remember that there are many other contributors to vertical movement at any site. GIA models are generally good at capturing the greater rates experienced in the past at high-latitude sites where there has been substantial rebound. However, the models are sensitive to small changes in parameterisation which may result in differences of up to 0.5 mm/year at far-field sites (Mitrovica *et al.*, 2010).

First-order GIA adjustments can be modelled across the globe by geodynamicists, but still more subtle land movements, including tectonic movement, seismic displacement, subsidence, and sediment compaction, and the local expression of ocean surface dynamics further complicate the sea-level trajectory on any coast. In view of these limitations, and recognising other contributions to vertical land movements, it should not be anticipated that correction for GIA will remove all vertical displacement and leave no further spatial variation. GIA corrections do not yet provide a single unique solution, and outputs are generally highly smoothed through adoption of discrete time steps. The early sea-level equation of Farrell and Clark (1976) continues to form the basis for computation, and modelling requires a space–time history for the load(s), represented by the history of late Pleistocene ice cover (see Figure 2.13). Whereas the initial disc-load style of ice volumes has been repeatedly refined, and further geological evidence has enabled better determination of former extent and timing of melt of most northern hemisphere icesheets, there is still need for further refinement. A second requirement concerns the governing viscoelastic rheological parameters capturing behaviour of the mantle. Modelling at the global scale generally adopts a symmetrical, spherical parameterisation developed by Peltier (1974). This does not adequately recognise three-dimensional variation in rheology. Inversions of GIA model outputs against calibrating sea-level observations have produced various inferences about mantle viscosity (Yokoyama *et al.*, 2012), and studies have adopted different viscosity values when fitting model and data in different regions – for example, Fennoscandia and the east coast of Australia. Slangen *et al.* (2011) have examined the effect of using the ICE5G GIA model compared with using that developed at ANU by Lambeck and others, and demonstrated that the latter produces a greater forebulge-response region around former icesheets, with further differences when using different mantle rheology. Comparison of anticipated GIA trends with GPS results indicates significant disparities, particularly greater rates of subsidence (Houston and Dean, 2012; King *et al.*, 2012a).

8.6 Sea level and climate models

In the first attempt to estimate sea-level rise explicitly addressing thermal expansion and land ice losses resulting from CO_2-related warming, Hoffman *et al.* (1983) warned that sea level will almost certainly rise in coming decades with the 'most likely' rise being between 1.44 and 2.17 m by 2100, but cautioned that a wider range of 0.56–3.45 m 'cannot be ruled out'.

Although observations can produce trends in the rate of sea-level rise, projections for the future are primarily derived from climate models. There are a range of models, from Earth Models of Intermediate Complexity (EMICs) to fully coupled Atmosphere–Ocean General Circulation Models (AOGCMs). The complexity of these models varies, and the number of atmospheric and ocean layers, and their spatial resolution have gradually increased in state-of-the-art models. For each model simulation, initial conditions have derived from control runs with a steady-state pre-industrial climate. Drift can amount to an equivalent sea-level adjustment of ~1 mm/year, so low-frequency drift observed in these control runs, in which no greenhouse gas forcing has been applied, is generally subtracted from subsequent transient simulations. Simulation of sea-level conditions is linked to the physics representing oceanographical processes capturing horizontal and vertical redistribution of heat (and to a lesser extent salinity) in the oceans. These aspects of dynamic oceanography are complex and there is considerable variation in the spatial distribution of outputs from different modelling groups (Pardaens *et al.*, 2011a,b). It is important to recognise that models are primarily representing steric, thermal expansion-driven components of sea level, reflecting underlying changes in ocean temperature and salinity that are determined in surface fluxes and ocean transport processes.

IPCC has reported sea-level rise projections through its successive assessment reports. Projections have not differed greatly; those in the most recent (AR4) have been interpreted as a narrower range of likely sea-level rise (Gehrels, 2010a), although incorporating different aspects than earlier reports. In the 1990 First Assessment Report (FAR) of the IPCC, Warrick and Oerlemans (1990) indicated three scenarios for sea-level rise: high, low, and best estimate based on business as usual (BAU). Sea-level projections for the years 2030, 2070, and 2100 were 0.29, 0.71, and 1.10 m higher than 1990 for the high, 0.18, 0.44, and 0.66 m for the best estimate, and 0.08, 0.21, and 0.31 m for the low, respectively. IPCC emission scenarios were revised in 1992, with a series of six scenarios being developed. The sea-level rise projections varied little from earlier estimates until 2050, after which they diverged. Best-case scenarios ranged between 0.22 and 1.15 m rise by 2100, whereas best-guess scenarios (which incorporated feedback effects in the carbon cycle) varied from 0.15 to 0.90 m by 2100 (Wigley and Raper, 1992). It was shown that as CO_2 continues to accumulate in the atmosphere, the rate of sea-level rise will be relatively insensitive to emissions; the sea will continue to rise beyond the twenty-first century whatever mitigation efforts are made to reduce greenhouse gas emissions (Warrick, 1993).

In the Second Assessment Report of the IPCC (SAR), it was anticipated that MSL would rise due to thermal expansion of the oceans and melting of glaciers and icesheets. The lowest of the emission scenarios (IS92c) produced an estimate of 0.15 m of sea-level rise by 2100; the highest (IS92e, combined with the upper values for climate and ice-melt sensitivity) indicated the sea could be 0.95 m above 1995 levels. Central values, using the IS92a scenario, and assuming 'best estimate' values of climate sensitivity and of ice-melt sensitivity to warming (including effects of future changes in aerosol concentrations) projected an increase in sea level of about 0.50 m between 1995 and 2100.

The Third Assessment by the IPCC (TAR), released in 2001, included further modelling using the IS92e scenario, which produced a range of 0.11–0.77 m rise by 2100, but provided results of simulations from an ensemble of coupled models using SRES (Special Report on Emissions Scenarios) greenhouse gas emission scenarios, with these ranging from 0.09 to 0.88 m, and a central value of 0.48 m rise by 2100 (Church *et al.*, 2001). There was considerable geographical variability generated by the different models (Gregory *et al.*, 2001). Patterns respond

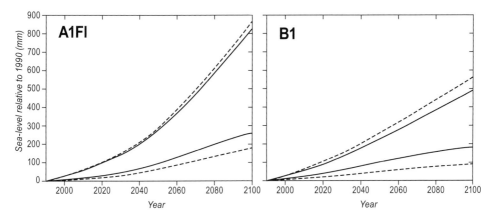

Figure 8.8 Projections of sea level derived by adjusting the TAR projections (solid line, 5 and 95 percentile) to correspond with the AR4 projections at 2095. The upper and lower dashed lines show the full range of TAR projections for the A1FI (fossil fuel-intensive) and B1 (world in which there is increasing global cooperation and convergence on sustainable technologies) scenarios (from Hunter, 2010, with kind permission from Springer Science + Business Media).

primarily to changes in surface heat, freshwater (particularly sea-ice melt in the Arctic), and wind stress (Lowe and Gregory, 2006).

The Fourth Assessment Report by the IPCC (AR4) provided projections of sea-level rise at 2095 (2090–2099) relative to 1990 (1980–1999; Meehl *et al.*, 2007). The rise projected by the models over this period ranged from 0.18 m for the low emission B1 scenario to 0.59 m for the high emission, A1FI (fossil fuel-intensive) scenario. AR4 simulations are reviewed by Church *et al.* (2011b) and extended to 2100 (whereas AR4 produced projections for the decade 2090–2099); A1FI and B1 projections are shown in Figure 8.8. The issue that could not be resolved in AR4 modelling was the dynamic behaviour of icesheets which had become apparent, but for which there were no physical models available; therefore, this was not included in projections but was addressed in a footnote, which indicated that ice dynamics might contribute up to a further 0.17 m by 2100. AR4 projections for sea-level rise did not give a best estimate or an upper bound, and larger rises could not be ruled out.

Figure 8.8 indicates that the sea-level projections are not particularly sensitive to emission scenario, particularly up to 2050; the main variation is between different models. Uncertainties related to the poor observational constraints on sea level and incomplete modelling of icesheet melt mean that it has not been possible to attribute probabilities to the outcomes. Pfeffer (2011) outlines the ice mass loss contributions used in successive IPCC assessments, and summarises the dilemma that arose in the AR4 when it was clear that rapid dynamics (acceleration) of icesheet behaviour (first known as rapid deglaciation, following Mercer (1978), but more recently termed collapse or disintegration) was insufficiently understood to be adequately modelled.

The steric sea-level changes in climate models are not spatially uniform and some regions are projected to experience more than twice the global average rate of sea-level rise with others experiencing considerably less (Landerer *et al.*, 2007). Comparing geographical patterns of sea-level variations generated by models using the IS92a scenario for the TAR with those generated using the A1B scenario for AR4 indicates negligible reduction of variability. Changes vary substantially in magnitude, and even sign, between models (Pardaens *et al.*, 2011a). The spatial root-mean square (RMS) of sea-level change in an ensemble of 13 AR4 models, which adopted the A1B scenario, is ~20% of the AR4 central estimate of global MSL change. Models of similar structure, based on similar formulations, tend to generate similar regional patterns, but substantial variation persists (Pardaens *et al.*, 2011a).

There are some common features between models; for example, sea-level rise that is less than the global average in the Southern Ocean, rise that exceeds the global average in the Arctic Ocean, and a similar enhancement of rise at the high-latitude ends of subtropical gyres. However, divergence between models is far larger for sea level than it is for temperature (which generally produces similar results between models), or precipitation (which produces marked differences between models), indicating relatively little reliability can be assigned to geographical variability shown by model output.

There are aspects of sea-level response that are not modelled. Some model outputs do not include the GIA component, and when they do, there remains disagreement about the best model parameters to use. Global models do not incorporate the many other vertical movements which are expressed at a range of scales across the globe. It is important to recognise that few models incorporate all aspects of ice melt, and even those that do, do not assign a geographical distribution to the water level increment. This is known to be a feature of rapid melt of a mass of ice, which incurs an isostatic response, but also incorporates a modification of the ocean surface through gravitational effects. For example, the anticipated geographical variation of water after a contribution of 1 mm equivalent to the global ocean from Greenland can be modelled. When rapid melt occurs, there is not only uplift beneath the reduced icesheet, but there is also reduced gravitational attraction from that icesheet, with a lowering of sea level near the ice margin, but a net increment slightly above average in the far-field (Tamisiea and Mitrovica, 2011).

Climate modelling undertaken for AR4 does not take mitigation measures into account. Emissions are modelled based on SRES scenarios but without reductions because of greenhouse gas mitigation policies, which is known as business-as-usual. A major aim of the Copenhagen Accord was to keep increases in global-mean surface air temperatures below 2°C. The effect of mitigation of this level on future sea level has been modelled by Pardaens *et al.* (2011b), using the HadCM3C and HadGEM2-AO models. Using two variants of the A1B scenario (A1B (IMAGE) which rises to a CO_2 equivalent concentration of 1050 ppmv by 2100 and A1B SRES which reaches 835 ppmv), modelling without dynamic acceleration of the icesheets, produced a range of 0.29–0.51 m sea-level rise by 2100, with a median of 0.39 m. In contrast, the mitigation modelling (E1, with emission reductions from 2010, peaking at 530 ppmv in 2045 and stabilising at 450 ppmv by the end of the century) reduced this to 0.17–0.34 m with a median of 0.26 m. The reduction resulted primarily from reduced thermal expansion; considerably larger sea-level rises of 0.72 m for BAU and 0.47 m for E1 in 2100 were obtained when accelerated icesheet dynamics were included. Such mitigation would reduce the number of people likely to experience coastal flooding by about 55%.

The relatively poor agreement between models, and the difference between model output and observed sea-level trends, may be reduced in future by higher-resolution modelling. A high-resolution iteration of the MIROC3.2 model produced a geographical pattern of steric sea-level changes at finer resolution, which captured features such as eddies that can be anticipated associated with known currents. However, substantial differences remain between observations and model estimates for the twentieth century and between various model projections for the twenty-first century (Church *et al.*, 2010b, p. 166; Suzuki and Ishii, 2011). In a simulation whereby a climate model was run for 500 years and compared with tide-gauge records, considerable 10–20-year biases were found in results. It was noted that, although sea-level changes are constrained by ocean mass (unlike other variables such as temperature that do not have such mass constraints), the 'global mean is a small number resulting as the sum of many large values of opposite sign' (Christiansen *et al.*, 2010, p. 4325).

Recently, Slangen *et al.* (2011) have modelled regional variability in future sea-level change, using a combination of spatial patterns of steric effects, land ice melt, and GIA obtained from 12 different AOGCM models for three of the SRES scenarios. They adopted many data sets used by AR4, with updated estimations of glacier and icesheet contributions. Although the steric component dominates the geographical pattern and variability in the sea-level anomaly (that is the deviation from global-mean value), the addition of mass (the land ice component) appears significant, but GIA is a major factor only in particular areas (Spada and Galassi, 2012). The steric component is the most variable because it contains density changes associated with ocean currents and freshwater additions; the land ice contributes to mass changes which do result in elastic solid earth deformation and self-gravitation effects; whereas the longer-term GIA is substantial only in areas close to the former icesheets or their forebulges (Slangen *et al.*, 2011).

8.6.1 Post-AR4 sea-level projections

Since AR4 there has been some criticism of the sea-level projections and a number of alternative approaches that indicate the possibility of much more rapid rise. Hansen (2007) insinuated that what he called 'scientific reticence' had produced a conservative estimate of likely sea-level rise. Shortly after release of AR4, Rahmstorf *et al.*

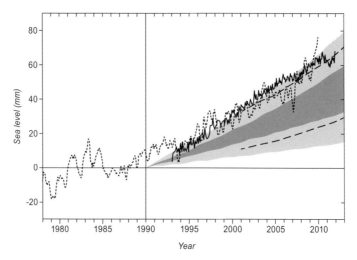

Figure 8.9 Comparison of observed and modelled sea level. The solid line represents the satellite altimetry data (see Figure 8.6), the dotted line is reconstructed global tide-gauge record (after Church and White, 2011). The TAR projections are shown (shaded, darker are central estimates and lighter is full range), and those inferred from AR4 (dashed), indicating that sea-level rise appears to continue at the upper limit of IPCC projections (after Rahmstorf *et al.*, 2012).

(2007) showed that observed sea-level rise derived from tide gauges (1990–2001) and altimetry (1993–2006) was tracking at the upper limit of projections from TAR (and hence by implication also AR4 as these were quite similar). This trend has continued with a longer data set (Church and White, 2011), as shown in Figure 8.9 (Rahmstorf *et al.*, 2012). The 'enigma' meant that until now observed sea-level rise could not be balanced with inferred sources, confounding development of projections for the future. As indicated above, this now appears achievable, and revision of projections in the fifth assessment by the IPCC (AR5) may eventuate.

Rahmstorf (2007) has also proposed an alternative semi-empirical method that could be used to project future sea-level rise which scales observed sea-level rise to globally averaged temperature. This method has been extended to imply 0.75–1.9 m rise by 2100 (Vermeer and Rahmstorf, 2009). Grinstead *et al.* (2010) undertook a similar analysis using non-linear relationships and response times. These methods produce sea levels that are 0.50–1.80 m higher than 1990 by 2100, but have been criticised on several grounds (Gregory *et al.*, 2013). First, there is no physical basis for the relationship; second, past relationships between temperature and sea level appear to have been non-linear (Siddall *et al.*, 2009); and, third, the relationship holds over only a short period of calibration, raising concerns about extrapolation decades into the future (Holgate *et al.*, 2007; von Storch *et al.*, 2008).

The observation that sea level was higher in the last interglacial has been used to tune models which indicate that Greenland could melt and contribute to sea level as

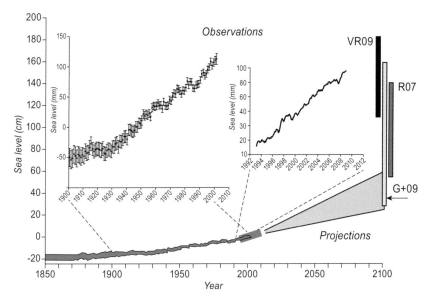

Figure 8.10 The trend in global mean sea level over the late nineteenth and twentieth centuries; the insets show the tide-gauge record for the twentieth century and the altimetry record since 1993. The range of AR4 IPCC projections for the twenty-first century is shown, together with columns capturing ranges projected by alternative studies since AR4, the dark column to the left is after Vermeer and Rahmstorf, 2009; the central column after Grinstead *et al.*, 2010; and the right-hand column after Rahmstorf, 2007 (redrawn from Nicholls and Cazenave, 2010, reprinted with permission from AAAS).

much as 6 m above present (Otto-Bliesner *et al.*, 2006; Hansen and Sato, 2012). However, there are physical constraints on how fast ice can melt in future, which limit the magnitude of sea-level rise to less than 2 m by 2100, and indicate that 0.8 m may be a more probable projection (Pfeffer *et al.*, 2008). Figure 8.10 summarises tide-gauge and altimeter records, IPCC projections, and later extrapolations (Nicholls and Cazenave, 2010). Although it remains uncertain just how fast the sea is rising, a majority of researchers consider that rise of several decimetres will be experienced during the coming decades (Perette *et al.*, 2012; Bamber and Aspinall, 2013). Future projections are likely to be based on representative concentration pathways (RCP), related to greenhouse gas concentrations or temperatures, rather than emissions scenarios (Moss *et al.*, 2010; van Vuuren *et al.*, 2011). The sea will rise beyond 2100; it has been suggested that by 2300 it might be ~1.5 m higher if the Earth is 1.5°C warmer, or ~2.7 m higher if it is 2°C warmer (Schaeffer *et al.*, 2012).

Vertical land movements remain a major complication. This can be illustrated with reference to variability around Britain, where there is reasonable correlation between records from tide gauges and regional GIA modelling. Uplift occurs in Scotland which was covered by an extensive icesheet during the last glaciation, and subsidence characterises southern England, with rapidly retreating coasts in low-lying areas such as East Anglia (Shennan and Woodworth, 1992; Shennan and Horton, 2002a). The broad pattern of regional tilting (see Figure 7.14) is also verified by GPS techniques (Woodworth *et al.*, 2009a; Bradley *et al.*, 2009), although there remain some differences in the way these corrections are applied by different researchers (Shennan *et al.*, 2009; Gehrels, 2010b).

In some cases there are site-specific compilations of the contributions to relative sea-level rise, several of which attribute most to subsidence of the land (Belperio, 1993; Paine, 1993). In the case of Manila Bay, Rodolfo and Siringan (2006) emphasise that compared with the threat of sea-level rise, inundation as a result of subsidence caused by groundwater extraction poses a much more imminent risk. Subsidence is a feature of megadeltas around the world and has required urgent action in Bangkok (see Figure 8.2) and Shanghai where groundwater extraction has been ceased beneath cities because it accentuated subsidence (Syvitski *et al.*, 2009).

8.6.2 The threat of catastrophic sea-level rise

Although the sea is rising at the majority of locations around the world, the rate of rise is not as fast as has been experienced during certain times in the past. The average rate of rise during much of the postglacial ice melt was of the order of 10 mm/year. Much more rapid rates occurred during meltwater pulses (Fairbanks, 1989), particularly MWP-1A, when it is inferred that decanting of ponded meltwater from the Laurentide Icesheet contributed a large volume to the ocean. Rapid future rises of this type are unlikely to occur because there are no longer the numerous large icesheets that could provide such meltwater sources compared with the late-glacial. There appear to have been periods in the instrumental record when the sea was rising at similar rates to those now determined from altimetry (Watson, 2011). However, recent accelerated melt observed in both Greenland and parts of Antarctica (Rignot *et al.*, 2011) raises prospects of faster melt from these sources than experienced during the twentieth century. In particular, there has been considerable debate about the stability of West Antarctica.

In contrast to the vast icesheet grounded over the major land mass of Antarctica, the West Antarctic Icesheet (WAIS) appears potentially less stable because it is flanked by vulnerable floating ice shelves that buttress inland ice streams. That it could melt rapidly, with fast flow of those ice streams, causing a rapid rise in sea level was first proposed by Mercer (1978). There is evidence for previous collapse of this icesheet during the Pliocene and in Pleistocene super-interglacials (Pollard and DeConto, 2009). It was believed that melt of the WAIS could contribute 5 m to global sea level, although the latest estimate revises that down to around 3.3 m (Bamber *et al.*, 2009). A rapid sea-level rise at the termination of the last interglacial has been suggested by several researchers and this is generally attributed to WAIS or Greenland sources (Hollin, 1977; Hearty *et al.*, 2007; Blanchon *et al.*, 2009). Other evidence supports catastrophic collapse of icesheets over a period of only a few centuries at the termination of the last interglacial (Grant *et al.*, 2012), raising the possibility of rapid WAIS melt in the future as a result of anthropogenic activities. It remains difficult to verify because the resolution of dating cannot resolve whether melt occurred over a century or a millennium (Hearty *et al.*, 2007). The past may preserve evidence of other potential changes, including erratic shifts in climate and oceanographic circulation, greater frequency and intensification of storms at subtropical latitudes, eventual melt back of the Greenland Icesheet, WAIS ice collapse and rapid sea-level rise.

8.7 Impacts and coastal vulnerability

Projections of future sea level imply that there will be significant impacts on many heavily populated coastal regions (Nicholls *et al.*, 2007). Impacts include erosion and retreat of erodible shorelines and inundation of low-lying areas (Church *et al.*, 2008). The gradual increment of sea-level rise will be imperceptible, but extreme high-water levels will have major impacts (Hunter, 2010; Lowe *et al.*, 2010). Woodworth and Blackman (2004) examined data from 141 tide gauges globally and found evidence for a slight increase in extreme high-water levels since 1975; this has been restated by Menéndez and Woodworth (2010), and there are also local studies which have demonstrated an increase (Haigh *et al.*, 2010). As the sea rises, inundation events that recurred only rarely become more frequent; for example, the 1 in 100-year event may be experienced each few decades, with increase in frequency being spatially variable as a function of the Gumbel scale parameter (Hunter, 2012).

Coastal erosion and inundation are anticipated to become widespread, but there are other risks, such as saltwater incursion into groundwater and freshwater wetlands (Mimura, 1999). A first-order estimate of areas that will be flooded can be derived by a 'bathtub' approach which presumes that those areas that lie within a contour interval (determined by amount of sea-level rise) above the present high-tide level will be subject to potential flooding. More sophisticated modelling may allow for set-up and run-up of waves and storm surges. However, a shortcoming of many of these approaches is that they do not recognise the natural dynamics associated with coastal environments. The sedimentary record provides ample evidence for complex depositional and erosional histories that coastal environments undergo (Woodroffe, 2003); a significant challenge for geoscientists is to decipher that history in order to build better models of how shorelines will respond to changes of sea level in the future.

It will be important to recognise that different processes within coastal systems operate at different timescales. This is synthesised in Figure 8.11, which combines timescales at which coastal systems have responded in the past (after Cowell and Thom, 1994; Woodroffe, 2003), with timescales at which coastal behaviour can be expected in the future. Instantaneous responses are seen in small-scale phenomena, such as ripples, based on the physics of fluid motion and shear stresses acting on individual sediment particles. Such physical understanding is good at the

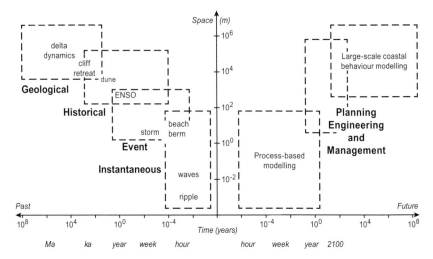

Figure 8.11 A framework for considering space and timescales in coastal systems, showing the past to the left (based on Cowell and Thom, 1994; Woodroffe, 2003), and the future to the right, with types of model that are appropriate (after Gelfenbaum and Kaminsky, 2010). Examples of characteristic geomorphological features or landforms and the scales at which they are best studied are proposed (from Woodroffe and Murray-Wallace, 2012, with permission from Elsevier).

instantaneous scale, but it cannot be effectively scaled up to determine aggregated behaviour of millions of irregular sand grains in a complex wave environment at longer timescales. There are sophisticated facies models that enable interpretation of the evolution of sedimentary landforms at millennial and longer 'geological' time-scales. Most complex, and of concern in the context of global climate change and policy development, is the intermediate timescale, termed historical in Figure 8.11, but also relevant for engineering works that must meet a planned design life. This decadal to century scale is the timescale at which managers and planners require forecasts of likely future changes to the coastal systems that they oversee (Mitsova and Esnard, 2012).

8.8 Relevance of Quaternary sea-level changes

There are lessons to be learned from the past that provide context and background for a clearer understanding of how sea level will change in the future (Gehrels, 2010a; Siddall and Milne, 2012). It will be important to continue to refine knowledge of patterns of past sea-level change. First, this provides an independent measure of past climate change (ocean volume is inversely related to ice volume). Second, it provides important geological analogues for times of higher sea level and rapid sea-level rise, enabling geoscientists to decipher how coastlines and ecosystems have responded to such events in the past. Third, the evidence of past shorelines can be used to assess the longer-term rates of vertical land movement at specific locations (e.g. if a past shoreline has been displaced vertically, inferences may be made about a significant component of land–sea interaction that can be incorporated into projections of future sea-level change at that location).

Reconstructions of Quaternary sea level initially focused on trying to construct a single 'eustatic' sea-level curve, as previous chapters have demonstrated, until it became evident that vertical movement of the land, most obviously through tectonic activity in active plate–margin settings, but more subtly through isostatic response of the Earth to redistribution of ice and water loads, meant that a 'eustatic' curve was largely unattainable. Attention has subsequently been focused on local reconstruction of relative sea-level changes, combining effects of vertical motion of land and associated changes of the sea surface at specific geographical locations, leading to compilation of atlases of relative sea-level curves (Bloom, 1977; Pirazzoli, 1991, 1996). In a similar way, climate scientists have endeavoured to derive a detectable 'global' trend in sea level. Observations that sea level has shown a net increase over recent decades provide important independent corroboration of climate change. It is also necessary to establish the contributions of different sources to this sea-level rise, and the proportion of change that can be attributed to each source, whether thermal expansion or additional meltwater (Meehl *et al.*, 2007). Although it will be essential that these estimates of global MSL rise are continually reassessed and updated, it is important to recognise that there is going to be considerable geographical variation in the pattern of sea-level rise in the future, just as there has been in the past. All too often, guidelines adopted for planning purposes are focused on the global-mean projection for sea level, with insufficient consideration of local or regional deviations. For example, the average global rate of sea-level rise is not the rate that is currently being observed at many recording sites around the coast of Australia where both satellite altimetry and high-resolution SeaFrame tide gauges, that have been operating concurrently over the past two decades, indicate considerable variation in sea-level trend (Deng *et al.*, 2010).

A broader understanding of past sea-level changes can contribute to the study of future sea level trends and their consequences in several ways. First, the nature of the evidence that palaeo-sea-level researchers use means that such studies may enable identification of other factors that need to be taken into account in determining the land–sea relationship at a site, and producing projections for what may happen in the future. Many coastlines experience subsidence; knowledge of this, and of the rates of change, will be important when planning for existing settlements or future infrastructure, and when managing resources at such locations. These studies also have the potential to provide evidence on which to determine long-term sediment budgets, and hence to formulate clearer models of future coastal behaviour for those coasts where erosion and re-deposition of sediment is a significant component of shoreline adjustment. These, and related challenges, will provide an ongoing focus for research by Quaternary coastal scientists that will have a direct relevance for society.

References

Abbey, E., Webster, J. M. & Beaman, R. J. (2011). Geomorphology of submerged reefs on the shelf edge of the Great Barrier Reef: the influence of oscillating Pleistocene sea-levels. *Marine Geology*, **288**, 61–78.

Abrahams, A. D. & Oak, H. L. (1975). Shore platform widths between Port Kembla and Durras Lake, New South Wales. *Australian Geographical Studies*, **13**, 190–194.

Adam, C. & Bonneville, A. (2005). Extent of the South Pacific Superswell. *Journal of Geophysical Research*, **110**(B09408): doi: 10.1029/2004JB003465.

Adam, P. (1990). *Saltmarsh ecology*. Cambridge University Press.

Adey, W. H. (1978). Coral reef morphogenesis: a multidimensional model. *Science*, **202**, 831–837.

Adhémar, J. A. (1842). *Revolutions de la Mer – Deluges Periodiques*. Carilian & V. Dalmont, Paris (2nd edition, 1870, 358 pp.).

Agassiz, L. (1840). *Etudes sur les glaciers*. Jent and Gassman, Neuchâtel.

Aguirre, E. & Pasini, G. (1985). The Pliocene–Pleistocene boundary. *Episodes*, **8**, 116–120.

Aharon, P. (1984). Implications of the coral-reef record from New Guinea concerning the astronomical theory of ice ages. *In* A. Berger, J. Imbrie, J. Hays, G. Kukla & B. Saltzman (Eds), *Milankovitch and climate: understanding the response to astronomical forcing*, pp. 379–389. Reidel, Dordrecht.

Aharon, P. & Chappell, J. (1986). Oxygen isotopes, sea level changes and temperature history of a coral reef environment in New Guinea over the last 10^5 years. *Palaeogeography, Palaeoclimatology, Palaeoecology*, **56**, 337–379.

Aharon, P., Chappell, J. & Compston, W. (1980). Stable isotope and sea-level data from New Guinea supports Antarctic ice-surge theory of ice ages. *Nature*, **283**, 649–651.

Airy, G. B. (1855). On the computation of the effect of the attraction of the mountain-masses, as disturbing the apparent astronomical latitude of stations in geodetic surveys. *Philosophical Transactions of the Royal Society of London, Series A*, **145**, 101–104.

Aitken, M. J. (1990). *Science-based dating in archaeology*. Longman, London, 274 pp.

Aitken, M. (1998). *An introduction to optical dating*. Oxford University Press, Oxford, 267 pp.

Alexander, C. S. (1968). The Marine Terraces of the Northeast Coast of Tanganyika. *Zeitschrift für Geomorphologie* Suppl. (Bd), **7**, 133–154.

Allen, J. R. L. (1987). Coal dust in the Severn Estuary, southwestern UK. *Marine Pollution Bulletin*, **18**, 169–174.

Allen, J. R. L. (1993). Muddy alluvial coasts of Britain: field criteria for shoreline position and movement in the recent past. *Proceedings of the Geologists Association*, **104**, 241–262.

Allen, J. R. L. (2000). Morphodynamics of Holocene salt marshes: a review sketch from the Atlantic and Southern North Sea coasts of Europe. *Quaternary Science Reviews*, **19**, 1155–1231.

Allen, J. R. L. & Rae, J. E. (1986). Time sequence of metal pollution, Severn

Estuary, southwestern UK. *Marine Pollution Bulletin*, **17**, 427–431.

Allen, J. R. L. & Haslett, S. K. (2002). Buried salt-marsh edges and tide-level cycles in the mid-Holocene of the Caldicot Level (Gwent), South Wales, UK. *The Holocene*, **12**, 303–324.

Alley, N. F. & Lindsay, J. M. (1995). Tertiary. *In* J.F. Drexel & W.V. Preiss (Eds), The Geology of South Australia, Vol. 2, The Phanerozoic, pp. 151–217. *South Australia Geological Survey Bulletin*, **54**.

Alvarez-Marrión, J., Hetzel, R., Niedermann, S., *et al.* (2007). Origin, structure and exposure history of a wave-cut platform more than 1 Ma in age at the coast of northern Spain: a multiple cosmogenic nuclide approach. *Geomorphology*, **93**, 316–334.

Amos, C. L. & Zaitlin, B. A. (1984). The effect of changes in tidal range on a sublittoral microtidal sequence, Bay of Fundy, Canada. *Geo-Marine Letters*, **4**, 161–169.

Anderson, E. C., Libby, W. F., Weinhouse, S., *et al.* (1947). Natural radiocarbon from cosmic radiation. *Physical Review*, **72**, 931–936.

Anderson, J. B., Shipp, S. A., Lowe, A. L., *et al.* (2002). The Antarctic ice sheet during the last glacial maximum and its subsequent retreat history: a review. *Quaternary Science Reviews*, **21**, 49–70.

Anderson, R. S., Densmore, A. L. & Ellis, M. A. (1999). The generation and degradation of marine terraces. *Basin Research*, **11**, 7–19.

Andréfouët, S., Cabioch, G., Flamand, B., *et al.* (2009). A reappraisal of the diversity of geomorphological and genetic processes of New Caledonian coral reefs: a synthesis from optical remote sensing, coring and acoustic multibeam observations. *Coral Reefs*, **28**, 691–707.

Angulo, R. J., Lessa, G. C. & de Souza, M. C. (2006). A critical review of mid- to late-Holocene sea-level fluctuations on the eastern Brazilian coastline. *Quaternary Science Reviews*, **25**, 486–506.

Antonioli, F., Ferranti, L. & Lo Schiavo, F. (1996). The submerged Neolithic burials of the Grotta Verde at Capo Caccia (Sardinia, Italy): implication for the Holocene sea-level rise. *Memorie Descrittive del Servizio Geologico Nazionale*, **52**, 329–336.

Antonioli, F., Chemello, R., Improta, S., *et al.* (1999). The *Dendropoma* (Mollusca Gastropoda, Vermetidae) intertidal reef formations and their paleoclimatological use. *Marine Geology*, **161**, 155–170.

Antonioli, F., Bard, E., Potter, E.-K., *et al.* (2004). 215-ka history of sealevel oscillations from marine and continental layers in Argenterola Cave speleothems (Italy). *Global and Planetary Change*, **43**, 57–78.

Antonioli, F., Ferranti, L., Lambeck, K., *et al.* (2006a). Late Pleistocene to Holocene record of changing uplift rates in southern Calabria and eastern Sicily (southern Italy, Central Mediterranean Sea). *Tectonophysics*, **422**, 23–40.

Antonioli, F., Kershaw, S., Renda, P., *et al.* (2006b). Elevation of the last interglacial highstand in Sicily (Italy): a benchmark of coastal tectonics. *Quaternary International*, **145–146**, 3–18.

Antonioli, F., D'Orefice, M., Ducci, S., *et al.* (2011). Palaeogeographic reconstruction of northern Tyrrhenian coast using archaeological and geomorphological markers at Pianosa Island (Italy). *Quaternary International*, **232**, 31–44.

Anzidei, M., Antonioli, F., Lambeck, K., *et al.* (2011). New insights on the relative sea level change during Holocene along the coasts of Tunisia and western Libya from archaeological and geomorphological markers. *Quaternary International*, **232**, 5–12.

Armijo, R., Meyer, B., King, C. P., *et al.* (1996). Quaternary evolution of the Corinth Rift and its implications for the Late Cenozoic evolution of the Aegean. *Geophysical Journal International*, **126**, 11–53.

Arnold, J. R. (1992). The early years with Libby at Chicago: a retrospective. *In* R. E. Taylor, A. Long & R. S. Kra (Eds), *Radiocarbon after four decades: an interdisciplinary perspective*, pp. 3–10. Springer-Verlag, New York.

Arnold, J. R. & Libby, W. (1949). Age determinations by radiocarbon content, checks with samples of known age. *Science*, **110**, 678–680.

Arnold, J. R. & Libby, W. F. (1951). Radiocarbon dates. *Science*, **113**, 111–120.

Arnold, L. D. (1995). Conventional radiocarbon dating. *In* N. W. Rutter & N. R. Catto (Eds), *Dating methods for Quaternary deposits*, pp. 107–115, Geotext 2. Geological Association of Canada, St Johns.

Arz, H. W., Lamy, F., Ganopolski, A., *et al.* (2007). Dominant Northern Hemisphere climate control over millennial-scale glacial sea-level variability. *Quaternary Science Reviews*, **26**, 312–321.

Ascough, P., Cook, G. & Dugmore, A. (2005). Methodological approaches to determining the marine radiocarbon reservoir effect. *Progress in Physical Geography*, **29**, 532–547.

Ase, L.-E. (1981). Studies of shores and shore displacement on the southern coast of Kenya: especially in Kilifi district. *Geografiska Annaler*, **63A**, 303–310.

Aubrey, D. G. & Emery, K. O. (1986). Australia – an unstable platform for tide-gauge measurements of changing sea levels. *Journal of Geology*, **94**, 699–712.

Aubrey, D. G., Emery, K. O. & Echupi, E. (1988). Changing coastal levels of South America and the Caribbean region from tide-gauge records. *Tectonophysics*, **154**, 269–284.

Austin, W. E. N. & Hibbert, F. D. (2012). Tracing time in the ocean: a brief review of chronological constraints (60–8 kyr) on North Atlantic marine event-based stratigraphies. *Quaternary Science Reviews*, **36**, 28–37.

Babbage, C. (1847). Observations on the Temple of Serapis, at Pozzuoli, near Naples, with remarks on certain causes which may produce geological cycles of great extent. *The Quarterly Journal of the Geological Society of London*, **3**, 186–217.

Bada, J. L. (1985). Racemization of amino acids. *In* G. C. Barrett (Ed.), *Chemistry and biochemistry of amino acids*, pp. 399–414. Chapman and Hall.

Bada, J. L. & Schroeder, R. A. (1976). Correction in the Glacial–Postglacial temperature difference computed from amino acid racemization. *Science*, **191**, 102–103.

Baeteman, C. (2008). Radiocarbon-dated sediment sequences from the Belgian coastal plain: testing the hypothesis of fluctuating or smooth late-Holocene relative sea-level rise. *The Holocene*, **18**, 1219–1228.

Baker, R. G. V. & Haworth, R. J. (1997). Further evidence from relic shell crust sequences for a late Holocene higher sea level for eastern Australia. *Marine Geology*, **141**, 1–9.

Baker, R. G. V. & Haworth, R. J. (2000a). Smooth or oscillating late Holocene sea-level curve? Evidence from cross-regional statistical regressions of fixed biological indicators. *Marine Geology*, **163**, 353–365.

Baker, R. G. V. & Haworth, R. J. (2000b). Smooth or oscillating late Holocene sea-level curve? Evidence from palaeozoology of fixed biological indicators in east Australia and beyond. *Marine Geology*, **163**, 367–386.

Baker, R. G. V., Haworth, R. J. & Flood, P. G. (2001a). Inter-tidal fixed indicators of former Holocene sea levels in Australia: a summary of sites and a review of methods and models. *Quaternary International*, **83–85**, 257–273.

Baker, R. G. V., Haworth, R. J. & Flood, P. G. (2001b). Warmer or cooler late Holocene marine palaeoenvironments? Interpreting southeast Australian and Brazilian sea-level changes using fixed biological indicators and their $\delta^{18}O$ composition. *Palaeogeography, Palaeoclimatology, Palaeoecology*, **168**, 249–272.

Baker, R. G. V., Davis, A. M., Aitchison, J. C. *et al.* (2003). Comment on mid-Holocene higher sea-level indicators from the south China coast. *Marine Geology*, **196**, 91–101.

Baker, R. G. V., Haworth, R. J. & Flood, P. G. (2005). An oscillating Holocene sea-level? Revisiting Rottnest Island, Western Australia and the Fairbridge eustatic hypothesis. *Journal of Coastal Research*, **42**, 3–14.

Balco, G., Schaefer, J. M. & LARISSA Group (2013). Exposure-age record of Holocene ice sheet and ice shelf change in the northeast Antarctic Peninsula. *Quaternary Science Reviews*, **59**, 101–111.

Ball, M. M. (1967). Carbonate sand bodies of Florida and the Bahamas. *Journal of Sedimentary Petrology*, **37**, 556–591.

Ball, R. (1891). *The cause of an ice age*. Kegan Paul, Trench, Trübner and Co., London, 180 pp.

Ballantyne, C. K., McCarroll, D., Nesje, A., *et al.* (1998). The last ice sheet in north-west Scotland: reconstruction and implications. *Quaternary Science Reviews*, **17**, 1149–1184.

Ballarini, M., Wallinga, J., Murray, A. S., *et al.* (2003). Optical dating of young coastal dunes on a decadal time scale. *Quaternary Science Reviews*, **22**, 1011–1017.

Ballu, V., Bouin, M.-N., Siméoni, P., *et al.* (2011). Comparing the role of absolute sea-level rise and vertical tectonic motions in coastal flooding, Torres Islands (Vanuatu). *Proceedings of the National Academy of Sciences*, **108**, 13019–13022.

Baltzer, F. (1969). *Les formations végétales associées au delta de al Dumbea (Nouvelle Calédonie) et leurs indications écologiques géomorphologiques et sédimentologiques, mises en évidence par la cartographie. Cahiers ORSTOM. Sér Géologie*, Paris 1, 59–84.

Bamber, J. L. & Aspinall, W. P. (2013). An assessment of expert opinion on future sea-level rise from melting of ice sheets. *Nature Climate Change*, **3**, 1–5.

Bamber, J. L., Riva, R. E. M., Vermeersen, B. L. A., *et al.* (2009). Reassessment of the potential sea-level rise from a collapse of the West Antarctic ice sheet. *Science*, **324**, 901–903.

Banerjee, D., Hildebrand, A. N., Murray-Wallace, C. V., *et al.* (2003). New quartz SAR-OSL ages from the stranded beach dune sequence in south-east South Australia. *Quaternary Science Reviews*, **22**, 1019–1025.

Banerjee, P. K. (2000). Holocene and Late Pleistocene relative sea level fluctuations along the east coast of India. *Marine Geology*, **167**, 243–260.

Banks, K. W., Riegl, B. M., Richards, V. P., *et al.* (2008). The reef tract of continental southeast Florida (Miami-Dade, Broward and Palm Beach counties, USA). *In* B. M. Riegl, & R. E. Dodge (Eds), *Coral reefs of the USA*, pp. 175–220. Springer, New York.

Barbante, C., Fischer, H., Masson-Delmotte, V., *et al.* (2010). Climate of the last million years: new insights from EPICA and other records. *Quaternary Science Reviews*, **29**, 1–7.

Bard, E. (1988). Correction of accelerator mass spectrometry ^{14}C ages measured in planktonic foraminifera: paleoceanographic implications. *Palaeoceanography*, **3**, 635–645.

Bard, E., Hamelin, B. & Fairbanks, R. G. (1990a). U-Th ages obtained by mass spectrometry in corals from Barbados: sea level during the past 130,000 years. *Nature*, **346**, 456–458.

Bard, E., Hamelin, B., Fairbanks, R. G., *et al.* (1990b). Calibration of the ^{14}C timescale over the last 30,000 years using mass spectrometric U-Th ages from Barbados corals. *Nature*, **345**, 405–410.

Bard, E., Hamelin, B., Arnold, M., *et al.* (1996a). Deglacial sea-level record from Tahiti corals and the timing of global meltwater discharge. *Nature*, **382**, 241–244.

Bard, E., Jouannic, C., Hamelin, B., *et al.* (1996b). Pleistocene sea levels and tectonic uplift based on dating of corals from Sumba Island, Indonesia. *Geophysical Research Letters*, **23**, 1473–1476.

Bard, E., Hamelin, B. & Delanghe-Sabatier, D. (2010). Deglacial Meltwater Pulse 1B and Younger Dryas sea levels revisited with boreholes at Tahiti. *Science*, **327**, 1235–1237.

Barendregt, R. W. (1995). Paleomagnetic dating methods. *In* N. W. Rutter & N. R. Catto (Eds), *Dating methods for Quaternary deposits*, pp. 29–49. Geotext 2. Geological Association of Canada, St Johns.

Barker, H. (1970). Critical assessment of radiocarbon dating. *Philosophical Transactions of the Royal Society of London, Series A*, **269**, 37–45.

Barlow, N. L. M., Shennan, I., Long, A. J., *et al.* (2013). Saltmarshes as late Holocene tide gauges. *Global and Planetary Change*, **106**, 90–110.

Barnes, J. (1984) (Ed.), *The complete works of Aristotle*, Bollingen Series LXXI.2 Vol. 1. Princeton University Press, Princeton, 1250 pp.

Barnes, J. W., Lang, E. J. & Potratz, H. A. (1956). Ratio of ionium to uranium in coral limestones. *Science*, **124**, 175–176.

Barnett, E. J., Harvey, N., Belperio, A. P., *et al.* (1997). Sea-level indicators from a

Holocene, tide-dominated coastal succession, Port Pirie, South Australia. *Royal Society of South Australia, Transactions*, **121**, 125–135.

Barnett, T. P. (1983). Recent changes in sea level and their possible causes. *Climatic Change*, **5**, 15–38.

Barrell, J. (1912). Criteria for the recognition of ancient delta deposits. *Geological Society of America Bulletin*, **23**, 377–446.

Barrell, J. (1917). Rhythms and measurement of geological time. *Geological Society of America Bulletin*, **28**, 745–904.

Barth, M. C. & Titus, J. G. (1984). *Greenhouse effect and sea level rise: a challenge for this generation*. van Nostrand Reinhold, New York, 325 pp.

Bartrum, J. A. (1916). High water rock platforms: a phase of shoreline erosion. *Transactions of the New Zealand Institute*, **48**, 132–134.

Bassinot, F. C., Labeyrie, L. D., Vincent, E., *et al.* (1994). The astronomical theory of climate and the age of the Brunhes–Matuyama magnetic reversal. *Earth and Planetary Science Letters*, **126**, 91–108.

Bateman, H. (1910). The solution of a system of differential equations occurring in the theory of radioactive transformations. *Proceedings of the Cambridge Philosophical Society*, **15**, 423–427.

Bateman, M. D., Holmes, P. J., Carr, A. S., *et al.* (2004). Aeolianite and barrier dune construction spanning the last two glacial–interglacial cycles from the southern Cape, South Africa. *Quaternary Science Reviews*, **23**, 1681–1698.

Bateman, M. D., Carr, A. S., Murray-Wallace, C. V., *et al.* (2008). A dating intercomparison study on Late Stone Age coastal midden deposits, South Africa. *Geoarchaeology*, **23**, 715–741.

Bateman, M. D., Carr, A. S., Dunajko, A. C., *et al.* (2011). The evolution of coastal barrier systems: a case study of the Middle–Late Pleistocene Wilderness barriers, South Africa. *Quaternary Science Reviews*, **30**, 63–81.

Bates, R. L. & Jackson, J. A. (1987). *Glossary of geology*. American Geological Institute, Alexandra, Virginia, 3rd edition, 788 pp.

Bauch, H. A., Erlenkeuser, H., Helmke, J. P., *et al.* (2000). A paleoclimatic evaluation of marine oxygen isotope stage 11 in the high-northern Atlantic (Nordic seas). *Global and Planetary Change*, **24**, 27–39.

Baulig, H. (1935). The changing sea level. Institute of British Geographers, Publication No. **3**, 46 pp.

Beach, D. K. & Ginsburg, R. N. (1980). Facies succession of Pliocene–Pleistocene carbonates, northwestern Great Bahama Bank. *American Association of Petroleum Geologists Bulletin*, **64**, 1634–1642.

Beaman, R., Larcombe, P. & Carter, R. M. (1994). New evidence for the Holocene sea-level high from the inner shelf, central Great Barrier Reef, Australia. *Sedimentary Research*, **A64**: 881–885.

Beaman, R. J., Webster, J. M. & Wust, R. A. J. (2008). New evidence for drowned shelf edge reefs in the Great Barrier Reef, Australia. *Marine Geology*, **247**, 17–34.

Beaton, J. M. (1985). Evidence for a coastal occupation time-lag at Princess Charlotte Bay (North Queensland) and implications for coastal colonization and population growth theories for Aboriginal Australia. *Archaeology in Oceania*, **20**, 1–20.

Beck, J. W., Récy, J., Taylor, F., *et al.* (1997). Abrupt changes in early Holocene tropical sea surface temperature derived from coral records. *Nature*, **385**, 705–707.

Becker, M., Meyssignac, B., Letetrel, C., *et al.* (2012). Sea level variations at tropical Pacific islands since 1950. *Global and Planetary Change*, **80**–81, 85–98.

Beckinsale, R. P. & Chorley, R. J. (1991). *The history of the study of landforms or development of geomorphology*. Vol. 3. Routledge, London, 496 pp.

Beesley, P. L., Ross, G. J. B. & Wells, A. (Eds) (1998). *Mollusca: the southern synthesis. Fauna of Australia*. Part A of Vol. 5. CSIRO Publishing. Melbourne, 563 pp.

Bekaroğlu, E. (2012). Comment on 'MIS 5a and MIS 3 relatively high sea-level stands on the Hatay-Samandağ coast, Eastern Mediterranean, Turkey'. U. Doğan, A. Koçyğit, B. Varol, İ. Özer, A. Molodkov, E. Zöhra. *Quaternary International* (2012), **262**, 65–79. *Quaternary International*, **262**, 80–83.

Belknap, D. F. & Kraft, J. C. (1977). Holocene relative sea-level changes and coastal stratigraphic units on the north-west flank of the Baltimore Canyon trough geosyncline. *Journal of Sedimentary Petrology*, **47**, 610–629.

Belperio, A. P. (1979). Negative evidence for a mid-Holocene high sea level along the coastal plain of the Great Barrier Reef province. *Marine Geology*, **32**, M1–M9.

Belperio, A. P. (1993). Land subsidence and sea level rise in the Port Adelaide estuary: implications for monitoring the greenhouse effect. *Australian Journal of Earth Sciences*, **40**, 359–368.

Belperio, A. P. (1995). Quaternary. *In* J. F. Drexel & W. V. Preiss (Eds), *The geology of South Australia*. Vol. 2, pp. 219–280, The Phanerozoic. Mines and Energy, South Australia, Bulletin 54.

Belperio, A. P., Hails, J. R. & Gostin, V. A. (1983). A review of Holocene sea levels in South Australia. *In* D. Hopley (Ed.), *Australian sea levels in the last 15000 years: a review*. Monograph Series, Occasional Paper, no. 3. Department of Geography, James Cook University of North Queensland, pp. 37–47.

Belperio, A. P., Hails, J. R., Gostin, V. A., *et al.* (1984a). The stratigraphy of coastal carbonate banks and Holocene sea levels of northern Spencer Gulf, South Australia. *Marine Geology*, **61**, 297–313.

Belperio, A. P., Smith, B. W., Polach, H. A., *et al.* (1984b). Chronological studies of the Quaternary marine sediments of northern Spencer Gulf, South Australia. *Marine Geology*, **61**, 265–296.

Belperio, A. P., Gostin, V. A., Cann, J. H., *et al.* (1988). Sediment–organism zonation and the evolution of Holocene tidal sequences in southern Australia. *In* P. L. de Boer, A. van Gelder & S. D. Nio (Eds), *Tide-influenced sedimentary environments and facies*, pp. 475–497. Reidel, Dordrecht.

Belperio, A. P., Harvey, N. & Bourman, R. P. (2002). Spatial and temporal variability in the Holocene palaeosea-level record around the South Australian coastline. *Sedimentary Geology*, **150**, 153–169.

Bender, M. L., Fairbanks, R. G., Taylor, F. W., *et al.* (1979). Uranium-series dating of the Pleistocene reefs tracts of Barbados, West Indies. *Geological Society of America Bulletin*, **90**, 577–594.

Benioff, H. (1949). Seismic evidence for the fault origin of oceanic ridges. *Bulletin, Geological Society of America*, **60**, 1837–1856.

Berger, A. & Loutre, M. F. (2002). An exceptionally long interglacial ahead? *Science*, **297**, 1287–1288.

Berggren, W. A., Burckle, L. H., Cita, M. B., *et al.* (1980). Towards a Quaternary time scale. *Quaternary Research*, **13**, 277–302.

Berkeley, A., Perry, C. T., Smithers, S. G., *et al.* (2008). The spatial and vertical distribution of living (stained) benthic foraminifera from a tropical, intertidal environment, north Queensland, Australia. *Marine Micropaleontology*, **69**, 240–261.

Berkeley, A., Perry, C. T., Smithers, S. G., *et al.* (2009). Foraminiferal biofacies across mangrove–mudflat environments at Cocoa Creek, north Queensland, Australia. *Marine Geology*, **263**, 64–86.

Bernard, H. A., Major, C. F., Parrott, B. S., *et al.* (1970). *Recent sediments of southeast Texas: a field guide to the Brazos alluvial and deltaic plains and the Galveston barrier island complex*. Bureau of Economic Geology, University of Texas, Austin. Guidebook No. 11.

Berryman, K. & Hull, A. (2003). Tectonic controls on late Quaternary shorelines: a review and prospects for future research. *In* J. R. Goff, S. L. Nichol & H. L. Rouse (Eds), *The New Zealand coast: Te Tai O Aotearoa*, pp. 25–58. Dunmore Press, Palmerston North.

Berryman, K. R., Ota, Y. & Hull, A. G. (1992). Holocene coastal evolution under the influence of episodic tectonic uplift: examples from New Zealand and Japan. *Quaternary International*, **15–16**, 31–45.

Beu, A. G., Griffin, M. & Maxwell, P. A. (1997). Opening of the Drake Passage gateway and Late Miocene to Pleistocene cooling reflected in Southern Ocean molluscan dispersal: evidence from New Zealand and Argentina. *Tectonophysics*, **281**, 83–97.

Bevin, K. (1996). Equifinality and uncertainty in geomorphological modelling. *In* B. L. Rhoads & C. E. Thorn (Eds),

The scientific nature of geomorphology, pp. 289–313. John Wiley & Sons, Chichester.

Bigarella, J. J. (1965). Sand-ridge structures from Parana coastal plain. *Marine Geology*, **3**, 269–278.

Billing, N. B. (1984). Palaeosol development in Quaternary marine sediments and palaeoclimatic interpretations, Spencer Gulf, Australia. *Marine Geology*, **61**, 315–343.

Bindoff, N., Willebrand, J., Artale, V., *et al.* (2007). Observations: oceanic climate change and sea level. *In* S. Solomon, D. Qin & M. Manning (Eds), *Climate Change 2007: the physical science basis.* Contribution of Working Group I to the Fourth Assessment Report of the Intergovernmental Panel on Climate Change, pp. 385–432. Cambridge University Press, Cambridge.

Bintanja, R., van de Wal, R. S. W. & Oerlemans, J. (2005). Modelled atmospheric temperatures and global sea levels over the past million years. *Nature*, **437**, 125–128.

Bird, E. C. F. (1988). The tubeworm *Galeolaria caespitosa* as an indicator of sea level rise. *Victorian Naturalist*, **10**, 98–105.

Bird, E. (2000). *Coastal geomorphology: an introduction.* John Wiley & Sons, Chichester, 322 pp.

Bird, E. C. F. & Dent, O. (1966). Shore platforms on the South Coast of New South Wales. *Australian Geographer*, **10**, 71–80.

Bird, M. I., Ayliffe, L. K., Turney, C. S. M., *et al.* (1999). Radiocarbon dating of 'old' charcoal using a wet oxidation stepped-combustion procedure. *Radiocarbon*, **41**, 127–140.

Bird, M. I., Fifield, L. K., Teh, T. S., *et al.* (2007). An inflection in the rate of early mid-Holocene eustatic sea-level rise: a new sea-level curve from Singapore. *Estuarine Coastal and Shelf Science*, **71**, 523–536.

Biton, E., Gildor, H. & Peltier, W. R. (2008). Red Sea during the Last Glacial Maximum: implications for sea level reconstruction. *Paleoceanography*, **23**, PA1214, doi:10.1029/2007PA001431.

Blackwell, B. (1995). Electron spin resonance dating. *In* N. W. Rutter & N. R. Catto (Eds), *Dating methods for Quaternary deposits*, pp. 209–268. Geotext 2, Geological Association of Canada, St Johns.

Blackwell, B. & Schwarcz, H. P. (1995). The uranium-series disequilibrium dating methods. *In* N. W. Rutter & N. R. Catto (Eds), *Dating methods for Quaternary deposits*, pp. 167–208. Geotext2. Geological Association of Canada, St Johns.

Blanchon, P. (2005). Comments on 'Corrected western Atlantic sea-level curve for the last 11,000 years based on calibrated ^{14}C dates from *Acropora palmata* framework and intertidal mangrove peat' by Toscano and Macintyre. *Coral Reefs (2003)* **22**, 257–270. *Coral Reefs* **24**, 183–186.

Blanchon, P. (2010). Reef demise and back-stepping during the last interglacial, northeast Yucatan. *Coral Reefs*, **29**, 481–498.

Blanchon, P. (2011a). Last Interglacial and reef development. *In* D. Hopley (Ed.), *Encyclopedia of modern coral reefs*, pp. 621–639. Springer, Dordrecht.

Blanchon, P. (2011b). Meltwater pulses. *In* Hopley, D. (Ed.), *Encyclopedia of Modern Coral Reefs*, pp. 683–690. Springer.

Blanchon, P. & Eisenhauer, A. (2005). Multistage reef development on Barbados during the last Interglaciation. *Quaternary Science Reviews*, **20**, 1093–1112.

Blanchon, P. & Shaw, J. (1995). Reef drowning during the last deglaciation: evidence for catastrophic sea-level rise and ice-sheet collapse. *Geology*, **23**, 4–8.

Blanchon, P., Jones, B. & Ford, D. C. (2002). Discovery of a submerged relic reef and shoreline off Grand Cayman: further support for an early Holocene jump in sea level. *Sedimentary Geology*, **147**, 253–270.

Blanchon, P., Eisenhauer, A., Fietzke, J., *et al.* (2009). Rapid sea-level rise and reef back-stepping at the close of the last interglacial highstand. *Nature*, **458**, 881–885.

Blatt, H., Middleton, G. & Murray, R. (1980). *Origin of sedimentary rocks*, 2nd Edition. Prentice Hall, New Jersey, 782 pp.

Blewitt, G., Altamimi, Z., Davis, J., *et al.* (2010). Geodetic observations and global reference frame contributions to understanding sea-level rise and variability. *In* J. A. Church, P. L. Woodworth,

T. Aarup & W. S. Wilson (Eds), *Understanding sea-level rise and variability*, pp. 256–284. Wiley-Blackwell, Chichester.

Blockley, S. P. E., Lane, C. S., Hardiman, M., *et al.* (2012). Synchronisation of palaeoenvironmental records over the last 60,000 years, and an extended INTIMATE event stratigraphy to 48,000 b2k. *Quaternary Science Reviews*, **36**, 2–10.

Blom, W. M. (1988). Late Quaternary sediments and sea levels in Bass Basin, southeastern Australia – a preliminary report. *Search*, **19**, 94–96.

Bloom, A. L. (1963). Late-Pleistocene fluctuations of sea level and postglacial crustal rebound in coastal Maine. *American Journal of Science*, **261**, 862–879.

Bloom, A. L. (1967). Pleistocene shorelines: a new test of isostasy. *Geological Society of America Bulletin*, **78**, 1477–1494.

Bloom, A. L. (1970). Paludal stratigraphy of Truk, Ponape, and Kusaie, Eastern Caroline Islands. *Geological Society of America Bulletin*, **81**, 1895–1904.

Bloom, A. L. (1971). Glacial-eustatic and isostatic controls of sea level since the last glaciation. *In* K. K. Turekian (Ed.), *The Late Cenozoic glacial ages*, pp. 355–379. Yale University Press, New Haven.

Bloom, A. L. (1977). *Atlas of sea-level curves*. IGCP 61. Cornell University, Ithaca.

Bloom, A. L. & Stuiver, M. (1963). Submergence of the Connecticut coast. *Science*, **139**, 332–334.

Bloom, A. L., Broecker, W. S., Chappell, J. M. A., *et al.* (1974). Quaternary sea level fluctuations on a tectonic coast: new ^{230}Th/^{234}U dates from the Huon Peninsula, New Guinea. *Quaternary Research*, **4**, 185–205.

Bloom, A. L. & Yonekura, N. (1985). Coastal terraces generated by sea-level change and tectonic uplift. *In* M. J. Woldenberg (Ed.), *Models in geomorphology*, pp. 139–154. Allen & Unwin, Boston.

Blum, M. D. & Törnqvist, T. E. (2000). Fluvial responses to climate and sea-level change: a review and look forward. *Sedimentology*, **47**, 2–48.

Blum, M. D., Misner, T. J., Collins, E. S., *et al.* (2001). Middle Holocene sea-level rise and highstand at +2 m, central Texas coast. *Journal of Sedimentary Research*, **71**, 581–588.

Blum, M. D., Tomkin, J. H., Purcell, A., *et al.* (2008). Ups and downs of the Mississippi Delta. *Geology*, **36**, 675–678.

Bond, G. H., Heinrich, H., Broecker, W., *et al.* (1992). Evidence for massive discharges of icebergs into the North Atlantic Ocean during the last glacial period. *Nature*, **360**, 245–249.

Bondevik, S., Svendsen, J. I. & Mangerud, J. (1998). Distinction between the Storegga tsunami and the Holocene marine transgression in coastal basin deposits of western Norway. *Journal of Quaternary Science*, **13**, 529–537.

Bonini, L., Di Bucci, D., Toscani, G., *et al.* (2011). Reconciling deep seismogenic and shallow active faults through analogue modelling: the case of the Messina Straits (southern Italy). *Journal of the Geological Society, London*, **168**, 191–199.

Bordoni, P. & Valensise, G. (1998). Deformation of the 125 ka marine terrace in Italy: tectonic implications. *In* I. Stewart & C. Vita-Finzi (Eds), *Coastal tectonics*, Geological Society, London, Special Publications, **146**, 71–110.

Boulton, G. S., Smith, G. D., Jones, A. S., *et al.* (1985). Glacial geology and glaciology of the last mid-latitude ice sheets. *Journal of the Geological Society of London*, **142**, 447–474.

Bourman, R. P. & Murray-Wallace, C. V. (1991). Holocene evolution of a sand spit at the mouth of a large river system: Sir Richard Peninsula and the Murray Mouth, South Australia. *Zeitschrift für Geomorphologie*, Suppl.-Bd. **81**, 63–83.

Bourman, R. P., Belperio, A. P., Murray-Wallace, C. V., *et al.* (1999). A last interglacial embayment fill at Normanville, South Australia, and its neotectonic implications. *Royal Society of South Australia, Transactions*, **123**, 1–15.

Bourman, R. P., Murray-Wallace, C. V., Belperio, A. P., *et al.* (2000). Rapid coastal geomorphic change in the River Murray Estuary of Australia. *Marine Geology*, **170**, 141–168.

Boutakoff, N. (1963). The geology and geomorphology of the Portland area. *Victoria Geological Survey Memoirs*, **22**, 172 pp.

Bowden, A. R. & Colhoun, E. A. (1984). Quaternary emergent shorelines of Tasmania. *In* B. G. Thom (Ed.), *Coastal geomorphology in Australia*, pp. 313–342. Academic Press, Sydney.

Bowen, D. Q. (1985). *Quaternary geology: a stratigraphic framework for multidisciplinary work.* Pergamon, London, 237 pp.

Bowen, D. Q. (2003). Uncertainty in oxygen isotope stage 11 sea-level: Great Britain. *In* A. W. Droxler, R. Z. Poore & L. H. Burkle (Eds), *Earth's climate and orbital eccentricity: the Marine Isotope Stage 11 question.* Geophysical Monograph, **137**, 131–144.

Bowen, D. Q., Rose, J., McCabe, A. M., *et al.* (1986). Correlation of Quaternary glaciations in England, Ireland, Scotland and Wales. *Quaternary Science Reviews*, **5**, 299–340.

Bowen, D. Q., Pillans, B., Sykes, G. A., *et al.* (1998). Amino acid geochronology of Pleistocene marine sediments in the Wanganui Basin: a New Zealand framework for correlation and dating. *Journal of the Geological Society, London*, **155**, 439–446.

Bowler, J. M. (1982). Aridity in the late Tertiary, Quaternary of Australia. *In* W. R. Barker & P. J. M. Greenslade (Eds), *Evolution of the flora and fauna of arid Australia*, pp. 35–45. Peacock Publications, Adelaide.

Bowler, J. M. (1986). Quaternary landform evolution. *In* D. N. Jeans (Ed.), *Australia – a geography.* Vol. 1, *The Natural Environment*, 2nd Ed., pp. 117–147. Sydney University Press, Sydney.

Bowler, J. M., Kotsonis, A. & Lawrence, C. R. (2006). Environmental evolution of the Mallee region, western Murray Basin. *Proceedings of the Royal Society of Victoria*, **118**, 161–210.

Bowman, G. & Harvey, N. (1983). Radiocarbon dating marine shells in South Australia. *Australian Archaeology*, **17**, 113–123.

Bowman, S. G. E. (1990). *Radiocarbon dating.* British Museum, London.

Boyd, R., Dalrymple, R. & Zaitlin, B. A. (1992). Classification of clastic coastal depositional environments. *Sedimentary Geology*, **80**, 139–150.

Boyd, R., Rumming, K., Goodwin, I., *et al.* (2008). Highstand transport of coastal sand to the deep ocean: a case study from Fraser Island, southeast Australia. *Geology*, **36**, 15–18.

Bradley, R. S. (1999). *Paleoclimatology: reconstructing climates of the Quaternary.* International Geophysics Series Vol. 64. Academic Press, San Diego, 613 pp.

Bradley, S. L., Milne, G. A., Teferle, F. N., *et al.* (2009). Glacial isostatic adjustment of the British Isles: new constraints from GPS measurements of crustal motion. *Geophysical Journal International*, **178**, 14–22.

Bradley, S. L., Milne, G. A., Shennan, I., *et al.* (2011). An improved glacial isostatic adjustment model of the British Isles. *Journal of Quaternary Science*, **26**, 541–552.

Braithwaite, C. J. R. (1984). Depositional history of the late Pleistocene Limestones of the Kenya coast. *Journal of the Geological Society, London*, **141**, 685–699.

Braithwaite, C. J. R., Taylor, J. D. & Kennedy, W. J. (1973). The evolution of an atoll: the depositional and erosional history of Aldabra. *Philosophical Transactions of the Royal Society of London, B*, **266**, 307–340.

Bretz, J. H. (1960). Bermuda: a partially drowned late mature Pleistocene karst. *Geological Society of America Bulletin*, **81**, 2523–2524.

Brice, W. R. (1982). Bishop Ussher, John Lightfoot and the age of creation. *Journal of Geological Education*, **30**, 18–24.

Brigham, J. K. (1983). Intrashell variations in amino acid concentrations and isoleucine epimerization ratios in fossil *Hiatella arctica. Geology*, **11**, 509–513.

Broecker, W. (2010). *The great ocean conveyor: discovering the trigger for abrupt climate change.* Princton University Press, Princeton, 154 pp.

Broecker, W. S. & van Donk, J. (1970). Insolation changes, ice volumes and ^{18}O record in deep-sea cores. *Reviews of Geophysics and Space Physics*, **8**, 169–198.

Broecker, W. S., Thurber, D. L., Goddard, J., *et al.* (1968). Milankovitch hypothesis supports the precise dating of coral reefs and deep sea sediments. *Science*, **159**, 297–300.

Broecker, W. S., Peng, T.-H., Ostlund, G., *et al.* (1985). The distribution of bomb radiocarbon in the ocean. *Journal of Geophysical Research*, **90**, 6953–6970.

Brooke, B. P. (2001). The distribution of carbonate eolianite. *Earth-Science Reviews*, **55**, 135–164.

Brooke, B. P., Young, R. W., Bryant, E. A., et al. (1994). A Pleistocene origin for shore platforms along the northern Illawarra coast, New South Wales. *Australian Geographer*, **25**, 178–185.

Brooke, B. P., Murray-Wallace, C. V., Woodroffe, C. D., et al. (2003a). Quaternary aminostratigraphy of eolianite on Lord Howe Island, southwest Pacific Ocean. *Quaternary Science Reviews*, **22**, 387–406.

Brooke, B. P., Woodroffe, C. D., Murray-Wallace, C. V., et al. (2003b). Quaternary calcarenite stratigraphy on Lord Howe Island, southwestern Pacific Ocean and the record of coastal carbonate deposition. *Quaternary Science Reviews*, **22**, 859–880.

Brooks, A. J., Bradley, S. L., Edwards, R. J., et al. (2008). Postglacial relative sea-level observations from Ireland and their role in glacial rebound modelling. *Journal of Quaternary Science*, **23**, 175–192.

Brothers, R. N. (1954). The relative Pleistocene chronology of the South Kaipara district, New Zealand. *Transactions of the Royal Society of New Zealand*, **82**, 677–694.

Brown, C. M. & Stephenson, A. E. (1991). *Geology of the Murray Basin, Southeastern Australia*. BMR Bulletin 235. Bureau of Mineral Resources, Geology and Geophysics, Canberra, 430 pp.

Brown, S. L. (1998). Sedimentation on a Humber saltmarsh. *In* K. S. Black, D. M. Paterson & A. Cramp (Eds), *Sedimentary processes in the intertidal zone*. Geological Society of London, *Special Publication*, **139**, pp. 69–83.

Bruggemann, J. H., Buffler, R. T., Guillaume, M. M., et al. (2004). Stratigraphy, palaeoenvironments and model for the deposition of the Abdur Reef Limestone: context for an important archaeological site from the last interglacial on the Red Sea coast of Eritrea. *Palaeogeography, Palaeoclimatology, Palaeoecology*, **203**, 179–206.

Bryan, W. B. & Stephens, R. S. (1993). Coastal bench formation at Hanauma Bay, Oahu, Hawaii. *Geological Society of America Bulletin*, **105**, 377–386.

Bryant, E. (1988). Sea-level variability and its impact within the greenhouse scenario. *In* G. I. Pearman (Ed.), *Greenhouse: planning for climate change*, pp. 135–146. CSIRO Division of Atmospheric Research, E. J. Brill, Leiden.

Bryant, E. A. (1993). The magnitude and nature of 'noise' in world sea-level records. *In* R. N. Chowdhury & S. M. Sivakumar (Eds), *Environmental management geo-water and engineering*, pp. 747–751. A. A. Balkema, Rotterdam.

Bryant, E. (2001). *Tsunami: the underrated hazard*. Cambridge University Press, Cambridge, 320 pp.

Bryant, E. A., Young, R. W., Price, D. M., et al. (1990). Thermoluminescence and uranium-thorium chronologies of Pleistocene coastal landforms of the Illawarra region, New South Wales. *Australian Geographer*, **21**, 101–112.

Buckland, W. (1823). *Reliquiae diluvianae; or observations on the organic remains contained in caves, fissures and diluvial gravel and on other geological phenomena attesting the action of an universal deluge.* John Murray, London.

Bull, W. B. & Cooper, A. F. (1986). Uplifted marine terraces along the alpine fault, New Zealand. *Science*, **234**, 1225–1228.

Bunt, J. S., Williams, W. T. & Bunt, E. D. (1985). Mangrove species distribution in relation to tide at the seafront and up rivers. *Australian Journal of Marine and Freshwater Research*, **36**, 481–492.

Burkle, L. H. (1993). Late Quaternary interglacial stages warmer than present. *Quaternary Science Reviews*, **12**, 825–831.

Burne, R. V. (1982). Relative fall of Holocene sea level and coastal progradation, northeastern Spencer Gulf, South Australia. *BMR Journal of Australian Geology and Geophysics*, **7**, 35–45.

Burroughs, W. J. (2005). *Climate change in prehistory – the end of the reign of chaos*. Cambridge University Press, Cambridge, 356 pp.

Burton, T. E. (1982). Mangrove development north of Adelaide, 1935–1982. *Royal Society of South Australia, Transactions*, **106**, 183–189.

Butzer, K. W. (1962). Coastal geomorphology of Majorca. *Annals of the Association of American Geographers*, **52**, 191–212.

Cabanes, C., Cazenave, A. & Le Provost, C. (2001). Sea level change from Topex-Poseidon altimetry for 1993–1999 and possible warming of the southern oceans. *Geophysical Research Letters*, **28**, 9–12.

Cabioch, G., Correge, T., Turpin, L., *et al.* (1999). Development patterns of fringing and barrier reefs in New Caledonia (southwest Pacific). *Oceanologica Acta*, **22**, 567–578.

Cabioch, G., Banks-Cutler, K., Beck, W. J., *et al.* (2003). Continuous reef growth during the last 23 ka in a tectonically active zone (Vanuatu, SouthWest Pacific). *Quaternary Science Reviews*, **22**, 1771–1786.

Cabioch, G., Montaggioni, L., Frank, N., *et al.* (2008a). Successive reef depositional events along the Marquesas foreslopes (French Polynesia) since 26 ka. *Marine Geology*, **254**, 18–34.

Cabioch, G., Montaggioni, L. F., Thouveny, N., *et al.* (2008b). The chronology and structure of the western New Caledonian barrier reef tracts. *Palaeogeography, Palaeoclimatology, Palaeoecology*, **268**, 91–105.

Callard, S. L., Gehrels, W. R., Morrison, B. V., *et al.* (2011). Suitability of saltmarsh foraminifera as proxy indicators of sea level in Tasmania. *Marine Micropaleontology*, **79**, 121–131.

Camoin, G. F., Colonna, M. & Montaggioni, L. F. (1997). Holocene sea level changes and reef development in the southwestern Indian Ocean. *Coral Reefs*, **16**, 247–259.

Camoin, G. F., Ebren, P., Eisenhauer, A., *et al.* (2001). A 300000-yr coral reef record of sea level changes, Mururoa atoll (Tuamotu Archipelago, French Polynesia). *Paleogeography, Palaeoclimatology, Palaeoecology*, **175**, 325–241.

Camoin, G. F., Montaggioni, L. F. & Braithwaite, C. J. R. (2004). Late glacial to post glacial sea levels in the Western Indian Ocean. *Marine Geology*, **206**, 119–146.

Camoin, G., Seard, C., Deschamps, P., *et al.* (2012). Reef response to sea-level and environmental changes during the last deglaciation: Integrated Ocean Drilling Program Expedition 310, Tahiti Sea Level. *Geology*, **40**, 643–646.

Campbell, J. F. (1986). Subsidence rates for the southeastern Hawaiian Islands determined from submerged terraces. *Geo-Marine Letters*, **6**, 139–146.

Cane, S. (2001). The Great Flood: eustatic change and cultural change in Australia during the late Pleistocene and Holocene. *In* A. Anderson, I. Lilley & S. O'Connor (Eds), *Histories of old age: essays in honour of Rhys Jones*, pp. 141–165. Pandanus Books, Australian National University, Canberra.

Cann, J. H. (1978). An exposed reference section for the Glanville Formation. *Quarterly Geological Notes, Geological Survey of South Australia*, **65**, 2–4.

Cann, J. H. & Clarke, J. D. A. (1993). The significance of *Marginopora vertebralis* (Foraminifera) in surficial sediments at Esperance, Western Australia, and in last interglacial sediments in northern Spencer Gulf, South Australia. *Marine Geology*, **111**, 171–187.

Cann, J. H. & Gostin, V. A. (1985). Coastal sedimentary facies and foraminiferal biofacies of the St Kilda Formation at Port Gawler, South Australia. *Royal Society of South Australia, Transactions*, **109**, 121–142.

Cann, J. H. & Murray-Wallace, C. V. (1999). Source of food items in an Aboriginal midden at Little Dip, near Robe, southeastern South Australia: implications for coastal geomorphic change. *Royal Society of South Australia, Transactions*, **123**, 43–51.

Cann, J. H., Belperio, A. P., Gostin, V. A., *et al.* (1988). Sea-level history, 45,000 to 30,000 yr B.P., inferred from benthic foraminifera, Gulf St. Vincent, South Australia. *Quaternary Research*, **29**, 153–175.

Cann, J. H., De Deckker, P. & Murray-Wallace, C. V. (1991). Coastal Aboriginal shell middens and their palaeoenvironmental significance, Robe Range, South Australia. *Royal Society of South Australia, Transactions*, **115**, 161–175.

Cann, J. H., Belperio, A. P., Gostin, V. A., *et al.* (1993). Contemporary benthic foraminifera in Gulf St. Vincent, South

Australia and a refined Late Pleistocene sea-level history. *Australian Journal of Earth Sciences*, **40**, 197–211.

Cann, J. H., Belperio, A. P. & Murray-Wallace, C. V. (2000). Late Quaternary paleosealevels and paleoenvironments inferred from foraminifera, northern Spencer Gulf, South Australia. *Journal of Foraminiferal Research*, **30**, 29–53.

Cann, J. H., Murray-Wallace, C. V., Riggs, N. J., *et al.* (2006). Successive foraminiferal faunas and inferred palaeoenvironments associated with the postglacial (Holocene) marine transgression, Gulf St. Vincent, South Australia. *The Holocene*, **16**, 224–234.

Caputo, M. & Pieri, L. (1976). Eustatic variation in the last 2000 years in the Mediterranean. *Journal of Geophysical Research*, **81**, 5787–5790.

Carey, S. W. (1976). *The expanding Earth*. Elsevier, Amsterdam, 488 pp.

Carlson, A. E., Legrande, A. N., Oppo, D. W., *et al.* (2008). Rapid early Holocene deglaciation of the Laurentide ice sheet. *Nature Geoscience*, **1**, 620–624.

Carozzi, A. V. (1965). Lavoisier's fundamental contribution to stratigraphy. *The Ohio Journal of Science*, **65**, 71–85.

Carozzi, A. V. (Ed.) (1968). *Telliamed or conversations between an Indian philosopher and a French Missionary on the diminution of the sea by Benoît de Maillet*. University of Illinois Press, Urbana, 465 pp.

Carozzi, A. V. (1969). de Maillett's Telliamed (1748): an ultra-neptunian theory of the Earth. *In*, C. J. Schneer (Ed.), *Toward a history of geology*, pp. 80–99. The MIT Press, Cambridge, Massachusetts.

Carozzi, A. V. (1992). De Maillett's Telliamed (1748): the diminution of the sea or the fall portion of a complete cosmic eustatic cycle. *In* R. H. Dott (Ed.), *Eustasy: the historical ups and downs of a major geological concept*. Geological Society of America Memoir 180, pp. 17–24. Boulder, Colorado.

Carr, A. P. & Graff, J. (1982). The tidal immersion factor and shore platform development. *Transactions of the Institute of British Geographers*, **7**, 240–245.

Carr, A. S., Bateman, M. D., Roberts, D. L., *et al.* (2010). The last interglacial sea-level high stand on the southern Cape coastline of South Africa. *Quaternary Research*, **73**, 351–363.

Carter, R. M. & Johnson, D. P. (1986). Sea-level controls on the post-glacial development of the Great Barrier Reef, Queensland. *Marine Geology*, **71**, 137–164.

Carter, R. M. & Naish, T. R. (1998). A review of Wanganui Basin, New Zealand: global reference section for shallow marine, Plio-Pleistocene (2.5–0 Ma) cyclostratigraphy. *Sedimentary Geology*, **122**, 37–52.

Carter, R. M., Carter, L. & Johnson, D. P. (1986). Submergent shorelines in the SW Pacific: evidence for an episodic postglacial transgression. *Sedimentology*, **33**, 629–649.

Carter, R. M., Abbott, S. T. & Naish, T. R. (1999). Plio-Pleistocene cyclothems from Wanganui Basin, New Zealand: type locality for an astrochronologic timescale, or template for recognizing ancient glacio-eustasy? *Philosophical Transactions of the Royal Society, London*, Series A, **357**, 1861–1872.

Carter, R. W. G. (1982). Sea-level changes in Northern Ireland. *Proceedings of the Geological Association*, **93**, 7–23.

Carter, R. W. G., Johnston, T. W., McKenna, J., *et al.* (1987). Sea-level, sediment supply and coastal changes: examples from the coast of Ireland. *Progress in Oceanography*, **18**, 79–101.

Catto, N. (1995). Other isotopic methods. *In* N. W. Rutter & N. R. Catto (Eds), *Dating methods for Quaternary deposits*, pp. 67–71. Geo Text2, Geological Association of Canada, St Johns.

Catuneanu, O. (2006). *Principles of sequence stratigraphy*. Elsevier, Amsterdam, 375 pp.

Cazenave, A. & Llovel, W. (2010). Contemporary sea level rise. *Annual Review of Marine Science*, **2**, 45–73.

Cazenave, A., Dominh, K., Allegre, C. J., *et al.* (1986). Global relationship between oceanic geoid and topography. *Journal of Geophysical Research*, **91**, 11439–11450.

Celsius, A. (1743). Anmärkningar om vatnets förminskande så i Östersjön som Vesterhafvet. *Kungl. Vet. Akad. Handl.* 1743.

Cerling, T. E. & Craig, H. (1994). Geomorphology and in-situ cosmogenic isotopes.

Annual Review of Earth and Planetary Science, **22**, 273–317.

Chamberlin, T. C. (1890). The method of multiple working hypotheses. *Science*, **15**, 92–96.

Chamberlin, T. C. (1897). The method of multiple working hypotheses. *Journal of Geology*, **5**, 837–848.

Chambers, R. (1848). *Ancient sea–margins, as memorials of changes in the relative level of sea and land.* W. & R. Chambers, Edinburgh, 338 pp.

Chao, B. F., Wu, Y. H. & Li, Y. S. (2008). Impact of artificial reservoir water impoundment on global sea level. *Science*, **320**, 212–214.

Chapman, D. M., Geary, M., Roy, P. S., et al. (1982). *Coastal evolution and coastal erosion in New South Wales.* Coastal Council of New South Wales, Sydney, 340 pp.

Chappell, J. (1974a). Geology of coral terraces, Huon Peninsula, New Guinea: a study of Quaternary tectonic movements and sea-level changes. *Geological Society of America Bulletin*, **85**, 553–570.

Chappell, J. (1974b). Upper mantle rheology in a tectonic region: evidence from New Guinea. *Journal of Geophysical Research*, **79**, 390–398.

Chappell, J. (1974c). Late Quaternary glacio- and hydro-isostasy on a layered Earth. *Quaternary Research*, **4**, 429–440.

Chappell, J. (1974d). The geomorphology and evolution of small valleys in dated coral reef terraces, New Guinea. *Journal of Geology*, **82**, 795–812.

Chappell, J. (1975). Upper Quaternary warping and uplift rates in the Bay of Plenty and West Coast, North Island, New Zealand. *New Zealand Journal of Geology and Geophysics*, **18**, 129–155.

Chappell, J. (1980). Coral morphology, diversity and reef growth. *Nature*, **286**, 249–252.

Chappell, J. (1982). Radiocarbon dating uncertainties and their effects on studies of the past. *In* W. Ambrose & P. Duerden (Eds), *Archaeometry: an Australasian perspective*, pp. 322–335. Australian National University Press, Canberra.

Chappell, J. (1983a). A revised sea-level record for the last 300,000 years from Papua New Guinea. *Search*, **14**, 99–101.

Chappell, J. (1983b). Sea level changes and coral reef growth. *In* D. J. Barnes (Ed.), *Perspectives on coral reefs*, pp. 46–55. Australian Institute of Marine Science, Brian Clouston Publisher, Canberra.

Chappell, J. (1983c). Aspects of sea levels, tectonics, and isostasy since the Cretaceous. *In* R. Garner & H. Scoging (Eds), *Mega-geomorphology*, pp. 56–72. Clarendon Press, Oxford.

Chappell, J. (1983d). Evidence for smoothly falling sea levels relative to north Queensland, Australia, during the past 6000 years. *Nature*, **302**, 406–408.

Chappell, J. (1987). Late Quaternary sea-level changes in the Australian region. *In* M. J. Tooley & I. Shennan (Eds), *Sea-level changes*, pp. 296–331. Basil Blackwell, London.

Chappell, J. (1993a). Contrasting Holocene sedimentary geologies of lower Daly River, northern Australia, and lower Sepik-Ramu, Papua New Guinea. *Sedimentary Geology*, **83**, 339–358.

Chappell, J. (1993b). Late Pleistocene coasts and human migrations in the Austral Region. *In* M. Spriggs, D. E. Yen, W. Ambrose, R. Jones, A. Thorne & A. Andrews (Eds), *A community of culture – the people and prehistory of the Pacific*, pp. 43–48. Occasional Papers in Prehistory No. 21, Department of Prehistory, Research School of Pacific Studies, The Australian National University, Canberra.

Chappell, J. (2002). Sea level changes forced ice breakouts in the Last Glacial cycle: new results from coral terraces. *Quaternary Science Reviews*, **21**, 1229–1240.

Chappell, J. & Grindrod, J. (1984). Chenier Plain Formation in Northern Australia. *In* B. G. Thom (Ed.), *Coastal geomorphology in Australia*, pp. 197–231. Academic Press, Sydney.

Chappell, J. & Polach, H. A. (1976). Holocene sea-level change and coral-reef growth at Huon Peninsula, Papua New Guinea. *Geological Society of America Bulletin*, **87**, 235–240.

Chappell, J. & Polach, H. A. (1991). Postglacial sea-level rise from a coral record at Huon Peninsula, Papua New Guinea. *Nature*, **349**, 147–149.

Chappell, J. & Shackleton, N. J. (1986). Oxygen isotopes and sea level. *Nature*, **324**, 137–140.

Chappell, J. & Thom, B. (1977). Sea levels and coasts. *In* J. Allen, J. Golson & B. Thom (Eds), *Sunda and Sahul*, pp. 275–291. Academic Press, New York.

Chappell, J. & Thom, B. G. (1986). Coastal morphodynamics in north Australia: review and prospect. *Australian Geographical Studies*, **24**, 110–127.

Chappell, J. & Veeh, H. H. (1978). Late Quaternary tectonic movements and sea-level changes at Timor and Atauro Island. *Geological Society of America Bulletin*, **89**, 356–368.

Chappell, J., Rhodes, E. G., Thom, B. G., et al. (1982). Hydro-isostasy and the sea-level isobase of 5500 B.P. in north Queensland, Australia. *Marine Geology*, **49**, 81–90.

Chappell, J., Chivas, A., Wallensky, E., et al. (1983). Holocene palaeo-environment changes, central to north Great Barrier Reef inner zone. *BMR Journal Australian Geology and Geophysics*, **8**, 223–235.

Chappell, J., Omura, A., Esat, T., et al. (1996a). Reconciliation of late Quaternary sea levels derived from coral terraces at Huon Peninsula with deep sea oxygen isotope records. *Earth and Planetary Science Letters*, **141**, 227–236.

Chappell, J., Ota, Y. & Berryman, K. (1996b). Late Quaternary coseismic uplift history of Huon Peninsula, Papua New Guinea. *Quaternary Science Reviews*, **15**, 7–22.

Chappell, J., Ota, Y. & Campbell, C. (1998). Decoupling post-glacial tectonism and eustasy at Huon Peninsula, Papua New Guinea. *In* I. Stewart & C. Vita-Finzi (Eds), *Coastal Tectonics*. Geological Society, London, Special Publications, **146**, 31–40.

Chaput, E. (1917). Recherches sur les terrasses alluviales de la Loire et de ses principaux affluents. *Annales de la Université de Lyon, N. Sér.*, **1**, 1–303.

Chaput, E. (1927). Les principales phases de l'evolution de la Vallée de la Seine. *Annales de Géographie*, **36**, 125–135.

Chen, J. H., Edwards, R. L. & Wasserburg, G. J. (1986). ^{238}U, ^{234}U and ^{232}Th in seawater. *Earth and Planetary Science Letters*, **80**, 241–251.

Chen, J. H., Curran, H. A., White, B., et al. (1991). Precise chronology of the last interglacial period: ^{234}U/^{230}Th data from fossil coral reefs in the Bahamas. *Geological Society of America Bulletin*, **103**, 82–97.

Chen, J. L., Wilson, C. R., Tapley, B. D., et al. (2005). Seasonal global mean sea level change from satellite altimeter, GRACE, and geophysical models. *Journal of Geodesy*, **79**, 532–539.

Chenhall, B. E., Yassini, I., Depers, A. M., et al. (1995). Anthropogenic marker evidence for accelerated sedimentation in Lake Illawarra, New South Wales, Australia. *Environmental Geology*, **26**, 124–135.

Chiu, T.-C., Fairbanks, R. G., Cao, L., et al. (2007). Analysis of the atmospheric ^{14}C record spanning the past 50,000 years derived from high-precision ^{230}Th/^{234}U/^{238}U, ^{231}Pa/^{235}U and ^{14}C dates on fossil corals. *Quaternary Science Reviews*, **26**, 18–36.

Chivas, A., Chappell, J., Polach, H., et al. (1986). Radiocarbon evidence for the timing and rate of island development, beach-rock formation and phosphatization at Lady Elliot Island, Queensland, Australia. *Marine Geology*, **69**, 273–287.

Chivas, A. R., García, A., van der Kaars, S., et al. (2001). Sea-level and environmental changes since the last interglacial in the Gulf of Carpentaria, Australia: an overview. *Quaternary International*, **83–85**, 19–46.

Choi, K. H., Seong, Y. B., Jung, P. M., et al. (2012). Using cosmogenic ^{10}Be dating to unravel the antiquity of a rocky shore platform on the west coast of Korea. *Journal of Coastal Research*, **28**, 641–657.

Chorley, R. J. & Kennedy, B. A. (1971). *Physical geography – a systems approach*. Prentice-Hall International Inc., London, 370 pp.

Chorley, R. J., Dunn, A. J. & Beckinsale, R. P. (1964). *The history of the study of landforms, or the development of geomorphology. Volume 1, Geomorphology before Davis*. Methuen & Co., London, 678 pp.

Christiansen, B., Schmith, T. & Thejll, P. (2010). A surrogate ensemble study of sea level reconstructions. *Journal of Climate*, **23**, 4306–4326.

Church, J. A. & White, N. J. (2006). A 20th century acceleration in global sea-level rise. *Geophysical Research Letters*, **33**, L01602.

Church, J. A. & White, N. J. (2011). Sea-level rise from the late 19th to the early 21st century. *Surveys in Geophysics*, **24**, 339–386.

Church, J. A., Gregory, J. M., Huybrechts, P., *et al.* (2001). Changes in sea level. *In* J. T. Houghton, Y. Ding, D. J. Griggs, M. Noguer, P. J. van der Linden & D. Xiaosu (Eds), *Climate change 2001. The scientific basis*, pp. 639–693. Cambridge University Press, Cambridge.

Church, J. A., White, N. J., Coleman, R., *et al.* (2004). Estimates of the regional distribution of sea-level rise over the 1950 to 2000 period. *Journal of Climatology*, **17**, 2609–2625.

Church, J. A., White, N. J. & Arblaster, J. M. (2005). Significant decadal-scale impact of volcanic eruptions on sea level and ocean heat content. *Nature*, **438**, 74–77.

Church, J. A., White, N. J. & Hunter, J. R. (2006). Sea-level rise at tropical Pacific and Indian Ocean islands. *Global and Planetary Change*, **53**, 155–168.

Church, J. A., White, N. J., Aarup, T., *et al.* (2008). Understanding global sea levels: past, present and future. *Sustainability Science*, **3**, 9–22.

Church, J. A., Roemimich, D., Domingues, C. M., *et al.* (2010a). Ocean temperature and salinity contributions to global and regional sea-level change. *In* J. A. Church, P. L. Woodworth, T. Aarup & W. S. Wilson (Eds), *Understanding sea-level rise and variability*, pp. 143–176. Wiley-Blackwell, Chichester.

Church, J. A., Aarup, T., Woodworth, P. L., *et al.* (2010b). Sea-level rise and variability: synthesis and outlook for the future. *In* J. A. Church, P. L. Woodworth, T. Aarup & W. S. Wilson (Eds), *Understanding sea-level rise and variability*, pp. 402–419. Wiley-Blackwell, Chichester.

Church, J. A., White, N. J., Konikow, L. F., *et al.* (2011a). Revisiting the Earth's sea-level and energy budgets from 1961 to 2008. *Geophysical Research Letters*, **38**, (L18601).

Church, J. A., Gregory, J. M., White, N. J., *et al.* (2011b). Understanding and projecting sea level change. *Oceanography*, **24**, 130–143.

Clark, J. A. & Lingle, C. S. (1979). Predicted relative sea-level changes (18,000 years B.P. to Present) caused by late-glacial retreat of the Antarctic ice sheet. *Quaternary Research*, **11**, 279–298.

Clark, J. A., Farrell, W. E. & Peltier, W. R. (1978). Global changes in postglacial sea level: a numerical calculation. *Quaternary Research*, **9**, 265–287.

Clark, P. U. & Mix, A. C. (2002). Ice sheets and sea level of the Last Glacial Maximum. *Quaternary Science Reviews*, **21**, 1–7.

Clark, P. U., Mitrovica, J. X., Milne, G. A., *et al.* (2002). Sea-level fingerprinting as a direct test for the source of global meltwater pulse IA. *Science*, **295**, 2438–2441.

Clark, P. U., Archer, D., Pollard, D., *et al.* (2006). The middle Pleistocene transition: characteristics, mechanisms and implications for long-term changes in atmospheric pCO_2. *Quaternary Science Reviews*, **25**, 3150–3184.

Clark, R. L. & Guppy, J. C. (1988). A transition from mangrove forest to freshwater wetland in the monsoon tropics of Australia. *Journal of Biogeography*, **15**, 665–684.

Clarke, M. F., Wasson, R. J. & Williams, M. A. J. (1979). Point Stuart chenier and Holocene sea levels in northern Australia. *Search*, **10**, 90–92.

Clarke, S. J. & Murray-Wallace, C. V. (2006). Mathematical expressions used in amino acid racemisation geochronology – a review. *Quaternary Geochronology*, **1**, 261–278.

CLIMAP Project Members (1976). The surface of the ice age Earth. *Science*, **191**, 1131–1137.

Clouard, V. & Bonneville, A. (2001). How many Pacific hotspots are fed by deep-mantle plumes? *Geology*, **29**, 695–698.

Cloud, P. E. (1952). Preliminary report on geology and marine environments of Onotoa Atoll, Gilbert Islands. *Atoll Research Bulletin*, **12**, 1–73.

Clough, B. F. (Ed.) (1982). *Mangrove ecosystems in Australia: structure, function and management*. Australian National University Press, Canberra, 302 pp.

Cobb, K. M., Charles, C. D., Cheng, H., *et al.* (2003). El Niño/Southern Oscillation and tropical Pacific climate during the last millennium. *Nature*, **424**, 271–276.

Cockburn, H. A. P. & Summerfield, M. A. (2004). Geomorphological applications of cosmogenic isotope analysis. *Progress in Physical Geography*, **28**, 1–42.

Cohen, M. C. L., Souza Filho, P. W. M., Lara, R. J., *et al.* (2005). A model of Holocene mangrove development and relative sea-level changes on Braganca Peninsula (northern Brazil). *Wetlands Ecology and Management*, **13**, 433–443.

Colby, N. D. & Boardman, M. R. (1989). Depositional evolution of a windward, high energy lagoon, Graham's Harbour, San Salvador, Bahamas. *Journal of Sedimentary Petrology*, **59**, 819–834.

Coles, B. J. (2000). Doggerland: the cultural dynamics of a shifting coastline. *In* K. Pye & J. R. L. Allen (Eds), *Coastal and estuarine environments: sedimentology, geomorphology and geoarchaeology.* Geological Society, London, Special Publication, **175**, 393–401.

Colhoun, E. A., Kiernan, K., Barrows, T. T., *et al.* (2010). Advances in Quaternary studies in Tasmania. *In* P. Bishop & B. Pillans (Eds), *Australian Landscapes.* Geological Society, London, Special Publications, **346**, 165–183.

Collins, L. B. (2010). Controls on morphology and growth history of coral reefs of Australia's western margin. *In* W. A. Morgan, A. D. George, P. M. Harris, J. A. Kupecz & J. F. Sarg (Eds), *Cenozoic carbonate systems of Australia.* SEPM Special Publication, **95**, 195–217.

Collins, L. B., Zhu, Z. R., Wyrwoll, K.-H., *et al.* (1993). Late Quaternary evolution of coral reefs on a cool-water carbonate margin: the Abrolhos Carbonate Platforms, southwest Australia. *Marine Geology*, **110**, 203–212.

Collins, L. B., Zhou, J.-X. & Freeman, H. (2006). A high-precision record of mid-late Holocene sea-level events from emergent coral pavements in the Houtman Abrolhos Islands, southwest Australia. *Quaternary International*, **145**–146, 78–85x.

Colman, S. M., Pierce, K. L. & Birkeland, P. W. (1987). Suggested terminology for Quaternary dating methods. *Quaternary Research*, **28**, 314–319.

Colonna, M., Casanova, J., Dullo, W. C., *et al.* (1996). Sea-level changes and $\delta^{18}O$ record for the past 30,000 yr from Mayotte Reef, Indian Ocean. *Quaternary Research*, **46**, 335–339.

Compton, J. S. (2001). Holocene sea-level fluctuations inferred from the evolution of depositional environments of the southern Langebaan Lagoon salt marsh, South Africa. *The Holocene*, **11**, 395–405.

Conybeare, W. D. & Phillips, W. (1822). *Outlines of the geology of England and Wales, with an introductory compendium of the general principles of that science, and comparative views of the structure of foreign countries.* Part 1, William Phillips, London, 470 pp.

Cook, P. J. & Polach, H. A. (1973). A chenier sequence at Broad Sound, Queensland, and evidence against a Holocene high sea level. *Marine Geology*, **14**, 253–268.

Cooper, J. A. G. (1993). Sedimentation in a river dominated estuary. *Sedimentology*, **40**, 979–1017.

Cotton, C. A. (1916). Fault coasts in New Zealand. *Geographical Review*, **1**, 20–47.

Cowell, P. J. & Thom, B. G. (1994). Morphodynamics of coastal evolution. *In* R. W. G. Carter & C. D. Woodroffe (Eds), *Coastal evolution – late Quaternary shoreline morphodynamics*, pp. 33–86. Cambridge University Press, Cambridge.

Cox, A. (Ed.) (1973). *Plate tectonics and geomagnetic reversals.* W. H. Freeman & Co., San Francisco, 702 pp.

Coyne, M. K., Jones, B. & Ford, D. (2007). Highstands during Marine Isotope Stage 5: evidence from the Ironshore Formation of Grand Cayman, British West Indies. *Quaternary Science Reviews*, **26**, 536–559.

Craig, H. (1953). The geochemistry of stable carbon isotopes. *Geochimica et Cosmochimica Acta*, **3**, 53–92.

Croll, J. (1875). *Climate and time in their geological relations: a theory of secular changes of the Earth's climate.* Daldy, Isbister, London, 577 pp.

Cronin, T. M. (2010). *Paleoclimates: understanding climate change past and present*. Columbia University Press, New York, 441 pp.

Cronin, T. M., Szabo, B. J., Ager, T. A., *et al.* (1981). Quaternary climates and sea levels of the U.S. Atlantic Coastal Plain. *Science*, **211**, 233–240.

Cronin, T. M., Vogt, P. R., Willard, D. A., *et al.* (2007). Rapid sea level rise and ice sheet response to 8,200-year climate event. *Geophysical Research Letters*, **34**, L20603.

Crowley, G. M. (1996). Late Quaternary mangrove distribution in Northern Australia. *Australian Systematic Botany*, **9**, 219–225.

Cullingford, R. A. & Smith, D. E. (1980). Late Devensian raised shorelines in Angus and Kincardineshire, Scotland. *Boreas*, **9**, 21–38.

Curran, J. M. (1899). *The geology of Sydney and the Blue Mountains*. Angus & Robertson, Sydney, 391 pp.

Curray, J. R. (1961). Late Quaternary sea level: a discussion. *Geological Society of America Bulletin*, **72**, 1707–1712.

Curray, J. R. (1964). Transgressions and regressions. *In* R. L. Miller (Ed.), *Papers in marine geology*, pp. 175–203. Macmillan, New York.

Curray, J. R. (1965). Late Quaternary history, continental shelves of the United States. *In* H. E. Wright Jr & D. G. Frey (Eds), *The Quaternary of the United States*, pp. 723–735. Princeton University Press, Princeton.

Curray, J. R., Emmel, F. J. & Crampton, P. J. S. (1969). Holocene history of a strand plain, lagoonal coast, Nayarit, Mexico. *In* A. A. Castanares & F. B. Phleger (Eds), *Coastal lagoons – a symposium*, pp. 63–100. Universidad Nacional Autonoma, Mexico City.

Curray, J. R., Shepard, F. P. & Veeh, H. H. (1970). Later Quaternary sea-level studies in Micronesia: Carmarsel expedition. *Geological Society of America Bulletin*, **81**, 1865–1880.

Cutler, K. B., Edwards, R. L., Taylor, F. W., *et al.* (2003). Rapid sea-level fall and deep-ocean temperature change since the last interglacial period. *Earth and Planetary Science Letters*, **206**, 253–271.

Dalrymple, G. B. (1991). *The age of the Earth*. Stanford University Press, Stanford, 474 pp.

Dalrymple, R. W., Zaitlin, B. A. & Boyd, R. (1992). Estuarine facies models: conceptual basis and stratigraphic implications. *Journal of Sedimentary Petrology*, **62**, 1130–1146.

Daly, J. F., Belknap, D. F., Kelley, J. T., *et al.* (2007). Late Holocene sea-level change around Newfoundland. *Canadian Journal of Earth Sciences*, **44**, 1453–1465.

Daly, R. A. (1915). The glacial-control theory of coral reefs. *Proceedings of the American Academy of Arts and Sciences*, **51**, 155–251.

Daly, R. A. (1920). A general sinking of sea-level in recent time. *Proceedings of the National Academy of Sciences*, **6**, 246–250.

Daly, R. A. (1925). Pleistocene changes of level. *American Journal of Science*, **10**, 281–313.

Daly, R. A. (1934). *The changing world of the Ice Age*. Yale University Press, New Haven, 271 pp.

Dana, J. D. (1849). *Geology. Report of the United States Exploring Expedition 1838–1842, 10*. C. Sherman, Philadelphia, 756 pp.

Dana, J. D. (1872). *Corals and coral islands*. Dodd and Mean, New York, 406 pp.

Dana, J. D. (1890). *Characteristics of volcanoes with contributions of facts and principles from the Hawaiian Islands*. Dodd, Mead, & Company, New York.

Dansgaard, W. & Oeschger, H. (1989). Past environmental long-term records from the Arctic. *In* H. Oeschger, & C. C. Langway Jr (Eds), *The environmental record in glaciers and ice sheets*, pp. 297–318. John Wiley and Sons, Chichester.

Dansgaard, W. S., Johnsen, S. J., Moller, I., *et al.* (1969). One thousand centuries of climatic record from Camp Century on the Greenland Ice Sheet. *Science*, **166**, 377–381.

Dansgaard, W. S., Clausen, H. B., Gundestrup, N., *et al.* (1982). A new Greenland deep ice core. *Science*, **218**, 1273–1277.

Dansgaard, W., Johnsen, S. J., Clausen, H. B., *et al.* (1984). North Atlantic climatic oscillations revealed by deep Greenland ice cores. *In* J. E. Hansen, & T. Takahashi (Eds), *Climate processes and climate sensitivity*, pp. 288–298. American Geophysical Union, Washington, DC.

Dansgaard, W., Johnsen, S. J., Clausen, H. B., *et al.* (1993). Evidence for general instability of climate from a 250-kyr ice-core record. *Nature*, **364**, 218–220.

Darby, D. J. & Beanland, S. (1992). Possible source models for the 1855 Wairarapa Earthquake, New Zealand. *Journal of Geophysical Research*, **97**, 12,375–12,389.

Darwin, C. (1839). Observations on the Parallel Roads of Glen Roy, and of other parts of Lochaber in Scotland, with an attempt to prove that they are of marine origin. *Philosophical Transactions of the Royal Society of London*, **129**, 39–81.

Darwin, C. (1842). *The structure and distribution of coral reefs*. Smith, Elder and Co., London, 214 pp.

Darwin, C. (1845). *Journal of researches into the natural history and geology of the countries visited during the voyage round the world of H.M.S. 'Beagle' under the command of Captain Fitz Roy, R.N.*, Second edition. John Murray, Albermarle Street, London.

Darwin, C. (1859). *The origin of species by means of natural selection or the preservation of favoured races in the struggle for life*. John Murray, London, 460 pp.

Darwin, C. (1896). *Geological observations on the volcanic islands and parts of South America visited during the voyage of H.M.S Beagle*, 3rd edition. D. Appleton and Company, New York, 648 pp.

David, T. W. E. & Sweet, G. (1904). The geology of Funafuti. *In* The Royal Society (Ed.), *The atoll of Funafuti*, pp. 61–124. The Royal Society, London.

Davies, B. J., Hambrey, M. J., Smeillie, J. L., *et al.* (2012). Antarctic Peninsula ice sheet evolution during the Cenozoic Era. *Quaternary Science Reviews*, **31**, 30–60.

Davies, J. L. (1961). Tasmanian beach ridge systems in relation to sea level change. *Papers and Proceedings of the Royal Society of Tasmania*, **95**, 35–41.

Davies, P. J. & Montaggioni, L. F. (1985). Reef growth and sea-level change: the environmental signature. *Proceedings of the 5th International Coral Reef Congress*, **3**, 477–515.

Davies, S. M., Abbott, P. M., Pearce, N. J. G., *et al.* (2012). Integrating the INTIMATE records using tephrochronology: rising to the challenge. *Quaternary Science Reviews*, **36**, 11–27.

Davies, W. D. & Treloar, F. E. (1977). The application of racemisation dating in archaeology: a critical review. *The Artefact*, **2**, 63–94.

Davis, A. M., Aitchison, J. C., Flood, P. G., *et al.* (2000). Late Holocene higher sea-level indicators from the South China coast. *Marine Geology*, **171**, 1–5.

Davis, R. & Schaeffer, O. A. (1955). Chlorine-36 in nature. *Annals of the New York Academy of Sciences*, **62**, 107–121.

Davis, W. M. (1899). The geographical cycle. *Geographical Journal*, **14**, 481–504.

Davis, W. M. (1928). *The coral reef problem*. Special Publication, 9. American Geographical Society, 596 pp.

Davis, W. M. (1933). Glacial episodes of the Santa Monica Mountains, California. *Geological Society of America Bulletin*, **44**, 1041–1133.

Dawson, A. G., Long, D. & Smith, D. E. (1988). The Storegga Slides: evidence from eastern Scotland for a possible tsunami. *Marine Geology*, **82**, 271–276.

Dawson, A. G., Hampton, S., Fretwell, P., *et al.* (2002). Defining the centre of glacio-isostatic uplift of the last Scottish ice-sheet: the Parallel Roads of Glen Roy, Scottish Highlands. *Journal of Quaternary Science*, **17**, 527–533.

Dawson, A. G., Bondevik, S. & Teller, J. T. (2011). Relative timing of the Storegga submarine slide, methane release, and climate change during the 8.2 ka cold event. *The Holocene*, **21**, 1167–1171.

Day, R. H. (1981). Estuarine ecology. *In* R. H. Day (Ed.), *Estuarine ecology in South Africa*, pp. 86–99. A. A. Balkema, Cape Town.

Dean, R. G. & Houston, J. R. (2013). Recent sea level trends and accelerations: comparison of tide gauge and satellite results. *Coastal Engineering*, **75**, 4–9.

De Boer, B., van de Wal, R. W. S., Bintanja, R., *et al.* (2010). Cenozoic global ice-volume and temperature simulations with 1-D ice-sheet models forced by benthic δ^{18}O records. *Annals of Glaciology*, **51**, 23–33.

De Geer, G. (1888). Om skandinaviens niva-forandringar under Quartarperioden. *Geologiske Forenhandel Stockholm Forhandlingar*, **10**, 366–79.

De Lamothe, L. J. D. (1899). Note sur les anciennes plages et terrasses du basin de L'isser (Départment D'Alger) et de Quelques Autres bassins de la Côte Algérienne. *Bulletin de la Société Geologique de France, Series* **3**, 27, 257–303.

De Lamothe, L. J. D. (1911). Les anciennes lignes de rivages du Sahel d'Alger et d'une partie de la côte algérienne. *Mémoi Société Geologique de France*, **1**, 1–288.

De Lamothe, L. J. D. (1918). Les anciennes nappes alluviales et lignes de rivage du bassin de la Somme et leurs rapports avec celles de la Méditerranée occidentale. *Bulletin de la Société Geologique de France*, **18**, 3–58.

De Laune, R. D., Patrick, W. H. J. & Buresh, R. J. (1978). Sedimentation rates determined by ^{137}Cs dating in a rapidly accreting salt marsh. *Nature*, **275**, 532–533.

Delibrias, G. & Guillier, M. T. (1971). The sea level on the Atlantic coast and the Channel for the last 10,000 years by the ^{14}C method. *Quaternaria*, **14**, 131–135.

De Maillett, B. (1748). *Telliamed or conversations between an Indian philosopher and a French missionary on the diminution of the sea.* (Translated and edited by A.V. Carozzi, 1968, University of Illinois Press, Urbana, 465 pp.)

Deng, X., Griffin, D. A., Ridgway, K. *et al.* (2010). Satellite altimetry for geodetic, oceanographic, and climate studies in the Australian region. *In* S. Vignudelli, A. G. Kostianoy, P. Cipollini & J. Benveniste (Eds), *Coastal altimetry*, pp. 473–508. Springer-Verlag, Berlin.

Denton, G. H. & Hughes, T. J. (Eds) (1981). *The last great ice sheets.* Wiley, New York, 484 pp.

Denton, G. H. & Hughes, T. J. (2002). Reconstructing the Antarctic ice sheet at the Last Glacial Maximum. *Quaternary Science Reviews*, **21**, 193–202.

Denton, G. H., Armstrong, R. L. & Stuiver, M. (1971). The late Cenozoic glacial history of Antarctica. *In* K. K. Turekian (Ed.), *Late Cenozoic glacial ages*, pp. 267–307. Yale University Press, New Haven.

Depéret, C. (1913). Observations sur l'histoire géologique Pliocène et Quaternaire du golfe et de l'isthme de Corinthe. *C.R. Academie Sciences Paris*, **156**, 427–431, 659–663, 1048–1052.

Depéret, C. (1918). Essai de coordination chronologique des temps quaternaires. *Compte Rendu Sommaire et Bulletin de la Société Geologique de France* **116**, 480, 636, 884.

Deschamps, P., Durand, N., Bard, E., *et al.* (2012). Ice-sheet collapse and sea-level rise at the Bølling warming 14,600 years ago. *Nature*, **483**, 559–564.

Deshayes, G. P. (1840). Nouvelles espèces de mollusques provenant des côtes de la Californie, du Mexique, du Kamtschatka et de la Nouvelle-Zélande, Revue Zoologique, par la Société Cuvierienne; Association Universelle por l'Avancement de Zoologie, de l'Anatomie Comparée et de la Paléontologie. *Journal Mensuel Publié sous la Direction de M.F.-E. Guérin-Méneville*. Année 1839, **2**, pp. 356–361, pl. 21.

Desnoyers, J. (1829). Observations sur un ensemble de depots marine. *Annales des Sciences Naturelles (Paris)*, 171–214, 402–491.

Devoy, R. J. N. (1979). Holocene sea level changes and vegetational history of the lower Thames Estuary. *Philosophical Transactions of the Royal Society*, **B285**, 355–407.

Devoy, R. J. N. (1982). Analysis of the geological evidence for Holocene sea-level movements in southeast England. *Proceedings of the Geologists' Association*, **93**, 65–90.

Dickinson, W. R. (1998). Geomorphology and geodynamics of the Cook–Austral island–seamount chain in the South Pacific Ocean: implications for hotspots and plumes. *International Geology Review*, **40**, 1039–1075.

Dickinson, W. R. (1999). Holocene sea-level record on Funafuti and potential impact of global warming on central

Pacific atolls. *Quaternary Research*, **51**, 124–132.

Dickinson, W. R. (2000). Hydro-isostatic and tectonic influences on emergent Holocene paleoshorelines in the Mariana Islands, western Pacific Ocean. *Journal of Coastal Research*, **16**, 735–746.

Dickinson, W. R. (2004). Impacts of eustasy and hydro-isostasy on the evolution and landforms of Pacific atolls. *Palaeogeography, Palaeoclimatology, Palaeoecology*, **213**, 251–269.

Dietz, R. S. (1961). Continental and ocean basin evolution by spreading of the sea floor. *Nature*, **190**, 854–857.

Digerfeldt, G. & Hendry, M. D. (1987). An 8000 year Holocene sea-level record from Jamaica: implications for interpretation of Caribbean reef and coastal history. *Coral Reefs*, **5**, 165–169.

Dillon, W. P. & Oldale, R. N. (1978). Late Quaternary sea-level curve: reinterpretation based on glaciotectonic influence. *Geology*, **6**, 56–60.

Dixon, R. G. (1979). Sedimentary facies in the Red Crag (Lower Pleistocene, East Anglia). *Proceedings of the Geologist's Association*, **90**, 117–132.

Dodge, R. E., Fairbanks, R. G., Benninger, L. K., *et al.* (1983). Pleistocene sea levels from raised coral reefs of Haiti. *Science*, **219**, 1423–1425.

Doğan, U., Koçyğit, A., Varol, B., *et al.* (2012a). MIS 5a and MIS 3 relatively high sea-level stands on the Hatay-Samandağ Coast, Eastern Mediterranean, Turkey. *Quaternary International*, **262**, 65–79.

Doğan, U., Koçyğit, A., Varol, B., *et al.* (2012b). Reply to the comments by Erdem Bekaroğlu on 'MIS 5a and MIS 3 relatively high sea-level stands on the Hatay-Samandağ Coast, Eastern Mediterranean, Turkey'. *Quaternary International*, **262**, 84–87.

Domingues, C. M., Church, J. A., White, N. J., *et al.* (2008). Improved estimates of upper-ocean warming and multi-decadal sea-level rise. *Nature*, **453**, 1090–1094.

Dominguez, L. L. & Mullins, H. T. (1988). Cat Island platform, Bahamas: an incipiently drowned Holocene carbonate shelf. *Sedimentology*, **35**, 805–819.

Donnelly, J. P., Cleary, P., Newby, P., *et al.* (2004). Coupling instrumental and geological records of sea-level change: evidence from southern New England of an increase in the rate of sea-level rise in the late 19th century. *Geophysical Research Letters*, **31**, art. no.-L05203.

Donner, J. (1995). *The Quaternary history of Scandinavia*. Cambridge University Press, Cambridge, 200 pp.

Donovan, D. T. & Jones, E. J. W. (1979). Causes of world-wide changes in sea level. *Journal of the Geological Society of London*, **136**, 187–192.

D'Orbingy, A. D. (1842). *Voyages dans l'Amérique méridionale execute pendant les années 1826, 1827, 1828, 1829, 1830, 1831, 1832 et 1833 III: Partie Géologique*. Paris.

Dott, R. H. (1988). An episodic view of shallow marine clastic sedimentation. *In* P. L. de Boer, A. van Gelder & S. D. Nio (Eds), *Tide-influenced sedimentary environments and facies*, pp. 3–12. D. Reidel Publishing Company, Dordrecht.

Dougherty, A. J. & Dickson, M. E. (2012). Sea level and storm control on the evolution of a chenier plain, Firth of Thames, New Zealand. *Marine Geology*, **307–310**, 58–72.

Douglas, B. C. (1991). Global sea level rise. *Journal of Geophysical Research*, **96**, 6981–6992.

Douglas, B. C. (1992). Global sea level acceleration. *Journal of Geophysical Research*, **97**, 12699–12706.

Douglas, B. C. (2008). Concerning evidence for fingerprints of glacial melting. *Journal of Coastal Research*, **24**, 218–227.

Droxler, A. W. & Farrell, J. W. (2000). Marine Isotope Stage 11 (MIS 11): new insights for a warm future. *Global and Planetary Change*, **24**, 1–5.

Dubois, G. (1925). Sur la nature des oscillations de type Atlantique des lignes de rivages Quaternaires. *Compte Rendu Sommaire et Bulletin de la Société Geologique de France*, **25**, 857–878.

Dubois, J., Launay, J. & Recy, J. (1974). Uplift movements in New Caledonia – Loyalty Islands area and their plate tectonics interpretation. *Tectonophysics*, **24**, 133–150.

Duggen, S., Hoernle, K., van den Bogaard, P., *et al.* (2003). Deep roots of the Messinian Salinity Crisis. *Nature*, **422**, 602–606.

Duke, N. C., Ball, M. C. & Ellison, J. C. (1998). Factors influencing biodiversity and distributional gradients in mangroves. *Global Ecology and Biogeography Letters*, **7**, 27–47.

Duller, G. A. T. & Wintle, A. G. (2012). A review of the thermally transferred optically stimulated luminescence signal from quartz for dating sediments. *Quaternary Geochronology*, **7**, 6–20.

Dullo, W.-C. (2005). Coral growth and reef growth: a brief review. *Facies*, **51**, 33–48.

Dutton, A. & Lambeck, K. (2012). Ice volume and sea level during the Last Interglacial. *Science*, **337**, 216–219.

Dutton, A., Bard, E., Antonioli, F., *et al.* (2009). Phasing and amplitude of sea-level and climate change during the penultimate interglacial. *Nature Geoscience*, **2**, 355–359.

Dutton, C. E. (1882). Review of physics of the Earth's crust by the Rev. Osmond Fisher. *American Journal of Science, 3rd Series*, **23**, 283–290.

Dutton, C. E. (1889). On some of the greater problems of physical geology. *Bulletin of the Philosophical Society of Washington*, **11**, 51–59.

Dyke, A. S., Andrews, J. T., Clark, P. U., *et al.* (2002). The Laurentide and Innuitan ice sheets during the Last Glacial Maximum. *Quaternary Science Reviews*, **21**, 9–31.

Easton, W. H. & Olson, E. A. (1976). Radiocarbon profile of Hanauma Reef, Oahu, Hawaii. *Geological Society of America Bulletin*, **87**, 711–719.

Edwards, A. B. (1941). Storm-wave platforms. *Journal of Geomorphology*, **4**, 223–236.

Edwards, R. L., Chen, J. H., Ku, T.-L., *et al.* (1987a). Precise timing of the Last Interglacial period from mass spectrometric determination of Thorium-230 in corals. *Science*, **236**, 1547–1553.

Edwards, R. L., Chen, J. H. & Wasserburg, G. J. (1987b). ^{238}U-^{234}U-^{230}Th-^{232}Th systematics and the precise measurement of time over the past 500,000 years. *Earth and Planetary Science Letters*, **81**, 175–192.

Eggins, S. M., Grün, R., McCulloch, M. T., *et al.* (2005). *In situ* U-series dating by laser-ablation multi-collector ICPMS: new prospects for Quaternary geochronology. *Quaternary Science Reviews*, **24**, 2523–2538.

Egler, F. E. (1952). Southeast saline everglades vegetation, Florida: and its management. *Vegetatio*, **3**, 213–265.

Ehlers, J. & Gibbard, P. L. (2004a). *Quaternary glaciations – extent and chronology. Part I: Europe. Developments in Quaternary science 2*. Elsevier, Amsterdam, 475 pp.

Ehlers, J. & Gibbard, P. L. (2004b). *Quaternary glaciations – extent and chronology. Part II: North America. Developments in Quaternary science 2*. Elsevier, Amsterdam, 440 pp.

Ehlers, J. & Gibbard, P. L. (2004c). *Quaternary glaciations – extent and chronology. Part III: South America, Asia, Africa, Australasia, Antarctica. Developments in Quaternary science*. Elsevier, Amsterdam, 380 pp.

Eisenhauer, A., Wasserburg, G. J., Chen, J. H., *et al.* (1993). Holocene sea-level determinations relative to the Australian continent: U/Th (TIMS) and ^{14}C (AMS) dating of coral cores from the Abrolhos Islands. *Earth and Planetary Science Letters*, **114**, 529–547.

Eisenhauer, A., Zhu, Z. R., Collins, L. B., *et al.* (1996). The Last Interglacial sea level change: new evidence from the Abrolhos islands, West Australia. *Geologische Rundschau*, **85**, 606–614.

Eisenhauer, A., Heiss, G. A., Sheppard, C. R. C., *et al.* (1999). Reef and island formation and Late Holocene sea-level changes in the Chagos islands. *In* C. R. C. Sheppard & M. R. D. Seaward (Eds), *Ecology of the Chagos Archipelago*. Linnean Society, Occasional Publications, London, pp. 21–33.

Elias, S. A. (2001). Beringian paleoecology: results from the 1997 workshop. *Quaternary Science Reviews*, **20**, 7–13.

Elias, S. A. & Brigham-Grette, J. (Eds) (2001). Beringian paleoenvironments festschrift in honour of D.M. Hopkins. *Quaternary Science Reviews*, **20**, 1–574.

Emery, K. O. & Aubrey, D. G. (1985). Glacial rebound and relative sea levels in Europe from tide-gauge records. *Tectonophysics*, **120**, 239–255.

Emery, K. O. & Aubrey, D. G. (1991). *Sea levels, land levels and tide gauges.* Springer-Verlag, New York, 237 pp.

Emery, K. O., Niino, H. & Sullivan, B. (1971). Post-Pleistocene levels of the East China Sea. *In* K. K. Turekian (Ed.), *The Late Cenozoic glacial ages*, pp. 381–390. Yale University Press, New Haven.

Emiliani, C. (1955). Pleistocene temperatures. *Journal of Geology*, **63**, 538–578.

Emmel, F. J. & Curray, J. R. (1982). A submerged late Pleistocene delta and other features related to sea level changes in the Malacca Strait. *Marine Geology*, **47**, 197–216.

Engel, M. H., Zumberge, J. E. & Nagy, B. (1977). Kinetics of amino acid racemization in *Sequoiadendron giganteum* heartwood. *Analytical Biochemistry*, **82**, 415–422.

Engelhart, S. E. & Horton, B. P. (2012). Holocene sea level database for the Atlantic coast of the United States. *Quaternary Science Reviews*, doi:10.1016/j.quascirev.2011.09.013.

Engelhart, S. E., Horton, B. P., Roberts, D. H., *et al.* (2007). Mangrove pollen of Indonesia and its suitability as a sea-level indicator. *Marine Geology*, **242**, 65–81.

Engelhart, S. E., Horton, B. P. & Kemp, A. C. (2011a). Holocene sea level changes along the United States' Atlantic Coast. *Oceanography*, **24**, 70–79.

Engelhart, S. E., Peltier, W. R. & Horton, B. P. (2011b). Holocene relative sea-level changes and glacial isostatic adjustment of the U.S. Atlantic coast. *Geology*, **39**, 751–754.

Engelkeimer, A. G., Hamill, W. H., Ingham, M. G., *et al.* (1949). The half-life of radiocarbon (^{14}C). *Physical Review*, **75**, 1825–1833.

Enos, P. & Perkins, R. D. (1979). Evolution of Florida Bay from island stratigraphy. *Geological Society of America Bulletin*, **90**, 59–83.

Eronen, M. (1983). Late Weichselian and Holocene shore displacement in Finland. *In* D. E. Smith & A. G. Dawson (Eds), *Shorelines and isostasy*, pp. 183–207. Academic Press, London.

Eronen, M., Glückert, G., Hatakka, L., *et al.* (2001). Rates of Holocene isostatic uplift and relative sea-level lowering of the Baltic in SW Finland based on studies of isolation contacts. *Boreas*, **30**, 17–30.

Esat, T. M., McCulloch, M. T., Chappell, J., *et al.* (1999). Rapid fluctuations in sea level recorded at Huon Peninsula during the Penultimate deglaciation. *Science*, **283**, 197–201.

Esteban, M. & Klappa, C. F. (1983). Subaerial exposure environment. *In* P. A. Scholle, D. G. Bebout & C. H. Moore (Eds), *Carbonate depositional environments*, American Association of Petroleum Geologists Memoir 33, pp. 1–54. Tulsa, Oklahoma.

Evelpidou, N., Pirazzoli, P. A., Saliegec, J.-F., *et al.* (2011). Submerged notches and doline sediment as evidence for Holocene subsidence. *Continental Shelf Research*, **31**, 1273–1281.

Evelpidou, N., Vassilopoulos, A. & Pirazzoli, P. A. (2012a). Submerged notches on the coast of Skyros Island (Greece) as evidence for Holocene subsidence. *Geomorphology*, **141**, 81–87.

Evelpidou, N., Kampolis, I., Pirazzoli, P.A., *et al.* (2012b). Global sea-level rise and the disappearance of tidal notches. *Global and Planetary Change*, **92–93**, 248–256.

Fairbanks, R. G. (1989). A 17,000 year glacio-eustatic sea level record: influence of glacial melting rates on Younger Dryas event and deep-ocean circulation. *Nature*, **342**, 637–642.

Fairbanks, R. G. (1990). The age and origin of the 'Younger Dryas Climate Event' in Greenland ice cores. *Paleoceanography*, **5**, 937–948.

Fairbanks, R. G., Mortlock, R. A., Chiu, T.-C., *et al.* (2005). Marine radiocarbon calibration curve spanning 10,000 to 50,000 years B.P. based on paired ^{230}Th/^{234}U/^{238}U and ^{14}C dates on pristine corals. *Quaternary Science Reviews*, **24**, 1781–1796.

Fairbridge, R. W. (1950). The geology and geomorphology of Point Peron, Western Australia. *Journal of the Royal Society of Western Australia*, **33**, 1–43.

Fairbridge, R. W. (1954). Quaternary eustatic data for Western Australia and adjacent states. *Proceedings of the Pan Indian Ocean Science Congress*, Perth, August 1954, Section F, pp. 64–84.

Fairbridge, R. W. (1960). The changing level of the sea. *Scientific American*, **202**, 70–79.

Fairbridge, R. W. (1961). Eustatic changes in sea level. *Physics and Chemistry of the Earth*, **4**, 99–185.

Fairbridge, R. W. (1981). The concept of neotectonics: an introduction. *Zeitschrift für Geomorphologie*. **Suppl.-Bd 40**, VII–XII.

Fairbridge, R. W. (1995). Eolianite and eustasy: early concepts on Darwin's voyage of HMS Beagle. *Carbonates and Evaporites*, **10**, 92–101.

Fairbridge, R. W. & Hillaire-Marcel, C. (1977). An 8,000-yr palaeoclimatic record of the 'Double-Hale' 45-yr solar cycle. *Nature*, **268**, 413–416.

Falkland, A. C. & Woodroffe, C. D. (1997). Geology and hydrogeology of the Tarawa and Christmas Island (Kiritimati), Kiribati, Central Pacific. *In* H. L. Vacher & T. M. Quinn (Eds), *Geology and hydrogeology of carbonate islands*, pp. 577–610. Elsevier, Amsterdam.

Fallon, S. (2011). Radiocarbon (14C): dating and corals. *In* D. Hopley (Ed.), *Encyclopedia of modern coral reefs*, pp. 829–834. Springer, Dordrecht.

Farrell, W. E. & Clark, J. A. (1976). On postglacial sea-level. *Geophysical Journal of the Royal Astronomical Society*, **46**, 647–667.

Faure, G. (1986). *Principles of isotope geology*. John Wiley & Sons, New York, 589 pp.

Faure, H., Fontes, J. C., Hebrard, L., *et al.* (1980). Geoidal change and shore-level tilt along Holocene estuaries: Senegal River area, West Africa. *Science*, **210**, 421–423.

Featherstone, W. E. (2006). Yet more evidence for a north–south slope in the Australian Height Datum. *Journal of Spatial Science*, **52**, 1–6.

Ferland, M. A. & Roy, P. S. (1997). Southeastern Australia: a sea-level dependent, cool-water carbonate margin. *In* N. P. James & J. D. A. Clarke (Eds), *Cool-water carbonates*, SEPM Special Publication No. 56, pp. 37–52.

Ferland, M. A., Roy, P. S. & Murray-Wallace, C. V. (1995). Glacial lowstand deposits on the outer continental shelf of southeastern Australia. *Quaternary Research*, **44**, 294–299.

Ferranti, L., Antonioli, F., Mauz, B., *et al.* (2006). Markers of the last interglacial sea-level high stand along the coast of Italy: tectonic implications. *Quaternary International*, **145–146**, 30–54.

Fiedel, S. J. (2011). The mysterious onset of the Younger Dryas. *Quaternary International*, **242**, 262–266.

Fiedler, J. W. & Conrad, C. P. (2010). Spatial variability of sea level rise due to water impoundment behind dams. *Geophysical Research Letters*, **37**, L12603.

Fisk, H. N. (1944). *Geological investigation of the Alluvial Valley of the lower Mississippi River*. Mississippi River Commission, Vicksburg, 170 pp.

Fleming, C. A. (1953). The geology of the Wanganui Subdivision, Waverley and Wanganui sheet districts (N137 and N138). *New Zealand Geological Survey Bulletin*, **52**, 362 pp.

Fleming, K., Johnston, P., Zwartz, D., *et al.* (1998). Refining the eustatic sea-level curve since the Last Glacial Maximum using far- and intermediate-field sites. *Earth and Planetary Science Letters*, **163**, 327–342.

Flemming, N. C. (1969). Archaeological evidence for eustatic change of sea level and earth movements in the Western Mediterranean during the last 2000 years. *Geological Society of America Special paper*, **109**, 1–125.

Flemming, N. C. (1986). A survey of the late Quaternary landscape of the Cootamundra Shoals, north Australia: a preliminary report. *In* N. C. Flemming, F. Marchetti & A. Stefanon (Eds), *Proceedings of the 7th International Diving Science Symposium of CMAS, 1983*, Padova, Italy, pp. 149–180.

Flemming, N. C. (1998). Archaeological evidence for vertical movement on the continental shelf during the Palaeolithic, Neolithic and Bronze Age periods. *In* I. S. Stewart & C. Vita-Finzi (Eds), *Coastal tectonics*. Geological Society, London, Special Publications, **146**, 129–146.

Flemming, N. C. & Roberts, D. G. (1973). Tectono-eustatic changes in sea level and seafloor spreading. *Nature*, **243**, 19–22.

Flessa, K. W. (1998). Well-traveled cockles: shell transport during the Holocene transgression of the southern North Sea. *Geology*, **26**, 187–190.

Fletcher, C. H. & Jones, A. T. (1996). Sea-level highstand recorded in Holocene shoreline deposits on Oahu, Hawaii. *Journal of Sedimentary Research*, **66**, 632–641.

Fletcher, C. H., Murray-Wallace, C. V. & Glenn, C., et al. (2005). Age and origin of late Quaternary eolianite Kaiehu Point, (Moomomi), Molokai, Hawaii. *Journal of Coastal Research*, **S142**, 97–112.

Fletcher, C. III, Bochicchio, C., Conger, C. L., et al. (2008). Geology of Hawaii reefs. *In* B. M. Riegl & R. E. Dodge (Eds), *Coral reefs of the USA*, pp. 435–487. Springer, New York.

Flinders, M. (1814). *A voyage to Terra Australis: undertaken for the purpose of completing the discovery of that vast country and, prosecuted in the years 1801, 1802 and 1803, in His Majesty's Ship, the Investigator, 2 Volumes and Atlas*. G. & W. Nicol, London.

Flint, R. F. (1961). *Glacial and Quaternary geology*. John Wiley & Sons, London, 538 pp.

Flint, R. F. (1971). *Glacial and Quaternary geology*. John Wiley & Sons, New York, 892 pp.

Flood, P. G. (1983). Holocene sea level data from the southern Great Barrier Reef and southeastern Queensland: a review. *In* D. Hopley (Ed.), *Australian sea levels in the last 15000 years: a review*, pp. 85–92. Australian Report for IGCP 61. Department of Geography, James Cook University of North Queensland.

Flood, P. G. & Frankel, E. (1989). Late Holocene higher sea level indicators from eastern Australia. *Marine Geology*, **90**, 193–195.

Florido, E., Auriemma, R., Faivre, S., et al. (2011). Istrian and Dalmatian fishtanks as sea-level markers. *Quaternary International*, **232**, 105–113.

Focke, J. W. (1978). Limestone cliff morphology on Curacao (Netherlands Antilles), with special attention to the origin of notches and vermetid/coralline algal surf benches ('cornices', 'trottoirs'). *Zeitschrift für Geomorphologie N.F.*, **22**, 329–349.

Fontaine, H. & Delibrias, G. (1974). Niveaux marins pendant le Quaternaire au Viet-Nam. *Archives Geologiques du Viet-Nam*, **17**, 35–44.

Forman, S. L., Pierson, J. & Lepper, K. (2000). Luminescence geochronology. *In* J. S. Noller, J. M. Sowers & W. R. Lettis (Eds), *Quaternary geochronology: methods and applications*, pp. 157–176. American Geophysical Union Reference Shelf 4, Washington, DC.

Frakes, L. A., Alley, N. F. & Deynoux, M. (1995). Early Cretaceous ice rafting and climate zonation in Australia. *International Geology Reviews*, **37**, 567–583.

Franke, J., Paul, A. & Schultz, M. (2008). Modeling variations of marine reservoir ages during the last 45000 years. *Climate of the Past*, **4**, 125–136.

Frazier, D. E. (1967). Recent deltaic deposits of the Mississippi River: their development and chronology. *Transactions of the Gulf Coast Association of Geological Societies*, **17**, 287–315.

Friedman, G. M., Sanders, J. E. & Kopaska-Merkel, D. C. (1992). *Principles of sedimentary deposits: stratigraphy and sedimentology*. Macmillan Publishing Company, New York, 717 pp.

Froggatt, P. C. (1983). Toward a comprehensive Upper Quaternary tephra and ignimbrite stratigraphy in New Zealand using electron microprobe analysis of glass shards. *Quaternary Research*, **19**, 188–200.

Fuchs, M., Kreutzer, S., Fischer, M., et al. (2012). OSL and IRSL dating of raised beach sand deposits along the southeastern coast of Norway. *Quaternary Geochronology*, **10**, 195–200.

Fujimoto, K., Miyagi, T., Kikuchi, T., et al. (1996). Mangrove habitat formation and response to Holocene sea-level changes on Kosrae Island, Micronesia. *Mangroves and Salt Marshes*, **1**, 47–57.

Funnell, B. M. (1995). Global sea-level and the (pen-)insularity of late Cenozoic Britain. *In* R. C. Preece (Ed.), *Island Britain: a Quaternary perspective*. Geological Society Special Publication, **96**, 3–13.

Gage, M. (1953). The study of Quaternary strandlines in New Zealand. *Transactions of the Royal Society of New Zealand*, **81**, 27–34.

Gale, S. J. (2009). Event chronostratigraphy: a high-resolution tool for dating the recent past. *Quaternary Geochronology*, **4**, 391–399.

Galloway, R. W. (1982). Distribution and physiographic patterns of Australian mangroves. *In* B. F. Clough (Ed.), *Mangrove ecosystems in Australia: structure, function and management*, pp. 31–54. Australian National University Press, Canberra.

Galloway, W. E. & Hobday, D. K. (1983). *Terrigenous–clastic depositional systems*. Springer-Verlag, New York, 423 pp.

Gallup, C. D., Edwards, R. L. & Johnson, R. D. (1994). The timing of high sea-levels over the past 200,000 years. *Science*, **263**, 786–800.

Gallup, C. D., Cheng, H., Taylor, F. W., *et al.* (2002). Direct determination of the timing of sea level change during Termination II. *Science*, **295**, 310–313.

Gaposchkin, E. M. & Lambeck, K. (1971). Earth's gravity field to the sixteenth degree and station coordinates from satellite and terrestrial data. *Journal of Geophysical Research*, **76**, 4855–4883.

Gardiner, J. S. (1902). The formation of the Maldives. *Symposium of the Zoological Society of London*, **19**, 277–301.

Gardiner, J. S. (Ed.) (1903). *The fauna and geography of the Maldive and Laccadive Archipelagoes, being an account of the work carried on and of collections made by an expedition during years 1899 and 1900*. Cambridge University Press, Cambridge, 118 pp.

Gardiner, J. S. (1931). *Coral reefs and atolls*. Macmillan, London, 182 pp.

Gardiner, J. S. (1936). The reefs of the western Indian Ocean. I Chagos Archipelago; II the Mascarene region. *Transactions of the Linnean Society of London II*, **19**, 393–436.

Gardner, J. V., Calder, B. R., Clarke, J. E., *et al.* (2007). Drowned shelf-edge deltas, barrier islands and related features along the outer continental shelf north of the head of De Soto Canyon, NE Gulf of Mexico. *Geomorphology*, **89**, 370–390.

Gehrels, W. R. (1999). Middle and late Holocene sea-level change in eastern Maine reconstructed from foraminiferal saltmarsh stratigraphy and AMS ^{14}C dates on basal peats. *Quaternary Research*, **52**, 350–359.

Gehrels, W. R. (2000). Using foraminiferal transfer functions to produce high-resolution sea-level records from saltmarsh deposits, Maine, USA. *The Holocene*, **10**, 367–376.

Gehrels, W. R. (2001). Discussion of: Nunn, Patrick D., 1998. Sea-level changes over the past 1,000 years in the Pacific. *Journal of Coastal Research*, **14**, 23–30. *Journal of Coastal Research*, **17**, 244–245.

Gehrels, W. R. (2010a). Sea-level changes since the Last Glacial Maximum: an appraisal of the IPCC Fourth Assessment Report. *Journal of Quaternary Science*, **25**, 26–38.

Gehrels, W. R. (2010b). Late Holocene land- and sea-level changes in the British Isles: implications for future sea-level predictions. *Quaternary Science Reviews*, **29**, 1648–1660.

Gehrels, W. R. & Woodworth, P. L. (2013). When did modern rates of sea-level rise start? *Global and Planetary Change*, **100**, 263–277.

Gehrels, W. R., Belknap, D. F. & Kelley, J. T. (1996). Integrated high precision analyses of Holocene relative sea-level changes: lessons from the coast of Maine. *Geological Society of America Bulletin*, **108**, 1073–1088.

Gehrels, W. R., Milne, G. A., Kirby, J. R., *et al.* (2004). Late Holocene sea-level changes and isostatic crustal movements in Atlantic Canada. *Quaternary International*, **120**, 79–89.

Gehrels, W. R., Kirby, J. R., Prokoph, A., *et al.* (2005). Onset of recent rapid sea-level rise in the Western Atlantic Ocean. *Quaternary Science Reviews*, **24**, 2083–2100.

Gehrels, W. R., Marshall, W. A., Gehrels, M. J., *et al.* (2006). Rapid sea-level rise in the North Atlantic Ocean since the first half of the 19th century. *The Holocene*, **16**, 948–964.

Gehrels, W. R., Hayward, B. W., Newnham, R. M., *et al.* (2008). A 20th century sea-level acceleration in New Zealand. *Geophysical Research Letters*, **35**, L02717.

Gehrels, W. R., Horton, B. P., Kemp, A. C., *et al.* (2011). Two millennia of sea level data; the key to predicting change. *EOS,*

Transactions of the American Geophysical Union, **92**, 289–291.

Gehrels, W. R., Callard, S. L., Moss, P. T., *et al.* (2012). Nineteenth and twentieth century sea-level changes in Tasmania and New Zealand. *Earth and Planetary Science Letters*, **315–316**, 94–102.

Geikie, J. (1894). *The Great Ice Age and its relation to the antiquity of Man*, 3rd edition. Edward Standford, London, 850 pp.

Gelfenbaum, G. & Kaminsky, G. M. (2010). Large-scale coastal change in the Columbia River littoral cell: an overview. *Marine Geology*, **273**, 1–10.

Geyh, M. A., Kudrass, H.-R. & Streif, H. (1979). Sea-level changes during the late Pleistocene and Holocene in the Strait of Malacca. *Nature*, **278**, 441–443.

Gibb, J. G. (1986). A New Zealand regional Holocene eustatic sea-level curve and its application to determination of vertical tectonic movements: a contribution to IGCP-Project 200. *Royal Society of New Zealand Bulletin*, **24**, 377–395.

Gibbard, P. L. (2003). Definition of the Middle–Upper Pleistocene boundary. *Global and Planetary Change*, **36**, 201–208.

Gignoux, M. (1913). Les formations marine Pliocènes et Quaternaires de l'Italie du Sud et de la Sicilie. *Annales de la Université de Lyon, N.S. I. Sc. Méd.* **36**, 1–693.

Gilbert, G. K. (1885). *The topographic features of lake shores*. US Geological Survey Annual Report, **5**, 75–123.

Gilbert, G. K. (1890). *Lake Bonneville*. United States Geological Survey, Monograph 1, 438 pp.

Gilbert, G. K. (1896). The origin of hypotheses: illustrated by the discussion of a topographic problem. *Science*, **3**, 1–13.

Gill, E. D. (1983). Australian sea levels in the last 15 000 years – Victoria, SE Australia. *In* D. Hopley (Ed.), *Australian sea levels in the last 15 000 years: a review*. Monograph Series Occasional Paper 3, Geography Department, James Cook University, Townsville, pp. 59–63.

Gill, E. D. & Hopley, D. (1972). Holocene sea levels in eastern Australia – a discussion. *Marine Geology*, **12**, 223–233.

Gill, E. D., Lang, J. G. & Boyd, S. E. (1982). The peak of the Flandrian Transgression in Victoria, S.E. Australia: faunas and sea level changes. *Proceedings of the Royal Society of Victoria*, **94**, 23–34.

Gillespie, R. (1990). The Australian marine shell correction factor. *In* R. Gillespie (Ed.), *Quaternary Dating Workshop 1990*, p. 15. Department of Biogeography and Geomorphology, Australian National University.

Gillespie, R. & Polach, H. A. (1979). The suitability of marine shells for radiocarbon dating of Australian prehistory. *In* R. Berger & H. Suess (Eds), *Radiocarbon dating: proceedings of the 9th international conference on radiocarbon dating*, pp. 404–421. University of California Press, Los Angeles.

Gillespie, R. & Temple, R. B. (1977). Radiocarbon dating of shell middens. *Archaeology and Physical Anthropology in Oceania*, **12**, 26–37.

Giresse, P. & Davies, O. (1980). High sea levels during the last Glaciation: one of the most puzzling problems of sea-level studies. *Quaternaria*, **22**, 211–236.

Gischler, E. (2006). Comment on 'Corrected western Atlantic sea-level curve for the last 11,000 years based on calibrated ^{14}C dates from *Acropora palmata* framework and intertidal mangrove peat' by Toscano and Macintyre. *Coral Reefs* **22**, 257–270 (2003), and their response in *Coral Reefs* **24**, 187–190 (2005). *Coral Reefs*, **25**, 273–279.

Gischler, E. (2008). Accretion patterns in Holocene tropical coral reefs: do massive coral reefs in deeper water with slowly growing corals accrete faster than shallower branched coral reefs with rapidly growing corals? *International Journal of Earth Sciences*, **97**, 851–859.

Gischler, E. & Hudson, J. H. (2004). Holocene development of the Belize Barrier Reef. *Sedimentary Geology*, **164**, 223–236.

Gischler, E., Hudson, J. H. & Pisera, A. (2008). Late Quaternary reef growth and sea level in the Maldives (Indian Ocean). *Marine Geology*, **250**, 104–113.

Godwin, H. (1962). Half-life of radiocarbon. *Nature*, **195**, 984.

Godwin, H. & Clifford, M. A. (1938). Studies of the postglacial history of British vegetation. I. Origin and stratigraphy of the Fenland deposits near Woodwalton,

Hunts. II. Origin and stratigraphy of deposits in southern Fenland. *Philosophical Transactions of the Royal Society of London*, **B229**, 323–406.

Godwin, H., Suggate, R. P. & Willis, E. H. (1958). Radiocarbon dating of the eustatic rise in ocean level. *Nature*, **181**, 1518–1519.

Goodbred, S. L. & Kuehl, S. A. (2000). Enormous Ganges–Brahmaputra sediment discharge during strengthened early Holocene monsoon. *Geology*, **28**, 1083–1086.

Goodfriend, G. A., Hare, P. E. & Druffel, E. R. M. (1992). Aspartic acid racemization and protein diagenesis in corals over the last 350 years. *Geochimica et Cosmochimica Acta*, **56**, 3847–3850.

Goodfriend, G. A., Kashgarian, M. & Harasewych, M. G. (1995). Use of aspartic acid racemization and post-bomb ^{14}C to reconstruct growth rate and longevity of the deep-water slit shell *Entemnotrochus adansonianus*. *Geochimica et Cosmochimica Acta*, **59**, 1125–1129.

Goodwin, I. D. (2003). Unravelling climatic influences on Late Holocene sea-level variability. *In* A. Mackay, R. Battarbee, J. Birks & F. Oldfield (Eds), *Global change in the Holocene*, pp. 406–421. Arnold, London.

Goodwin, I. D. & Harvey, N. (2008). Subtropical sea-level history from coral microatolls in the Southern Cook Islands, since 300 AD. *Marine Geology*, **253**, 14–25.

Gordon, J. E. & Harkness, D. D. (1992). Magnitude and geographic variation of the radiocarbon content in Antarctic marine life: implications for reservoir corrections in radiocarbon dating. *Quaternary Science Reviews*, **11**, 697–708.

Gornitz, V. (1991). Global coastal hazards from future sea level rise. *Palaeogeography, Palaeoclimatology, Palaeoecology*, **89**, 379–398.

Görür, N., Çagatay, M. N., Emre, Ö., *et al.* (2001). Is the abrupt drowning of the Black Sea shelf at 7150 yr BP a myth? *Marine Geology*, **176**, 65–73.

Gosse, J. C. & Phillips, F. M. (2001). Terrestrial in situ cosmogenic nuclides: theory and application. *Quaternary Science Reviews*, **20**, 1475–1560.

Gostin, V. A., Hails, J. R. & Belperio, A. P. (1984). The sedimentary framework of northern Spencer Gulf, South Australia. *Marine Geology*, **61**, 111–138.

Gould, H. R. & McFarlan, E. (1959). Geologic history of the Chenier Plain, southwestern Louisiana. *Transactions Gulf Coast Association of Geological Societies*, **9**, 261–270.

Gould, S. J. (2000). Chapter 5: The proof of Lavoisier's Plates. *In* S. J. Gould, *The lying stones of Marrakech: penultimate reflections in natural history*, pp. 91–114. Jonathan Cape, London.

Grabau, A. W. (1940). *The rhythm of the ages*. Robert E. Krieger Publishing Co., New York. (Reprint 1978, 561 pp. +XXV plates.)

Gradstein, F., Ogg, J. & Smith, A. Q. (2004). *A geologic time scale 2004*. Cambridge University Press, Cambridge, 589 pp.

Gradstein, F. M., Ogg, J. G., Schmitz, M. D., *et al.* (Eds) (2012). *The geologic time scale 2012*. 2 Volumes. Elsevier, Oxford, 1144 pp.

Graf, M.-T. & Chmura, G. L. (2010). Reinterpretation of past sea-level variation of the Bay of Fundy. *The Holocene*, **20**, 7–11.

Granger, D. E. (2007). Cosmogenic nuclides – landscape evolution. *In* S. A. Elias (Ed.), *Encyclopedia of Quaternary sciences*, pp. 445–452. Elsevier, Amsterdam.

Grant, D. R. (1970). Recent coastal submergence of the Maritime Provinces, Canada. *Canadian Journal of Earth Sciences*, **7**, 676–689.

Grant, D. R. (1980). Quaternary sea-level change in Atlantic Canada as an indication of crustal delevelling. *In* N.-A. Mörner (Ed.), *Earth rheology, isostasy and eustasy*, pp. 201–214. Wiley, London.

Grant, K. M., Rohling, E. J., Matthews, M. B., *et al.* (2012). Rapid coupling between ice volume and polar temperature over the past 150,000 years. *Nature*, **491**, 744–747.

Gray, S. C., Hein, J. R., Hausmann, R., *et al.* (1992). Geochronology and subsurface stratigraphy of Pukapuka and Rakahanga atolls, Cook Islands: late Quaternary reef growth and sea level history. *Palaeogeography, Palaeoclimatology, Palaeoecology*, **91**, 377–394.

Greensmith, J. T. & Tucker, E. V. (1973). Holocene transgressions and regressions on the Essex coast, outer Thames estuary. *Geologie en Mijnbouw*, **52**, 193–202.

Greenstein, B. J. & Pandolfi, J. M. (2008). Escaping the heat: range shifts of reef coral taxa in coastal Western Australia. *Global Change Biology*, **14**, 513–528.

Gregoire, L. J., Payne, A. J. & Valdes, P. J. (2012). Deglacial rapid sea level rises caused by ice-sheet saddle collapses. *Nature*, **487**, 219–223.

Gregory, J. M., Church, J. A., Boer, G. J. *et al.* (2001). Comparison of results from several AOGCMs for global and regional sea level change 1900–2100. *Climate Dynamics*, **18**, 225–240.

Gregory, J. M., White, N. J., Church, J. A., *et al.* (2013). Twentieth-century global-mean sea-level rise: is the whole greater than the sum of the parts? *Journal of Climate*, doi: 10.1175/JCLI-D-12-00319.1.

Grigg, R. W. (1982). Darwin point: a threshold for atoll formation. *Coral Reefs*, **1**, 29–34.

Grigg, R. W. (1998). Holocene coral reef accretion in Hawaii: a function of wave exposure and sea level history. *Coral Reefs*, **17**, 263–272.

Grigg, R. W. & Jones, A. T. (1997). Uplift caused by lithospheric flexure in the Hawaiian Archipelago as revealed by elevated coral deposits. *Marine Geology*, **141**, 11–25.

Grindrod, J. (1985). The palynology of mangroves on a prograded shore, Princess Charlotte Bay, North Queensland, Australia. *Journal of Biogeography*, **12**, 323–348.

Grindrod, J. (1988). The palynology of Holocene mangrove and saltmarsh sediments, particularly in northern Australia. *Review of Palaeobotany and Palynology*, **55**, 229–245.

Grindrod, J. & Rhodes, E. G. (1984). Holocene sea level history of a tropical estuary: Missionary Bay, North Queensland. *In* B. G. Thom (Ed.), *Coastal geomorphology in Australia*, pp. 151–178. Academic Press, Sydney.

Grinstead, A., Moore, J. C. & Jevrejeva, S. (2010). Reconstructing sea level from paleo and projected temperatures 200 to 2100 AD. *Climate Dynamics*, **34**, 461–472.

Gröger, M. & Plag, H.-P. (1993). Estimations of a global sea level trend: limitations from the structure of the PSMSL global sea level data set. *Global and Planetary Change*, **8**, 161–179.

Grossman, E. E. & Fletcher, C. H. (1998). Sea level higher than present 3500 years ago on the northern main Hawaiian Islands. *Geology*, **26**, 363–366.

Grossman, E. E., Fletcher, C. H. III & Richmond, B. M. (1998). The Holocene sea-level highstand in the equatorial Pacific: analysis of the insular paleosea-level database. *Coral Reefs*, **17**, 309–327.

Grossman, E. L. (2012). Oxygen isotope stratigraphy. *In* F. M. Gradstein, G. G. Ogg, M. D. Schmitz & G. M. Ogg (Eds), *The geologic time scale 2012*. Volume 1, pp. 181–206. Elsevier, Oxford.

Grün, R., Radtke, U. & Omura, A. (1992). ESR and U-Series analyses on corals from Huon Peninsula, New Guinea. *Quaternary Science Reviews*, **11**, 197–202.

Guilcher, A. (1969). Pleistocene and Holocene sea level changes. *Earth-Science Reviews*, **5**, 69–97.

Guilcher, A. (1988). *Coral reef geomorphology*. Wiley, Chichester, 228 pp.

Guilderson, T. P., Fallon, S., Moore, M. D., *et al.* (2009). Seasonally resolved surface water $\Delta^{14}C$ variability in the Lombok Strait: a coralline perspective. *Journal of Geophysical Research*, **114**, C7, doi: 10.1029/2008JC004876.

Guilin, S. B., Polikarpov, G. G. & Egorov, V. N. (2003). The age of microbial structures grown at methane seeps in the Black Sea with an implication of dating the seeping methane. *Marine Chemistry*, **84**, 67–72.

Guillou, H., Singer, B. S., Laj, C., *et al.* (2004). On the age of the Laschamp geomagnetic excursion. *Earth and Planetary Science Letters*, **227**, 331–343.

Guirand, F. (Ed.) (1996). *The Larousse encyclopedia of mythology*. Chancellor Press, London, 500 pp.

Gulliver, F. (1899). Shoreline topography. *Proceedings of the American Academy of Arts and Sciences*, **34**, 151–258.

Gupta, S. K. & Polach, H. A. (1985). *Radiocarbon dating practices at ANU*. Radiocarbon Laboratory, Research School of

Pacific Studies, The Australian National University, Canberra, 176 pp.

Guttenberg, B. (1941). Changes in sea level, postglacial uplift, and mobility of the earth's interior. *Geological Society of America Bulletin*, **52**, 721–772.

Haigh, I., Nicholls, R. & Wells, N. (2010). Assessing changes in extreme sea levels: application to the English Channel, 1900–2006. *Continental Shelf Research*, **30**, 1042–1055.

Hails, J. R. & Gostin, V. A. (Eds) (1984). The Spencer Gulf region. *Marine Geology*, **61**, 111–422.

Hails, J. R., Belperio, A. P., Gostin, V. A., et al. (1984a). The submarine Quaternary stratigraphy of Northern Spencer Gulf, South Australia. *Marine Geology*, **61**, 345–372.

Hails, J. R., Belperio, A. P. & Gostin, V. A. (1984b). Quaternary sea levels, northern Spencer Gulf, Australia. *Marine Geology*, **61**, 373–389.

Hailwood, E. A. (1989). *Magnetostratigraphy*. Geological Society Special Report No. 19, Blackwell Scientific Publications, Oxford, 84 pp.

Haines-Young, R. H. & Petch, J. R. (1983). Multiple working hypotheses: equifinality and the study of landforms. *Transactions of the Institute of British Geographers*, **8**, 458–466.

Häkansson, S. (1983). A reservoir age for the coastal waters of Iceland. *Geologiska Föreningens i Stockholm Förhandlingar*, **105**, 64–67.

Hall, J. (1815). On the revolutions of the Earth's surface. *Transactions of the Royal Society of Edinburgh*, **7**, 139–211.

Hall, K. (1982). Rapid deglaciation as an initiator of volcanic activity: an hypothesis. *Earth Surface Processes and Landforms*, **7**, 45–51.

Hallam, A. (1983). *Great geological controversies*. Oxford University Press, Oxford, 182 pp.

Hallam, A. (1984). Pre-Quaternary sea-level changes. *Annual Review of Earth and Planetary Science*, **12**, 205–243.

Hallam, A. (1992). *Phanerozoic sea-level changes*. Perspectives in paleobiology and Earth History Series. Columbia University Press, New York, 266 pp.

Halley, R. B., Shinn, E. A., Hudson, J. H., et al. (1977). Recent and relict topography of Boo Bee patch reef, Belize. *Proceedings of 3rd International Coral Reef Symposium*, **2**, 29–35.

Hamburger, M. W. & Everingham, I. B. (1988). Active tectonism within the Fiji platform, southwest Pacific. *Geology*, **16**, 237–241.

Hamilton, W. B. (1988). Plate tectonics and island arcs. *Geological Society of America Bulletin*, **100**, 1503–1527.

Han, W., Meehl, G. A., Rajagopalan, B., et al. (2010). Patterns of Indian Ocean sea-level change in a warming climate. *Nature Geoscience*, **3**, 546–550.

Hanebuth, T., Stattegger, K. & Grootes, P. M. (2000). Rapid flooding of the Sunda Shelf: a late-glacial sea-level record. *Science*, **288**, 1033–1035.

Hanebuth, T. J. J., Stattegger, K. & Bojanowski, A. (2009). Termination of the Last Glacial Maximum sea-level lowstand: the Sunda Shelf data revisited. *Global and Planetary Change*, **66**, 76–84.

Hansen, J. E. (2007). Scientific reticence and sea level rise. *Environmental Research Letters*, **2**, 1–6.

Hansen, J. E. & Sato, M. (2012). Paleoclimate implications for human-made climate change. *In* A. Berger, F. Mesinger & D. Šijački (Eds), *Climate change: inferences from paleoclimate and regional aspects*, pp. 21–48. Springer, Dordrecht.

Hantoro, W. S. (1996). Quaternary sea level variation in the Pacific-Indian Ocean gateways: response and impact. *Quaternary International*, **37**, 73–80.

Haq, B. U., Berggren, W. A. & van Couvering, J. A. (1977). Corrected age for the Pliocene/Pleistocene boundary. *Nature*, **269**, 483–488.

Haq, B. U., Hardenbol, J. & Vail, P. R. (1987). Chronology of fluctuating sea levels since the Triassic. *Science*, **235**, 1156–1167.

Hare, P. E. & Abelson, P. H. (1968). Racemization of amino acids in fossil shells. *Carnegie Institution of Washington Yearbook*, **66**, 526–528.

Harkness, D. D. (1983). The extent of natural 14C deficiency in the coastal environment of the United Kingdom. *PACT, Journal of the European Study Group on*

Physical, Chemical and Mathematical Techniques applied to Archaeology, **8**, 351–364.

Harmon, R. S., Schwarcz, H. P. & Ford, D. C. (1978). Late Pleistocene sea level history of Bermuda. *Quaternary Research*, **9**, 205–218.

Harmon, R. S., Land, L. S., Mitterer, R. M., *et al.* (1981). Bermuda sea level during the last interglacial. *Nature*, **289**, 481–483.

Harmon, R. S., Mitterer, R. M., Kriausakul, N., *et al.* (1983). U-series and amino-acid racemization geochronology of Bermuda: implications for eustatic sea-level fluctuation over the past 250,000 years. *Palaeogeography, Palaeoclimatology, Palaeoecology*, **44**, 41–70.

Harris, P. T. (1999). Sequence architecture during the Holocene transgression: an example from the Great Barrier Reef shelf, Australia – comment. *Sedimentary Geology*, **125**, 235–239.

Harris, P. T. & Heap, A. D. (2003). Environmental management of clastic coastal depositional environments: inferences from an Australian geomorphic database. *Ocean and Coastal Management*, **46**, 457–478.

Harris, P. T., Heap, A. D., Bryce, A. D., *et al.* (2002). Classification of Australian clastic coastal depositional environments based upon a quantitative analysis of wave, tidal, and river power. *Journal of Sedimentary Research*, **72**, 858–870.

Harris, P. T., Heap, A., Marshall, J. F., *et al.* (2007). Submerged coral reefs and benthic habitats of the southern Gulf of Carpentaria. *Geoscience Australia Record*, **2007**/02, 134 pp.

Harris, P. T., Heap, A. D., Marshall, J. F., *et al.* (2008). A new coral reef province in the Gulf of Carpentaria, Australia: colonisation, growth and submergence during the early Holocene. *Marine Geology*, **251**, 85–97.

Harrison, D. E. & Cane, M. A. (1984). Changes in the Pacific during the 1982–83 event. *Oceanus*, **27**, 21–28.

Hartley, R. A., Roberts, G. G., White, N., *et al.* (2011). Transient convective uplift of an ancient buried landscape. *Nature Geoscience*, **4**, 562–565.

Harvey, N., Barnett, E. J., Bourman, R. P., *et al.* (1999). Holocene sea-level change at Port Pirie, South Australia: a contribution to global sea level rise estimates from tide gauges. *Journal of Coastal Research*, **15**, 607–615.

Haslett, S. K. (2001). The palaeoenvironmental implications of the distribution of intertidal Foraminifera in a tropical Australian estuary: a reconnaissance study. *Australian Geographical Studies*, **39**, 67–74.

Haug, E. (1900). Les geosynclinaux et les aires continentals; contribution a l'etude des transgressions et des regressions inconnues. *Bulletin de la Société Geologique de la France, Series 3*, **28**, 617–711.

Hayne, M. & Chappell, J. (2001). Cyclone frequency during the last 5000 years at Curacao Island, north Queensland, Australia. *Palaeogeography, Palaeoclimatology, Palaeoecology*, **168**, 207–219.

Hays, J. D., Imbrie, J. & Shackleton, N. J. (1976). Variations in the Earth's orbit: pacemaker of the Ice Ages. *Science*, **194**, 1121–1132.

Head, M. J. & Gibbard, P. L. (2005). Early–Middle Pleistocene transitions: an overview and recommendations for the defining boundary. *In* M. J. Head. & P. L. Gibbard (Eds), *Early–middle Pleistocene transitions: the land–ocean evidence.* Geological Society, London, Special Publications, 247, pp. 1–18.

Heap, A. D. & Harris, P. T. (2008). Geomorphology of the Australian margin and adjacent seafloor. *Australian Journal of Earth Sciences*, **55**, 555–585.

Heap, A. D., Bryce, S. & Ryan, D. A. (2004). Facies evolution of Holocene estuaries and deltas: a large-sample statistical study from Australia. *Sedimentary Geology*, **168**, 1–17.

Hearty, P. J. (1986). An inventory of Last Interglacial (sensu lato) age deposits from the Mediterranean basin: a study of isoleucine epimerization and U-series dating. *Zeitschrift für Geomorphologie N.F. Suppl.-Bd.*, **62**, 51–69.

Hearty, P. J. (1998). The geology of Eleuthera Island, Bahamas: a Rosetta Stone of Quaternary stratigraphy and sea-level history. *Quaternary Science Reviews*, **17**, 333–355.

Hearty, P. J. (2002). The Kaena highstand on Oahu, Hawaii: further evidence of

Antarctic Ice collapse during the middle Pleistocene. *Pacific Science*, **56**, 65–82.

Hearty, P. J. & Aharon, P. (1988). Amino acid chronostratigraphy of Late Quaternary coral reefs – Huon Peninsula, New Guinea and the Great Barrier Reef, Australia. *Geology*, **16**, 579–583.

Hearty, P. J. & Kindler, P. (1993). New perspectives on Bahamian geology: San Salvador Island, Bahamas. *Journal of Coastal Research*, **9**, 577–594.

Hearty, P. J. & Neumann, A. C. (2001). Rapid sea level and climate change at the close of the Last Interglaciation (MIS 5e): evidence from the Bahama Islands. *Quaternary Science Reviews*, **20**, 1881–1895.

Hearty, P. J. & O'Leary, M. J. (2008). Carbonate eolianites, quartz sands and Quaternary sea-level cycles, Western Australia: a chronostratigraphic approach. *Quaternary Geochronology*, **3**, 26–55.

Hearty, P. J., Miller, G. H., Stearns, C. E., et al. (1986). Aminostratigraphy of Quaternary shorelines in the Mediterranean basin. *Bulletin Geological Society of America*, **97**, 850–858.

Hearty, P. J., Vacher, H. L. & Mitterer, R. M. (1992). Aminostratigraphy and ages of the Pleistocene limestones of Bermuda. *Geological Society of America Bulletin*, **104**, 471–480.

Hearty, P. J., Kindler, P., Cheng, H., et al. (1999). A +20 m middle Pleistocene sea-level highstand (Bermuda and the Bahamas) due to partial collapse of Antarctic ice. *Geology*, **27**, 375–378.

Hearty, P. J., Kaufman, D. S., Olson, S. L., et al. (2000). Stratigraphy and whole-rock amino acid geochronology of key Holocene and Last Interglacial carbonate deposits in the Hawaiian Islands. *Pacific Science*, **544**, 423–442.

Hearty, P. J., O'Leary, M. J., Kaufman, D. S., et al. (2004). Amino acid geochronology of individual foraminifer (*Pulleniatina obliquiloculata*) tests, north Queensland margin, Australia: a new approach to correlating and dating Quaternary tropical marine sediment cores. *Paleoceanography*, **19**, doi: 10.1029/2004PA001059.

Hearty, P. J., Hollin, J. T., Neumann, A. C., et al. (2007). Global sea-level fluctuations duirng the Last Interglaciation (MIS 5e). *Quaternary Science Reviews*, **26**, 2090–2112.

Hedges, R. E. M. (1992). Sample treatment strategies in radiocarbon dating. *In* R. E. Taylor, A. Long & R. S. Kra (Eds), *Radiocarbon after four decades – an interdisciplinary perspective*, pp. 165–183. Springer-Verlag, New York.

Heinrich, H. (1988). Origin and consequences of cyclic ice rafting in the northeast Atlantic Ocean during the past 130,000 years. *Quaternary Research*, **29**, 142–152.

Hendry, M. D. (1987). Tectonic and eustatic control on late Cenozoic sedimentation within an active plate boundary zone, west coast margin, Jamaica. *Geological Society of America Bulletin*, **99**, 718–728.

Hernández-Mendiola, E., Bernal, J. P., Lounejeva, E., et al. (2011). U-series dating of carbonates using inductively coupled plasma-quadrupole mass spectrometry. *Quaternary Geochronology*, **6**, 564–573.

Herweijer, J. P. & Focke, J. W. (1978). Late Pleistocene depositional and denudational history of Aruba, Bonaire and Curacao (Netherlands Antilles). *Geologie En Mijnbouw*, **57**, 177–187.

Hesp, P. A. (1984). Foredune formation in southeast Australia. *In* B. G. Thom (Ed.), *Coastal geomorphology in Australia*, pp. 69–97. Academic Press, Sydney.

Hesp, P. A., Hung, C. C., Hilton, M., et al. (1998). A first tentative Holocene sea-level curve for Singapore. *Journal of Coastal Research*, **14**, 308–314.

Heteren, S. V., Huntley, D. J., van de Plassche, O., et al. (2000). Optical dating of dune sand for the study of sea-level change. *Geology*, **28**, 411–414.

Hey, R. W. (1978). Horizontal Quaternary shorelines of the Mediterranean. *Quaternary Research*, **10**, 197–203.

Heyworth, A. & Kidson, C. (1982). Sea-level changes in England and Wales. *Proceedings of the Geologists Association*, **93**, 91–111.

Hill, P. J., De Deckker, P., Von der Borch, C., et al. (2009). Ancestral Murray River on the Lacepede Shelf, southern Australia: late Quaternary migrations of a major river outlet and strandline development. *Australian Journal of Earth Sciences*, **56**, 135–157.

Hillaire-Marcel, C., Karrow, O., Causse, C., et al. (1986). Th/U dating of *Strombus bubonius*-bearing marine terraces in southeastern Spain. *Geology*, **14**, 613–616.

Hille, P. (1979). An open system model for uranium series dating. *Earth and Planetary Science Letters*, **42**, 138–142.

Hills, E. S. (1949). Shore platforms. *Geological Magazine*, **86**, 137–152.

Hine, A. C. & Neumann, A. C. (1977). Shallow carbonate–bank–margin growth and structure, Little Bahama Bank, Bahamas. *American Association of Petroleum Geologists Bulletin*, **61**, 376–406.

Hinton, A. C. (1995). Holocene tides of The Wash, U.K.: the influence of water-depth and coastline-shape changes on the record of sea-level change. *Marine Geology*, **124**, 87–111.

Hodell, D. A., Charles, C. D. & Ninnemann, U. S. (2000). Comparison of interglacial stages in the South Atlantic sector of the southern ocean for the past 450 kyr: implications for Marine Isotope Stage (MIS) 11. *Global and Planetary Change*, **24**, 7–26.

Hoffman, J. S., Keyes, D. & Titus, J. G. (1983). *Projecting future sea level rise*. US Environmental Protection Agency, Washington, DC, 121 pp.

Hoffmeister, J. E. & Multer, H. G. (1968). Geology and origin of the Florida Keys. *Geological Society of America Bulletin*, **79**, 1487–1502.

Hoffmeister, J. E., Stockman, K. W. & Multer, H. G. (1967). Miami limestone of Florida and its recent Bahamian counterpart. *Geological Society of America Bulletin*, **78**, 175–190.

Holdgate, G. R., Wagstaff, B. & Gallagher, S. J. (2011). Did Port Phillip Bay nearly dry up between ~2800 and 1000 cal. yr BP? Bay floor channelling evidence, seismic and core dating. *Australian Journal of Earth Sciences*, **58**, 157–175.

Holgate, S. & Woodworth, P. L. (2004). Evidence for enhanced coastal sea level rise during the 1990s. *Geophysical Research Letters*, **31**, L07305.

Holgate, S., Jevrejeva, S., Woodworth, P., et al. (2007). Comment on 'A semi-empirical approach to projecting future sea-level rise'. *Science*, **317**, 1866.

Hollin, J. T. (1977). Thames interglacial sites, Ipswichian sea levels and Antarctic ice surges. *Boreas*, **1**, 33–52.

Holmes, A. (1944). *Principles of physical geology*. Thomas Nelson & Sons, Edinburgh.

Hopley, D. (1982). *The geomorphology of the Great Barrier Reef*. Wiley Interscience, New York, 453 pp.

Hopley, D. (1984). The Holocene 'high energy window' on the central Great Barrier Reef. *In* B. G. Thom (Ed.), *Coastal geomorphology in Australia*, pp. 135–150. Academic Press, Sydney.

Hopley, D. (1986a). Beachrock as a sea-level indicator. *In* O. van de Plassche (Ed.), *Sea-level research: a manual for the collection and evaluation of data*, pp. 157–173. GeoBooks, Norwich.

Hopley, D. (1986b). Corals and reefs as indicators of paleo-sea levels with special reference to the Great Barrier Reef. *In* O. Van de Plassche (Ed.), *Sea-level research: a manual for the collection and evaluation of data*, pp. 195–228. GeoBooks, Norwich.

Hopley, D. (1987). Holocene sea-level changes in Australasia and the Southern Pacific. *In* R. J. N. Devoy (Ed.), *Sea surface studies: a global review*, pp. 375–408. Croom Helm, London.

Hopley, D. (2006). Coral reef growth on the shelf margin of the Great Barrier reef with special reference to the Pompey Complex. *Journal of Coastal Research*, **22**, 150–174.

Hopley, D. & Isdale, P. (1977). Coral micro-atolls, tropical cyclones and reef flat morphology: a North Queensland example. *Search*, **8**, 79–81.

Hopley, D., Muir, F. J. & Grant, C. R. (1984). Pleistocene foundations and Holocene growth of Redbill Reef, south central Great Barrier Reef. *Search*, **15**, 288–289.

Hopley, D., Smithers, S. G. & Parnell, K. (2007). *Geomorphology of the Great Barrier Reef: development, diversity and change*. Cambridge University Press, Cambridge, 546 pp.

Hori, K. & Saito, Y. (2007). An early Holocene sea-level jump and delta initiation. *Geophysical Research Letters*, **34**, doi:10.1029/2007GL031029.

Horsfield, W. T. (1975). Quaternary vertical movements in the Greater Antilles. *Geological Society of America Bulletin*, **86**, 933–938.

Horton, B. P. & Edwards, R. J. (2006). Quantifying Holocene sea-level change using intertidal foraminifera: lessons from the British Isles. *Cushman Foundation for Foraminiferal Research, Special Publication*, **40**, 1–97.

Horton, B. P. & Shennan, I. (2009). Compaction of Holocene strata and the implications for relative sea-level change on the east coast of England. *Geology*, **37**, 1083–1086.

Horton, B. P., Edwards, R. J. & Lloyd, J. M. (1999). UK intertidal foraminiferal distributions: implications for sea-level studies. *Marine Micropalaeontology*, **36**, 205–223.

Horton, B. P., Corbett, R., Culver, S. J., *et al.* (2006). Modern saltmarsh diatom distributions of the Outer Banks, North Carolina, and the development of a transfer function for high resolution reconstructions of sea level. *Estuarine Coastal and Shelf Science*, **69**, 381–394.

Horton, B. P., Peltier, W. R., Culver, S. J., *et al.* (2009). Holocene sea-level changes along the North Carolina coastline and their implications for glacial isostatic adjustment models. *Quaternary Science Reviews*, **28**, 1725–1736.

Hossfeld, P. A. (1950). The Late Cainozoic history of the south-east of South Australia. *Royal Society of South Australia, Transactions*, **73**, 232–279.

Houston, J. R. & Dean, R. G. (2011). Sea-level acceleration based on U.S. tide gauges and extensions of previous global-gauge analyses. *Journal of Coastal Research*, **27**, 409–417.

Houston, J. R. & Dean, R. G. (2012). Comparisons at tide-gauge locations of Glacial Isostatic Adjustment predictions with Global Positioning System measurements. *Journal of Coastal Research*, **28**, 739–744.

Hsü, K. J. & Winterer, E. L. (1980). Discussion on causes of world-wide changes in sea level. *Journal of the Geological Society, London*, **137**, 509–510.

Hsü, K. J., Ryan, W. B. F. & Cita, M. B. (1973). Late Miocene desiccation of the Mediterranean. *Nature*, **242**, 240–244.

Hsü, K. J., Montadert, L., Bernoulli, D., *et al.* (1977). History of the Mediterranean salinity crisis. *Nature*, **267**, 399–403.

Hsu, J. T., Leonard, E. M. & Wehmiller, J. F. (1989). Aminostratigraphy of Peruvian and Chilean Quaternary marine terraces. *Quaternary Science Reviews*, **8**, 255–262.

Hua, Q. (2009). Radiocarbon: a chronological tool for the recent past. *Quaternary Geochronology*, **4**, 378–390.

Hua, Q., Woodroffe, C. D., Barbetti, M., *et al.* (2004). Marine reservoir correction for the Cocos (Keeling) Islands, Indian Ocean. *Radiocarbon*, **46**, 603–610.

Hubbard, D. K. (1997). Reefs as dynamic systems. *In* C. Birkeland (Ed.), *Life and death of coral reefs*, pp. 43–67. Chapman & Hall, New York.

Hubbard, D. K., Zankl, H., Van Heerden, I., *et al.* (2005). Holocene reef development along the northeastern St. Croix Shelf, Buck Island, U.S. Virgin Islands. *Journal of Sedimentary Research*, **75**, 97–113.

Hubbard, D., Burke, R., Gill, I., *et al.* (2008). Coral-reef geology: Puerto Rico and the U.S. Virgin Islands. *In* B. M. Riegl & R. E. Dodge (Eds), *Coral reefs of the USA*, pp. 263–302. Springer, Dordrecht.

Hughen, K., Lehman, S., Southon, J., *et al.* (2004). ^{14}C activity and global carbon changes over the past 50,000 years. *Science*, **303**, 202–207.

Hunter, J. (2010). Estimating sea-level extremes under conditions of uncertain sea-level rise. *Climatic Change*, **99**, 331–350.

Hunter, J. (2012). A simple technique for estimating an allowance for uncertain sea-level rise. *Climatic Change*, **113**, 239–252.

Hunter, J., Coleman, R. & Pugh, D. (2003). The sea level at Port Arthur, Tasmania, from 1841 to the present. *Geophysical Research Letters*, **30**, art. no.-1401.

Huntley, D. J. & Prescott, J. R. (2001). Improved methodology and new thermoluminescence ages for the dune sequence in south-east South Australia. *Quaternary Science Reviews*, **20**, 687–699.

Huntley, D. J., Godfrey-Smith, D. I. & Thewalt, M. L. W. (1985). Optical dating of sediments. *Nature*, **313**, 105–107.

Huntley, D. J., Hutton, J. T. & Prescott, J. R. (1993). The stranded beach-dune sequence of south-east South Australia: a test of thermoluminescence dating, 0–800 ka. *Quaternary Science Reviews*, **12**, 1–20.

Huntley, D. J., Hutton, J. T. & Prescott, J. R. (1994). Further thermoluminescence dates from the dune sequence in southeast of South Australia. *Quaternary Science Reviews*, **13**, 201–207.

Huxley, T. H. (1878). *Physiography: an introduction to the study of nature*. Macmillan, London, 384 pp.

Huybrechts, P. (2002). Sea-level changes at the LGM from ice-dynamic reconstructions of the Greenland and Antarctic ice sheets during the glacial cycles. *Quaternary Science Reviews*, **21**, 203–231.

Idnurm, M. & Cook, P. J. (1980). Palaeomagnetism of beach ridges in South Australia and the Milankovitch theory of ice ages. *Nature*, **286**, 699–702.

Ikehara, K., Danhara, T., Yamashita, T., *et al.* (2011). Paleoceanographic control on a large marine reservoir effect offshore of Tokai, south of Japan, NW Pacific, during the last glacial maximum-deglaciation. *Quaternary International*, **246**, 213–221.

Iliffe, J. & Lott, R. (2008). *Datums and map projections for remote sensing, GIS and surveying*, 2nd ed. Whittles Publishing, Dunbeath.

Imbrie, J. & Imbrie, K. P. (1979). *Ice ages – solving the mystery*. Harvard University Press, Cambridge, Massachusetts, 224 pp.

Imbrie, J. J. D., Hays, D. G., Martinson, D. G., *et al.* (1984). The orbital theory of Pleistocene climate: support from a revised chronology of the marine 18O record. *In* A. Berger, J. Imbrie, J. Hays, G. Kukla & B. Saltzman (Eds), *Milankovitch and climate: understanding the response to astronomical forcing*, pp. 269–305. D. Reidel, Dordrecht.

Ingram, B. L. & Southon, J. R. (1996). Reservoir ages in Eastern Pacific coastal and estuarine waters. *Radiocarbon*, **38**, 571–582.

Inman, D. L. & Nordstrom, C. E. (1971). On the tectonic and morphologic classification of coasts. *Journal of Geology*, **79**, 1–21.

Isacks, B., Oliver, J. & Sykes, L. R. (1968). Seismology and the new global tectonics. *Journal of Geophysical Research*, **73**, 5855–5899.

Isla, F. I. (1989). Holocene sea-level fluctuation in the southern hemisphere. *Quaternary Science Reviews*, **8**, 359–368.

Israelson, C. & Wohlfarth, B. (1999). Timing of the Last-Interglacial high sea level on the Seychelles Islands, Indian Ocean. *Quaternary Research*, **51**, 306–316.

Issel, A. (1914). Lembi fossiliferi Quaternari e recenti nella Sardegna meridonale. *Academia Nazionale dei Lincei e Series*, **523**, 759–770.

Ivey, M. L., Breyer, J. A. & Britton, J. C. (1980). Sedimentary facies and depositional history of the Swan Islands, Honduras. *Sedimentary Geology*, **27**, 195–212.

Jacobs, Z. (2008). Luminescence chronologies for coastal and marine sediments. *Boreas*, **37**, 508–535.

Jacobs, Z. & Roberts, R. G. (2009). Last interglacial age for aeolian and marine deposits and the Nahoon fossil human footprints, Southeast Coast of South Africa. *Quaternary Geochronology*, **4**, 160–169.

Jacobs, Z., Roberts, R. G., Lachlan, T. J., *et al.* (2011). Development of the SAR TT-OSL procedure for dating Middle Pleistocene dune and shallow marine deposits along the southern Cape coast of South Africa. *Quaternary Geochronology*, **6**, 491–513.

James, N. P. & Bone, Y. (2011). *Neritic carbonate sediments in a temperate realm*. Springer, Dordrecht, 254 pp.

James, N. P., Bone, Y., Carter, R. M., *et al.* (2006). Origin of the Late Neogene Roe Plains and their calcarenite veneer: implications for sedimentology and tectonics in the Great Australian Bight. *Australian Journal of Earth Sciences*, **53**, 407–419.

Jamieson, T. F. (1863). On the Parallel Roads of Glen Roy, and their place in the history of the glacial period. *Quarterly Journal of the Geological Society of London*, **19**, 235–259.

Jamieson, T. F. (1865). On the history of the last geological changes in Scotland. *Quarterly Journal of the Geological Society of London*, **21**, 161–203.

Jamieson, T. F. (1908). On changes of level and the production of raised beaches. *Geological Magazine*, **5**, 22–25.

Jansen, H. S. (1984). *Radiocarbon dating for contributors*. Institute of Nuclear Sciences Publication INS-R-328. Department of Scientific and Industrial Research, Lower Hutt, New Zealand, 55 pp.

Janssen, V. (2009). Understanding coordinate reference systems, datums and transformations. *International Journal of Geoinformatics*, **5**, 41–53.

Jelgersma, S. (1961). Holocene sea-level changes in the Netherlands. *Mededelingen Geologische Stichting*, ser. **C6**, no. 7, 100 pp.

Jelgersma, S. (1966). Sea-level changes during the last 10,000 years. *Proceedings of the International Symposium on World Climates from 8000 B.C. to 0 B.C.*, Imperial College, London, 18–19 April 1966. Royal Meteorological Society, London, pp. 54–71.

Jelgersma, S. (1979). Sea level changes in the North Sea Basin. *In* E. Oele, R. T. E. Schüttenhelm & A. J. Wiggers (Eds), The Quaternary history of the North Sea. *Acta Universitas Upsaliensis Symposium. Universitas Upsaliensis Annum Quingentesimum Celebrantis*, **2**, 233–248.

Jennings, J. N. (1975). Desert dunes and estuarine fill in the Fitzroy estuary, northwestern Australia. *Catena*, **2**, 215–262.

Jennings, J. N. (1985). *Karst geomorphology*. Blackwell, Oxford, 280 pp.

Jevrejeva, S., Grinsted, A., Moore, J. C., *et al.* (2006). Nonlinear trends and multiyear cycles in sea level records. *Journal of Geophysical Research*, **111**, C09012. doi:10.1029/2005JC003229.

Jevrejeva, S., Moore, J. C., Grinsted, A., *et al.* (2008). Recent global sea level acceleration started over 200 years ago? *Geophysical Research Letters*, **35**, L08715.

Jevrejeva, S., Grinsted, A. & Moore, J. C. (2009). Anthropogenic forcing dominates sea level rise since 1850. *Geophysical Research Letters*, **36**, L20706.

Johnson, D. L. (1978). The origin of Island Mammoths and the Quaternary land bridge history of the Northern Channel Islands, California. *Quaternary Research*, **10**, 204–225.

Johnson, D. W. (1911). Botanical evidence of coastal subsidence. *Science*, **33**, 300–302.

Johnson, D. W. (1919). *Shore processes and shoreline development*. Wiley, New York, 584 pp.

Johnson, D. W. (1925). *The New England–Acadian shoreline*. Wiley, New York, 608 pp.

Johnson, D. W. (1929). Base level. *Journal of Geology*, **37**, 775–782.

Johnson, G. C. & Wijffels, S. E. (2011). Ocean density change contributions to sea level rise. *Oceanography*, **24**, 112–121.

Johnston, P. J. & Lambeck, K. (1999). Postglacial rebound and sea-level contributions to changes in the geoid and the Earth's rotation axis. *Geophysical Journal International*, **136**, 537–558.

Jones, B. & Hunter, I. G. (1990). Pleistocene paleogeography and sea levels on the Cayman Islands, British West Indies. *Coral Reefs*, **9**, 81–91.

Jones, B. G., Young, R. W. & Eliot, I. G. (1979). Stratigraphy and chronology of receding barrier-beach deposits on the northern Illawarra coast of New South Wales. *Journal of the Geological Society of Australia*, **26**, 255–264.

Jones, B. G., Woodroffe, C. D. & Martin, G. R. (2003). Deltas in the Gulf of Carpentaria, Australia: forms, processes and products. *In* F. H. Sidi, D. Nummedal, P. Imbert, H. Darmen & H. W. Posamentier (Eds), *Tropical deltas of southeast Asia: sedimentology, stratigraphy and petroleum geology*, pp. 21–43. SEPM Special Publication, Tulsa, OK.

Jones, D. K. C. (1981). *Southeast and southern England*. Methuen, London, 332 pp.

Jones, D. K. C. (1999). Evolving models of the Tertiary evolutionary geomorphology of southern England, with special reference to the Chalklands. *In* B. J. Smith, W. B. Whalley & P. A. Warke (Eds), *Uplift, erosion and stability: perspectives on long-term landscape development*. Geological Society, London, Special Publications, **162**, 1–23.

Jones, M. R. & Torgersen, T. (1988). Late Quaternary evolution of Lake Carpentaria on the Australia–New Guinea continental shelf. *Australian Journal of Earth Sciences*, **35**, 313–324.

Jongsma, D. (1970). Eustatic sea level changes in the Arafura Sea. *Nature*, **228**, 150–151.

Jouannic, C., Hantoro, W. S., Hoang, C. T., *et al.* (1988). Quaternary raised reef terraces at Cape Laundi, Sumba, Indonesia: geomorphological analysis and first radiometric Th/U and ^{14}C determinations. *Proceedings of the 6th Coral Reef Symposium*, **3**, 441–447.

Jouzel, J., Masson-Delmotte, V., Cattani, O., *et al.* (2007). Orbital and millennial Antarctic climate variability over the past 800,000 years. *Science*, **317**, 793–796.

Joyce, E. B. (1975). Quaternary volcanism and tectonics in southeastern Australia. *In* R. P. Suggate & M. M. Cresswell (Eds), Quaternary studies. *Bulletin, Royal Society of New Zealand*, **13**, 169–176.

Jutson, J. T. (1939). Shore platforms near Sydney, New South Wales. *Journal of Geomorphology*, **2**, 237–250.

Katupotha, J. (1988). Evidence of high sea level during the mid-Holocene on the southwest coast of Sri Lanka. *Boreas*, **17**, 209–213.

Katupotha, J. & Fujiwara, K. (1988). Holocene sea level change on the southwest and south coasts of Sri Lanka. *Palaeogeography, Palaeoclimatology, Palaeoecology*, **68**, 189–203.

Kaufman, A., Broecker, W. S., Ku, T.-L., *et al.* (1971). The status of U-series methods of mollusk dating. *Geochimica et Cosmochimica Acta*, **35**, 1155–1183.

Kayanne, H., Ishii, T., Matsumoto, E., *et al.* (1993). Late Holocene sea-level change on Rota and Guam, Mariana Islands and its constraint on geophysical predictions. *Quaternary Research*, **40**, 189–200.

Kaye, C. A. & Barghoorn, E. S. (1964). Late Quaternary sea-level change and crustal rise at Boston, Massachusetts, with notes on the autocompaction of peat. *Geological Society of America Bulletin*, **75**, 63–80.

Kearey, P. & Vine, F. J. (1996). *Global tectonics*, 2nd edition. Blackwell Science, Oxford, 333 pp.

Kearey, P., Klepeis, K. A. & Vine, F. J. (2009). *Global tectonics*, 3rd edition. Wiley-Blackwell, Oxford, 482 pp.

Keigwin, L. D., Donelly, J. P., Cook, M. S., *et al.* (2006). Rapid sea-level rise and Holocene climate in the Chukchi Sea. *Geology*, **34**, 861–864.

Kelletat, D. (2006). Beachrock as sea-level indicator? Remarks from a geomorphological point of view. *Journal of Coastal Research*, **22**, 1558–1564.

Kelley, J. T., Belknap, D. F. & Claesson, S. (2010). Drowned coastal deposits with associated archaeological remains from a sea-level slowstand: northeastern Gulf of Maine, USA. *Geology*, **38**, 695–698.

Kelley, J. T., Belknap, D. F., Kelley, A. R., *et al.* (2013). A model for drowned terrestrial habitats with associated archaeological remains in the northwestern Gulf of Maine, USA. *Marine Geology*, doi: 10.1016/j.margeo.2012.10.016.

Kemp, A. C., Horton, B. P., Culver, S. J., *et al.* (2009). Timing and magnitude of recent accelerated sea-level rise (North Carolina, United States). *Geology*, **37**, 1035–1038.

Kemp, A. C., Horton, B. P., Donnelly, J. P. (2011). Climate related sea-level variations over the past two millennia. *Proceedings of the National Academy of Sciences*, **108**, 11017–11022.

Kemp, A. C., Horton, B. P., Vann, D. R., *et al.* (2012). Quantitative vertical zonation of salt-marsh foraminifera for reconstructing former sea level; an example from New Jersey, USA. *Quaternary Science Reviews*, **54**, 26–39.

Kench, P. S., Smithers, S. G., McLean, R. F., *et al.* (2009). Holocene reef growth in the Maldives: evidence of a mid-Holocene sea-level highstand in the central Indian Ocean. *Geology*, **37**, 455–458.

Kendall, R. A. & Mitrovica, J. X. (2007). Radial resolving power of far-field differential sea-level highstands in the inference of mantle viscosity. *Geophysical Journal International*, **171**, 881–889.

Kendall, R. A., Mitrovica, J. X. & Milne, G. A. (2008). The sea-level fingerprint of the 8.2 ka climate event. *Geology*, **36**, 423–426.

Kendrick, G. W., Wyrwoll, K.-H. & Szabo, B. J. (1991). Pliocene–Pleistocene coastal events and history along the western margin of Australia. *Quaternary Science Reviews*, **10**, 419–439.

Kennedy, D. M. (2010). Geological control on the morphology of estuarine shore platforms: Middle Harbour, Sydney, Australia. *Geomorphology*, **114**, 71–77.

Kennedy, D. M. & Woodroffe, C. D. (2002). Fringing reef growth and morphology: a review. *Earth-Science Reviews*, **57**, 257–279.

Kennedy, D. M., Paulik, R. & Dickson, M. E. (2011). Subaerial weathering versus wave processes in shore platform development: reappraising the Old Hat Island evidence. *Earth Surface Process and Landforms*, **36**, 686–694.

Kennedy, D. M., Marsters, H., Woods, J., et al. (2012). Shore platform development on an uplifting limestone island over multiple sea level cycles, Niue, South Pacific. *Geomorphology*, **141–142**, 170–182.

Kennedy, G. L., Lajoie, K. R. & Wehmiller, J. F. (1982). Aminostratigraphy and faunal correlations of late Quaternary marine terraces, Pacific Coast, USA. *Nature*, **299**, 545–547.

Keraudren, B. & Sorel, D. (1987). The terraces of Corinth (Greece): a detailed record of eustatic sea-level variations during the last 500,000 years. *Marine Geology*, **77**, 99–107.

Kershaw, S. & Guo, L. (2001). Marine notches in coastal cliffs: indicators of relative sea-level change, Perachora Peninsula, central Greece. *Marine Geology*, **179**, 213–228.

Kessler, J. D., Reeburgh, W. S. & Southon, J. (2006). Basin-wide estimates of the input of methane from seeps and clathrates to the Black Sea. *Earth and Planetary Science Letters*, **243**, 366–375.

Kidson, C. (1982). Sea level changes in the Holocene. *Quaternary Science Reviews*, **1**, 121–151.

Kidson, C. (1986). Sea-level changes in the Holocene. *In* O. van de Plassche (Ed.), *Sea-level research: a manual for the collection and evaluation of data*, pp. 27–64. GeoBooks, Norwich.

Kikuchi, T. (1990). *Late Quaternary uplift rate and Paleosea-levels in the Huon Peninsula, New Guinea*. Tokyo Metropolitan University. Department of Geography, Geographical Reports, 25, 139–153.

Kindler, P. & Hearty, P. J. (1997). Geology of the Bahamas: architecture of Bahamian Islands. *In* H. L. Vacher and T. Quinn (Eds), *Geology and hydrogeology of carbonate islands*. Developments in Sedimentology **54**, pp. 141–160. Elsevier, Amsterdam.

King, L. C. (1962). *Morphology of the Earth*, Oliver and Boyd, Edinburgh, 699 pp.

King, M. A., Keshin, M., Whitehouse, P. L., et al. (2012a). Regional biases in absolute sea-level estimates from tide gauge data due to residual unmodeled vertical land movement. *Geophysical Research Letters*, **39**, (L14604).

King, M. A., Bingham, R. J., Moore, P., et al. (2012b). Lower satellite-gravimetry estimates of Antarctic sea-level contribution. *Nature*, **491**, 586–589.

King-Hele, D. G. (1972). Heavenly harmony and Earthly harmonics. *Quarterly Journal of the Royal Astronomical Society*, **13**, 374–395.

Klein, J., Lerman, J. C., Damon, P. E., et al. (1982). Calibration of radiocarbon dates: tables based on the consensus of data of the workshop on calibrating the radiocarbon time scale. *Radiocarbon*, **24**, 103–150.

Knight, J. (2007). Beachrock reconsidered. Discussion of Kelletat, D. 2006. Beachrock as sea-level indicator? Remarks from a geomorphological point of view. *Journal of Coastal Research*, **23**, 1074–1078.

Koelling, M., Webster, J. M., Camoin, G., et al. (2009). SEALEX – internal reef chronology and virtual drill logs from a spreadsheet-based reef growth model. *Global and Planetary Change*, **66**, 149–159.

Kolb, C. R. & van Lopik, J. R. (1966). Depositional environments of the Mississippi River deltaic plain, southeastern Louisiana. *In* M. L. Shirley & J. A. Ragsdale (Eds), *Deltas*, pp. 17–62. Texas Geological Society, Houston.

Kong, P., Na, C. G. & Fink, D. (2007). Cosmogenic ^{10}Be inferred lake-level changes in Sumxi Co Basin, Western Tibet. *Journal of Asian Earth Sciences*, **29**, 698–703.

Komroff, M. (Ed.) (1947). *The history of Herodotus*. Tudor Publishing Company, New York, 544 pp.

Konishi, K., Schlanger, S. O. & Omura, A. (1970). Neotectonic rates in the central Ryukyu Islands derived from ^{230}Th coral ages. *Marine Geology*, **9**, 225–240.

Konishi, K., Omura, A. & Nakamichi, O. (1974). Radiometric coral ages and sea level records from the late Quaternary reef complexes of the Ryukyu Islands. *Proceedings of the Second International Coral Reef Symposium*, **2**, 595–613.

Kopp, R. E., Simons, F. J., Mitrovica, J. X., et al. (2009). Probabilistic assessment of sea level during the last interglacial stage. *Nature*, **462**, 863–868.

Kosnik, M. A. & Kaufman, D. S. (2008). Identifying outliers and assessing the accuracy of amino acid racemization measurements for geochronology: II. Data screening. *Quaternary Geochronology*, **3**, 328–341.

Kosnik, M. A., Kaufman, D. S. & Hua, Q. (2008). Identifying outliers and assessing the accuracy of amino acid racemization measurements for geochronology: I. Age calibration curves. *Quaternary Geochronology*, **3**, 308–327.

Kotsonis, A. (1999). Tertiary shorelines of the western Murray Basin: weathering, sedimentology, exploration potential. *In* R. Stewart (Ed.), Murray Basin Mineral Sands Conference, *Australian Institute of Geoscientists Bulletin*, 26, 57–63.

Kovanen, D. J. & Easterbrook, D. J. (2002). Paleodeviations of radiocarbon marine reservoir values for the northeast Pacific. *Geology*, **30**, 243–246.

Kraft, J. C. (1971). Sedimentary facies patterns and geologic history of a Holocene marine transgression. *Geological Society of America Bulletin*, **82**, 2131–2158.

Kraft, J. C., Rapp, G. R., Kayan, I., *et al.* (2003). Harbor areas at ancient Troy: sedimentology and geomorphology complement Homer's *Iliad*. *Geology*, **31**, 163–166.

Krige, A. V. (1927). An examination of Tertiary and Quaternary changes of sea-level in South Africa, with special stress on the evidence in favour of a recent world-wide sinking of ocean level. *Annals of the University of Stellenbosch*, Capetown, Vol. **5**, Section A, No. 1, pp. 1–81 & 5 plates.

Ku, T.-L. (2000). Uranium-series methods. *In* J. S. Noller, J. M. Sowers & W. R. Lettis (Eds), *Quaternary geochronology: methods and applications*, pp. 101–114. American Geophysical Union, Reference Shelf 4, Washington, DC.

Ku, T. L., Kimmel, M. A., Easton, W. H., *et al.* (1974). Eustatic sea level 120,000 years ago on Oahu, Hawaii. *Science*, **183**, 959–962.

Kuchar, J., Milne, G., Hubbard, A., *et al.* (2012). Evaluation of a numerical model of the British–Irish ice sheet using relative sea-level data: implications for the interpretation of trimline observations. *Journal of Quaternary Science*, doi: 10.1002/jqs.2552.

Kuehl, S. A., Hariu, T. M. & Moore, W. S. (1989). Shelf sedimentation off the Ganges–Brahmaputra river system: evidence for sediment bypassing to the Bengal fan. *Geology*, **17**, 1132–1135.

Kukkuri, J. (1997). Postglacial deformation of the Fennoscandian Coast. *Geophysica*, **33**, 99–109.

Kukla, G. J., Matthews, R. K. & Mitchell, J. M. (1972). The end of the present interglacial. *Quaternary Research*, **2**, 261–269.

Labeyrie, L. D., Duplessy, J. C. & Blanc, P. L. (1987). Variations in mode of formation and temperature of oceanic deep waters over the past 125,000 years. *Nature*, **327**, 477–482.

Laborel, J. & Laborel-Deguen, F. (1994). Biological indicators of relative sea-level variations and of co-seismic displacements in the Mediterranean region. *Journal of Coastal Research*, **10**, 395–415.

Laborel, J., Morhange, C., Lafont, R., *et al.* (1994). Biological evidence of sea-level rise during the last 4500 years on the rocky coasts of continental southwestern France and Corsica. *Marine Geology*, **120**, 203–223.

Laborel, J., Morhange, C., Collina-Girard, J., *et al.* (1999). Shoreline bioerosion, a tool for the study of sea level variations during the Holocene. *Danish Geological Journal*, **45**, 144–148.

Lajoie, K. (1987). Coastal tectonics. *In* R. E. Wallace (Ed.), *Active tectonics*, Studies in Geophysics, pp. 95–124. National Academy Press, Washington, DC.

Lambeck, K. (1995). Late Devensian and Holocene shorelines of the British Isles and North Sea from models of glacio-hyrdo-isostatic rebound. *Journal of the Geological Society, London*, **152**, 437–448.

Lambeck, K. (1996a). Shoreline reconstructions for the Persian Gulf since the last glacial maximum. *Earth and Planetary Science Letters*, **142**, 43–57.

Lambeck, K. (1996b). Sea-level change and shore-line evolution in Aegean Greece since Upper Palaeolithic time. *Antiquity*, **70**, 588–611.

Lambeck, K. (1997). Sea-level change along the French Atlantic and Channel coasts since the time of the last glacial maximum. *Palaeogeography, Palaeoclimatology, Palaeoecology*, **129**, 1–22.

Lambeck, K. (2002). Sea level change from Mid-Holocene to Recent time: an Australian example with global implications. *In* J. X. Mitrovica & L. L. A. Vermeersen (Eds), *Glacial isostatic adjustment and the Earth system: sea level, crustal deformation, gravity and rotation*, pp. 33–50. AGU Monograph, Geodynamics Series, Vol. 29. American Geophysical Union, Washington, DC.

Lambeck, K. & Bard E. (2000). Sea-level change along the French Mediterranean coast since the time of the Last Glacial Maximum. *Earth and Planetary Science Letters*, **175**, 202–222.

Lambeck, K. & Chappell, J. (2001). Sea level change through the Last Glacial Cycle. *Science*, **292**, 679–686.

Lambeck, K. & Johnston, P. (1995) Land subsidence and sea-level change: contributions from the melting of the last great ice sheets and the isostatic adjustment of the Earth. *In* F. B. J. Barends, F J. J. Brouwer & F. H. Schröder (Eds), *Land subsidence*, pp. 3–18. Balkema, Rotterdam.

Lambeck, K. & Johnston, P. (1998). The viscosity of the mantle: evidence from analysis of glacial-rebound phenomena. *In* I. Jackson (Ed.), *The Earth's mantle*, pp. 461–502. Cambridge University Press, Cambridge.

Lambeck, K. & Nakada, M. (1990). Late Pleistocene and Holocene sea-level change along the Australian coast. *Palaeogeography, Palaeoclimatology, Palaeoecology (Global and Planetary Change Section)*, **89**, 143–176.

Lambeck, K. & Nakada, M. (1992). Constraints on the age and duration of the last interglacial period and on sea-level variations. *Nature*, **357**, 125–128.

Lambeck, K. & Nakiboglu, S. M. (1980). Seamount loading and stress in the ocean lithosphere. *Journal of Geophysical Research*, **85**, 6403–6418.

Lambeck, K. & Nakiboglu, S. M. (1981). Seamount loading and stress in the ocean lithosphere 2. Viscoelastic and elastic–viscoelastic models. *Journal of Geophysical Research*, **86**, 6961–6984.

Lambeck, K. & Purcell, A. (2005). Sea-level change in the Mediterranean Sea since the LGM: model predictions for tectonically stable areas. *Quaternary Science Reviews*, **24**, 1969–1988.

Lambeck, K., Smither, C. & Johnson, P. (1998). Sea-level change, glacial rebound and mantle viscosity for northern Europe. *Geophysics Journal International*, **134**, 102–144.

Lambeck, K., Yokoyama, Y. & Purcell, T. (2002a). Into and out of the Last Glacial Maximum: sea-level change during Oxygen Isotope Stages 3 and 2. *Quaternary Science Reviews*, **21**, 343–360.

Lambeck, K., Esat, T. M. & Potter, E.-M. (2002b). Links between climate and sea levels for the past three million years. *Science*, **419**, 199–206.

Lambeck, K., Anzidei, M., Antonioli, F., *et al.* (2004). Sea level in Roman time in the central Mediterranean and implications for modern sea level rise. *Earth and Planetary Science Letters*, **224**, 563–575.

Lambeck, K., Woodroffe, C. D., Antonioli, F., *et al.* (2010). Paleoenvironmental records, geophysical modeling, and reconstruction of sea-level trends and variability on centennial and longer timescales. *In* J. A. Church, P. L. Woodworth, T. Aarup & W. S. Wilson (Eds), *Understanding sea-level rise and variability*, pp. 61–121. Wiley-Blackwell, Chichester.

Lambeck, K., Purcell, A., Flemming, N. C., *et al.* (2011). Sea level and shoreline reconstructions for the Red Sea: isostatic and tectonic considerations and implications for hominin migration out of Africa. *Quaternary Science Reviews*, **30**, 3542–3574.

Lambeck, K., Purcell, A. & Dutton, A. (2012). The anatomy of interglacial sea levels: the relationship between sea levels and ice volumes during the Last Interglacial. *Earth and Planetary Science Letters*, **315–316**, 4–11.

Lamprell, K. & Whitehead, T. (1992). *Bivalves of Australia. Vol. 1.* Crawford House Press, Bathurst, Australia, 182 pp.

Land, L. S., Mackenzie, F. T. & Gould, S. J. (1967). The Pleistocene history of Bermuda. *Geological Society of America Bulletin*, **78**, 993–1006.

Landerer, F. W. & Volkov, D. L. (2013). The anatomy of recent large sea level fluctuations in the Mediterranean Sea. *Geophysical Research Letters*, **40**, 553–557.

Landerer, F. W., Jungclaus, J. H. & Marotzke, J. (2007). Regional dynamic and steric sea level change in response to the IPCC-A1B scenario. *Journal of Physical Oceanography*, **37**, 296–312.

Landerer, F. W., Jungclaus, J. H. & Marotzke, J. (2008). El Niño–Southern Oscillation signals in sea level, surface mass redistribution, and degree-two geoid coefficients. *Journal of Geophysical Research*, **113**, (C08014).

Larcombe, P. & Woolfe, K. J. (1999). Terrigenous sediments as influences upon Holocene nearshore coral reefs, central Great Barrier Reef, Australia. *Australian Journal of Earth Sciences*, **46**, 141–154.

Larcombe, P., Carter, R. M., Dye, J., *et al.* (1995). New evidence for episodic postglacial sea-level rise in the central Great Barrier Reef, Australia. *Marine Geology*, **127**, 1–44.

Lavoisier, A. L. (1789). *Observations générales sur les couches modernes horizontales, qui ont été déposées par la mer, et sur les conséquences qu'on peut tirer de leurs dispositions, relativement à l'ancienneté du globe terrestre.* France, Acadamie Royale des Sciences, Memoir.

Lea, D. W., Martin, P. A., Pak, D. K., *et al.* (2002). Reconstructing a 350 ky history of sea level using planktonic Mg/Ca and oxygen isotope records from a Cocos Ridge core. *Quaternary Science Reviews*, **21**, 283–293.

Le Breton, E., Dauteuil, O. & Biessy, G. (2010). Post-glacial rebound of Iceland during the Holocene. *Journal of the Geological Society, London*, **167**, 417–432.

Le Grand, H. E. (1988). *Drifting continents and shifting theories.* Cambridge University Press, Cambridge, 313 pp.

Lemieux-Dudon, B., Blayo, E., Petit, J.–R., *et al.* (2010). Consistent dating for Antarctic and Greenland ice cores. *Quaternary Science Reviews*, **29**, 8–20.

Leonard, E. M. & Wehmiller, J. F. (1991). Geochronology of marine terraces at Caleta Michilla, Northern Chile; implications for Late Pleistocene and Holocene uplift. *Revista Geológica de Chile*, **18**, 81–86.

Leonard, E. M. & Wehmiller, J. F. (1992). Low uplift rates and terrace reoccupation inferred from mollusc aminostratigraphy, Coquimbo Bay area, Chile. *Quaternary Research*, **38**, 246–259.

Leopold, L. B. & Bull, W. B. (1979). Base level, aggradation and grade. *Proceedings of the American Philosophical Society*, **123**, 168–202.

Leopold, L. B., Wolman, M. G. & Miller, J. P. (1964). Fluvial processes in geomorphology. W. H. Freeman, San Francisco, 522 pp.

Leorri, E., Horton, B. P. & Cearreta, A. (2008). Development of a foraminifera-based transfer function in the Basque marshes, N. Spain: implications for sea-level studies in the Bay of Biscay. *Marine Geology*, **251**, 60–74.

Le Pichon, X., Francheteau, J. & Bonnin, J. (1973). *Plate tectonics*, Developments in Geotectonics 6. Elsevier Scientific Publishing Company, Amsterdam, 300 pp.

Lessa, G. C. & Angulo, R. J. (1998). Oscillations or not oscillations, that is the question – reply. *Marine Geology*, **150**, 189–196.

Leuliette, E. W. & Willis, J. K. (2011). Balancing the sea level budget. *Oceanography*, **24**, 122–129.

Leuliette, E. W., Nerem, R. S. & Mitchum, G. T. (2004). Calibration of TOPEX/Poseidon and Jason altimeter data to construct a continuous record of mean sea level change. *Marine Geodesy*, **27**, 79–94.

Levitus, S., Antonov, J. I., Boyer, T. P., *et al.* (2012). World ocean heat content and thermosteric sea level change. *Geophysical Research Letters*, **39** (L10603), doi: 10.1029/2012GL051106.

Lewis, S. E., Wüst, R. A. J., Webster, J. M., *et al.* (2008). Mid–Late Holocene sea-level variability in eastern Australia. *Terra Nova*, **20**, 74–81.

Lewis, S. E., Sloss, C. R., Murray-Wallace, C. V., *et al.* (2013). Post-glacial sea-level changes around the Australian margin: a review. *Quaternary Science Reviews*, **74**, 115–138.

Li, W.-X., Lundberg, J., Dickin, A. P., *et al.* (1989). High precision mass spectrometric uranium-series dating of cave deposits and implications for paleoclimate studies. *Nature*, **339**, 534–536.

Li, W.-X., Törnqvist, T., Nevitt, J. M., *et al.* (2012). Synchronizing a sea-level jump, final Lake Agassiz drainage, and abrupt cooling 8200 years ago. *Earth and Planetary Science Letters*, **315–316**, 41–50.

Lian, O. B. (2006). Luminescence dating: thermoluminescence. *In* S. A. Elias (Ed.), *Encyclopedia of Quaternary science*, pp. 1480–1491. Elsevier, Amsterdam.

Lian, O. B. & Huntley, D. J. (2001). Luminescence dating. *In* W. M. Last & J. P Smol (Eds), *Tracking environmental change using lake sediments: basin analysis, coring, and chronological techniques*, vol. 1, pp. 261–282. Kluwer Academic Publishing, Dordrecht.

Lian, O. B. & Roberts, R. G. (2006). Dating the Quaternary: progress in luminescence dating of sediments. *Quaternary Science Reviews*, **25**, 2449–2468.

Libby, W. A. (1955). *Radiocarbon dating*, 2nd edition. The University of Chicago Press, Chicago, 175 pp.

Liew, P. M., Pirazzoli, P. A. & Hsieh, M. L., *et al.* (1993). Holocene tectonic uplift deduced from elevated shorelines, eastern Coastal Range of Taiwan. *Tectonophysics*, **222**, 55–68.

Lighty, R. G., Macintyre, I. G. & Stuckenrath, R. (1978). Submerged early Holocene barrier reef south-east Florida shelf. *Nature*, **276**, 59–60.

Lighty, R. G., Macintyre, I. G. & Stuckenrath, R. (1982). *Acropora palmata* reef framework: a reliable indicator of sea level in the western Atlantic for the past 10,000 years. *Coral Reefs*, **1**, 125–130.

Linné, C. von (1745). *Öländska och Gothländska resa på Riksens häglofiga Ständers befallning förrättad*. 1741, Stockholm.

Lisiecki, L. E. & Raymo, M. E. (2005). A Pliocene–Pleistocene stack of 57 globally distributed benthic $\delta^{18}O$ records. *Palaeoceanography*, **20**, PA1003, pp. 1–17.

Lisitzin, E. (1974). *Sea-level changes*. Elsevier, Oceanography Series 8, Amsterdam, 286 pp.

Lloyd, J. (2000). Combined Foraminiferal and Thecamoebian environmental reconstruction from an isolation basin in NW Scotland: implications for sea-level studies. *Journal of Foraminiferal Research*, **30**, 294–305.

Lloyd, J. M., Norddahl, H. & Bentley, M. J., *et al.* (2009). Lateglacial to Holocene relative sea-level changes in the Bjarkarlundur area near Reykhólar, North West Iceland. *Journal of Quaternary Science*, **24**, 816–831.

Long, A. J., Roberts, D. H. & Dawson, S. (2006). Early Holocene history of the West Greenland Ice Sheet and the GH-8.2 event. *Quaternary Science Reviews*, **25**, 904–922.

Lorbacher, K., Marsland, S. J., Church, J. A., *et al.* (2012). Rapid barotrophic sea-level rise from ice-sheet melting scenarios. *Journal of Geophysical Research*, **117**, C06003.

Lowe, J. A. & Gregory, J. M. (2006). Understanding projections of sea level rise in a Hadley Centre coupled climate model. *Journal of Geophysical Research*, **111**, C11014.

Lowe, J. A., Woodworth, P. L., Knutson, T., *et al.* (2010). Past and future changes in extreme sea levels and waves. *In* J. A. Church, P. L. Woodworth, T. Aarup & W. S. Wilson (Eds), *Understanding sea-level rise and variability*, pp. 326–375. Wiley-Blackwell, Chichester.

Lowe, J. J. & Walker, M. J. C. (1997). *Reconstructing Quaternary environments*, 2nd Ed. Longman, London.

Lowry, D. C. (1970). Geology of the Western Australian part of the Eucla Basin. *Geological Survey of Western Australia, Bulletin*, **122**, 201 pp.

Ludbrook, N. H. (1958). The stratigraphic sequence in the western portion of the Eucla Basin. *Royal Society of South Australia, Transactions*, **41**, 108–114.

Ludbrook, N. H. (1978). Quaternary molluscs of the western part of the Eucla Basin. *Geological Survey of Western Australia Bulletin*, **125**, 286 pp.

Ludbrook, N. H. (1983). Molluscan faunas of the Early Pleistocene Point Ellen Formation and Burnham Limestone, South Australia. *Royal Society of South Australia, Transactions*, **107**, 37–49.

Ludbrook, N. H. (1984). *Quaternary molluscs of South Australia*. South Australia, Department of Mines and Energy, Handbook No. 9, 327 pp.

Ludwig, K. R., Szabo, B. J., Moore, J. G., *et al.* (1991). Crustal subsidence rates off Hawaii determined from $^{234}U/^{238}U$ ages of drowned coral reefs. *Geology*, **19**, 171–174.

Ludwig, K. R., Muhs, D. R., Simmons, K. R., *et al.* (1996). Sea-level records at ~80 ka from tectonically stable platforms: Florida and Bermuda. *Geology*, **24**, 211–214.

Lugg, A. (1978). Over determined problems in science. *Studies in History and Philosophy of Science*, **9**, 1–18.

Lundberg, N. & Dorsey, R. J. (1990). Rapid Quaternary emergence, uplift, and denudation of the coastal range, eastern Taiwan. *Geology*, **18**, 638–641.

Lundberg, J. & Ford, D. C. (1994). Late Pleistocene sea level change in the Bahamas from mass spectrometric U-series dating of submerged speleothem. *Quaternary Science Reviews*, **13**, 1–14.

Lundberg, J., Ford, D. C., Schwarcz, H. P., *et al.* (1990). Dating sea level in caves. *Nature*, **343**, 217–218.

Lyell, C. (1839). *Noveaux elements de Géologie*. Pitois-Levrault, Paris, 648 pp.

Lyell, C. (1853). *Principles of geology; or, the modern changes of the Earth and its inhabitants considered as illustrative of geology*. 9th Edition. John Murray, London, 835 pp.

Lyell, C. (1855). *A manual of elementary geology or, the ancient changes of the Earth and its inhabitants as illustrated by geological monuments*. John Murray, London, 655 pp.

Lyman, J. M., Good, S. A., Gouretski, V. V., *et al.* (2010). Robust warming of the global upper ocean. *Nature*, **465**, 334–337.

Lyon, C. J. & Goldthwait, J. W. (1934). An attempt to crossdate trees in drowned forests. *Geographical Review*, **24**, 605–614.

Macintyre, I. G. (1967). Submerged coral reefs, west coast of Barbados, West Indies. *Canadian Journal of Earth Sciences*, **4**, 461–474.

Macintyre, I. G. (1971). Some submerged coral reefs in the Caribbean. *Transactions of the Fifth Caribbean Geological Conference Bulletin*, **5**, 49–54.

Macintyre, I. G. (1988). Modern coral reefs of Western Atlantic: new geological perspective. *American Association of Petroleum Geologists Bulletin*, **72**, 1360–1369.

Macintyre, I. G. & Glynn, P. W. (1976). Evolution of modern Caribbean fringing reef, Galeta Point, Panama. *The American Association of Petroleum Geologists Bulletin*, **60**, 1054–1072.

Macintyre, I. G., Rutzler, K., Norris, J. N., *et al.* (1991). An early Holocene reef in the western Atlantic: submersible investigations of a deep relict reef off the west coast of Barbados, W.I. *Coral Reefs*, **10**, 167–174.

Macintyre, I. G., Littler, M. M. & Littler, D. S. (1995). Holocene history of Tobacco Range, Belize, Central America. *Atoll Research Bulletin*, **430**, 1–18.

Macintyre, I. G., Toscano, M. A., Lighty, R. G., *et al.* (2004). Holocene history of the mangrove islands of Twin Cays, Belize, Central America. *Atoll Research Bulletin*, **510**, 1–16.

Mackin, J. H. (1948). Concept of a graded river. *Geological Society of America Bulletin*, **59**, 463–511.

Maclaren, C. (1842). The Glacial Theory of Prof. Agassiz. *American Journal of Science*, **42**, 346–655.

Macnae, W. (1968). A general account of the fauna and flora of mangrove swamps and forests in the Indo-West-Pacific region. *Advances in Marine Biology*, **6**, 73–270.

Maeda, Y., Sirringan, F., Omura, A., *et al.* (2004). Higher-than-present Holocene mean sea levels in Ilocos, Palawan and Samar, Philippines. *Quaternary International*, **115–116**, 15–26.

Mallett, C. W. (1982). Late Pliocene planktonic foraminifera from subsurface shell beds, Jandakot, near Perth, Western Australia. *Search*, **13**, 35–36.

Mallinson, D. J., Culver, S. J., Riggs, S. R., *et al.* (2010). Regional seismic stratigraphy and controls on Quaternary evolution of the Cape Hatteras region of the Atlantic passive margin, USA. *Marine Geology*, **268**, 16–33.

Maroukian, H., Gaki-Papanastassiou, K., Karymbalis, E., *et al.* (2008). Morphotectonics control on drainage network evolution in the Perachora Peninsula, Greece. *Geomorphology*, **102**, 81–92.

Marshall, J. F. & Davies, P. J. (1984). Last interglacial reef growth beneath modern reefs in the southern Great Barrier Reef. *Nature*, **307**, 44–46.

Marshall, J. F. & Jacobson, G. (1985). Holocene growth of a mid-plate atoll: Tarawa, Kiribati. *Coral Reefs*, **4**, 11–17.

Marshall, J. F. & Thom, B. G. (1976). The sea level in the last interglacial. *Nature*, **263**, 120–121.

Martin, L. & Suguio, K. (1992). Variation of coastal dynamics during the last 7000 years recorded in beach-ridge plains associated with river mouths: example from the central Brazilian coast. *Palaeogeography, Palaeoclimatology, Palaeoecology*, **99**, 119–140.

Martin, L., Flexor, J.-M., Blitzkow, D., *et al.* (1985). Geoid change indications along the Brazilian coast during the last 7000 years. *Proceedings of the 5th International Coral Reef Congress*, **3**, 85–90.

Martin, L., Suguio, K. & Flexor, J. M. (1986). Relative sea-level reconstruction during the last 7000 years along the States of Parana and Santa Catarina coastal plains: additional information derived from shell-middens. *Quaternary of South America and Antarctic Peninsula*, **4**, 219–236.

Martin, L., Bittencourt, A. C. S. P., Dominguez, J. M. L., *et al.* (1998). Oscillations or not oscillations, that is the question: comment on Angulo, R. J. and Lessa, G. C. 'the Brazilian sea-level curves: a critical review with emphasis on the curves from the Paranagua and Cananéia regions'. *Marine Geology*, **150**, 179–187.

Martinson, D. G., Pisias, N. G., Hays, J. D., *et al.* (1987). Age dating and the orbital theory of the ice ages: development of a high resolution 0 to 300,000-year chronostratigraphy. *Quaternary Research*, **27**, 1–29.

Maslin, M. A. & Ridgwell, A. J. (2005). Mid-Pleistocene revolution and the 'eccentricity myth'. *In* M. J. Head & P. L. Gibbard (Eds), *Early–Middle Pleistocene Transitions: the land–ocean evidence*. Geological Society, London, Special Publication, **247**, 19–34.

Massey, A. C., Gehrels, W. R., Charman, D. J., *et al.* (2006). An intertidal foraminifera-based transfer function for reconstructing Holocene sea-level change in southwest England. *Journal of Foraminiferal Research*, **36**, 215–232.

Massey, A. C., Gehrels, W. R., Charman, D. J., *et al.* (2008). Relative sea-level change and postglacial isostatic adjustment along the coast of south Devon, United Kingdom. *Journal of Quaternary Science*, **23**, 415–433.

Masson-Delmotte, V., Stenni, B., Pol, K., *et al.* (2010). EPICA Dome C record of glacial and interglacial intensities. *Quaternary Science Reviews*, **29**, 113–128.

Mastronuzzi, G. & Sansò, P. (2003). *Quaternary coastal morphology and sea level changes*. Puglia 2003 – Final conference, IGCP Project 437, Field Guide, 180 pp.

Matchan, E. & Phillips, D. (2011). New $^{40}Ar/^{39}Ar$ ages for selected young (<1 Ma) basalt flows of the Newer Volcanic Province, southeastern Australia. *Quaternary Geochronology*, **6**, 356–368.

Mather, R. S., Rizos, C. & Coleman, R. (1979). Remote sensing of surface ocean circulation with satellite altimetry. *Science*, **205**, 11–17.

Matthews, R. K. (1969). Tectonic implications of glacio-eustatic sea-level fluctuations. *Earth and Planetary Science Letters*, **5**, 459–462.

Matthews, R. K. (1990). Quaternary sea-level change. *In Sea-level change*, pp. 88–103. National Research Council Studies in Geophysics, National Academy Press, Washington, DC.

Mayer, W. (2008). Early geological investigations of the Pleistocene Tamala Limestone, Western Australia. *In* R. H. Grapes, D. Oldroyd & A. Grigelis (Eds), *History of geomorphology and Quaternary geology*. Geological Society, London, Special Publications, **301**, 279–293.

McBride, R. A., Taylor, M. J. & Byrnes, M. R. (2007). Coastal morphodynamics and chenier-plain evolution in southwestern Louisiana, USA: a geomorphic model. *Geomorphology*, **88**, 367–422.

McCartan, L., Owens, J. P., Blackwelder, B. W., *et al.* (1982). Comparison of amino acid racemization geochronometry with lithostratigraphy, biostratigraphy, uranium-series coral dating and magnetostratigraphy in the Atlantic Coastal Plain of southeastern United States. *Quaternary Research*, **18**, 337–359.

McCoy, W. D. (1987). The precision of amino acid geochronology and palaeo-thermometry. *Quaternary Science Reviews*, **6**, 43–54.

McCulloch, M. T. & Esat, T. (2000). The coral record of last interglacial sea levels and sea surface temperatures. *Chemical Geology*, **169**, 107–129.

McCulloch, M. T. & Mortimer, G. E. (2008). Applications of the ^{238}U–^{230}Th decay series to dating of fossil and modern corals using MC-ICPMS. *Australian Journal of Earth Sciences*, **55**, 955–965.

McCulloch, M., Mortimer, G., Esat, T., *et al.* (1996). High resolution windows into early Holocene climate: Sr/Ca coral records from the Huon Peninsula. *Earth and Planetary Science Letters*, **138**, 169–178.

McCulloch, M. T., Tudhope, A. W., Esat, T. M., *et al.* (1999). Coral record of equatorial sea-surface temperatures during the Penultimate deglaciation at Huon Peninsula. *Science*, **283**, 202–204.

McDougall, I. & Duncan, R. A. (1980). Linear volcanic chains: recording plate motions? *Tectonophysics*, **63**, 275–295.

McGowran, B. (2005). *Biostratigraphy – microfossils and geological time*. Cambridge University Press, Cambridge, 459 pp.

McGregor, H. V., Gagan, M. K., McCulloch, M. T., *et al.* (2008). Mid-Holocene variability in marine ^{14}C reservoir age for northern coastal Papua New Guinea. *Quaternary Geochronology*, **3**, 213–225.

McGregor, H. V., Hellstrom, J., Fink, D., *et al.* (2011). Rapid U-series dating of young fossil corals by laser ablation MC-ICPMS. *Quaternary Geochronology*, **6**, 195–206.

McGuire, W. J., Howarth, R. J., Firth, C., *et al.* (1997). Correlation between rate of sea-level change and frequency of explosive volcanism in the Mediterranean. *Nature*, **389**, 473–476.

McKee, E. D. & Ward, W. C. (1983). Eolian environment. *In* P. A. Scholle, D. G. Bebout & C. H. Moore (Eds), *Carbonate depositional environments*, pp. 131–170. American Association of Petroleum Geologists Memoir 33, Tulsa, Oklahoma.

McKee, K. L. & Patrick, W. H. (1988). The relationship of smooth cordgrass (*Spartina alterniflora*) to tidal datums: a review. *Estuaries*, **11**, 143–151.

McLaren, S. J. & Rowe, P. J. (1996). The reliability of uranium series mollusc dates from the western Mediterranean Basin. *Quaternary Science Reviews*, **15**, 709–717.

McLaren, S., Wallace, M. W., Gallagher, S. J., *et al.* (2011). Palaeogeographic, climatic and tectonic change in southeastern Australia: the Late Neogene evolution of the Murray Basin. *Quaternary Science Reviews*, **30**, 1086–1111.

McLean, R. & Shen, J.-S. (2006). From fore-shore to foredune: foredune development over the last 30 years at Moruya Beach, New South Wales, Australia. *Journal of Coastal Research*, **22**, 28–36.

McLean, R. F. & Hosking, P. L. (1991). Geo-morphology of reef islands and atoll motu in Tuvalu. *South Pacific Journal of Natural Science*, **11**, 167–189.

McLean, R. F., Stoddart, D. R., Hopley, D., *et al.* (1978). Sea level change in the Holocene on the northern Great Barrier Reef. *Philosophical Transactions of the Royal Society, London*, **A291**, 167–186.

McManus, J., Oppo, D., Cullen, J., *et al.* (2003). Marine isotope stage 11 (MIS 11) analog for Holocene and future climate change? *In* A. W. Droxler, R. Z. Poore & H. Lloyd Droxler (Eds), *Earth's climate and orbital eccentricity: the marine stage 11 question*. Geophysical Monographs, **137**, 69–85, American Geophysical Union, Washington, DC.

McMaster, R. L., Lachance, T. P. & Ashraf, A. (1970). Continental shelf geomorphic features off Portugese Guinea, Guinea and Sierra Leone (West Africa). *Marine Geology*, **9**, 203–213.

McMenamin, M. A. S., Blunt, D., Kvenvolden, K. A., *et al.* (1982). Amino acid geochemistry of fossil bones from Rancho La Brea asphalt deposits, California. *Quaternary Research*, **18**, 174–183.

McMillan, A. A., Hamblin, R. J. O. & Merritt, J. W. (2011). A lithostratigraphical framework for onshore Quaternary and Neogene (Tertiary) superficial deposits of Great Britain and the Isle of Man. *British Geological Survey Report*, **RR/10/03**, 343 pp.

McMurty, G. M., Campbell, J. F., Fryer, G. J., *et al.* (2010). Uplift of Oahu, Hawaii, during the past 500 k.y. as recorded by elevated reef deposits. *Geology*, **38**, 27–30.

McNutt, M. & Menard, H. W. (1978). Lithospheric flexure and uplifted atolls. *Journal of Geophysical Research*, **83**, 1206–1212.

Meehl, G. A., Stocker, T. F., Collins, W., *et al.* (2007). Global climate projections. *In* S. Solomon, D. Qin & M. Manning (Eds), *Climate Change 2007: the physical science basis*. Contribution of Working Group I to the Fourth Assessment Report of the Intergovernmental Panel on Climate Change, pp. 747–845. Cambridge University Press, Cambridge.

Meek, N. (1989). Geomorphic and hydrologic implications of the rapid incision of Afton Canyon, Mojave Desert, California. *Geology*, **17**, 7–10.

Meier, M. E. (1984). Contribution of small glaciers to global sea level. *Science*, **226**, 1418–1421.

Meltzner, A. J., Sieh, K., Abrams, M., *et al.* (2006). Uplift and subsidence associated with the great Aceh–Andaman earthquake of 2004. *Journal of Geophysical Research*, **111**, B02407, doi: 10.1029/2005JB003891.

Meltzner, A. J., Sieh, K., Chiang, H.-W., *et al.* (2010). Coral evidence for earthquake recurrence and an A.D. 1390–1455 cluster at the south end of the 2004 Aceh–Andaman rupture. *Journal of Geophysical Research*, **115**.

Menard, H. W. (1986a). *The ocean of truth: a personal history of global tectonics*. Princeton University Press, Princeton, 353 pp.

Menard, H. W. (1986b). *Islands*. Scientific American Library, New York, 230 pp.

Menard, Y. (1981). Étude de la variabilité de la topographie dynamique des océans liéé à la circulation océanique, à l'ouest de la dorsale médio-Atlantique entre 30°N et 55°S en latitude. *Annales Géophysique*, **37**, 99–106.

Menéndez, M. & Woodworth, P. L. (2010). Changes in extreme high water levels based on a quasi-global tide-gauge data set. *Journal of Geophysical Research*, **115**, C10011.

Mercer, J. H. (1978). West Antarctic Ice Sheet and CO_2 greenhouse effect: a threat of disaster. *Nature*, **271**, 321–325.

Merrifield, M. A., Merrifield, S. T. & Mitchum, G. T. (2009). An anomalous recent acceleration of global sea level rise. *Journal of Climate*, **22**, 5772–5781.

Mesolella, K. J. (1967). Zonation of uplifted Pleistocene coral reefs on Barbados, West Indies. *Science*, **156**, 638–640.

Mesolella, K. J. (1968). *The uplifted reefs of Barbados: physical stratigraphy, facies relationship and absolute chronology*. PhD dissertation, Brown University, Philadelphia, Parts 1 & 2, 736 pp.

Mesolella, K. J., Matthews, R. K., Broecker, W. S., *et al.* (1969). The astronomical theory of climatic change: Barbados data. *Journal of Geology*, **77**, 250–274.

Meyssignac, B. & Cazenave, A. (2012). Sea level: a review of present-day and recent-past changes and variability. *Journal of Geodynamics*, **58**, 96–109.

Miall, A. D. (1990). *Principles of sedimentary basin analysis*, 2nd edition. Springer-Verlag, New York, 668 pp.

Milankovitch, M. (1920). *Théorie mathématique des phénomènes thermiques produits par la radiation solaire*. Ganthier-Villars, Paris, 339 pp.

Milankovitch, M. (1941). *Kanon der Erdbestrahlung und seine Anwendung auf das Eiszeitproblem*. Royal Serbian Sciences, Special Publications 132, Section of Mathematical and Natural Sciences, Belgrade (Canon of Insolation and the Ice Age Problem, English translation by Israel Program for Scientific Translation, published for the US Department of Commerce and the National Science Foundation, Washington, DC, 1969).

Miller, G. H. & Brigham-Grette, J. (1989). Amino acid geochronology: resolution and precision in carbonate fossils. *Quaternary International*, **1**, 111–128.

Miller, G. H. & Hare, P. E. (1980). Amino acid geochronology: integrity of the carbonate matrix and potential of molluscan fossils. *In* P. E. Hare, T. C. Hoering & K. King (Eds), *Biogeochemistry of amino acids*, pp. 415–444. Wiley, New York.

Miller, G. H. & Mangerud, J. (1985). Aminostratigraphy of European marine interglacial deposits. *Quaternary Science Reviews*, **4**, 215–278.

Miller, G. H., Hollin, J. T. & Andrews, J. T. (1979). Aminostratigraphy of UK Pleistocene deposits. *Nature*, **281**, 539–543.

Miller, G. H., Magee, J. W. & Jull, A. J. T. (1997). Low-latitude glacial cooling in the Southern Hemisphere from aminoacid racemization in emu eggshells. *Nature*, **385**, 241–244.

Miller, K. G., Sugarman, P. J., Browning, J. V., *et al.* (2009). Sea-level rise in New Jersey over the past 5000 years: implications to anthropogenic changes. *Global and Planetary Change*, **66**, 10–18.

Milliken, K. T., Anderson, J. B., & Rodriguez, A. B. (2008). A new composite Holocene sea-level curve for the northern Gulf of Mexico. *In* J. B. Anderson & A. B. Rodriguez (Eds), *Response of Upper Gulf Coast estuaries to Holocene climate change and sea-level rise*. Geological Society of America Special Paper, **443**, pp. 1–11.

Milliman, J. D. & Emery, K. O. (1968). Sea levels during the past 35,000 years. *Science*, **162**, 1121–1123.

Milliman, J. D., Freile, D., Steinen, R. P., *et al.* (1993). Great Bahama bank aragonitic muds: mostly inorganically precipitated, mostly exported. *Journal of Sedimentary Petrology*, **63**, 589–595.

Milly, P. C. D., Cazenave, A., Famiglietti, J. S., *et al.* (2010). Terrestrial water-storage contributions to sea-level rise and variability. *In* J. A. Church, P. L. Woodworth, T. Aarup & W. S. Wilson (Eds), *Understanding sea-level rise and variability*, pp. 228–255. Wiley-Blackwell, Oxford.

Milne, G. A. & Mitrovica, J. X. (1998). Postglacial sea-level change on a rotating Earth. *Geophysical Journal International*, **133**, 1–19.

Milne, G. A. & Mitrovica, J. X. (2008). Searching for eustasy in deglacial sea-level histories. *Quaternary Science Reviews*, **27**, 2292–2302.

Milne, G. A. & Shennan, I. (2007). Isostasy. *In* S. A. Elias (Ed.), *Encyclopedia of Quaternary science*, Vol. 4, pp. 3043–3051. Elsevier, Oxford.

Milne, G. A., Mitrovica, J. X. & Schrag, D. P. (2002). Estimating past continental ice volumes from sea-level data. *Quaternary Science Reviews*, **21**, 361–376.

Milne, G. A., Mitrovica, J. X., Scherneck, H.-G., *et al.* (2004). Continuous GPS measurements of postglacial adjustment in Fennoscandia. *Science*, **291**, 2381–2385.

Milne, G. A., Long, A. J. & Bassett, S. E. (2005). Modelling Holocene relative sea-level observations from the Caribbean and South America. *Quaternary Science Reviews*, **24**, 1183–1202.

Milne, G. A., Gehrels, W. R., Hughes, C. W., *et al.* (2009). Identifying the causes of sea-level change. *Nature Geoscience*, **2**, 471–478.

Milnes, A. R., Ludbrook, N. H., Lindsay, J. M., *et al.* (1983). The succession of Cainozoic marine sediments on Kangaroo Island, South Australia. *Royal Society of South Australia, Transactions*, **107**, 1–35.

Mimura, N. (1999). Vulnerability of island countries in the South Pacific to sea level rise and climate change. *Climate Research*, **12**, 137–143.

Mitchum, G. T., Nerem, R. S., Merrifield, M. A., *et al.* (2010). Modern sea-level changes. *In* J. A. Church, P. L. Woodworth, T. Aarup, & W. S. Wilson (Eds), *Understanding sea-level rise and variability*, pp. 122–142. Wiley-Blackwell, Oxford.

Mitrovica, J. X. & Davis, J. L. (1995). Present-day post-glacial sea level change far from the Late Pleistocene ice sheets: implications for recent analyses of tide gauge records. *Geophysical Research Letters*, **22**, 2529–2532.

Mitrovica, J. X. & Milne, G. A. (2002). On the origin of late Holocene sea-level highstands within equatorial ocean basins. *Quaternary Science Reviews*, **21**, 2179–2190.

Mitrovica, J. X. & Peltier, W. R. (1991). On postglacial geoid subsidence over the equatorial oceans. *Journal of Geophysical Research*, **96**, 20053–20071.

Mitrovica, J. X., Gomez, N. & Clark, P. U. (2009). The sea-level fingerprint of West Antarctic collapse. *Science*, **323**, 753. doi: 10.1126/science.1166510.

Mitrovica, J. X., Tamisiea, M. E., Ivins, E. R., *et al.* (2010). Surface mass loading on a dynamic earth: complexity and contamination in the geodetic analysis of global sea-level trends. *In* J. A. Church, P. L. Woodworth, T. Aarup & W. S. Wilson (Eds), *Understanding sea-level rise and variability*, pp. 285–325. Wiley-Blackwell, Oxford.

Mitsova, D. & Esnard, A.-M. (2012). Holding back the sea: an overview of shore zone planning and management. *Journal of Planning Literature*, **27**, 446–459.

Mitterer, R. M. & Kriausakul, N. (1989). Calculation of amino acid racemization ages based on apparent parabolic kinetics, *Quaternary Science Reviews*, **8**, 353–357.

Mix, A. C., Bard, E. & Schneider, R. (2001). Environmental processes of the ice age: land, oceans, glaciers (EPILOG). *Quaternary Science Reviews*, **20**, 627–657.

Molengraaff, G. (1921). Modern deep-sea research in the East Indian archipelago. *Geographical Journal*, **57**, 95–121.

Montaggioni, L. F. (2005). History of Indo-Pacific coral reef systems since the last glaciation: development patterns and controlling factors. *Earth-Science Reviews*, **71**, 1–75.

Montaggioni, L. F. & Braithwaite, C. J. R. (2009). *Quaternary coral reef systems: history, development processes and controlling factors*. Elsevier, Amsterdam, 532 pp.

Montaggioni, L. F. & Pirazzoli, P. A. (1984). The significance of exposed coral conglomerates from French Polynesia (Pacific Ocean) as indications of recent sea-level changes. *Coral Reefs*, **3**, 29–42.

Montaggioni, L. F., Richard, G., Bourrouilh-Le Jan, F., *et al.* (1985). Geology and marine biology of Makatea, an uplifted atoll, Tuamotu Archipelago, central Pacific Ocean. *Journal of Coastal Research*, **1**, 165–171.

Montaggioni, L. F., Cabioch, G., Camoinau, G. F., *et al.* (1997). Continuous record of reef growth over the past 14 k.y. on the mid-Pacific island of Tahiti. *Geology*, **25**, 555–558.

Moore, J. G. (1970). Relationship between subsidence and volcanic load. *Bulletin of Volcanology*, **34**, 562–576.

Moore, J. G. & Clague, D. A. (1992). Volcanic growth and evolution of the island of Hawaii. *Geological Society of America Bulletin*, **104**, 1471–1484.

Moore, J. G. & Moore, G. W. (1984). Deposit from a giant wave on the island of Lana'I Hawai'i. *Science*, **226**, 1312–1315.

Moore, P. R. (1987). Age of the raised beach ridges at Turakirae Head, Wellington: a reassessment based on new radiocarbon dates. *Journal of the Royal Society of New Zealand*, **17**, 313–324.

Moore, W. S. (1982). Late Pleistocene sea-level history. *In* M. Ivanovich & R. S. Harmon (Eds), *Uranium series disequilibrium: applications to environmental problems*, pp. 481–496. Clarendon Press, Oxford.

Morhange, C. & Pirazzoli, P. A. (2005). Mid-Holocene emergence of southern Tunisian coasts. *Marine Geology*, **220**, 205–213.

Morhange, C., Bourcier, M., Laborel, J., *et al.* (1999). New data on historical relative sea level movements in Pozzuoli, Phlegraean Fields, southern Italy. *Physics and Chemistry of the Earth*, **24**, 349–354.

Morhange, C., Laborel, J. & Hesnard, A. (2001). Changes of relative sea level during the past 5000 years in the ancient harbour of Marseilles, southern France. *Palaeogeography, Palaeoclimatology, Palaeoecology*, **166**, 319–329.

Morhange, C., Marriner, N., Laborel, J., *et al.* (2006). Rapid sea-level movements and noneruptive crustal deformation in the Phlegrean Fields caldera, Italy. *Geology*, **34**, 93–96.

Moriwaki, H., Chikamori, M., Okuno, M., *et al.* (2006). Holocene changes in sea level and coastal environments on Rarotonga, Cook Islands, South Pacific Ocean. *The Holocene*, **16**, 839–848.

Mörner, N.-A. (1969). *The Late Quaternary history of the Kattegatt Sea and the Swedish west coast.* Sveriges Geolo. Under., Ser. C., No. 640, 487 pp.

Mörner, N.-A. (1976a). Eustasy and geoid changes. *Journal of Geology*, **84**, 123–151.

Mörner, N.-A. (1976b). Eustatic changes during the last 8,000 years in view of radiocarbon calibration and new information from the Kattegat region and other northwestern European coastal areas. *Palaeogeography, Palaeoclimatology, Palaeoecology*, **19**, 63–85.

Mörner, N.-A. (1979). The Fennoscandian uplift and Late Cenozoic geodynamics: geological evidence. *Geojournal*, **3**, 287–318.

Mörner, N.-A. (1980). The northwest European 'sea-level laboratory' and regional Holocene eustasy. *Palaeogeography, Palaeoclimatology, Palaeoecology*, **29**, 281–300.

Mörner, N.-A. (1983). Sea levels. *In* R. Gardner & H. Scoging (Eds), *Mega-geomorphology*, pp. 73–91, Clarendon Press, Oxford.

Mörner, N.-A. (2003). *Paleoseismicity of Sweden: a novel paradigm.* A contribution to INQUA from its subcommission on Paleoseismology, Paleogeophysics and Geodynamics, Stockholm University, 320 pp.

Morton, R. A., Paine, J. G. & Blum, M. D. (2000). Responses of stable bay–margin and barrier–island systems to Holocene sea-level highstands, western Gulf of Mexico. *Journal of Sedimentary Research*, **70**, 478–490.

Moses, C. (2012). Tropical rock coasts: cliff, notch and platform erosion dynamics. *Progress in Physical Geography*, doi: 10.1177/0309133312460073.

Moslow, T. F. & Colquhoun, D. J. (1981). Influence of sea level change on barrier island evolution. *Oceanis*, **7**, 439–454.

Moss, R. H., Edmonds, J. A., Hibbard, K. A., *et al.* (2010). The next generation of scenarios for climate change research and assessment. *Nature*, **463**, 747–756.

Mudge, B. F. (1858). The salt marsh formations of Lynn. *Essex Institute Proceedings*, **2**, 117–119.

Muhs, D. R. (1992). The last interglacial – glacial transition in North America: evidence from uranium-series dating of coastal deposits. *In* P. U. Clark & P. D. Lea (Eds), *The last interglacial–glacial transition in North America*, pp. 31–51. Geological Society of America Special Paper 270, Boulder, Colorado.

Muhs, D. R. (2002). Evidence for the timing and duration of the last interglacial period from high-precision uranium-series ages of corals on tectonically stable coastlines. *Quaternary Research*, **58**, 36–40.

Muhs, D. R. & Szabo, B. J. (1994). New uranium-series ages of the Waimanalo limestone, Oahu, Hawaii: implications for sea level during the last interglacial period. *Marine Geology*, **118**, 315–326.

Muhs, D. R., Simmons, K. R. & Steinke, S. (2002). Timing and warmth of the last interglacial period: new U-series evidence from Hawaii and Bermuda and a new fossil compilation for North America. *Quaternary Science Reviews*, **21**, 1355–1383.

Muhs, D. R., Simmons, K. R., Schumann, R. R., *et al.* (2012). Sea-level history during the last interglacial complex on San Nicholas Island, California: implications for glacial isostatic adjustment processes, paleozoogeography and tectonics. *Quaternary Science Reviews*, **37**, 1–25.

Multer, H. G. & Hoffmeister, J. E. (1968). Subaerial laminated crusts of the Florida Keys. *Geological Society of America Bulletin*, **79**, 183–192.

Multer, H. G., Gischler, E., Lundberg, J., *et al.* (2002). Key Largo Limestone revisited: Pleistocene shelf-edge facies, Florida Keys, USA. *Facies*, **46**, 229–271.

Munk, W. (2002). Twentieth century sea level: an enigma. *Proceedings of the National Academy of Sciences of the United States of America*, **99**, 6550–6555.

Murchison, R. I. (1839). *The Silurian System, founded on geological researches in the Counties of Salop, Hereford, Radnor, Montgomery, Caermarthen, Brecon, Pembroke, Monmouth, Gloucestor, Worcestor, and Stafford; with descriptions of the coalfields and overlying formations.* John Murray, London.

Murray, A. S. & Olley, J. M. (2002). Precision and accuracy in the optically stimulated luminescence dating of sedimentary quartz: a status review. *Geochronometria*, **21**, 1–16.

Murray, A. S. & Roberts, R. G. (1997). Determining the burial time of single grains of quartz using optically stimulated luminescence. *Earth and Planetary Science Letters*, **152**, 163–180.

Murray-Wallace, C. V. (1995). Aminostratigraphy of Quaternary coastal sequences in southern Australia – an overview. *Quaternary International*, **26**, 69–86.

Murray-Wallace, C. V. (2000). Quaternary coastal aminostratigraphy: Australian data in a global context. *In* G. A. Goodfriend, M. J. Collins, M. L. Fogel, S. A. Macko & J. F. Wehmiller (Eds), *Perspectives in amino acid and protein geochemistry*, pp. 279–300. Oxford University Press, New York.

Murray-Wallace, C. V. (2002). Pleistocene coastal stratigraphy, sea-level highstands and neotectonism of the southern Australian passive continental margin – a review. *Journal of Quaternary Science*, **17**, 469–489.

Murray-Wallace, C. V. (2007). Sea level studies: eustatic sea-level changes, glacial-interglacial cycles. *In* S. A. Elias (Ed.), *Encyclopedia of Quaternary Science*, pp. 3024–3034. Elsevier, Oxford.

Murray-Wallace, C. V. & Belperio, A. P. (1991). The last interglacial shoreline in Australia – a review. *Quaternary Science Reviews*, **10**, 441–461.

Murray-Wallace, C. V. & Belperio, A. P. (1994). Identification of remanié fossils using amino acid racemisation. *Alcheringa*, **18**, 219–227.

Murray-Wallace, C. V. & Goede, A. (1991). Aminostratigraphy and electron spin resonance studies of late Quaternary sea level change and coastal neotectonics in Tasmania, Australia. *Zeitschrift für Geomorphologie*, **35**, 129–149.

Murray-Wallace, C. V. & Goede, A. (1995). Aminostratigraphy and electron spin resonance dating of Quaternary coastal neotectonism in Tasmania and the Bass Strait islands. *Australian Journal of Earth Sciences*, **42**, 51–67.

Murray-Wallace, C. V. & Kimber, R. W. L. (1989). Quaternary marine aminostratigraphy: Perth Basin, Western Australia. *Australian Journal of Earth Sciences*, **36**, 553–568.

Murray-Wallace, C. V., Kimber, R. W. L., Gostin, V. A., et al. (1988). Amino acid racemisation dating of the 'Older Pleistocene marine beds', Redcliff, Northern Spencer Gulf, South Australia. *Royal Society of South Australia, Transactions*, **112**, 51–55.

Murray-Wallace, C. V., Belperio, A. P., Picker, K., et al. (1991). Coastal aminostratigraphy of the Last Interglaciation in southern Australia. *Quaternary Research*, **35**, 63–71.

Murray-Wallace, C. V., Belperio, A. P., Gostin, V. A., et al. (1993). Amino acid racemization and radiocarbon dating of interstadial marine strata (oxygen isotope stage 3), Gulf St. Vincent, South Australia. *Marine Geology*, **110**, 83–93.

Murray-Wallace, C. V., Belperio, A. P., Cann, J. H., et al. (1996a). Late Quaternary uplift history, Mount Gambier region, South Australia. *Zeitschrift für Geomorphologie N.F.*, Suppl.-Bd. **106**, 41–56.

Murray-Wallace, C. V., Ferland, M. A., Roy, P. S., et al. (1996b). Unravelling patterns of reworking in lowstand shelf deposits using amino acid racemisation and radiocarbon dating. *Quaternary Science Reviews*, **15**, 685–697.

Murray-Wallace, C. V., Belperio, A. P. & Cann, J. H. (1998). Quaternary neotectonism and intra-plate volcanism: the Coorong to Mount Gambier Coastal Plain, Southeastern Australia: a review. *In* I. S. Stewart & C. Vita-Finzi (Eds), *Coastal tectonics*. Geological Society, London, Special Publication, **146**, 255–267.

Murray-Wallace, C. V., Belperio, A. P., Bourman, R. P., et al. (1999). Facies architecture of a last interglacial barrier: a model for Quaternary barrier development from the Coorong to Mount Gambier Coastal Plain, southeastern Australia. *Marine Geology*, **158**, 177–195.

Murray-Wallace, C. V., Beu, A. G., Kendrick, G. W., et al. (2000). Palaeoclimatic implications of the occurrence of the arcoid bivalve *Anadara trapezia* (Deshayes) in the Quaternary of Australasia. *Quaternary Science Reviews*, **19**, 559–590.

Murray-Wallace, C. V., Brooke, B. P., Cann, J. H., *et al.* (2001). Whole-rock amino-stratigraphy of the Coorong Coastal Plain, South Australia: towards a 1 million year record of sea-level highstands. *Journal of the Geological Society, London*, **158**, 111–124.

Murray-Wallace, C. V., Banerjee, D., Bourman, R. P., *et al.* (2002). Optically stimulated luminescence dating of Holocene relict foredunes, Guichen Bay, South Australia. *Quaternary Science Reviews*, **21**, 1077–1086.

Murray-Wallace, C. V., Ferland, M. A. & Roy, P. S. (2005). Further amino acid racemisation evidence for glacial age, multiple lowstand deposition on the New South Wales outer continental shelf, southeastern Australia. *Marine Geology*, **214**, 235–250.

Murray-Wallace, C. V., Bourman, R. P., Prescott, J. R., *et al.* (2010). Aminostratigraphy and thermoluminescence dating of coastal aeolianites and the later Quaternary history of a failed delta: the River Murray mouth region, South Australia. *Quaternary Geochronology*, **5**, 28–49.

Naish, T. (1997). Constraints on the amplitude of late Pliocene eustatic sea-level fluctuations: new evidence from the New Zealand shallow-marine sediment record. *Geology*, **25**, 1139–1142.

Nakada, M. (1986). Holocene sea levels in oceanic islands: implications for the rheological structure of the Earth's mantle. *Tectonophysics*, **121**, 263–276.

Nakada, M. & Lambeck, K. (1987). Glacial rebound and relative sea-level variations: a new appraisal. *Geophysical Journal*, **90**, 171–224.

Nakada, M. & Lambeck, K. (1988). The melting history of the late Pleistocene Antarctic ice sheet. *Nature*, **333**, 36–40.

Nakada, M. & Lambeck, K. (1989). Late Pleistocene and Holocene sea-level change in the Australian region and mantle rheology. *Geophysical Journal*, **96**, 497–517.

Nakada, M. & Yokose, H. (1992). Ice age as a trigger of active Quaternary volcanism and tectonism, *Tectonophysics*, **212**, 321–329.

Naylor, L. A., Stephenson, W. J. & Trenhaile, A. S. (2010). Rock coast geomorphology: recent advances and future research directions. *Geomorphology*, **114**, 3–11.

Negrete, J., Soibelzon, E., Tonni, E. P., *et al.* (2011). Antarctic radiocarbon reservoir: the case of the mummified crabeater seals (*Lobodon carcinophaga*) in Bodman Cape, Seymour Island, Antarctica. *Radiocarbon*, **53**, 161–166.

Nerem, R. S., Chambers, D. P., Choe, C., *et al.* (2010). Estimating mean sea level change from the TOPEX and Jason altimeter missions. *Marine Geodesy*, **33**, 435–446.

Neumann, A. C. (1971). Quaternary sea-level data from Bermuda. *Quaternaria*, **14**, 41–43.

Neumann, A. C. & Macintyre, I. (1985). Reef response to sea level rise: keep-up, catch-up or give-up. *Proceedings of the 5th International Coral Reef Congress*, **3**, 105–110.

Neumann, A. C. & Moore, W. S. (1975). Sea level events and Pleistocene coral ages in the northern Bahamas. *Quaternary Research*, **5**, 215–224.

Newell, N. D. (1961). Recent terraces of tropical limestone shores. *Zeitschrift für Geomorphologie N.F.* Suppl-Bd., **3**, 87–106.

Newell, N. D. & Bloom, A. L. (1970). The reef flat and 'two-meter eustatic terrace' of some Pacific atolls. *Geological Society of America, Bulletin*, **81**, 1881–1894.

Newell, N. D. & Rigby, J. K. (1957). Geological studies on the Great Bahama Bank. *Society of Economic Paleonologists and Mineralogists, Special Publication*, **5**, 15–79.

Newman, W. S., Cinquemani, L. J., Pardi, R. R., *et al.* (1980). Holocene delevelling of the United States' east coast. *In* N.-A. Morner (Ed.), *Earth rheology, isostasy and eustasy*, pp. 449–463. Wiley, Chichester.

Ngo-Duc, T., Laval, K. & Polcher, J. (2005). Contribution of continental water to sea level variations during the 1997–1998 El Niño–Southern Oscillation event: comparison between Atmospheric Model Intercomparison Project simulations and TOPEX//Poseidon satellite data. *Journal of Geophysical Research*, **110**, D09103.

Nguyen, V. L., Ta, T. K. O. & Tateishi, M. (2000). Late Holocene depositional environments and coastal evolution of the Mekong River Delta, southern Vietnam. *Journal of Asian Earth Sciences*, **18**, 427–439.

Nichol, S. L. & Murray-Wallace, C. V. (1992). A partially preserved last interglacial estuarine fill: Narrawallee Inlet, New South Wales. *Australian Journal of Earth Sciences*, **39**, 545–553.

Nicholas, W. A., Chivas, A. R., Murray-Wallace, C. V., et al. (2011). Prompt transgression and gradual salinisation of the Black Sea during the early Holocene constrained by amino acid racemization and radiocarbon dating. *Quaternary Science Reviews*, **30**, 3769–3790.

Nicholls, R. J. & Cazenave, A. (2010). Sea-level rise and its impact on coastal zones. *Science*, **328**, 1517–1520.

Nicholls, R. J., Wong, P. P., Burkett, V. R., et al. (2007). Coastal systems and low-lying areas. *In* M. L. Parry, O. F. Canziani, J. P. Palutikof, P. J. Van Der Linden, & C. E. Hanson (Eds), *Climate Change 2007: impacts, adaptation and vulnerability. Contribution of Working Group II to the Fourth Assessment Report of the Intergovernmental Panel on Climate Change (IPCC)*, pp. 315–357. Cambridge University Press, Cambridge.

Nield, T. (2007). *Supercontinent – ten billion years in the life of our planet*. Granta Books, London, 288 pp.

Nishiizumi, K., Winterer, E. L., Kohl, C. P., et al. (1989). Cosmic ray production rates of ^{10}Be and ^{26}Al in quartz from glacially polished rocks. *Journal of Geophysical Research*, **94**, 17907–17915.

Nishiizumi, K., Kohl, C. P., Arnold, J. R., et al. (1993). Role of *in situ* cosmogenic nuclides ^{10}Be and ^{26}Al in the study of diverse geomorphic processes. *Earth Surface Processes and Landforms*, **18**, 407–425.

Niskanen, E. (1939). On the upheaval of land in Fennoscandia. *Annales Academiae Scientiarum Fennicae*, **Series A 53**, 10, 30 pp.

Noller, J. S., Sowers, J. M. & Lettis, W. R. (Eds) (2000). *Quaternary geochronology: methods and applications*. American Geophysical Union, Reference Shelf 4, Washington, DC, 581 pp.

Nott, J. (1996). Late Pleistocene and Holocene sea-level highstands in northern Australia. *Journal of Coastal Research*, **12**, 907–910.

Nott, J. (2012). Storm tide recurrence intervals – a statistical approach using beach ridge plains in Northern Australia. *Geographical Research*, **50**, 368–376.

Nunn, P. D. (1986). Implications of migrating geoid anomalies for the interpretation of high-level fossil coral reefs. *Geological Society of America Bulletin*, **97**, 946–952.

Nunn, P. D. (1990). Coastal processes and landforms Fiji: their bearing on Holocene sea-level changes in the south and west Pacific. *Journal of Coastal Research*, **6**, 279–310.

Nunn, P. D. (1994). *Oceanic Islands*. Blackwell, Oxford, 413 pp.

Nunn, P. D. (1998). *Pacific Island landscapes: landscape and geological development of southwest Pacific Islands, especially Fiji, Samoa and Tonga*. Institute of Pacific Studies, The University of the South Pacific, 318 pp.

Nunn, P. D. & Britton, J. M. R. (2004). The long-term evolution of Niue Island. *In* J. P. Terry, W. E. Murray (Eds), *Niue Island: geographic perspectives on the rock of Polynesiai*, pp. 31–73. Insula, Paris.

Nunn, P. D. & Peltier, W. R. (2001). Far-field test of the ICE-4G model of global isostatic response to deglaciation using empirical and theoretical Holocene sea-level reconstructions for the Fiji Islands, southwestern Pacific. *Quaternary Research*, **55**, 203–214.

Nunn, P. D., Ollier, C., Hope, G., et al. (2002). Late Quaternary sea-level and tectonic changes in northeast Fiji. *Marine Geology*, **187**, 299–311.

Oerlemans, J. (1989). A projection of future sea level. *Climatic Change*, **15**, 151–174.

Oerlemans, J. & Fortuin, J. P. F. (1992). Sensitivity of glaciers and small ice caps to greenhouse warming. *Science*, **258**, 115–117.

Oeschger, H., Beer, J., Siegenthaler, U., et al. (1984). Late glacial climate history from ice cores. *In* J. E. Hansen & T. Takahashi (Eds), *Climate processes and climate sensitivity*, pp. 299–306. American Geophysical Union, Washington, DC.

Ogg, J., Ogg, G. & Gradstein, F. M. (2008). *The concise geologic time scale.* Cambridge University Press, Cambridge, 177 pp.

Ohde, S., Greaves, M., Masuzawa, T., *et al.* (2002). The chronology of Funafuti Atoll: revisiting an old friend. *Proceedings of the Royal Society of London*, **A458**, 2289–2306.

O'Leary, M. J., Hearty, P. J. & McCulloch, M. T. (2008). Geomorphic evidence of major sea-level fluctuations during marine isotope substage-5e, Cape Cuvier, Western Australia. *Geomorphology*, **102**, 595–602.

Oldroyd, D. (1996). *Thinking about the Earth: a history of ideas in Geology.* Athlone, London, 410 pp.

Olson, S. L. & Hearty, P. J. (2009). A sustained +21 m sea-level highstand during MIS 11 (400 ka): direct fossil and sedimentary evidence from Bermuda. *Quaternary Science Reviews*, **28**, 271–285.

Olsson, I. U. (1974). Some problems in connection with the evaluation of ^{14}C dates. *Geologiska Föreningens i Stockholm Förhandlingar*, **96**, 311–320.

Omura, A., Maeda, Y., Kawana, T., *et al.* (2004). U-Series of Pleistocene corals and their implications to the paleo-sea levels and the vertical displacement in the Central Philippines. *Quaternary International*, **115–116**, 3–13.

Oppenheimer, S. (1998). *Eden in the east – the drowned continent of southeast Asia.* Weidenfeld and Nicolson, London, 560 pp.

Oreskes, N., Shrader-Frechette, K. & Belitz, K. (1994). Verification, validation, and confirmation of numerical models in the earth sciences. *Science*, **263**, 641–646.

Orme, A. R. (1998). Late Quaternary tectonism along the Pacific coast of the Californias: a contrast of style. *In* I. Stewart & C. Vita-Finzi (Eds), *Coastal tectonics*, Geological Society, London, Special Publications, **146**, 179–197.

Ortlieb, L. (1995). *Late Quaternary coastal changes in northern Chile.* Guidebook for a fieldtrip (Antofagasta – Iquique, 23–25 November 1995) organised during the 1995 Annual Meeting of the International Geological Correlation Program Project 367 (19–28 November 1995, Antofagasta, Chile), Orstom, Antofagasta, Chile, 175 pp.

Ortlieb, L., Zazo, C., Goy, J. L., *et al.* (1996). Coastal deformation and sea-level changes in the northern Chile subduction area (23°S) during the last 330 KY. *Quaternary Science Reviews*, **15**, 819–831.

Ortlieb, L., Vargas, G. & Saliège, J.-F. (2011). Marine radiocarbon reservoir effect along the northern Chile–southern Peru coast (14–24°S) throughout the Holocene. *Quaternary Research*, **75**, 91–103.

Ota, Y. (1986). Marine terraces as reference surfaces in late Quaternary tectonics studies: examples from the Pacific rim. *Royal Society of New Zealand Bulletin*, **24**, 357–375.

Ota, Y. (1992). Coastal evolution and coastal tectonics in the Japanese Islands and some parts of the Western Pacific Rim (IGCP-274). *Japan Contribution to the IGCP*, 1992, pp. 41–46.

Ota, Y. (1994). Last interglacial shorelines in the Western Pacific Rim. *Journal of Geography*, **103**, 809–827.

Ota, Y. (1996). Various aspects of coral reef studies with special references to tectonic geomorphology. *In* S. B. McCann & D. C. Ford (Eds), *Geomorphology sans frontières*, pp. 115–136. John Wiley and Sons Ltd, Chichester.

Ota, Y. & Kaizuka, S. (1991). Tectonic geomorphology at active plate boundaries: examples from the Pacific rim. *Zeitschrift für Geomorphologie* Suppl.-Bd, **82**, 119–146.

Ota, Y. & Machida, H. (1987). Quaternary sea-level changes in Japan. *In* M. J. Tooley, & I. Shennan (Eds), *Sea-level changes*, The Institute of British Geographers Special Publication Series, 20, Basil Blackwell, London, pp. 182–224.

Ota, Y, & Omura, A. (1991). Late Quaternary shorelines in the Japanese Islands. *The Quaternary Research*, **30**, 175–186.

Ota, Y, & Omura, A. (1992). Contrasting styles and rates of tectonic uplift of coral reef terraces in the Ryukyu and Daito Islands, Southwestern Japan. *Quaternary International*, **15/16**, 17–29.

Ota, Y., Berryman, K. R., Hull, A. G., *et al.* (1988). Age and height distribution of Holocene transgressive deposits in eastern North Island, New Zealand. *Palaeogeography, Palaeoclimatology, Palaeoecology*, **68**, 135–151.

Ota, Y., Berryman, K. R., Brown, L. J., *et al.* (1989). Holocene sediments and vertical tectonic downwarping near Wairoa, Northern Hawke's Bay, New Zealand. *New Zealand Journal of Geology and Geophysics*, **32**, 333–341.

Ota, Y., Hull, A. G. & Berryman, K. R. (1991). Coseismic uplift of Holocene marine terraces in the Pakarae River area, Eastern North Island, New Zealand. *Quaternary Research*, **35**, 331–346.

Ota, Y., Miyauchi, T., Paskoff, R., *et al.* (1995). Plio-Quaternary terraces and their deformation along the Altos de Talinay, north-central Chile. *Revista Geologica de Chile*, **22**, 89–102.

Otto-Bliesner, B. L., Marshall, S. J., Overpeck, J. T., *et al.* (2006). Simulating arctic climate warmth and icefield retreat in the Last Interglaciation. *Science*, **311**, 1751–1753.

Otvos, E. G. (2000). Beach ridges – definitions and significance. *Geomorphology*, **32**, 83–108.

Otvos, E. G. (2012). Coastal barriers – nomenclature, processes, and classification issues. *Geomorphology*, **139–140**: 39–52.

Otvos, E. G. & Price, W. A. (1979). Problems of chenier genesis and terminology – an overview. *Marine Geology*, **31**, 251–263.

Owen, M., Day, S. & Maslin, M. (2007). Late Pleistocene submarine mass movements: occurrence and causes. *Quaternary Science Reviews*, **26**, 958–978.

Ozawa, S., Nishimura, T., Suito, H., *et al.* (2011). Coseismic and postseismic slip of the 2011 magnitude -9 Tohoku-Oki earthquake. *Nature*, **475**, 373–377.

Paine, J. G. (1993). Subsidence of the Texas coast: inferences from historical and late Pleistocene sea levels. *Tectonophysics*, **222**, 445–458.

Pandolfi, J. (1996). Limited membership in Pleistocene reef coral assemblages from the Huon Peninsula, Papua New Guinea: constancy during global change. *Paleobiology*, **22**, 152–176.

Pandolfi, J. M., Best, M. M. R. & Murray, S. P. (1994). Coseismic event of May 15, 1992, Huon Peninsula, Papua New Guinea: comparison with Quaternary tectonic history. *Geology*, **22**, 239–242.

Pandolfi, J. M., Llewellyn, G. & Jackson, J. B. C. (1999). Pleistocene reef environments, constituent grains, and coral community structure: Curacao, Netherlands Antilles. *Coral Reefs*, **18**, 107–122.

Pandolfi, J. M., Tudhope, A. W., Burr, G. S., *et al.* (2006). Mass mortality following disturbance in Holocene coral reefs from Papua New Guinea. *Geology*, **34**, 949–952.

Pardaens, A. K., Gregory, J. M. & Lowe, J. A. (2011a). A model study of factors influencing projected changes in regional sea level over the twenty-first century. *Climate Dynamics*, **36**, 2015–2033.

Pardaens, A. K., Lowe, J. A., Brown, S., *et al.* (2011b). Sea-level rise and impacts projections under a future scenario with large greenhouse gas emission reductions. *Geophysical Research Letters*, **38**, L12604.

Parkinson, R. W. (1989). Decelerating Holocene sea-level rise and its influence on southwest Florida coastal evolution: a transgressive/regressive stratigraphy. *Journal of Sedimentary Petrology*, **59**, 960–972.

Paskoff, R. (1970). *Recherches Géomorphologiques dans le Chili semi-aride*. Biscaye Frères, Bordeaux, 420 pp.

Paskoff, R. (1995). *Field meeting in the La Serena – Coquimbo Bay area (Chile)*. Guidebook for a fieldtrip (27–28 November 1995) organized during the 1995 Annual meeting of International Geological Correlation Program Project 367, (Antofagasta, Chile), Orstom, Chile, 69 pp.

Pavlis, N. K., Holmes, S. A., Kenyon, S. C., *et al.* (2008). An Earth gravitational model to degree 2160: EGM2008, presented at the 2008 General Assembly of the European Geosciences Union, Vienna, Austria, 13–18 April 2008.

Pearson, H. W. (1901). Oscillations in the sea-level. *The Geological Magazine*, **8**, 167–174 (Part I), 223–231 (Part II), and 253–265 (Part III).

Pedoja, K., Husson, L., Regard, V., *et al.* (2011). Relative sea-level fall since the last interglacial stage: are coasts uplifting worldwide? *Earth-Science Reviews*, **108**, 1–15.

Peltier, W. R. (1974). The impulse response of a Maxwell Earth. *Reviews of Geophysics*, **12**, 649–669.

Peltier, W. R. (1987). Mechanisms of relative sea-level change and the geophysical responses to ice-water loading. *In* R. J. N. Devoy (Ed.), *Sea surface studies – a global view*, pp. 57–94. Croom Helm, London.

Peltier, W. R. (1996). Global sea level rise and glacial isostatic adjustment: an analysis of data from the east coast of North America. *Geophysical Research Letters*, **23**, 717–720.

Peltier, W. R. (1998). Global glacial isostatic adjustment and coastal tectonics. *In* I. S. Stewart & C. Vita-Finzi (Eds), *Coastal tectonics*, Geological Society, London, Special Publications, **146**, 1–29.

Peltier, W. R. (2000). Global glacial isostatic adjustment and modern instrumental records of relative sea-level history. *In* B. C. Douglas, M. S. Kearney & S. P. Leatherman (Eds), *Sea level rise: history and consequences*, pp. 65–95. Academic Press, San Diego.

Peltier, W. R. (2002a). On eustatic sea level history: Last Glacial Maximum to Holocene. *Quaternary Science Reviews*, **21**, 377–397.

Peltier, W. R. (2002b). Global glacial isostatic adjustment: palaeogeodetic and space-geodetic tests of the ICE-4G (VM2) model. *Journal of Quaternary Science*, **17**, 491–510.

Peltier, W. R. (2004). Global glacial isostasy and the surface of the Ice-Age Earth: the ICE-5G (VM2) Model and GRACE. *Annual Review of Earth and Planetary Sciences*, **32**, 111–149.

Peltier, W. R. (2005). On the hemispheric origins of meltwater pulse 1a. *Quaternary Science Reviews*, **24**, 1655–1671.

Peltier, W. R. (2009). Closure of the budget of global sea level rise over the GRACE era: the importance and magnitudes of the required corrections for global glacial isostatic adjustment. *Quaternary Science Reviews*, **28**, 1658–1674.

Peltier, W. R. & Andrews, J. T. (1976). Glacial isostatic adjustment – 1. The forward problem. *Geophysical Journal of the Royal Astronomical Society*, **46**, 605–646.

Peltier, W. R. & Drummond, R. (2008). Rheological stratification of the lithosphere: a direct inference based upon the geodetically observed pattern of the glacial isostatic adjustment of the North American continent. *Geophysical Research Letters*, **35**. doi:10.1029/2008GL034586.

Peltier, W. R. & Fairbanks, R. G. (2006). Global glacial ice volume and Last Glacial Maximum duration from an extended Barbados sea level record. *Quaternary Science Reviews*, **25**, 3322–3337.

Peltier, W. R. & Tushingham, A. M. (1989). Global sea level rise and the greenhouse effect: might they be connected. *Science*, **244**, 806–810.

Peltier, W. R. & Tushingham, A. M. (1991). Influence of glacial isostatic adjustment on tide gauge measurements of secular sea level change. *Journal of Geophysical Research*, **96**, 6779–6796.

Penck, A. & Brückner, E. (1909). *Die Alpen im Eiszeitalter*. Leipzig, Tauchnitz, 3 volumes, 1199 pp.

Penkman, K. E. H., Kaufman, D. S., Maddy, D., *et al.* (2008). Closed-system behaviour of the intra-crystalline fraction of amino acids in mollusc shells. *Quaternary Geochronology*, **3**, 2–25.

Penland, S. & Suter, J. R. (1989). The geomorphology of the Mississippi River chenier plain. *Marine Geology*, **90**, 231–258.

Penland, S., Boyd, R. & Suter, J. R. (1988). Transgressive depositional systems of the Mississippi Delta plain: a model for barrier shoreline and shelf sand development. *Journal of Sedimentary Petrology*, **58**, 932–949.

Perrette, M., Landerer, F., Riva, R., *et al.* (2012). Probabilistic projection of sea-level change along the world's coastlines. *Earth System Dynamics Discussions*, **3**, 357–389.

Perry, C. T. & Smithers, S. G. (2011). Cycles of coral reef 'turn-on', rapid growth and 'turn-off' over the past 8500 years: a context for understanding modern ecological states and trajectories (refer to supplementary material). *Global Change Biology*, **17**, 76–86.

Petchey, F. & Clark, G. (2011). Tongatapu hardwater: investigation into the ^{14}C marine reservoir offset in lagoon, reef and open ocean environments of a limestone island. *Quaternary Geochronology*, **6**, 539–549.

Petchey, F., Anderson, A., Zondervan, A., et al. (2008). New marine ΔR values for the South Pacific subtropical gyre region. *Radiocarbon*, **50**, 373–397.

Petersen, K. S. (1986). Marine molluscs as indicators of former sea-level stands. *In* O. van de Plassche (Ed.), *Sea-level research: a manual for the collection and evaluation of data*, pp. 129–155. Geo Books, Norwich.

Pethick, J. (1984). *An introduction to coastal geomorphology*. Edward Arnold, London, 260 pp.

Petley, D. N. & Reid, S. (1999). Uplift and landscape stability at Taroko, eastern Taiwan. *In* B. J. Smith, W. B. Whalley & P. A. Warke (Eds), *Uplift, erosion and stability: perspectives on long-term landscape development*. Geological Society, London, Special Publications, **162**, 169–181.

Petzelberger, B. E. M. (2000). Coastal development and human activities in NW Germany. *In* K. Pye & J. R. L. Allan (Eds), *Coastal and estuarine environments: sedimentology, geomorphology and geoarchaeology*. Geological Society, London, Special Publications, **175**, 365–376.

Pfeffer, W. T. (2011). Land ice and sea level rise: a thirty-year perspective. *Oceanography*, **24**, 94–111.

Pfeffer, W. T., Harper, J. & O'Neel, S. (2008). Kinematic constraints on glacier contribution to 21st century sea-level rise. *Science*, **321**, 1340–1343.

Phelan, M. B. (1999). A Delta R correction value for Samoa from known-age marine shells. *Radiocarbon*, **41**, 99–101.

Phillips, F. M. (1995). Cosmogenic chlorine-36 accumulation: a method for dating Quaternary landforms. *In* N. W. Rutter & N. R. Catto (Eds), *Dating methods for Quaternary deposits*, pp. 61–66, Geo Text 2, Geological Association of Canada, St Johns.

Phipps, C. V. G. (1970). Dating of eustatic events from cores taken in the Gulf of Carpentaria and samples from the New South Wales continental shelf. *Australian Journal of Science*, **32**, 329–330.

Pickett, J. W., Ku, T. L., Thompson, C. H., et al. (1989). A review of age determination of Pleistocene corals in eastern Australia. *Quaternary Research*, **31**, 392–395.

Pierre, C., Belanger, P., Saliège, J. F., et al. (1999). Paleoceanography of the Western Mediterranean during the Pleistocene: oxygen and carbon isotope records at Site 975. *In* R. Zahn, M. C. Comas & A. Klaus (Eds), *Proceedings of the Ocean Drilling Program, Scientific Results*, **161**, 481–488.

Pillans, B. (1982). Amino acid racemisation dating: a review. *In* W. Ambrose & P. Duerden (Eds), *Archaeometry: an Australasian perspective*, pp. 228–235. Australian National University, Canberra.

Pillans, B. (1983). Upper Quaternary marine terrace chronology and deformation, South Taranaki, New Zealand. *Geology*, **11**, 292–297.

Pillans, B. (1990). Pleistocene marine terraces in New Zealand: a review. *New Zealand Journal of Geology and Geophysics*, **33**, 219–231.

Pillans, B. (1991). New Zealand Quaternary stratigraphy: an overview. *Quaternary Science Reviews*, **10**, 405–418.

Pillans, B. (1994). Direct marine–terrestrial correlations, Wanganui Basin, New Zealand: the last 1 million years. *Quaternary Science Reviews*, **13**, 189–200.

Pillans, B. & Bourman, R. P. (1997). The Brunhes/Matuyama polarity transition (0.78 Ma) as a chronostratigraphic marker in Australian regolith studies. *AGSO Journal of Australian Geology and Geophysics*, **13**, 289–294.

Pillans, B. & Naish, T. (2004). Defining the Quaternary. *Quaternary Science Reviews*, **23**, 2271–2282.

Pillans, B. J., Roberts, A. P., Wilson, G. S., et al. (1994). Magnetostratigraphic, lithostratigraphic and tephrostratigraphic constraints on Lower and Middle Pleistocene sea-level changes, Wanganui Basin, New Zealand. *Earth and Planetary Science Letters*, **121**, 81–98.

Pinet, P. R. (1992). *Oceanography: an introduction to the planet Oceanus*. West Publishing Co., St Paul, USA, 572 pp.

Pirazzoli, P. A. (1976). Sea level variation in the northwest Mediterranean during Roman times. *Science*, **194**, 519–521.

Pirazzoli, P. A. (1991). *World atlas of Holocene sea-level changes*, Elsevier

Oceanography Series, 58. Elsevier, Amsterdam, 300 pp.

Pirazzoli, P. A. (1993). Global sea-level changes and their measurement. *Global and Planetary Change*, **8**, 135–148.

Pirazzoli, P. A. (1994). Tectonic shorelines. *In* W. Carter & C. D. Woodroffe (Eds), *Coastal evolution*, pp. 451–476. Cambridge University Press, Cambridge.

Pirazzoli, P. A. (1996). *Sea-level changes: the last 20,000 years*. John Wiley and Sons, Chichester, 211 pp.

Pirazzoli, P. A. (2005). A review of possible eustatic, isostatic and tectonic contributions in eight late-Holocene relative sea-level histories from the Mediterranean area. *Quaternary Science Reviews*, **24**, 1989–2001.

Pirazzoli, P. A. (2007). Sea level studies: geomorphological indicators. *In* S. A. Elias (Ed.), *Encyclopedia of Quaternary science*, pp. 2974–2983. Elsevier, Oxford.

Pirazzoli, P. A. & Montaggioni, L. F. (1988). Holocene sea-level changes in French Polynesia. *Palaeogeography, Palaeoclimatology, Palaeoecology*, **68**, 153–175.

Pirazzoli, P. A., Delibrias, G., Kawana, T., *et al.* (1985). The use of barnacles to measure and date relative sea-level changes in the Ryukyu Islands, Japan. *Palaeogeography, Palaeoclimatology, Palaeoecology*, **49**, 161–174.

Pirazzoli, P. A., Koba, M., Montaggioni, L. F., *et al.* (1988). Anaa (Tuamotu Islands, Central Pacific): an incipient rising atoll? *Marine Geology*, **82**, 261–269.

Pirazzoli, P. A., Kaplin, P. A. & Montaggioni, L. F. (1990). Differential vertical crustal movement deduced from late Holocene coral-rich conglomerates: Farqhar and St Joseph atolls (Seychelles, western Indian Ocean). *Journal of Coastal Research*, **6**, 381–389.

Pirazzoli, P. A., Arnold, M., Giresse, P., *et al.* (1993a). Marine deposits of late glacial times exposed by tectonic uplift on the east coast of Taiwan. *Marine Geology*, **110**, 1–6.

Pirazzoli, P. A., Radtke, U., Hantoro, W. S., *et al.* (1993b). A one million-year-long sequence of marine terraces on Sumba Island, Indonesia. *Marine Geology*, **109**, 221–236.

Pirazzoli, P. A., Stiros, S. C., Laborel, J., *et al.* (1994a). Late-Holocene shoreline changes related to palaeoseismic events in the Ionian Islands, Greece. *The Holocene*, **4**, 397–405.

Pirazzoli, P. A., Stiros, S. C., Arnold, M., *et al.* (1994b). Episodic uplift deduced from Holocene shorelines in the Perachora Peninsula, Corinth area, Greece. *Tectonophysics*, **229**, 201–209.

Pirazzoli, P. A., Stiros, S. C., Fontugne, M., *et al.* (1994c). Holocene and Quaternary uplift in the central part of the southern coast of the Corinth Gulf (Greece). *Marine Geology*, **212**, 35–44.

Pirazzoli, P. A., Laborel, J. & Stiros, S. C. (1996). Coastal indicators of rapid uplift and subsidence: examples from Crete and other eastern Mediterranean sites. *Zeitschrift für Geomorphologie N.F.* Suppl.-Bd., **102**, 21–35.

Pirazzoli, P. A., Mastronuzzi, G. Saliège, J. F., *et al.* (1997). Late Holocene emergence in Calabria, Italy. *Marine Geology*, **141**, 61–70.

Pitman, W. C. (1978). Relationship between eustacy [*sic*] and stratigraphic sequences of passive margins. *Geological Society of America Bulletin*, **89**, 1389–1403.

Plafker, G. (1965). Tectonic deformation associated with the 1964 Alaska Earthquake. *Science*, **148**, 1675–1687.

Plaziat, J.-C., Baltzer, F., Chouki, A., *et al.* (1998). Quaternary marine and continental sedimentation in the northern Red Sea and Gulf of Suez (Egyptian coast): influences of rift tectonics, climatic changes and sea-level fluctuations. *In* B. H. Purser & D. W. J. Bosence (Eds), *Sedimentation and tectonics in Rift Basins: Red Sea–Gulf of Aden*, pp. 537–573. Chapman and Hall, London.

Plint, A. G., Eyles, N., Eyles, C. H., *et al.* (1992). Control of sea level change. *In* R. G. Walker & N. P. James (Eds), *Facies models: response to sea level change*, pp. 15–25. Geological Association of Canada, St Johns.

Pokhrel, Y. N., Hanasaki, N., Yeh, P. J.-F., *et al.* (2012). Model estimates of sea-level change due to anthropogenic impacts on terrestrial water storage. *Nature Geoscience*, **6**, 3–4.

Pollard, D. & De Conto, R. M. (2009). Modelling West Antarctic ice sheet growth and collapse through the past five million years. *Nature*, **458**, 329–332.

Poole, A. J., Shimmield, G. B. & Robertson, A. H. F. (1990). Late Quaternary uplift of the Troodos ophiolite, Cyprus: uranium-series dating of Pleistocene coral. *Geology*, **18**, 894–897.

Poore, H., White, N. & Maclennan, J. (2011). Ocean circulation and mantle melting controlled by radial flow of hot pulses in the Iceland plume. *Nature Geoscience*, **4**, 558–561.

Porqueddu, A., Antonioli, F., Rubens, D., et al. (2006). Relative sea level change in Olbia Gulf (Sardinia, Italy), a historically important Mediterranean harbour. *Quaternary International*, **232**, 21–30.

Posamentier, H. W., James, D. P. & Allen, G. P. (1990). Aspects of sequence stratigraphy: recent and ancient examples of forced regressions. *American Association of Petroleum Geologists Bulletin*, **74**, 742 pp.

Potter, E.-K. & Lambeck, K. (2003). Reconciliation of sea-level observations in the Western North Atlantic during the last glacial cycle. *Earth and Planetary Science Letters*, **217**, 171–181.

Powell, J. W. (1875). *Exploration of the Colorado River of the west and its tributaries*. Government Printing Office, Washington, DC.

Prandi, P., Cazenave, A. & Becker, M. (2009). Is coastal mean sea level rising faster than the global mean? A comparison between tide gauges and satellite altimetry. *Geophysical Research Letters*, **36**, L05602.

Pratt, B. R., James, N. P. & Cowan, C. A. (1992). Peritidal carbonates. *In* R. G. Walker & N. P. James (Eds), *Facies models: response to sea level change*, pp. 303–322. Geological Association of Canada, St Johns.

Pratt, J. H. (1855). On the attraction of the Himalaya Mountains, and of the elevated regions beyond them, upon the plumb-line in India. *Philosophical Transactions of the Royal Society of London, Series A*, **145**, 53–100.

Preiss, W. (1993). *Geology of South Australia, Vol. 1, Mines and Energy*. South Australia, Adelaide.

Prell, W. L., Imbrie, J., Martinson, D. G., et al. (1986). Graphic correlation of oxygen isotope stratigraphy: application to the Late Quaternary. *Paleoceanography*, **1**, 137–162.

Prescott, J. R. & Hutton, J. T. (1994). Cosmic ray contributions to dose rates for luminescence and ESR dating: large depths and long-term variations. *Radiation Measurements*, **23**, 497–500.

Prescott, J. R., Huntley, D. J. & Hutton, J. T. (1993). Estimation of equivalent dose in thermoluminescence dating – the Australian slide method. *Ancient TL*, **11**, 1–5.

Price, D. (1994). TL signatures of quartz grains of different origin. *Radiation Measurements*, **23**, 413–417.

Price, R. C., Nicholls, I. A. & Gray, C. M. (2003). Chapter 12: Cainozoic igneous activity. *In* W. D. Birch (Ed.), *Geology of Victoria*, pp. 361–375. Geological Society of Australia, Special Publication 23.

Pritchard, H. D., Ligtenberg, S. R. M. & Fricker, H. A. (2012). Antarctic ice-sheet loss driven by basal melting of ice shelves. *Nature*, **484**, 502–505.

Pugh, D. T. (1987). *Tides, surges and mean sea-level*. Wiley, Chichester, 471 pp.

Pugh, D. (2004). *Changing sea levels: effects of tides, weather and climate*, Cambridge University Press, Cambridge, 265 pp.

Purdy, E. G. (1974). Karst-determined facies patterns in British Honduras: Holocene carbonate sedimentation model. *The American Association of Petroleum Geologists Bulletin*, **58**, 825–855.

Purdy, E. G. & Gischler, E. (2005). The transient nature of the empty bucket model of reef sedimentation. *Sedimentary Geology*, **175**, 35–47.

Purdy, E. G. & Winterer, E. L. (2001). Origin of atoll lagoons. *Geological Society of America Bulletin*, **113**, 837–854.

Purdy, E. G. & Winterer, E. L. (2006). Contradicting Barrier Reef relationships for Darwin's evolution of reef types. *International Journal of Earth Sciences*, **95**, 143–167.

Quigley, M. C., Clark, D. & Sandiford, M. (2010). Tectonic geomorphology of Australia. *In* P. Bishop & B. Pillans (Eds), *Australian landscapes*. Geological Society, London, Special Publications, **346**, 243–265.

Quinlan, G. (1985). A numerical model of postglacial relative sea level change near Baffin Island. *In* J. T. Andrews (Ed.), *Quaternary environments, eastern Canadian Arctic, Baffin Island and western Greenland*, pp. 560–584. Allen and Unwin, London.

Quinlan, G. & Beaumont, C. (1982). The deglaciation of Atlantic Canada as reconstructed from the postglacial relative sea-level record. *Canadian Journal of Earth Sciences*, **19**, 2232–2246.

Radtke, U., Grün, R., Omura, A., *et al.* (1996). The Quaternary coral reef tracts of Hateruma, Ryukyu Islands, Japan. *Quaternary International*, **31**, 61–70.

Rahmstorf, S. (2007). A semi-empirical approach to projecting future sea-level rise. *Science*, **315**, 368–370.

Rahmstorf, S. & Vermeer, M. (2011). Discussion of: Houston, J. R. & Dean, R. G. 2011. Sea-level acceleration based on U.S. tide gauges and extensions of previous global-gauge analyses. *Journal of Coastal Research*, **27**, 409–417. *Journal of Coastal Research*, **27**, 784–787.

Rahmstorf, S., Cazenave, A., Church, J. A., *et al.* (2007). Recent climate observations compared to projections. *Science*, **316**, 709.

Rahmstorf, S., Foster, G. & Cazenave, A. (2012). Comparing climate projections to observations up to 2011. *Environmental Research Letters*, **7**, doi: 10.1088/1748-9326/7/4/044035.

Rampino, M. R. & Sanders, J. E. (1980). Holocene transgression in south-central Long Island, New York. *Journal of Sedimentary Petrology*, **50**, 1063–1080.

Ramsay, P. J. & Cooper, J. A. G. (2002). Late Quaternary sea-level change in South Africa. *Quaternary Research*, **57**, 82–90.

Rashid, T., Suzuki, S., Sato, H., *et al.* (2013). Relative sea-level changes during the Holocene in Bangladesh. *Journal of Asian Earth Sciences*, **64**, 136–150.

Raup, D. M. & Stanley, S. M. (1978). *Principles of paleontology*, 2nd edition. W. H. Freeman & Co., San Francisco, 481 pp.

Raymo, M. E. (1992). Global climate change: a three million year perspective. *In* G. J. Kukla & E. Went (Eds), *Start of a glacial*, pp. 207–223. Springer-Verlag, Berlin.

Raymo, M. E. & Mitrovica, J. X. (2012). Collapse of polar ice sheets during the stage 11 interglacial. *Nature*, **483**, 453–456.

Redfield, A. C. (1967). Postglacial change in sea level in the western North Atlantic Ocean. *Science*, **157**, 687–691.

Redfield, A. C. (1972). Development of a New England salt marsh. *Ecological Monographs*, **42**, 201–237.

Redfield, A. C. & Rubin, M. (1962). The age of salt marsh peat and its relation to recent changes in sea level at Barnstable, Massachusetts. *Proceedings of the National Academy of Science*, **48**, 1728–1735.

Reid, C. (1913). *Submerged forests*. The Cambridge Manuals of Science and Literature. Cambridge University Press, Cambridge, 129 pp.

Reimer, P. J., Baillie, M. G. L., Bard, E., *et al.* (2009). INTCAL09 and MARINE09 radiocarbon age calibration curves, 0–50,000 years Cal BP. *Radiocarbon*, **51**, 1111–1150.

Renfrew, C. (1973). *Before civilization – the radiocarbon revolution and prehistoric Europe*. Penguin Books, England, 320 pp.

Rhodes, E. G. (1982). Depositional model for a chenier plain, Gulf of Carpentaria, Australia. *Sedimentology*, **29**, 201–221.

Rhodes, E. G., Polach, H. A., Thom, B. G., *et al.* (1980). Age structure of Holocene coastal sediments: Gulf of Carpentaria, Australia. *Radiocarbon*, **22**, 718–727.

Rial, J. A., Oh, J. & Reischmann, E. (2013). Synchronization of the climate system to eccentricity forcing and the 100,000-year problem. *Nature Geoscience*, **6**, 289–293.

Richards, D. A., Smart, P. L. & Edwards, R. L. (1994). Maximum sea levels for the last glacial period from U-series ages of submerged speleothems. *Nature*, **367**, 357–360.

Richter, J. P. (Ed.) (1970). *The notebooks of Leonardo da Vinci*, Vol. II. Dover Publications, New York, 499 pp.

Riggs, S. R., York, L. L., Wehmiller, J. F., *et al.* (1992). Depositional patterns resulting from high-frequency Quaternary sea-level fluctuations in northeastern North Carolina. *In* C. H. Fletcher III & J. F. Wehmiller (Eds), *Quaternary coasts of the United States: marine and lacustrine systems.* SEPM (Society for Sedimentary Geology), Special Publication, No. 48, pp. 141–153.

Rignot, E. & Kanagaratnam, P. (2006). Changes in the velocity structure of the Greenland Ice Sheet. *Science*, **311**, 986–990.

Rignot, E., Casassa, G., Gogineni, P., *et al.* (2004). Accelerated ice discharge from the Antarctic Peninsula following the collapse of Larsen B ice shelf. *Geophysical Research Letters*, **31**, L18401.

Rignot, E., Velicogna, I., van den Broeke, M. R., *et al.* (2011). Acceleration of the contribution of the Greenland and Antarctic ice sheets to sea level rise. *Geophysical Research Letters*, **38**, L05503.

Rittner, S. (2013). *The efficacy of electron spin resonance for the dating of quartz.* Unpublished PhD thesis, University of New South Wales, ADFA, Canberra, Australia.

Riva, R. E., Bamber, M. J. L., Lavallée, D. A., *et al.* (2010). Sea level fingerprint of continental water and ice mass change from GRACE. *Geophysical Research Letters*, **37**, L19605.

Roberts, A. & Mountford, C. P. (1969). *The dawn of time: Australian Aboriginal myths in paintings.* Rigby Ltd, Adelaide, 79 pp.

Roberts, A. P., Tauxe, L. & Heslop, D. (2013). Magnetic paleointensity stratigraphy and high-resolution Quaternary geochronology: successions and future challenges. *Quaternary Science Reviews*, **61**, 1–16.

Roberts, D. L., Bateman, M. D., Murray-Wallace, C. V., *et al.* (2009). West coast dune plumes: climate driven contrasts in dunefield morphogenesis along the western and southern South African coasts. *Palaeogeography, Palaeoclimatology, Palaeoecology*, **271**, 24–38.

Roberts, H. H. (1998). Delta switching: early reponses to the Atchafalaya River diversion. *Journal of Coastal Research*, **14**, 882–899.

Rodolfo, K. S. & Siringan, F. (2006). Global sea-level rise is recognised, but flooding from anthropogenic land subsidence is ignored around northern Manila Bay, Philippines. *Disasters*, **30**, 118–139.

Rodriguez, A. B., Anderson, J. B., Banfield, L. A., *et al.* (2000). Identification of a − 15 m middle Wisconsin shoreline on the Texas inner continental shelf. *Palaeogeography, Palaeoclimatology, Palaeoecology*, **158**, 25–43.

Rodriguez, A. B., Anderson, J. B., Siringan, F. P., *et al.* (2004). Holocene evolution of the east Texas coast and inner continental shelf: along-strike variability in coastal retreat rates. *Journal of Sedimentary Research*, **74**, 405–421.

Rogers, J. J. W. & Santosh, M. (2004). *Continents and supercontinents.* Oxford University Press, Oxford, 289 pp.

Rohling, E. J. (1994). Glacial conditions in the Red Sea. *Paleoceanography*, **9**, 653–660.

Rohling, E. J., Fenton, M., Jorissen, F. J., *et al.* (1998). Magnitudes of sea-level lowstands of the past 500,000 years. *Nature*, **394**, 162–165.

Rohling, E. J., Grant, K., Hemleben, C. H., *et al.* (2008). High rates of sea-level rise during the last interglacial period. *Nature Geoscience*, **1**, 38–42.

Rohling, E. J., Grant, K., Bolshaw, M., *et al.* (2009). Antarctic temperature and global sea level closely coupled over the past five glacial cycles. *Nature Geoscience*, **2**, 500–504.

Rohling, E. J., Braun, K., Grant, K., *et al.* (2010). Comparison between Holocene and Marine Isotope Stage-11 sea-level histories. *Earth and Planetary Science Letters*, **291**, 97–105.

Rohling, E. J., Medina-Elizalde, M., Shepherd, J. G., *et al.* (2012). Sea surface and high-latitude temperature sensitivity to radiative forcing of climate over several glacial cycles. *Journal of Climate*, **25**, 1635–1656.

Rotzoll, K. & Fletcher, C. H. (2012). Assessment of groundwater inundation as a consequence of sea-level rise. *Nature Climate Change*, doi: 10.1038/NCLIMATE1725.

Roveri, M. & Manzi, V. (2006). The Messinian salinity crisis: looking for a new paradigm? *Palaeogeography, Palaeoclimatology, Palaeoecology*, **238**, 386–398.

Roy, P. S. (1984). New South Wales estuaries: their origin and evolution. *In* B. G. Thom (Ed.), *Coastal geomorphology in Australia*, pp. 99–121. Academic Press, Sydney.

Roy, P. S. (1994). Holocene estuary evolution – stratigraphic studies from southeastern Australia. *In* R. W. Dalrymple, R. Boyd & B. A. Zaitlin (Eds), *Incised-valley systems: origin and sedimentary sequences.* SEPM Special Publication No. 51, pp. 241–263.

Roy, P. S., Cowell, P. J., Ferland, M. A., *et al.* (1994). Wave-dominated coasts. *In* R. W. G. Carter & C. D. Woodroffe (Eds), *Coastal evolution: late Quaternary shoreline morphodynamics*, pp. 121–186. Cambridge University Press, Cambridge.

Roy, P. S., Whitehouse, J., Cowell, P. J., *et al.* (2000). Mineral sands occurrences in the Murray Basin, Southeastern Australia. *Economic Geology*, **95**, 1107–1128.

Roy, P. S, Williams, R. J., Jones, A. R., *et al.* (2001). Structure and function of southeast Australian estuaries. *Estuarine Coastal and Shelf Science*, **53**, 351–384.

Ruddiman, W. F. (2003). The anthropogenic greenhouse era began thousands of years ago. *Climatic Change*, **61**, 261–293.

Ruddiman, W. F. (2006). Orbital changes and climate. *Quaternary Science Reviews*, **25**, 3092–3112.

Rudwick, M. J. S. (1978). Charles Lyell's dream of a statistical palaeontology. *Palaeontology*, **21**, 225–244.

Rudwick, M. J. S. (2005). *Bursting the limits of time: the reconstruction of geohistory in the age of revolution.* University of Chicago Press, Chicago, 708 pp.

Russell, R. J. & Howe, H. V. (1935). Cheniers of Southwestern Louisiana. *Geographical Review*, **25**, 449–461.

Rutter, N. W. & Blackwell, B. (1995). Amino acid racemization dating. *In* N. W. Rutter & N. R. Catto (Eds), *Dating methods for Quaternary deposits*, pp. 125–164. Geotext 2, Geological Association of Canada, St Johns.

Rutter, N. W. & Catto, N. R. (Eds) (1995). *Dating methods for Quaternary deposits.* Geotext 2, Geological Association of Canada, St Johns, 308 pp.

Ryan, W. & Pitman, W. (1998). *Noah's flood: the new scientific discoveries about the event that changed history.* Simon & Schuster, New York, 319 pp.

Ryan, W. B. F., Pitman, W. C., Major, C. O., *et al.* (1997). An abrupt drowning of the Black Sea Shelf. *Marine Geology*, **138**, 119–126.

Sahagian, D. L., Schwartz, F. W. & Jacobs, D. K. (1994). Direct anthropogenic contributions to sea level rise in the twentieth century. *Nature*, **367**, 54–57.

Salvador, A. S. (1994). *International stratigraphic guide*, 2nd edition. Geological Society of America, Boulder, Colorado, 214 pp.

Samoilov, I. V. (1956). *Die Flussmündungen.* H. Haack, Gotha, 647 pp.

Sandgren, B. E., Barnekow, P., Hannon, L., *et al.* (2005). Early Holocene history of the Baltic Sea, as reflected in coastal sediments in Blekinge, southeastern Sweden. *Quaternary International*, **130**, 111–139.

Sandiford, M. (2003). Neotectonics of southeastern Australia: linking the Quaternary faulting record with seismicity and in situ stress. *In* R. R. Hillis & R. D. Müller (Eds), *Evolution and dynamics of the Australian plate*, pp. 107–119. Geological Society of Australia Special Publication 22.

Sandiford, M. (2007). The tilting continent: a new constraint on the dynamic topographic field from Australia. *Earth and Planetary Science Letters*, **261**, 152–163.

Santamaría-Gómez, A., Gravelle, M., Collilieux, X., *et al.* (2012). Mitigating the effects of vertical land motion in tide gauge records using a state-of-the-art GPS velocity field. *Global and Planetary Change*, **98–99**, 6–17.

Sarnthein, M., Grootes, P. M., Kennett, J. P., *et al.* (2007). ^{14}C reservoir ages show deglacial changes in ocean currents. *In* A. Schmittner, J. Chiang & S. Hemming (Eds), *Ocean circulation: mechanisms and impacts*, pp. 175–197. Geophysical Monograph Series 173, American Geophysical Union, Washington, DC.

Sasaki, K., Omura, A., Murakami, K., *et al.* (2004). Interstadial coral reef terraces and relative sea-level changes during marine oxygen isotope stages 3–4, Kikai Island, central Ryukyus, Japan. *Quaternary International*, **120**, 51–64.

Sayles, R. W. (1931). Bermuda during the ice age. *Proceedings of the American Academy of Arts and Sciences*, **66**, 381–467.

Schaeffer, M., Hare, W., Rahmstorf, S., *et al.* (2012). Long-term sea-level rise implied by 1.5°C and 2°C warming levels. *Nature Climate Change*, **2**, 867–870.

Schellmann, G. & Radtke, U. (2004a). *The marine Quaternary of Barbados.* Kölner Geographische Arbeiten, Heft 81, Geographisches Institut der Universität zu Köln, 137 pp.

Schellmann, G. & Radtke, U. (2004b). A revised morpho- and chronostratigraphy of the Late and Middle Pleistocene coral reef terraces on Southern Barbados (West Indies). *Earth-Science Reviews*, **64**, 157–187.

Schellmann, G., Radtke, U., Scheffers, A., *et al.* (2004). ESR dating of coral reef terraces on Curaçao (Netherlands Antilles) with estimates of Younger Pleistocene sea level elevations. *Journal of Coastal Research*, **20**, 947–957.

Schilt, A., Baumgartner, M., Blunier, T., *et al.* (2010). Glacial–interglacial and millennial-scale variations in the atmospheric nitrous oxide concentration during the last 800,000 years. *Quaternary Science Reviews*, **29**, 182–192.

Schmidt, S., De Deckker, P., Etcheber, H., *et al.* (2010). Are the Murray Canyons offshore southern Australia still active for sediment transport? *In* P. Bishop & B. Pillans (Eds), *Australian landscapes*, Geological Society, London, Special Publications, **346**, 43–55.

Schofield, J. C. (1959). The geology and hydrogeology of Niue Island, South Pacific. *New Zealand Geological Survey Bulletin*, **62**, 1–28.

Schofield, J. C. (1960). Sea level fluctuations during the last 4,000 years as recorded by a chenier plain, Firth of Thames, New Zealand. *New Zealand Geology and Geophysics*, **3**, 467–485.

Schofield, J. C. (1970). Notes on late Quaternary sea levels, Fiji and Rarotonga. New Zealand. *New Zealand Geology and Geophysics*, **13**, 199–206.

Schofield, J. C. (1977a). Late Holocene sea level, Gilbert and Ellice Islands, west Central Pacific Ocean. *New Zealand Journal of Geology and Geophysics*, **20**, 503–529.

Schofield, J. C. (1977b). Effect of Late Holocene sea-level fall on atoll development. *New Zealand Journal of Geology and Geophysics*, **20**, 531–536.

Scholl, D. W. (1964). Recent sedimentary record in mangrove swamps and rise in sea level over the southwestern coast of Florida: Part 1. *Marine Geology*, **1**, 344–366.

Scholl, D. W. & Stuiver, H. (1967). Recent submergence of southern Florida: reply. *Geological Society of America Bulletin*, **78**, 1195–1198.

Scholl, D. W., Craighead, F. C. & Stuiver, M. (1969). Florida submergence curve revised: its relation to coastal sedimentation rates. *Science*, **163**, 562–564.

Schroeder, R. A. & Bada, J. L. (1976). A review of the geochemical applications of the amino acid racemization reaction. *Earth-Science Reviews*, **12**, 347–391.

Schumm, S. A. (1991). *To interpret the Earth: ten ways to be wrong.* Cambridge University Press, Cambridge.

Schumm, S. A. (1993). River response to baselevel change: implications for sequence stratigraphy. *Journal of Geology*, **101**, 279–294.

Scicchitano, G., Lo Presti, V., Spampinato, C. R., *et al.* (2011). Millstones as indicators of relative sea-level changes in northern Sicily and southern Calabria coastlines, Italy. *Quaternary International*, **232**, 92–104.

Scoffin, T. P. (1987). *An introduction to carbonate sediments and rocks.* Chapman & Hall, New York, 274 pp.

Scoffin, T. P. (1992). Taphonomy of coral reefs: a review. *Coral Reefs*, **11**, 57–77.

Scoffin, T. P. & Le Tissier, M. D. A. (1998). Late Holocene sea level and reef-flat progradation, Phuket, South Thailand. *Coral Reefs*, **17**, 273–276.

Scoffin, T. P. & Stoddart, D. R. (1978). The nature and significance of micro atolls. *Philosophical Transactions of the Royal Society, London*, **B284**, 99–122.

Scoffin, T. P., Stoddart, D. R., Tudhope, A. W., *et al.* (1985). Exposed limestone of Suwarrow Atoll. *Proceedings of the 5th International Coral Reef Congress*, **3**, 137–140.

Scott, D. B. & Medioli, F. S. (1978). Vertical zonations of marsh foraminifera as accurate indicators of former sea-levels. *Nature*, **272**, 528–531.

Scott, D. B. & Medioli, F. S. (1986). Foraminifera as sea-level indicators. *In* O. van de

Plassche (Ed.), *Sea-level research: a manual for the collection and evaluation of data*, pp. 435–456. Geo-Books, Norwich.

Scott, G. A. J. & Rotondo, G. M. (1983). A model to explain the differences between Pacific Plate island atoll types. *Coral Reefs*, **1**, 139–150.

Scourse, J. D., Austin, W. E. N., Sejrup, H. P., *et al*. (1999). Foraminiferal isoleucine epimerization determinations from the Nar Valley Clay, Norfolk, UK: implications for Quaternary correlations in the southern North Sea basin. *Geological Magazine*, **136**, 543–560.

Semeniuk, V. (1980). Quaternary stratigraphy of the tidal flats, King Sound, Western Australia. *Journal of the Royal Society of Western Australia*, **63**, 65–78.

Semeniuk, V. (1981). Sedimentology and the stratigraphic sequence of a tropical tidal flat, north-western Australia. *Sedimentary Geology*, **29**, 195–221.

Semeniuk, V. (1982). Geomorphology and Holocene history of the tidal flats, King Sound, north-western Australia. *Journal of the Royal Society of Western Australia*, **65**, 47–68.

Semeniuk, V. (1983). Mangrove distribution in northwestern Australia in relationship to regional and local freshwater seepage. *Vegetatio*, **53**, 11–31.

Semeniuk, V. (2008). Holocene sedimentation, stratigraphy, biostratigraphy, and history of the Canning Coast, north-western Australia. *Journal of the Royal Society of Western Australia*, Supplement to volume **91**, 53–148.

Semeniuk, V. & Searle, D. J. (1986). Variability of Holocene sea level history along the southwestern coast of Australia: evidence for the effect of significant local tectonism. *Marine Geology*, **72**, 47–58.

Sewell, R. B. S. (1935). Studies on coral and coral formations in Indian waters. *Memoirs of the Royal Asiatic Society of Bengal*, **9**, 461–540.

Sewell, R. B. S. (1936). An account of Addu Atoll. *John Murray Expedition to the Indian Ocean, 1933–1934, Scientific Reports*, **1**, 63–93.

Shackleton, N. J. (1969). The last interglacial in the marine and terrestrial records. *Proceedings of the Royal Society of London, Series B*. **174**, 135–154.

Shackleton, N. J. (1987). Oxygen isotopes, ice volume and sea level. *Quaternary Science Reviews*, **6**, 183–190.

Shackleton, N. J. (2000). The 100,000-year ice-age cycle identified and found to lag temperature, carbon dioxide, and orbital eccentricity. *Science*, **289**, 1897–1902.

Shackleton, N. J. (2006). Formal Quaternary stratigraphy – What do we expect and need? *Quaternary Science Reviews*, **25**, 3458–3461.

Shackleton, N. J. & Matthews, R. K. (1977). Oxygen isotope stratigraphy of Late Pleistocene coral terraces in Barbados. *Nature*, **268**, 618–620.

Shackleton, N. J. & Opdyke, N. D. (1973). Oxygen isotope and palaeomagnetic stratigraphy of equatorial Pacific core V28–238: Oxygen isotope temperatures and ice volumes on a 10^5-year and 10^6-year scale. *Quaternary Research*, **3**, 39–55.

Shackleton, N. J., Berger, A., & Peltier, W. R. (1990). An alternative astronomical calibration of the lower Pleistocene timescale based on ODP Site 677. *Transactions of the Royal Society of Edinburgh; Earth Sciences*, **81**, 251–261.

Shackleton, N. J., Chapman, M., Sánchez-Goñi, M. F., *et al*. (2002). The classic marine isotope substage 5e. *Quaternary Research*, **58**, 14–16.

Shackleton, N. J., Sánchez-Goñi, M. F., Pailler, D., *et al*. (2003). Marine Isotope Substage 5e and the Eemian Interglacial. *Global and Planetary Change*, **36**, 151–155.

Sheard, M. J. (1990). A guide to Quaternary volcanoes in the lower south-east of South Australia. *Mines and Energy Review*, **157**, 40–50.

Shennan, I. (1986a). Flandrian sea-level changes in the Fenland I. The geographical setting and evidence of relative sea-level changes. *Journal of Quaternary Science* **1**, 119–154.

Shennan, I. (1986b). Flandrian sea-level changes in the Fenland II. Tendencies of sea-level movement, altitudinal changes and local and regional factors. *Journal of Quaternary Science*, **1**: 155–179.

Shennan, I. (1989). Holocene crustal movements and sea-level changes in Great Britain. *Journal of Quaternary Science*, **4**, 77–89.

Shennan, I. (1999). Global meltwater discharge and the deglacial sea-level record from northwest Scotland. *Journal of Quaternary Science*, **14**, 715–719.

Shennan, I. (2007). Sea level studies: overview. *In* E. S. Alias (Ed.), *Encyclopedia of Quaternary science*, Vol. 4, pp. 2967–2974. Elsevier, Amsterdam.

Shennan, I. (2009). Late Quaternary sea-level changes and palaeoseismology of the Bering Glacier region, Alaska. *Quaternary Science Reviews*, **28**, 1762–1773.

Shennan, I. & Horton, B. P. (2002a). Relative sea-level changes and crustal movements of the UK. *Journal of Quaternary Science*, **16**, 511–526.

Shennan, I. & Horton, B. P. (2002b). Holocene land- and sea-level changes in Great Britain. *Journal of Quaternary Science*, **17**, 511–526.

Shennan, I. & Milne, G. (2003). Sea-level observations around the Last Glacial Maximum from the Bonaparte Gulf, NW Australia. *Quaternary Science Reviews*, **22**, 1543–1547.

Shennan, I. & Woodworth, P. L. (1992). A comparison of late Holocene and twentieth-century sea-level trends from the UK and North Sea region. *Geophysical Journal International*, **109**, 96–105.

Shennan, I., Tooley, M. J., Davis, M. J., *et al.* (1983). Analysis and interpretation of Holocene sea-level data. *Nature*, **302**: 404–406.

Shennan, I., Innes, J. B., Long, A. J., *et al.* (1995). Late Devensian and Holocene relative sea-level changes in northwestern Scotland: new data to test existing models. *Quaternary International*, **26**, 97–123.

Shennan, I., Tooley, M., Green, F., *et al.* (1999). Sea level, climate change and coastal evolution in Morar, northwest Scotland. *Geologie en Mijnbouw*, **77**, 247–262.

Shennan, I., Lambeck, K., Horton, B., *et al.* (2000a). Late Devensian and Holocene records of relative sea-level changes in northwest Scotland and their implications for glacio-hydro-isostatic modelling. *Quaternary Science Reviews*, **19**, 1103–1135.

Shennan, I, Lambeck, K., Flather, R. *et al.* (2000b). Modelling western North Sea palaeogeographies and tidal changes during the Holocene. *In* I. Shennan & J. Andrews (Eds), *Holocene land–ocean interaction and environmental change around the north sea*. Geological Society, London, Special Publications, **166**, 299–319.

Shennan, I., Bradley, S., Milne, G., *et al.* (2006a). Relative sea-level changes, glacial isostatic modelling and ice sheet reconstructions from the British Isles since the last Glacial Maximum. *Journal of Quaternary Science*, **21**, 585–599.

Shennan, I., Hamilton, S., Hillier, C., *et al.* (2006b). Relative sea-level observations in western Scotland since the Last Glacial Maximum for testing models of glacial isostatic land movements and ice-sheet reconstructions. *Journal of Quaternary Science*, **21**, 601–613.

Shennan, I., Milne, G. & Bradley, S. L. (2009). Late Holocene relative land- and sea-level changes: providing information for stakeholders. *GSA Today*, **19**, 52–53.

Shennan, I., Milne, G. & Bradley, S. (2012). Late Holocene vertical land motion and relative sea-level changes: lessons from the British Isles. *Journal of Quaternary Science*, **27**, 64–70.

Shennan, I., Long, A. J. & Horton, B. P. (2014). *Handbook of sea-level Research*. Wiley-Blackwell, Chichester.

Shepard, F. P. (1960). Rise of sea level along northwest Gulf of Mexico. *In* F. P. Shepard, F. B. Phleger & Tj. H. Van Andel (Eds), *Recent sediments, northwest Gulf of Mexico*, pp. 338–381. American Association of Petroleum Geologists, Tulsa.

Shepard, F. P. (1961). Sea level rise during the past 20000 years. *Zeitschrift für Geomorphologie* Suppl.-Bd., **3**, 30–35.

Shepard, F. P. (1963). Thirty-five thousand years of sea level. *In* T. Clements (Ed.), *Essays in marine geology in honour of K.O. Emery*, pp. 1–10. University of Southern California Press, Los Angeles.

Shepard, F. P., Curray, J. R., Newman, W. A., *et al.* (1967). Holocene changes in sea level: evidence in Micronesia. *Science*, **157**, 542–544.

Shepherd, A., Wingham, D., Wallis, D., *et al.* (2010). Recent loss of floating ice and the consequent sea level contribution. *Geophysical Research Letters*, **37**, L13503.

Shepherd, M. J. (1990). Relict and contemporary foredunes as indicators of coastal processes. *In* G. Brierley & J. Chappell (Eds), *Applied Quaternary studies*. Papers presented at a workshop at the Australian National University, 2–3 July, 1990, pp. 17–24.

Shepherd, S. A. (1983). Benthic communities of upper Spencer Gulf, South Australia. *Royal Society of South Australia, Transactions*, **107**, 69–85.

Sherman, C. E., Fletcher, C. H. & Rubin, K. H. (1999). Marine and meteoric diagenesis of Pleistocene carbonates from a nearshore submarine terrace, Oahu, Hawaii. *Journal of Sedimentary Research*, **69**, 1083–1097.

Sherwood, J., Barbetti, M., Ditchburn, R., et al. (1994). A comparative study of Quaternary dating techniques applied to sedimentary deposits in southwest Victoria, Australia. *Quaternary Science Reviews*, **13**, 95–110.

Short, A. D. (1989). Chenier research on the Australian coast. *Marine Geology*, **90**, 345–351.

Short, S. A., Lowson, R. T., Ellis, J., et al. (1989). Thorium–uranium disequilibrium dating of Late Quaternary ferruginous concretions and rinds. *Geochimica et Cosmochimica Acta*, **53**, 1379–1389.

Siddall, M. & Milne, G. A. (2012). Understanding sea-level change is impossible without both insights from paleo studies and working across disciplines. *Earth and Planetary Science Letters*, **315–316**, 2–3.

Siddall, M., Rohling, E. J., Almogi-Labin, A., et al. (2003). Sea-level fluctuations during the last glacial cycle. *Nature*, **423**, 853–858.

Siddall, M., Smeed, D. A., Hemleben, C., et al. (2004). Understanding the Red Sea response to sea level. *Earth and Planetary Science Letters*, **225**, 421–434.

Siddall, M., Chappell, J. & Potter, E.-K. (2006). Eustatic sea level during past interglacials. *In* F. Sirocko, T. Litt, M. Claussen & M.-F. Sanchez-Goni (Eds), *The climate of past interglacials*, pp. 75–92. Elsevier, Amsterdam.

Siddall, M., Stocker, T. F. & Clark, P. U. (2009). Constraints on future sea-level rise from past sea-level change. *Nature Geoscience*, **2**, 571–575.

Siddiquie, H. N. (1980). The ages of the storm beaches of the Lakshadweep (Laccadives). *Marine Geology*, **38**, M11–M20.

Siegert, M. J. & Dowdeswell, J. A. (2004). Numerical reconstructions of the Eurasia Ice Sheet and climate during the Late Weichselian. *Quaternary Science Reviews*, **23**, 1273–1283.

Siegrist, H. G. & Randall, R. H. (1989). Sampling implications of stable isotope variation in Holocene reef corals from the Mariana Islands. *Micronesica*, **22**, 173–189.

Símonarson, L. A. & Leifsdóttir, Ó. E. (2002). Late-Holocene sea-level changes in south and southwest Iceland reconstructed from littoral molluscan stratigraphy. *The Holocene*, **12**, 149–158.

Simpson, G. G. (1980). *Why and how – some problems and methods in historical biology*. Pergamon Press, Oxford, 263 pp.

Sirocko, F. (1996). Classics in physical geography revisited. *Progress in Physical Geography*, **20**, 447–452.

Sissons, J. B. (1963). Scottish raised shoreline heights with particular reference to the Forth valley. *Geografiska Annaler*, **45**, 180–185.

Sissons, J. B. (1972). Dislocation and non-uniform uplift of raised shorelines in the western part of the Forth valley. *Transactions of the Institute of British Geographers*, **55**, 149–159.

Sissons, J. B. (1974). Late-glacial marine erosion in Scotland. *Boreas*, **3**, 41–48.

Sissons, J. B., Cullingford, R. A. & Smith, D. E. (1966). Late-glacial and post-glacial shorelines in south-east Scotland. *Transactions of the Institute of British Geographers*, **39**, 9–18.

Sivan, D., Lambeck, K., Toueg, R., et al. (2004). Ancient coastal wells of Caesarea Maritima, Israel, an indicator for sea level changes during the last 2000 years. *Earth and Planetary Science Letters*, **222**, 315–330.

Slangen, A. B. A., Katsman, C. A., van de Wal, R. S. W., et al. (2011). Towards regional projections of twenty-first century sea-level change based on IPCC SRES scenarios. *Climate Dynamics*, **38**, 1191–1209.

Sloss, C. R., Murray-Wallace, C. V. & Jones, B. G. (2004). Aspartic acid racemisation dating of mid-Holocene to recent estuarine sedimentation in New South Wales, Australia: a pilot study. *Marine Geology*, **212**, 45–59.

Sloss, C. R., Murray-Wallace, C. V. & Jones, B. G. (2007). Holocene sea-level change on the southeast coast of Australia: a review. *The Holocene*, **17**, 999–1014.

Sloss, L. L. (1963). Sequences in the cratonic interior of North America. *Geological Society of America Bulletin*, **74**, 93–114.

Smart, J. (1976). The nature and origin of beach ridges, western Cape York Peninsula, Queensland. *BMR Journal of Australian Geology and Geophysics*, **1**, 211–218.

Smart, J. (1977). Late Quaternary sea-level changes, Gulf of Carpentaria, Australia. *Geology*, **5**, 755–759.

Smith, D. E., Sissons, J. B. & Cullingford, R. A. (1969). Isobases for the Main Perth raised shoreline in South East Scotland as determined by trend-surface analysis. *Institute of British Geographers Transactions*, **46**, 45–52.

Smith, D. E., Shi, S., Cullingford, R. A., *et al.* (2004). The Holocene Storegga Slide tsunami in the United Kingdom. *Quaternary Science Reviews*, **23**, 2291–2321.

Smith, D. E., Davies, M. H., Brooks, C. L., *et al.* (2010). Holocene relative sea levels and related prehistoric activity in the Forth lowland, Scotland, United Kingdom. *Quaternary Science Reviews*, **29**, 2382–2410.

Smith, D. E., Harrison, S., Firth, C. R., *et al.* (2011). The early Holocene sea level rise. *Quaternary Science Reviews*, **30**, 1846–1860.

Smith, D. E., Hunt, N., Firth, C. R., *et al.* (2012). Patterns of Holocene relative sea level change in the North of Britain and Ireland. *Quaternary Science Reviews*, doi:10.1016/j.quascirev.2012.02.007.

Smith, J. (1838). On the last changes in the relative levels of the land and sea in the British Islands. *Memoirs of the Wernian Natural History Society*, **8**, 49–88.

Smithers, S. G. (2011). Sea-level indicators. *In* D. Hopley (Ed.), *Encyclopedia of Modern coral reefs*, pp. 978–991. Springer, Dordrecht.

Smithers, S. G. & Woodroffe, C. D. (2000). Microatolls as sea-level indicators on a mid-ocean atoll. *Marine Geology*, **168**, 61–78.

Smithers, S. G. & Woodroffe, C. D. (2001). Coral microatolls and 20th century sea level in the eastern Indian Ocean. *Earth and Planetary Science Letters*, **191**, 173–184.

Southon, J., Kashgarian, M., Fontugne, M., *et al.* (2002). Marine reservoir corrections for the Indian Ocean and Southeast Asia. *Radiocarbon*, **44**, 167–180.

Spackman, W., Dolsen, C. P. & Riegel, W. (1966). Phytogenic organic sediments and sedimentary environments in the everglades – mangrove complex. *Paleontographica Abteilung*, **B117**, 135–152.

Spada, G. & Galassi, G. (2012). New estimates of secular sea level rise from tide gauge data and GIA modelling. *Geophysical Journal International*, **191**, 1067–1094.

Spada, G., Bamber, J. L. & Hurkmans, R. T. W. L. (2012). The gravitationally consistent sea-level fingerprint of future terrestrial ice loss. *Geophysical Research Letters*, **40**, 1–5.

Spencer, T. (1988). Limestone coastal morphology: the biological contribution. *Progress in Physical Geography*, **12**, 66–101.

Spencer, T., Stoddart, D. R. & Woodroffe, C. D. (1987). Island uplift lithospheric flexure: observations and cautions from the South Pacific. *Zeitschrift für Geomorphologie, N.F.* Suppl.-Bd., **63**, 87–102.

Spencer, T., Tudhope, A. W., French, J. R., *et al.* (1997). Reconstructing sea level change from coral microatolls, Tongareva (Penrhyn) Atoll, northern Cook Islands. *Proceedings of the 8th International Coral Reef Symposium*, **1**, 489–494.

Sprigg, R. C. (1947). Submarine canyons of the New Guinea and South Australian Coasts. *Royal Society of South Australia, Transactions*, **71**, 296–310.

Sprigg, R. C. (1948). Stranded Pleistocene sea beaches of South Australia and aspects of the theories of Milankovitch and Zeuner. Report of the 18th session of the Geological Congress, *Great Britain*, pp. 226–237.

Sprigg, R. C. (1952). The geology of the south-east province, South Australia, with special reference to Quaternary coast-line migrations and modern beach developments. *Geological Survey of South Australia, Bulletin* **29**, 120 pp.

Stanley, D. J. & Warne, A. G. (1993). Sea level and initiation of Predynastic culture in the Nile Delta. *Nature*, **363**, 435–438.

Stanley, D. J. & Warne, A. G. (1994). Worldwide initiation of Holocene marine deltas by deceleration of sea-level rise. *Science*, **265**, 228–231.

Stearns, C. E. (1984). Uranium-series dating and the history of sea level. *In* W. C. Mahaney (Ed.), *Quaternary dating methods*, pp. 53–66. Elsevier, Amsterdam.

Stearns, H. T. (1935). Shore benches on the island of Oahu, Hawaii. *Geological Society of America Bulletin*, **46**, 1467–1482.

Stearns, H. T. (1974). Submerged shoreline and shelves in the Hawaii Islands and a revision of some of the eustatic emerged shorelines. *Geological Society of America Bulletin*, **85**, 795–804.

Stearns, H. T. (1978). *Quaternary Shorelines in the Hawaiian Islands*. Bernice P. Bishop Museum Bulletin 237. Bishop Museum Press, Honolulu, Hawaii, 57 pp.

Stearns, H. T. (1985). *Geology of the State of Hawaii*, 2nd edition. Pacific Books, Palo Alto, 325 pp.

Steffen, K., Thomas, R. H., Rignot, E., *et al.* (2010). Cryospheric contributions to sea-level rise and variability. *In* J. A. Church, P. L. Woodworth, T. Aarup & W. S. Wilson (Eds), *Understanding sea-level rise and variability*, pp. 177–225. Wiley-Blackwell, Oxford.

Stein, M., Wasserburg, G. J., Lajoie, K. R., *et al.* (1991). U-series ages of solitary corals from the California coast by mass spectrometry. *Geochimica et Cosmochimica Acta*, **55**, 3709–3722.

Stein, M., Wasserburg, G. J., Aharon, P., *et al.* (1993). TIMS U-series dating and stable isotopes of the last interglacial event in Papua New Guinea. *Geochimica et Cosmochimica Acta*, **57**, 2541–2554.

Steinen, R. P., Harrison, R. S. & Matthews, R. K. (1973). Eustatic low stand of sea level between 125,000 and 105,000 BP: evidence from the subsurface of Barbados, West Indies. *Geological Society of America Bulletin*, **84**, 63–70.

Stephenson, W. J. & Kirk, R. M. (2000). Development of shore platforms on Kaikoura Peninsula, South Island, New Zealand: II: the role of subaerial weathering. *Geomorphology*, **32**, 43–56.

Stephenson, W. J., Kirk, R. M., Kennedy, D. M., *et al.* (2012). Long term shore platform surface lowering rates: revisiting Gill and Lang after 32 years. *Marine Geology*, **299–302**, 90–95.

St-Hilaire-Gravel, D., Bell, T. J. & Forbes, D. L. (2010). Raised gravel beaches as proxy indicators of past sea-ice and wave conditions, Lowther Island, Canadian Arctic Archipelago. *Arctic*, **63**, 213–226.

Stewart, I. & Morhange, C. (2009). Coastal geomorphology and sea-level change. *In* J. C. Woodward (Ed.), *The physical geography of the Mediterranean*, pp. 385–413. Oxford University Press, Oxford.

Stewart, I. & Vita-Finzi, C. (Eds) (1998). *Coastal tectonics*. Geological Society, London, Special Publications, **146**, 378 pp.

Stirling, C. H., Esat, T. M., McCulloch, M. T., *et al.* (1995). High precision U-series dating of corals from Western Australia and implications for the timing and duration of the Last Interglacial. *Earth and Planetary Science Letters*, **135**, 115–130.

Stirling, C. H., Esat, T. M., Lambeck, K., *et al.* (1998). Timing and duration of the Last Interglacial: evidence for a restricted interval of widespread coral reef growth. *Earth and Planetary Science Letters*, **160**, 745–762.

Stirling, C. H., Esat, T. M., Lambeck, K., *et al.* (2001). Orbital forcing of the marine isotope stage 9 interglacial. *Science*, **291**, 290–293.

Stiros, S. C. & Pirazzoli, P. A. (1995). Palaeoseismic studies in Greece: a review. *Quaternary International*, **25**, 57–63.

Stiros, S. C. & Pirazzoli, P. A. (1998). *Late Quaternary coastal changes in the Gulf of Corinth – tectonics, earthquakes and aechaeology*. Guidebook for the Gulf of Corinth Field Trip, 14–16 September 1998, UNESCO-IUGS IGCP-367, Geodesy Laboratory, Department of Civil Engineering, Patras University, 49 pp.

Stiros, S. C., Pirazzoli, P. A., Laborel, J., *et al.* (1994). The 1953 earthquake in

Cephalonia (Western Hellenic Arc): coastal uplift and halotectonic faulting. *Geophysics Journal International*, **117**, 834–849.

Stoddart, D. R. (1971). Geomorphology of Diego Garcia Atoll. *Atoll Research Bulletin*, **149**, 7–26.

Stoddart, D. R. (1990). Coral reefs and islands and predicted sea-level rise. *Progress in Physical Geography*, **14**, 521–536.

Stoddart, D. R. & Scoffin, T. P. (1979). Microatolls: review of form, origin and terminology. *Atoll Research Bulletin*, **224**, 1–17.

Stoddart, D. R., Davies, P. S. & Keith, A. (1966). Geomorphology of Addu Atoll. *Atoll Research Bulletin*, **116**, 13–41.

Stoddart, D. R., McLean, R. F. & Hopley, D. (1978). Geomorphology of reef islands, northern Great Barrier Reef. *Philosophical Transactions of the Royal Society London*, **B284**, 39–61.

Stokes, C. R. & Clark, C. D. (2001). Palaeo-ice streams. *Quaternary Science Reviews*, **20**, 1437–1457.

Stone, G. W. & McBride, R. A. (1998). Louisiana barrier islands and their importance in wetland protection: forecasting shoreline change and subsequent response of wave climate. *Journal of Coastal Research*, **14**, 900–915.

Stone, J. O. (2000). Air pressure and cosmogenic isotope production. *Journal of Geophysical Research*, **105**: 23753–23759.

Stone, J., Lambeck, K., Fifield, L. K., *et al.* (1996). A late glacial age for the Main Rock Platform, western Scotland. *Geology*, **24**, 707–710.

Stroeve, J., Holland, M. M., Meier, W., *et al.* (2007). Arctic sea ice decline: faster than forecast. *Geophysical Research Letters*, **34**, L09501.

Stuiver, M. (1982). A high-precision calibration of the AD radiocarbon timescale. *Radiocarbon*, **24**, 1–26.

Stuiver, M. & Polach, H. (1977). Reporting of ^{14}C data. *Radiocarbon*, **19**, 355–363.

Stuiver, M., Reimer, P. J., Bard, E., *et al.* (1998). INTCAL98 radiocarbon age calibration, 24,000–0 cal BP. *Radiocarbon*, **40**, 1041–1084.

Suess, E. (1888). *Das Antlitz der Erde*. F. Tempsky, Vienna, 776 pp.

Summerfield, M. (1991). *Global geomorphology*. Longman, London, 537 pp.

Summerfield, M. A. (Ed.) (2000). *Geomorphology and global tectonics*. Wiley, Chichester, 367 pp.

Sunamura, T. (1992). *Geomorphology of rocky coasts*. John Wiley & Sons, Chichester, 302 pp.

Suter, J. R. (1994). Deltaic coasts. *In* R. W. G. Carter & C. D. Woodroffe (Eds), *Coastal evolution: late Quaternary shoreline morphodynamics*, pp. 87–120. Cambridge University Press, Cambridge.

Suter, J. R. & Berryhill, H. L. (1985). Late Quaternary shelf-margin deltas, northwest Gulf of Mexico. *American Association of Petroleum Geologists Bulletin*, **69**, 77–91.

Sutherland, F. L. & Kershaw, R. C. (1971). The Cainozoic geology of Flinders Island, Bass Strait. *Papers and Proceedings of the Royal Society of Tasmania*, **105**, 151–175.

Suzuki, T. & Ishii, M. (2011). Regional distribution of sea level changes resulting from enhanced greenhouse warming in the Model for Interdisciplinary Research on Climate version 3.2. *Geophysical Research Letters*, **38**: L02601.

Svendsen, J. I., Alexanderson, H., Astakhov, V. I., *et al.* (2004). Late Quaternary ice sheet history of northern Eurasia. *Quaternary Science Reviews*, **23**, 1229–1271.

Svensson, A., Anderson, K. K., Bigler, M., *et al.* (2008). A 60 000 year Greenland stratigraphic ice core chronology. *Climate of the Past*, **4**, 47–57.

Syvitski, J. P. M. & Kettner, A. (2011). Sediment flux and the Anthropocene. *Philosophical Transactions of the Royal Society A – Mathematical, Physical and Engineering Sciences*, **369**, 957–975.

Syvitski, J. P. M., Kettner, A. J., Overeem, I., *et al.* (2009). Sinking deltas due to human activites. *Nature Geoscience*, **2**, 681–686.

Szabo, B. J. (1979). Th, Pa, and open system dating of fossil corals and shells. *Journal of Geophysical Research*, **87**, 4927–4929.

Szabo, B. J. (1985). Uranium-series dating of fossil corals from marine sediments of southeastern United States Atlantic Coastal Plain. *Geological Society of America Bulletin*, **96**, 398–406.

Szabo, B. J. & Rosholt, J. N. (1969). Uranium-series dating of Pleistocene molluscan shells from southern California – an open system model. *Journal of Geophysical Research*, **74**, 3253–3260.

Szabo, B. J., Ward, W. C., Weidie, A. E., *et al.* (1978). Age and magnitude of the late Pleistocene sea-level rise on the eastern Yucatan Peninsula. *Geology*, **6**, 713–715.

Szabo, B. J., Tracey, J. I. & Goter, E. R. (1985). Ages of subsurface stratigraphic intervals in the Quaternary of Eniwetak Atoll, Marshall Islands. *Quaternary Research*, **23**, 54–61.

Szabo, B. J., Ludwig, K. R., Muhs, D. R., *et al.* (1994). Thorium-230 ages of corals and duration of the last interglacial sea-level high stand on Oahu, Hawaii. *Science*, **266**, 93–96.

Tamisiea, M. E. & Mitrovica, J. X. (2011). The moving boundaries of sea level change understanding of the origins of geographic variability. *Oceanography*, **24**, 24–39.

Tamura, T. (2012). Beach ridges and prograded beach deposits as palaeoenvironment records. *Earth-Science Reviews*, **114**, 279–297.

Tamura, T., Saito, Y., Sieng, S., *et al.* (2009). Initiation of the Mekong River delta at 8 ka: evidence from the sedimentary succession in the Cambodian lowland. *Quaternary Science Reviews*, **28**, 327–344.

Tanabe, S., Ta, T. K. O., Nguyen, V. L., *et al.* (2003). Delta evolution model inferred from the Mekong Delta, southern Vietnam. *In* F. H. Sidi, D. Nummedal, P. Imbert, H. Darman & H. W. Postamentier (Eds), *Tropical deltas of Southeast Asia – sedimentology, stratigraphy, and petroleum geology*. SEPM Special Publication no. 76, pp. 175–188.

Tatumi, S. H., Kowata, E. A., Gozzi, G., *et al.* (2003). Optical dating results of beachrock, aeolian dunes and sediments applied to sea-level changes study. *Journal of Luminescence*, **102–103**, 562–565.

Taylor, F. W. & Bloom, A. L. (1977). Coral reefs on tectonic blocks, Tonga Island arc. *Proceedings of the Third International Coral Reef Symposium*, pp. 275–281. Rosenstiel School of Marine and Atmospheric Science, University of Miami, Miami, Florida.

Taylor, M. & Stone, G. W. (1996). Beach ridges: a review. *Journal of Coastal Research*, **12**, 612–621.

Taylor, R. E. (1987). *Radiocarbon dating – an archaeological perspective*. Academic Press, London, 212 pp.

Teichert, C. (1950). Late Quaternary sea-level changes at Rottnest Island, Western Australia. *Proceedings of the Royal Society of Victoria*, **59**, 63–79.

Ters, M. (1973). Les variations du niveau marin depuis 10,000 ans le long du littoral Atlantique Français. Le Quaternaire: geodynamique, stratigraphie et environment, tavaux Français récents. *9th Congrès International de l'INQUA, Christchurch, New Zealand*, pp. 114–135.

Thom, B. G. (1973). The dilemma of high interstadial sea levels during the last glaciation. *Progress in Physical Geography*, **5**, 170–246.

Thom, B. G. (1982). Mangrove ecology – a geomorphological perspective. *In* B. F. Clough (Ed.), *Mangrove ecosystems in Australia: structure, function and management*, pp. 3–17. Australian National University Press, Canberra.

Thom, B. G. (1984). Transgressive and regressive stratigraphies of coastal sand barriers in eastern Australia. *Marine Geology*, **56**, 137–158.

Thom, B. G. & Chappell, J. (1975). Holocene sea levels relative to Australia. *Search*, **6**, 90–93.

Thom, B. G. & Murray-Wallace, C. V. (1988). Last interglacial (Stage 5e) estuarine sediments at Largs, New South Wales. *Australian Journal of Earth Sciences*, **35**, 571–574.

Thom, B. G. & Roy, P. (1983). Sea level change in New South Wales over the past 15000 years. *In* D. Hopley (Ed.), *Australian sea levels in the last 15 000 years: a review*, pp. 64–84. Monograph Series Occasional Paper 3. Geography Department, James Cook University, Townsville.

Thom, B. G. & Roy, P. S. (1985). Relative sea levels and coastal sedimentation in southeast Australia in the Holocene. *Journal of Sedimentary Petrology*, **55**, 257–264.

Thom, B. G., Hails, J. R. & Martin, A. R. H. (1969). Radiocarbon evidence against higher postglacial sea levels in eastern Australia. *Marine Geology*, **7**, 161–168.

Thom, B. G., Hails, J. R., Martin, A. R. H., et al. (1972). Post-glacial sea levels in eastern Australia – a reply. *Marine Geology*, **12**, 233–242.

Thom, B. G., Wright, L. D. & Coleman, J. M. (1975). Mangrove ecology and deltaic–estuarine geomorphology, Cambridge Gulf-Ord River, Western Australia. *Journal of Ecology*, **63**, 203–222.

Thom, B. G., Orme, G. R. & Polach, H. A. (1978). Drilling investigations of Bewick and Stapleton Islands. *Philosophical Transactions of the Royal Society, London A*, **291**, 37–54.

Thom, B. G., Bowman, G. M., Gillespie, R., et al. (1981). *Radiocarbon dating of Holocene beach ridge sequences in south-east Australia*. Monograph No. 11, Department of Geography, Royal Military College, University of New South Wales, Duntroon, 36 pp.

Thom, B. G., Shepherd, M., Ly, C. K., et al. (1992). *Coastal geomorphology and Quaternary geology of the Port Stephens – Myall Lakes Area*. ANU Monograph No. 6. Department of Biogeography and Geomorphology, Australian National University, Canberra, 407 pp.

Thomas, R., Krabill, F. E., Manizade, W., et al. (2006). Progressive increase in ice loss from Greenland. *Geophysical Research Letters*, **33**, L10503.

Thommeret, J. & Thommeret, Y. (1978). ^{14}C datings of some Holocene sea levels on the north coast of the island of Java (Indonesia). *Modern Quaternary Research in Southeast Asia*, **4**, 51–56.

Thompson, W. G. & Goldstein, S. L. (2005). Open-system coral ages reveal persistent suborbital sea-level cycles. *Science*, **308**, 401–404.

Thompson, W. G., Spiegelman, M. W., Goldstein, S. L., et al. (2003). An open-system model for U-series age determinations of fossil corals. *Earth and Planetary Science Letters*, **210**, 365–381.

Thompson, W. G., Curran, H. A., Wilson, M. A., et al. (2011). Sea-level oscillations during the last interglacial highstand recorded by Bahamas corals. *Nature Geoscience*, **4**, 684–687.

Thorn, C. E. (1988). *Introduction to theoretical geomorphology*. Unwin Hyman, Boston, 247 pp.

Thornton, L. E. & Stephenson, W. J. (2006). Rock strength: a control of shore platform elevation. *Journal of Coastal Research*, **22**, 224–231.

Thurber, D. L., Broecker, W. S., Blanchard, R. L., et al. (1965). Uranium-series ages of Pacific atoll coral. *Science*, **149**, 55–58.

Titus, J. G. (1986). Greenhouse effect, sea level rise, and coastal zone management. *Coastal Zone Management Journal*, **14**, 147–171.

Tjia, H. D. (1977). Sea level variations during the last six thousand years in Peninsular Malaysia. *Sains Malaysia*, **6**, 171–183.

Tjia, H. D. (1980). The Sunda Shelf, Southeast Asia. *Zeitschrift für Geomorphologie N.F.*, **24**: 405–427.

Tjia, H. D. (1996). Sea-level changes in the tectonically stable Malay–Thai peninsula. *Quaternary International*, **31**, 95–101.

Tooley, M. J. (1978). *Sea-level changes: northwest England during the Flandrian stage*. Oxford University Press, Oxford.

Tooley, M. J. (1985). Sea levels. *Progress in Physical Geography*, **9**, 113–120.

Tooley, M. J. & Smith, D. E. (2005). Relative sea-level change and evidence for the Holocene Storegga Slide tsunami from a high-energy coastal environment: Cocklemill Burn, Fife, Scotland, UK. *Quaternary International*, **133–134**: 107–119.

Torgersen, T., Jones, M. R., Stephens, A. W., et al. (1985). Late Quaternary hydrological changes in the Gulf of Carpentaria. *Nature*, **313**, 785–787.

Torgersen, T., Luly, J., De Deckker, P., et al. (1988). Late Quaternary environments of the Carpentaria Basin, Australia. *Palaeogeography Palaeoclimatology and Palaeoecology*, **67**, 245–261.

Törnqvist, T. E. (1994). Middle and late Holocene avulsion history of the River Rhine (Rhine–Meuse delta, Netherlands). *Geology*, **22**, 711–714.

Törnqvist, T. & Hijma, M. P. (2012). Links between early Holocene ice-sheet decay, sea-level rise and abrupt climate change. *Nature Geoscience*, **5**, 601–606.

Törnqvist, T. E., González, J. L., Newsom, L. A., et al. (2004). Deciphering Holocene sea-level history on the U.S. Gulf Coast: a high-resolution record from the Mississippi Delta. *Geological Society of America Bulletin*, **116**, 1026–1039.

Törnqvist, T. E., Wallace, D. J., Storms, J. E. A., *et al.* (2008). Mississippi delta subsidence primarily caused by compaction of Holocene strata. *Nature Geoscience*, **1**, 173–176.

Toscano, M. A. & Lundberg, J. (1999). Submerged Late Pleistocene reefs on the tectonically-stable S.E. Florida margin: high-precision geochronology, stratigraphy, resolution of Substage 5a sea-level elevation, and orbital forcing. *Quaternary Science Reviews*, **18**, 753–767.

Toscano, M. A. & Macintyre, I. G. (2003). Corrected western Atlantic sea-level curve for the last 11,000 years based on calibrated ^{14}C dates from *Acropora palmata* framework and intertidal mangrove peat. *Coral Reefs*, **22**, 257–270.

Toscano, M. A., Peltier, W. R. & Drummond, R. (2011). ICE-5G and ICE-6G models of postglacial relative sea-level history applied to the Holocene coral reef record of northeastern St Croix, U.S.V.I.: investigating the influence of rotational feedback on GIA processes at tropical latitudes. *Quaternary Science Reviews*, **30**, 3032–3042.

Trenhaile, A. S. (1974). The geometry of shore platforms in England and Wales. *Transactions of the Institute of British Geographers*, **46**, 129–142.

Trenhaile, A. S. (1997). *Coastal dynamics and landforms*. Clarendon Press, Oxford, 366 pp.

Trenhaile, A. S. (2002). Rock coasts, with particular emphasis on shore platforms. *Geomorphology*, **48**, 7–22.

Trenhaile, A. S. & Layzell, M. G. J. (1981). Shore platform morphology and the tidal factor. *Transactions of the Institute of British Geographers, New Series*, **6**, 82–102.

Trimmer, J. (1850). Generalizations respecting the erratic Tertiaries of Norfolk Drift, founded on the mapping of the superficial deposits of a large portion of Norfolk with a description of the freshwater deposits of the Gaytonthorpe Valley; and a note on the contorted strata of Cromer Cliffs. *Quarterly Journal of the Geological Society of London*, **7**, 19–31.

Trumbone, S. E. (2000). Radiocarbon geochronology. *In* J. S. Noller, J. M. Sowers & W. R. Lettis (Eds), *Quaternary geochronology: methods and applications*, pp. 41–60. American Geophysical Union, Reference Shelf 4, Washington, DC.

Tushingham, A. M. & Peltier, W. R. (1991). Ice-3G: a new global model of late Pleistocene deglaciation based upon geophysical predictions of post-glacial relative sea level change. *Journal of Geophysical Research*, **96**, 4497–4523.

Twidale, C. R. (1976). *Analysis of landforms*. Jacaranda Wiley, Brisbane, 572 pp.

Twidale, C. R., Bourne, J. A. & Twidale, N. (1977). Shore platforms and sea level changes in the Gulfs Region of South Australia. *Royal Society of South Australia, Transactions*, **101**, 63–74.

Tzedakis, P. C., Channell, J. E. T., Hodell, D. A., *et al.* (2012). Determining the natural length of the current interglacial. *Nature Geoscience*, **5**(2), 138–141.

Ulm, S. (2002). Marine and estuarine reservoir effects in central Queensland, Australia: determination of Delta R values. *Geoarchaeology – An International Journal*, **17**, 319–348.

Umbgrove, J. H. F. (1929). The amount of the maximal lowering of sea level in the Pleistocene. *Proceedings of the 4th Pacific Scientific Congress*, **2**, 105–113, Java.

Urey, H. C. (1948). Oxygen isotopes in nature and in the laboratory. *Science*, **108**, 489–496.

Ussher, J. (1658). *The Annals of the World. Deduced from the origin of time, and continued to the beginning of the Emperour Vespasian's reign, and the totall destruction and abolition of the temple and commonwealth of the Jews. Containing the historie of the Old and New Testament, with that of the Macchabees. Also all the most memorable affairs of Asia and Egypt, and the rise of the empire of the Roman caesars, under C. Julius, and Octavianus. Collected from all history, as well sacred, as prophane and methodically digested by the Most Reverend James Ussher, Archbishop of Armagh, and Primate of Ireland.* Printed by E. Taylor for J. Crook and G. Bell, London.

Vacher, H. L. & Hearty, P. (1989). History of Stage 5 sea level in Bermuda: review with new evidence of a brief rise to present sea level during Substage 5a. *Quaternary Science Reviews*, **8**, 159–168.

Vacher, H. L. & Quinn, T. M. (Eds) (1997). *Geology and hydrogeology of carbonate islands. Developments in Sedimentology*, Vol. 54. Elsevier, Amsterdam, 948 pp.

Vacher, H. L. & Rowe, M. P. (1997). Geology and hydrogeology of Bermuda. *In* H. L. Vacher & T. M. Quinn (Eds), *Geology and hydrogeology of carbonate islands*, pp. 35–90. Elsevier, Amsterdam.

Valentin, H. (1953). Present vertical movements of the British Isles. *Geographical Journal*, **119**, 299–305.

Van Andel, T. H. & Veevers, J. J. (1967). Morphology and sediments of the Timor Sea. *Bulletin of the Bureau of Mineral Resources, Geology and Geophysics, Australia*, **83**, 1–173.

van de Plassche, O. (1982). Sea-level change and water-level movements in the Netherlands during the Holocene. *Mededelingen Rijks Geologische Dienst*, **36** (4), 1–93.

van de Plassche, O. (1986). *Sea-level research: a manual for collection and evaluation of data*. GeoBooks, Norwich, 618 pp.

van de Plassche, O. (1990). Mid-Holocene sea-level change on the eastern shore of Virginia. *Marine Geology*, **91**, 149–154.

van de Plassche, O. & Roep, T. B. (1989). Sea-level changes in the Netherlands during the last 6,500 years: basal peat vs coastal barrier data. *In* D. B. Scott, P. A. Pirazzoli & C. A. Honig (Eds), *Late Quaternary sea-level correlation and applications*. Kluwer, Dordrecht, NATO ASI series C, 256, pp. 41–56.

van de Plassche, O., Van der Borg, K. & De Jong, A. F. M. (1998). Sea level-climate correlation during the past 1400 yr. *Geology*, **26**, 319–322.

Van der Pluijm, B. A. & Marshak, S. (2004). *Earth structure*, 2nd edition. W.W. Norton & Co., New York, 656 pp.

Van Straaten, L. M. J. U. (1954). Radiocarbon datings and changes of sea level at Velzen (Netherlands). *Geologie en Mijnbouw*, **16**, 247–253.

Van Vuuren, D. P., Edmonds, J. A., Kainuma, M., *et al.* (2011). The representative concentration pathways: an overview. *Climatic Change*, **109**, 5–31.

Van Wagoner, J. C., Posamentier, H. W., Mitchum, R. M., *et al.* (1988). An overview of the fundamentals of sequence stratigraphy and key definitions. *In* C. K. Wilgus, B. S. Hastings, C. A. Ross, H. Posamentier, J. Van Wagoner, & C. G. St. C.Kendall (Eds), *Sea-level changes: an integrated approach*. Society of Economic Paleontologists and Mineralogists, pp. 39–45, Special Publication No. 42, Tulsa, Oklahoma, USA.

Van Wagoner, J. C., Mitchum, R. M., Campion, K. M., *et al.* (1990). Siliciclastic sequence stratigraphy in well-logs, cores, and outcrops. *American Association of Petroleum Geologists, Methods in Exploration Series*, **7**, 55 pp.

Veeh, H. H. (1966). Th^{230}/U^{238} and U^{234}/U^{238} ages of Pleistocene high sea level stand. *Journal of Geophysical Research*, **71**, 3379–3386.

Veeh, H. H. & Chappell, J. (1970). Astronomical theory of climatic change: support from New Guinea. *Science*, **167**, 862–865.

Veeh, H. H. & Veevers, J. J. (1970). Sea level at −175 m off the Great Barrier Reef 13,600 to 17,000 years ago. *Nature*, **226**, 536–537.

Veevers, J. J. (2000). *Billion-year history of Australia and neighbours in Gondwanaland*. Gemoc Press, Sydney, 388 pp.

Veevers, J. J. & Eittreim, S. L. (1988). Reconstruction of Antarctica and Australia at breakup (95 ± 5 Ma) and before rifting (160 Ma). *Australian Journal of Earth Sciences*, **35**, 355–362.

Veevers, J. J., Powell, C. Mc.A. & Roots, S. R. (1991). Review of sea floor spreading around Australia. I. Synthesis of patterns of spreading. *Australian Journal of Earth Sciences*, **38**, 373–389.

Velicogna, I. & Wahr, J. (2006a). Measurements of time-variable gravity show mass loss in Antarctica. *Science*, **311**, 1754–1756.

Velicogna, I. & Wahr, J. (2006b). Acceleration of Greenland ice mass loss in spring 2004. *Nature*, **443**, 329–331.

Ventris, P. A. (1996). Hoxnian interglacial freshwater and marine deposits in North-West Norfolk, England and their implications for sea-level reconstruction. *Quaternary Science Reviews*, **15**, 437–450.

Vermeer, M. & Rahmstorf, S. (2009). Global sea level linked to global temperature. *Proceedings of the National Academy of Sciences*, **106**, 21527–21532.

Veron, J. E. N. (1995). *Corals in space and time: the biogeography and evolution of the Scleractinia*. University of New South Wales Press, Sydney, 321 pp.

Vézina, J. L., Jones, B. & Ford, D. (1999). Sea-level highstands over the last 500,000 years: evidence from the Ironshore Formation on Grand Cayman, British West Indies. *Journal of Sedimentary Research*, **69**, 317–327.

Villemant, B. & Feuillet, N. (2003). Dating open systems by the ^{238}U–^{234}U–^{230}Th method: application to Quaternary reef terraces. *Earth and Planetary Science Letters*, **210**, 105–118.

Vine, F. J. & Matthews, D. H. (1963). Magnetic anomalies over oceanic ridges. *Nature*, **199**, 947–949.

Vita-Finzi, C. (1973). *Recent Earth history*. Macmillan, London, 138 pp.

Vita-Finzi, C. (1986). *Recent Earth movements: an introduction to neotectonics*. Academic Press, London, 226 pp.

Vita-Finzi, C. (1987). Opinion: precision, accuracy and usefulness. *Geology Today*, November–December, p. 190.

Von Grafenstein, R., Zahn, R., Tiedemann, R., *et al.* (1999). Planktonic δ18O records at Sites 976 and 977, Alboran Sea: stratigraphy, forcing, and paleoceanographic implications. *In* R. Zahn, M. C. Comas & A. Klaus (Eds), *Proceedings of the Ocean Drilling Program, Scientific Results*, **161**, 469–479.

Von Richthofen, F. (1886). *Führer für Forschungsreisende*. Jänecke, Hannover, 745 pp.

von Storch, H., Zorita, E. & González-Rouco, J. F. (2008). Relationship between global mean sea-level and global mean temperature and heat-flux in a climate simulation of the past millennium. *Ocean Dynamics*, **58**, 227–236.

Voris, H. K. (2000). Maps of Pleistocene sea levels in Southeast Asia: shorelines, river systems and time durations. *Journal of Biogeography*, **27**, 1153–1167.

Vousdoukas, M. I., Velegrakis, A. F. & Plomaritis, T. A. (2007). Beachrock occurrence, characteristics, formation mechanisms and impacts. *Earth-Science Reviews*, **85**, 23–46.

Waelbroeck, C., Labeyrie, L., Michel, E., *et al.* (2002). Sealevel and deep water temperature changes derived from benthonic foraminifera isotopic records. *Quaternary Science Reviews*, **21**, 295–305.

Wagner, C. A. & Cheney, R. E. (1992). Global sea level change from satellite altimetry. *Journal of Geophysical Research*, **97**, 15607–15615.

Wagner, G. A. (1998). *Age determination of young rocks and artifacts*. Springer-Verlag, Berlin, 466 pp.

Walcott, R. I. (1972). Past sea levels, eustasy and deformation of the Earth. *Quaternary Research*, **2**, 1–14.

Walker, M. (2005). *Quaternary dating methods*. John Wiley and Sons, Chichester, 286 pp.

Walker, R. G. (1992). Facies, facies models and modern stratigraphic concepts. *In* R. G. Walker & N. P. James (Eds), *Facies models: response to sea level change*, pp. 1–14. Geological Association of Canada, St Johns.

Walker, R. G. & James, N. P. (Eds) (1992). *Facies models: response to sea level change*. Geological Association of Canada, St Johns, 409 pp.

Walker, R. G. & Plint, A. G. (1992). Wave- and storm-dominated shallow marine systems. *In* R. G Walker & N. P. James (Eds), *Facies models – response to sea level change*, pp. 219–238, Geological Association of Canada, St Johns.

Wallace, D. J. & Anderson, J. B. (2013). Unprecedented erosion of the upper Texas coast: response to accelerated sea-level rise and hurricane impacts. *Geological Society of America Bulletin*, doi: 10.1130/B30725.1.

Wallman, P. C., Mahood, G. A. & Pollard, D. D. (1988). Mechanical models for correlation of ring-fracture eruptions at Pantelleria, Strait of Sicily, with glacial sea-level drawdown. *Bulletin of Volcanology*, **50**, 327–339.

Walsh, J. P. & Nittrouer, C. A. (2009). Understanding fine-grained river-sediment dispersal on continental margins. *Marine Geology*, **263**, 34–45.

Ward, J. M. (1967). Studies in ecology on a shell barrier beach. *Vegetatio*, **14**, 241–342.

Ward, W. T. (1966). *Geology, geomorphology, and soils of the south-western part of County Adelaide, South Australia.* Soil Publication No. 23, Commonwealth Scientific and Industrial Research Organization, Australia, 115 pp.

Ward, W. T. & Jessup, R. W. (1965). Changes of sea-level in southern Australia. *Nature,* **205,** 791–792.

Ward, W. T. & Little, I. P. (2001). Sea-rafted pumice on the Australian east coast: numerical classification and stratigraphy. *Australian Journal of Earth Sciences,* **47,** 95–109.

Warrick, R. A. (1993). Slowing global warming and sea-level rise: the rough road from Rio. *Transactions of the Institute of British Geographers N.S.,* **18,** 140–148.

Warrick, R. & Oerlemans, J. (1990). Sea level rise. *In* J. T. Houghton, G. J. Jenkins & J. J. Ephraum (Eds), *Climate change: the IPCC scientific assessment,* pp. 257–281. Cambridge University Press, Cambridge.

Watson, J. G. (1928). Mangrove forests of the Malay Peninsula. *Malayan Forestry Records,* **6,** 1–275.

Watson, P. J. (2011). Is there evidence yet of acceleration in mean sea level rise around mainland Australia? *Journal of Coastal Research,* **27,** 368–377.

Watts, A. B. (1982). Tectonic subsidence, flexure and global changes of sea level. *Nature,* **297,** 469–474.

Watts, A. B. and ten Brink, U. S. (1989). Crustal structure, flexure, and subsidence history of the Hawaiian islands. *Journal of Geophysical Research,* **94,** 10473–10500.

Wegemann, E. (1969). Changing ideas about moving shorelines. *In* C. J. Sheer (Ed.), *Toward a history of geology,* pp. 386–414. The MIT Press, Cambridge, Massachusetts.

Wehmiller, J. F. (1981). Kinetic model options for interpretation of amino acid enantiomeric ratios in Quaternary mollusks: comments on a paper by Kvenvolden *et al.* (1979). *Geochimica et Cosmochimica Acta,* **45,** 261–264.

Wehmiller, J. F. (1982). A review of amino acid racemization studies in Quaternary mollusks: stratigraphic and chronological applications in coastal and interglacial sites, Pacific and Atlantic coasts, United States, United Kingdom, Baffin Island and Tropical Islands. *Quaternary Science Reviews,* **1,** 83–120.

Wehmiller, J. F. (1984). Relative and absolute dating of Quaternary molluscs with amino acid racemization: evaluation, applications and questions. *In* W. C. Mahaney (Ed.), *Quaternary dating methods,* pp. 171–193. Elsevier Science Publishers, Amsterdam.

Wehmiller, J. F. (1993). Applications of organic geochemistry for Quaternary research aminostratigraphy and aminochronology. *In* M. H. Engel & S. A. Macko (Eds), *Organic geochemistry,* pp. 735–783. Plenum Press, New York.

Wehmiller, J. F. & Belknap, D. F. (1978). Alternative kinetic models for the interpretation of amino acid enantiomer ratios in Pleistocene mollusks: examples from California, Washington and Florida. *Quaternary Research,* **9,** 330–348.

Wehmiller, J. F. & Belknap, D. F. (1982). Amino acid age estimates, Quaternary Atlantic coastal plain: comparison with U-series dates, biostratigraphy and paleomagnetic control. *Quaternary Research,* **18,** 311–336.

Wehmiller, J. F. & Miller, G. H. (2000). Aminostratigraphic dating methods in Quaternary geology. *In* J. S. Noller, J. M. Sowers & W. R. Lettis (Eds), *Quaternary geochronology: methods and applications,* pp. 187–222. American Geophysical Union, Reference Shelf 4, Washington, DC.

Wehmiller, J. F., York, L. L. & Bart, M. L. (1995). Amino acid racemization geochronology of reworked Quaternary molluscs on U.S. Atlantic coast beaches: Implications for chronostratigraphy, taphonomy and coastal sediment transport. *Marine Geology,* **124,** 303–337.

Wehmiller, J. F., Belknap, D. F., Boutin, B. S., *et al.* (1988). A review of the aminostratigraphy of Quaternary mollusks from United States Atlantic Coastal Plain sites. *In* D. J. Easterbrook (Ed.), *Dating Quaternary sediments,* pp. 69–110, Geological Society of America, Special Paper 227.

Wehmiller, J. F., Simmons, K. R., Cheng, H., *et al.* (2004). Uranium-series coral ages from the US Atlantic Coastal Plain – the

'80 ka problem' revisited. *Quaternary International*, **120**, 3–14.

Wehmiller, J. F., Thieler, E. R., Miller, D., *et al.* (2010). Aminostratigraphy of surface and subsurface Quaternary sediments, North Carolina coastal plain, USA. *Quaternary Geochronology*, **5**, 459–492.

Wehmiller, J. F., Harris, W. B., Boutin, B. S., *et al.* (2012). Calibration of amino acid racemization (AAR) kinetics in United States mid-Atlantic Coastal Plain Quaternary molluscs using $^{87}Sr/^{86}Sr$ analyses: evaluation of kinetic models and estimation of regional Late Pleistocene temperature history. *Quaternary Geochronology*, **7**, 21–36.

Wellner, R. W., Ashley, G. M. & Sheridan, R. E. (1993). Seismic stratigraphic evidence for a submerged middle Wisconsin barrier – implications for sea-level history. *Geology*, **21**, 109–112.

Wentworth, C. K. (1931). Geology of the Pacific equatorial islands. *B.P. Bishop Museum Occasional Papers*, **9**, 1–25.

Wentworth, C. K. (1938). Marine bench-forming processes, water level weathering. *Journal of Geomorphology*, **1**, 6–32.

Wescott, W. A. (1993). Geomorphic thresholds and complex response of fluvial systems – some implications for sequence stratigraphy. *American Association of Petroleum Geologists Bulletin*, **77**, 1208–1218.

Westaway, K. E. & Roberts, R. G. (2006). A dual-aliquot regenerative-dose protocol (DAP) for thermoluminescence (TL) dating of quartz sediments using the light-sensitive and isothermally-stimulated red emissions. *Quaternary Science Reviews*, **25**, 2513–2528.

Westaway, R. (2002). The Quaternary evolution of the Gulf of Corinth, central Greece: coupling between surface processes and flow in the lower continental crust. *Tectonophysics*, **348**, 269–318.

Whewell, W. (1832). [Review of] Art IV. Principles of Geology, being an attempt to explain the former changes of the Earth's surface by reference to causes now in operation. Vol. II. *The Quarterly Review*, **47**, 103–132.

Wigley, T. M. L. & Raper, S. C. B. (1992). Implications for climate and sea level of revised IPCC emission scenarios. *Nature*, **357**, 293–300.

Williams, K. M. & Smith, G. G. (1977). A critical evaluation of the application of amino acid racemization to geochronology and geothermometry. *Origins of Life*, **8**, 91–144.

Williams, M., Dunkerley, D., De Deckker, P., *et al.* (1998). *Quaternary environments*, 2nd edition. Arnold, London, 329 pp.

Williams, P. W. (1982). Speleothem dates, Quaternary terraces and uplift rates in New Zealand. *Nature*, **298**, 257–260.

Williams, P. W. (1991). Tectonic geomorphology, uplift rates and geomorphic response in New Zealand. *Catena*, **18**, 439–452.

Wilson, J. T. (1963). A possible origin of the Hawaiian Islands. *Canadian Journal of Physics*, **41**, 863–870.

Wilson, R. C. L., Drury, S. A. & Chapman, J. L. (2000). *The Great Ice Age: climate change and life*. Routledge, London, 267 pp.

Wintle, A. G. & Huntley, D. J. (1982). Thermoluminescence dating of sediments. *Quaternary Science Reviews*, **1**, 31–53.

Wintle, A. G. & Murray, A. S. (2006). A review of quartz optically stimulated luminescence characteristics and their relevance in single aliquot regeneration dating protocols. *Radiation Measurements*, **41**, 369–391.

Woodroffe, C. D. (1981). Mangrove swamp stratigraphy and Holocene transgression, Grand Cayman Island, West Indies. *Marine Geology*, **41**, 271–294.

Woodroffe, C. D. (1983). Development of Mangrove swamps behind beach ridges, Grand Cayman Island, West Indies. *Bulletin of Marine Science*, **33**, 864–880.

Woodroffe, C. D. (1988a). Mangroves and sedimentation in reef environments: indicators of past sea-level changes and present sea-level trends? *Proceedings 6th International Coral Reef Congress, Townsville*, **3**, 535–539.

Woodroffe, C. D. (1988b). Vertical movement of isolated oceanic islands at plate margins: evidence from emergent reefs in Tonga (Pacific Ocean) Cayman Islands (Caribbean Sea) and Christmas Island (Indian Ocean). *Zeitschrift für geomorphologie* Suppl. Bd, **69**, 17–37.

Woodroffe, C. D. (1993a). Morphology and evolution of reef islands in the Maldives. *Proceedings of the 7th International Coral Reef Symposium*, pp. 1217–1226.

Woodroffe, C. D. (1993b). Late Quaternary evolution of coastal and lowland riverine plains of Southeast Asia and northern Australia: an overview. *Sedimentary Geology*, **83**, 163–173.

Woodroffe, C. D. (1995). Response of tide-dominated mangrove shorelines in northern Australia to anticipated sea-level rise. *Earth Surface Processes and Landforms*, **20**, 65–85.

Woodroffe, C. D. (1996). Late Quaternary infill of macrotidal estuaries in northern Australia. *In* K. F. Nordstrom & C. J. Roman (Eds), *Estuarine shores: evolution, environments and human alterations*, pp. 89–114. Wiley, Chichester.

Woodroffe, C. D. (2000). Deltaic and estuarine environments and their Late Quaternary dynamics on the Sunda and Sahul shelves. *Journal of Asian Earth Sciences*, **18**, 393–413.

Woodroffe, C. D. (2003). *Coasts: form, process and evolution.* Cambridge University Press, Cambridge, 623 pp.

Woodroffe, C. D. (2005). Late Quaternary sea-level highstands in the central and eastern Indian Ocean: a review. *Global and Planetary Change*, **49**, 121–138.

Woodroffe, C. D. (2008). Reef-island topography and the vulnerability of atolls to sea-level rise. *Global and Planetary Change*, **62**, 77–96.

Woodroffe, C. D. & Chappell, J. (1993). Holocene emergence and evolution of the McArthur River Delta, southwestern Gulf of Carpentaria, Australia. *Sedimentary Geology*, **83**, 303–317.

Woodroffe, C. D. & Gagan, M. K. (2000). Coral microatolls from the central Pacific record late Holocene El Niño. *Geophysical Research Letters*, **27**, 1511–1514.

Woodroffe, C. D. & Grime, D. (1999). Storm impact and evolution of a mangrove-fringed chenier plain, Shoal Bay, Darwin, Australia. *Marine Geology*, **159**, 303–321.

Woodroffe, C. D. & Grindrod, J. (1991). Mangrove biogeography: the role of Quaternary environmental and sea-level fluctuations. *Journal of Biogeography*, **18**, 479–492.

Woodroffe, C. D. & McLean, R. F. (1990). Microatolls and recent sea level change on coral atolls. *Nature*, **344**, 531–534.

Woodroffe, C. D. & McLean, R. F. (1998). Pleistocene morphology and Holocene emergence of Christmas (Kiritimati) Island, Pacific Ocean. *Coral Reefs*, **17**, 235–248.

Woodroffe, C. D. & Morrison, R. J. (2001). Reef-island accretion and soil development, Makin Island, Kiribati, central Pacific. *Catena*, **44**, 245–261.

Woodroffe, C. D. & Murray-Wallace, C. V. (2012). Sea-level rise and coastal change: the past as a guide to the future. *Quaternary Science Reviews*, **54**, 4–11.

Woodroffe, C. D., Stoddart, D. R., Harmon, R. S., *et al.* (1983a). Coastal morphology and late Quaternary history, Cayman Islands, West Indies. *Quaternary Research*, **19**, 64–84.

Woodroffe, C. D., Curtis, R. J. & McLean, R. F. (1983b). Development of a chenier plain, Firth of Thames, New Zealand. *Marine Geology*, **53**, 1–22.

Woodroffe, C. D., Thom, B. G. & Chappell, J. (1985). Development of widespread mangrove swamps in mid-Holocene times in northern Australia. *Nature*, **317**, 711–713.

Woodroffe, C. D., Thom, B. G., Chappell, J., *et al.* (1987). Relative sea level in the South Alligator River region, North Australia, during the Holocene. *Search*, **18**, 198–200.

Woodroffe, C. D., Chappell, J. & Thom, B. G. (1988). Shell middens in the context of estuarine development, South Alligator River, Northern Territory. *Archaeology in Oceania*, **23**, 95–103.

Woodroffe, C. D., Chappell, J., Thom, B. G., *et al.* (1989). Depositional model of a macrotidal estuary and floodplain, South Alligator River, Northern Australia. *Sedimentology*, **36**, 737–756.

Woodroffe, C. D., Stoddart, D. R., Spencer, T., *et al.* (1990a). Holocene emergence in the Cook Islands, South Pacific. *Coral Reefs*, **9**, 31–39.

Woodroffe, C. D., McLean, R. F., Polach, H., *et al.* (1990b). Sea level and coral atolls: Late Holocene emergence in the Indian Ocean. *Geology*, **18**, 62–66.

Woodroffe, C. D., Short, S., Stoddart, D. R., *et al.* (1991a). Stratigraphy and chronology of Pleistocene reefs and the southern Cook Islands, South Pacific. *Quaternary Research*, **35**, 246–263.

Woodroffe, C. D., Veeh, H. H., Falkland, A., *et al.* (1991b). Last interglacial reef and subsidence of the Cocos (Keeling) Islands, Indian Ocean. *Marine Geology*, **96**, 137–143.

Woodroffe, C. D., Bryant, E. A., Price, D. M., *et al.* (1992). Quaternary inheritance of coastal landforms, Cobourg Peninsula, Northern Territory. *Australian Geographer*, **23**, 101–115.

Woodroffe, C. D., Mulrennan, M. E. & Chappell, J. (1993). Estuarine infill and coastal progradation, southern van Diemen Gulf, northern Australia. *Sedimentary Geology*, **83**, 257–275.

Woodroffe, C. D., McLean, R. F. & Wallensky, E. (1994). Geomorphology of the Cocos (Keeling) Islands. *Atoll Research Bulletin*, **402**, 1–33.

Woodroffe, C. D., Murray-Wallace, C. V., Bryant, E. A., *et al.* (1995). Late Quaternary sea-level highstands in the Tasman Sea: evidence from Lord Howe Island. *Marine Geology*, **125**, 61–72.

Woodroffe, C. D., Kennedy, D. M., Hopley, D., *et al.* (2000). Holocene reef growth in Torres Strait. *Marine Geology*, **170**, 331–346.

Woodroffe, C. D., Nicholls, R. J., Saito, Y., *et al.* (2006). Landscape variability and the response of Asian megadeltas to environmental change. *In* N. Harvey (Ed.), *Global change and integrated coastal management: the Asia-Pacific region*, pp. 277–314. Springer, Berlin.

Woodroffe, C. D., Brooke, B. P., Linklater, M., *et al.* (2010). Response of coral reefs to climate change: expansion and demise of the southernmost Pacific coral reef. *Geophysical Research Letters*, **37**, L15602, doi:10.1029/2010GL044067.

Woodroffe, C. D., Cowell, P. J., Callaghan, D. P., *et al.* (2012a). *Approaches to risk assessment on Australian coasts: a model framework for assessing risk and adaptation to climate change on Australian coasts.* National Climate Change Adaptation Research Facility, Gold Coast, 203 pp.

Woodroffe, C. D., McGregor, H., Lambeck, K., *et al.* (2012b). Mid-Pacific microatolls record sea-level stability over the past 5000 yr. *Geology*, **40**, 951–954.

Woodroffe, S. A. (2009). Testing models of mid to late Holocene sea-level change, North Queensland, Australia. *Quaternary Science Reviews*, **28**, 2474–2488.

Woodroffe, S. A. & Horton, B. P. (2005). Holocene sea-level changes in the Indo-Pacific. *Journal of Asian Earth Sciences*, **25**, 29–43.

Woodroffe, S. A. & Long, A. J. (2009). Salt marshes as archives of recent relative sea-level change in West Greenland. *Quaternary Science Reviews*, **28**, 1750–1761.

Woods, J. E. (1862). *Geological observations in South Australia: Principally in the district south-east of Adelaide.* Longman, Green, Longman, Roberts & Green, Melbourne, Victoria, 404 pp.

Woodworth, P. L. (2012). A note on the nodal tide in sea level records. *Journal of Coastal Research*, **28**, 316–323.

Woodworth, P. L. & Blackman, D. L. (2004). Evidence for systematic changes in extreme high waters since the mid-1970s. *Journal of Climate*, **17**, 1190–1197.

Woodworth, P. L. & Player, R. (2003). The Permanent Service for Mean Sea Level: an update to the 21st century. *Journal of Coastal Research*, **19**, 287–295.

Woodworth, P. L., Teferle, F. N., Bingley, R. M., *et al.* (2009a). Trends in UK mean sea level revisited. *Geophysical Journal International*, **176**, 19–30.

Woodworth, P. L., White, N. J., Jevrejeva, S., *et al.* (2009b). Evidence for the accelerations of sea level on multi-decade and century timescales. *International Journal of Climatology*, **29**, 777–789.

Woodworth, P. L., Gehrels, W. R. & Nerem, R. S. (2011). Nineteenth and twentieth century changes in sea level. *Oceanography*, **24**, 80–93.

Wooldridge, S. W. & Linton, D. L. (1939). *Structure, surface and drainage in south-east England.* Institute of British Geographers, Publication 10.

Wooldridge, S. W. & Linton, D. L. (1955). *Structure, surface and drainage in south-east England.* Philip, London.

Woolfe, K. J., Larcombe, P., Naish, T., *et al.* (1998). Lowstand rivers need not incise the shelf: an example from the Great

Barrier Reef, Australia, with implications for sequence stratigraphic models. *Geology*, **26**, 75–78.

Wöppelmann, G., Letetrel, C., Santamaria, A., *et al.* (2009). Rates of sea-level change over the past century in a geocentric reference frame. *Geophysical Research Letters*, **36**, L12607.

Wouters, B., Riva, R. E. M., Lavallée, D. A., *et al.* (2011). Seasonal variations in sea level induced by continental water mass: first results from GRACE, *Geophysical Research Letters*, **38**, L03303.

Wright, A. J., Edwards, R. J. & van der Plassche, O. (2011). Reassessing transfer-function performance in sea-level reconstruction based on benthic salt-marsh foraminifera from the Atlantic Coast of NE North America. *Marine Micropaleontology*, **81**, 43–62.

Wright, L. D. (1985). River deltas. *In* R. A. Davis (Ed.), *Coastal sedimentary environments*, pp. 1–76. Springer-Verlag, New York.

Wright, L. D., Coleman, J. M. & Thom, B. G. (1972). Emerged tidal flats in the Ord River estuary, Western Australia. *Search*, **3**, 339–341.

Wright, W. B. (1914). *The Quaternary ice age*. Macmillan, London, 464 pp.

Wyrtki, K. (1985). Sea level fluctuations in the Pacific during the 1982–83 El Niño. *Geophysical Research Letters*, **12**, 125–128.

Yim, W.W.-S. & Huang, G. (2002). Middle Holocene higher sea-level indicators from the south China coast. *Marine Geology*, **182**, 225–230.

Yokoyama, Y. & Esat, T. M. (2011). Global climate and sea level: enduring variability and rapid fluctuations over the past 150,000 years. *Oceanography*, **24**, 54–69, doi:10.5670/oceanog.2011.27.

Yokoyama, Y., Lambeck, K., De Deckker, P., *et al.* (2000). Timing of the Last Glacial Maximum from observed sea-level minima. *Nature*, **406**, 713–716.

Yokoyama, Y., De Deckker, P., Lambeck, K., *et al.* (2001a). Sea-level at the Last Glacial Maximum: evidence from northwestern Australia to constrain ice volumes for oxygen isotope stage 2. *Palaeogeography, Palaeoclimatology, Palaeoecology*, **165**, 281–297.

Yokoyama, Y., Purcell, A., Lambeck, K., *et al.* (2001b). Shore-line reconstruction around Australia during the Last Glacial Maximum and Late Glacial Stage. *Quaternary International*, **83**, 9–18.

Yokoyama, Y., Okuno, J., Miyairi, Y., *et al.* (2012). Holocene sea-level change and Antarctic melting history derived from geological observations and geophysical modeling along the Shimokita Peninsula, northern Japan. *Geophysical Research Letters*, **39**, L13502.

Yonekura, N., Ishii, T., Saito, Y., *et al.* (1988). Holocene fringing reefs and sea-level change in Mangaia Island, southern Cook Islands. *Palaeogeography, Palaeoclimatology and Palaeoecology*, **68**, 177–188.

Young, R. W. & Bryant, E. A. (1993). Coastal rock platforms and ramps of Pleistocene and Tertiary age in southern New South Wales, Australia. *Zeitschrift für Geomorphologie N.F.*, **37**, 257–272.

Young, R. W., Bryant, E. A. & Price, D. (1993a). Last interglacial sea levels on the South Coast of New South Wales. *Australian Geographer*, **24**, 72–75.

Young, R. W., Bryant, E. A., Price, D. M., *et al.* (1993b). Theoretical constraints and chronological evidence of Holocene coastal development in central and southern New South Wales, Australia. *Geomorphology*, **7**, 317–329.

Yu, K.-F. & Zhao, J. (2010). U-series dates of Great Barrier Reef corals suggest at least +0.7 m sea level ~7000 years ago. *The Holocene*, **20**, 161–168.

Yu, K.-F., Zhao, J.-X., Done, T., *et al.* (2009). Microatoll record for large century-scale sea-level fluctuations in the mid-Holocene. *Quaternary Research*, **71**, 354–360.

Yu, S.-Y., Berglund, B. E. & Sandgren, P. (2007). Evidence for a rapid sea-level rise 7600 yr ago. *Geology*, **35**, 891–894.

Zachariasen, J., Sieh, K., Taylor, F. W., *et al.* (1999). Submergence and uplift associated with the giant 1833 Sumatran subduction earthquake: evidence from coral microatolls. *Journal of Geophysical Research*, **104**, 895–919.

Zachariasen, J., Siringan, F., Taylor, F. W., *et al.* (2000). Modern vertical deformation above the Sumatran Subduction Zone: paleogeodetic insights from coral microatolls. *Bulletin of the Seismological Society of America*, **90**, 897–913.

Zachos, J., Pagani, M., Sloan, L., *et al.* (2001). Trends, rhythms and aberrations in global climate 65 Ma to present. *Science*, **292**, 686–693.

Zalasiewicz, J., Williams, M., Steffen, W., *et al.* (2010). The new world of the Anthropocene. *Environment, Science and Technology*, **44**, 2228–2231.

Zar, J. H. (1984). *Biostatistical analysis*, 2nd edition. Prentice-Hall, New Jersey, 718 pp.

Zazo, C., Silva, P. G., Goy, J. L., *et al.* (1999). Coastal uplift in continental collision plate boundaries: data from Last Interglacial marine terraces of the Gibraltar Strait area (south Spain). *Tectonophysics*, **301**, 95–109.

Zeuner, F. E. (1945). *The Pleistocene Period*. Ray Society, London, 322 pp.

Zeuner, F. E. (1952). Pleistocene shore-lines. *Geologische Rundschau*, **40**, 39–50.

Zhang, X. & Church, J. A. (2012). Sea level trends, interannual and decadal variability in the Pacific Ocean. *Geophysical Research Letters*, **39**, L21701.

Zhao, J., Yu, K. & Feng, Y. (2009). High-precision ^{238}U–^{234}U–^{230}Th disequilibrium dating of the recent past: a review. *Quaternary Geochronology*, **4**, 423–433.

Zhu, Z. R., Wyrwoll, K.-H., Collins, L. B. *et al.* (1993). High precision U-series dating of Last Interglacial events by mass spectrometry: Houtman Abrolhos Islands, Western Australia. *Earth and Planetary Science Letters*, **118**, 281–293.

Zong, Y. (2007). Sea level studies: tropics. *In* S. A. Elias (Ed.), *Encyclopedia of Quaternary science*, Vol. 4, pp. 3087–3094. Elsevier, Amsterdam.

Zong, Y. & Tooley, M. J. (1996). Holocene sea-level changes and crustal movements in Morecambe Bay, northwest England. *Journal of Quaternary Science*, **11**, 43–58.

Zoppi, U., Albani, A., Ammerman, A. J. (2001). Preliminary estimate of the reservoir age in the Lagoon of Venice. *Radiocarbon*, **43**, 489–494.

Zwally, H. J., Giovinetto, M. B., Li, J., *et al.* (2005). Mass changes of the Greenland and Antarctic ice sheets and shelves and contributions to sea-level rise: 1992–2002. *Journal of Glaciology*, **51**, 509–527.

Index